AF847067

B&SS – Bocconi & Springer Series

Series Editors: Lorenzo Peccati • Sandro Salsa (Editors-in-Chief)
Carlo A. Favero • Peter Müller • Eckhard Platen • Wolfgang J. Runggaldier

Volume 8

More information about this series at http://www.springer.com/series/8762

Kęstutis Kubilius • Yuliya Mishura •
Kostiantyn Ralchenko

Parameter Estimation in Fractional Diffusion Models

BOCCONI
UNIVERSITY
PRESS

Springer

Kęstutis Kubilius
Institute of Data Science and Digital Technologies
Vilnius University
Vilnius, Lithuania

Yuliya Mishura
Department of Probability Theory, Statistics and Actuarial Mathematics
Taras Shevchenko National University of Kyiv
Kyiv, Ukraine

Kostiantyn Ralchenko
Department of Probability Theory, Statistics and Actuarial Mathematics
Taras Shevchenko National University of Kyiv
Kyiv, Ukraine

ISSN 2039-1471 ISSN 2039-148X (electronic)
B&SS – Bocconi & Springer Series
ISBN 978-3-319-71029-7 ISBN 978-3-319-71030-3 (eBook)
https://doi.org/10.1007/978-3-319-71030-3

Library of Congress Control Number: 2017964371

© Springer International Publishing AG 2017
This work is subject to copyright. All rights are reserved by the Publisher, whether the whole or part of the material is concerned, specifically the rights of translation, reprinting, reuse of illustrations, recitation, broadcasting, reproduction on microfilms or in any other physical way, and transmission or information storage and retrieval, electronic adaptation, computer software, or by similar or dissimilar methodology now known or hereafter developed.
The use of general descriptive names, registered names, trademarks, service marks, etc. in this publication does not imply, even in the absence of a specific statement, that such names are exempt from the relevant protective laws and regulations and therefore free for general use.
The publisher, the authors and the editors are safe to assume that the advice and information in this book are believed to be true and accurate at the date of publication. Neither the publisher nor the authors or the editors give a warranty, express or implied, with respect to the material contained herein or for any errors or omissions that may have been made. The publisher remains neutral with regard to jurisdictional claims in published maps and institutional affiliations.

Printed on acid-free paper

This Springer imprint is published by Springer Nature
The registered company is Springer International Publishing AG
The registered company address is: Gewerbestrasse 11, 6330 Cham, Switzerland

Preface

The present book is devoted to parameter estimation in diffusion continuous-time models involving fractional Brownian motion and related processes. Our models extend and generalize standard diffusion models involving a Wiener process.

On the one hand, standard stochastic diffusion models are widely used in various fields of application. There is an extended literature devoted to standard and nonstandard approaches to parameter estimation in the diffusion models, for both discrete and continuous observations. We mention only the books [26, 69, 75, 87, 106, 112, 145] and references therein. Many complicated models have been studied. However, there was a curious situation in the case of the simplest homogeneous diffusion model: there were no conditions for the strong consistency of the standard maximum likelihood estimator that are close to being necessary and are sufficiently mild. We have filled the gap in this book and have proved that, in some sense, the standard maximum likelihood estimator is always strongly consistent unless the drift coefficient is identically zero.

On the other hand, the substantial limitation in the use of stochastic diffusion models with a Brownian motion is caused by the fact that it has independent increments, and, therefore, the random noise generated by this motion is "white", i.e., uncorrelated. However, many processes in natural sciences, computer networks, financial markets, etc. have long-term dependences, i.e., the correlations of random noise in these processes decrease slowly with time. In particular, the models of financial markets demonstrate various kinds of memory (short or long) and the simplest approach is to model this memory by fractional Brownian diffusion. To the best of our knowledge, there are only a very few books [24, 146] devoted to parameter estimation in fractional diffusions, though some fractional models are also considered in [26] and [120]. Note that many practical recommendations concerning the statistical methods for long-memory processes, especially for autoregression models, are proposed in [19] and [20]. Of course, there are hundreds of articles on statistical estimation for short- and long-memory processes, some of which are quoted in our bibliography, albeit without any claim to completeness. Thus, our goal is to systematize known methods already presented in various articles and to describe the new ones.

The book is the result of years of work by the authors in the area of statistical inference for processes with memory and is based, in particular, on the papers [14, 48, 49, 92, 95–105, 121–123, 128, 129, 143]. Moreover, the book is partially based on the courses "Statistics of stochastic processes" and "Fractional and multifractional processes" delivered at Taras Shevchenko National University of Kyiv between 2012 and 2017.

The book consists of six chapters.

In Chap. 1 we review basic properties of the fractional Brownian motion and Wiener integrals with respect to it. Then we consider the pathwise integration using the fractional calculus. Also, we give the existence and uniqueness theorems for stochastic differential equations. We consider Itô SDEs driven by a Wiener process, general SDEs driven by an fBm with $H > 1/2$, and some particular SDEs driven by an fBm with $H > 1/2$, namely, SDEs with polynomial drift and the fractional Gompertz diffusion process. The fractional Ornstein–Uhlenbeck process is described in detail. Further we consider mixed SDEs, involving both a Wiener process and an fBm with $H > 1/2$, and conclude with the description of the multifractional, bifractional, and subfractional Brownian motions. In this chapter we concentrate all "non-statistical" results concerning stochastic processes and stochastic differential equations that are used in the subsequent chapters to construct and investigate statistical estimators.

Chapter 2 is devoted to the consideration of the asymptotic behavior of the quadratic variations, as well as to a review of the known estimators of the Hurst index together with the construction of new ones. We consider the convergence rate of quadratic and weighted quadratic variations. Moreover, we recall some asymptotic normality results for quadratic variations of an fBm.

Chapter 3 is devoted to the Hurst parameter estimation from the solutions of SDEs driven by an fBm. Our goal is to construct strongly consistent and asymptotically normal estimators of the Hurst parameter H based on discrete observations of the underlying processes. First, we propose the Hurst index estimators for a general fractional diffusion model. Then we apply this estimator to two transformable SDEs. One is an SDE driven by an fBm with polynomial drift and another is the fractional Gompertz diffusion process. Moreover, we compare the asymptotic behavior of some estimators of the Hurst index and of the diffusion coefficient of the fractional Gompertz diffusion process with the aid of computer simulations. Also, we consider the limit behavior of the second-order quadratic variations of the solution of an SDE driven by an fBm along an arbitrary partition of the interval and derive strongly consistent Hurst index estimators.

In Chap. 4 we investigate the following stochastic process

$$M_t^H = aB_t^H + bW_t, \quad t \geq 0,$$

where B^H is an fBm, W is a Wiener process, a and b are some non-zero coefficients. Such a process is frequently called a mixed fractional Brownian motion. We consider the statistical identification of the model. Attention is devoted principally to the estimation of H, though we will also present estimators for a and b. The estimators are partially based on both weak and almost sure asymptotic behavior of mixed power variations that are related to the components of the model. Weak and strong limit theorems for the mixed variations are studied in detail, with the calculations of moments and other numerical characteristics of the limit distributions. These results are of independent interest.

Chapter 5 is devoted to the drift parameter estimation in the diffusion, fractional diffusion and mixed Brownian-fractional Brownian diffusion models. We start with the diffusion model and give a quite general condition for the strong consistency of the maximum likelihood estimator. Then we consider the parameter estimation in the fractional diffusion model. We introduce both the maximum likelihood estimator and the so-called transferred maximum likelihood estimator. Also, we consider a mixed diffusion model, which includes both standard and fractional Brownian motion. The maximum likelihood estimator for this model does not exist. Therefore we construct an analog of the transferred maximum likelihood estimator and prove its strong consistency. Then we discretize observations and establish the strong consistency of the corresponding estimators. We investigate in detail the fractional Ornstein–Uhlenbeck process, for which we consider the maximum likelihood estimator and two simpler estimators together with their discretized versions. Since in this model the asymptotic properties of the estimators depend on the drift parameter sign, we study hypothesis testing of this sign. Also, we construct the maximum likelihood estimator for a simple linear model involving two independent fractional Brownian motions and prove its strong consistency. Finally, we study multifractional generalizations of linear and Ornstein–Uhlenbeck models and construct the drift parameter estimators.

Chapter 6 is devoted to the introduction of the Orey index concept for a second order stochastic process, construction of the estimators and study of the asymptotic properties of these estimators. This index allows consideration of a class of Gaussian processes which may not have stationary increments and the trajectories of these processes are Hölder continuous up to the Orey index. The fBm is an example of this class of processes. The Orey index of an fBm is equal to H.

To make the book as self-contained as possible, we added Appendices A and B, devoted to non-random and stochastic objects that we need to consider when constructing estimators. Details of the sections and descriptions of the previous related results with the references are given at the beginning of each chapter. The most informative sections are also equipped with introductions.

Concerning the authorship of the results of chapters, Chaps. 2, 3 and 6 are written by K. Kubilius on the basis of his personal and collaborative work with co-authors, Chaps. 4 and 5 are written by Y. Mishura, K. Ralchenko and their coauthors, and Chap. 1 and the Appendices are the work of all the authors.

The book is intended for a wide variety of readers. We have tried to combine the most common statements with specific models and examples. The simplest methods, for example, Hurst estimation and drift parameter estimation for diffusion and fractional diffusion equations, are available for study and obtainment of numerical results by undergraduate students. More involved results will be interesting for graduate students, researchers and practitioners.

We are grateful to everyone who contributed to the creation and improvement of the content of this book. Our personal thanks are due to Prof. Yury Kutoyants and Prof. Marina Kleptsyna with whom we discussed the results and received useful advice during visits to the University of Maine and to the numerous Statistique Asymptotique des Processus Stochastiques (SAPS) workshops. Prof. Marco Dozzi was always hospitable during our visits to the University of Lorraine and made a significant contribution to the joint work. Y. Mishura is grateful to Prof. Oleg Seleznjev for his constant hospitality at Umeå University and to him and Prof. Mounir Zili for fruitful discussions of statistical problems. We are thankful to Prof. Georgiy Shevchenko for his contribution to the parameter estimation in the linear mixed fractional models, Prof. Yuriy Kozachenko for estimation of the rate of asymptotic growth of fractional and multifractional processes, and Assoc. Prof. Viktor Skorniakov for his contribution to the construction and consideration of asymptotic properties of the Hurst index estimators.

Our special thanks go to Springer Milan and the editorial board of the Bocconi & Springer Series for their helpful comments and recommendations which contributed significantly in improving the book's presentation.

Vilnius, Lithuania — Kęstutis Kubilius
Kyiv, Ukraine — Yuliya Mishura
Kyiv, Ukraine — Kostiantyn Ralchenko
July 2017

Contents

1 **Description and Properties of the Basic Stochastic Models** 1
1.1 Fractional Brownian Motion 2
1.1.1 Definition, Basic Properties and Integral Representations... 2
1.1.2 Wiener Integration with Respect to fBm 5
1.1.3 Elements of Fractional Calculus: Path-Wise Integration with Respect to fBm 7
1.2 Homogeneous Diffusion Model Involving a Wiener Process 11
1.2.1 Existence and Uniqueness of Weak and Strong Solutions... 12
1.2.2 The Distribution of the Local Time of the Recurrent and Transient Diffusion Processes 14
1.2.3 Integral Functionals of a Transient Diffusion Processes..... 18
1.3 Stochastic Differential Equations Involving fBm 22
1.3.1 Existence and Uniqueness of Solution............ 23
1.3.2 Fractional Ornstein–Uhlenbeck Process............ 24
1.3.3 SDE with Polynomial Drift 31
1.3.4 Fractional Gompertz Diffusion Process 35
1.4 Mixed SDE with Wiener Process and Fractional Brownian Motion............ 38
1.5 Sub-fractional, Bifractional and Multifractional Brownian Motions............ 39
1.5.1 Sub-fractional Brownian Motion 39
1.5.2 Bifractional Brownian Motion 40
1.5.3 Multifractional Brownian Motion............ 41

2 **The Hurst Index Estimators for a Fractional Brownian Motion** 45
2.1 Quadratic Variations of a Fractional Brownian Motion 46
2.1.1 The Rate of Convergence of Quadratic Variations of an fBm............ 47
2.1.2 Asymptotic Behavior of Weighted Quadratic Variations 54
2.1.3 CLT for Quadratic Variation of an fBm 57

2.2 The Hurst Index Estimators ... 67
2.2.1 Some Known Statistics ... 67
2.2.2 A New Class of Statistics for Estimation of the Hurst Index ... 68
2.2.3 Examples ... 71

3 Estimation of the Hurst Index from the Solution of a Stochastic Differential Equation ... 75
3.1 Strong Consistency of the Hurst Index Estimators Constructed from a Solution of SDE ... 75
3.1.1 Stochastic Differential Equation Involving a fBm: Bounds for the Increments of the Solution ... 76
3.1.2 Strongly Consistent Estimators of the Hurst Index ... 78
3.1.3 The Rate of Convergence of the Hurst Index Estimator Constructed from a SDE ... 80
3.1.4 On Euler–Peano Approximation of SDE and Consistency of Respective Estimator ... 82
3.2 Strongly Consistent and Asymptotically Normal Estimators of the Hurst Index Constructed from a SDE ... 84
3.2.1 Estimators of the Hurst Index of the Solution of SDE Driven by a fBm Under Additional Restrictions on Diffusion Coefficient: Asymptotic Normality ... 84
3.3 Estimation of the Hurst Index and of Diffusion Coefficient for Transformable SDEs ... 92
3.3.1 Estimators of the Hurst Index and the Diffusion Coefficient for Fractional Diffusion Processes with Polynomial Drift ... 93
3.3.2 Estimators of the Hurst Index and the Diffusion Coefficient for Fractional Gompertz Diffusion Process ... 103
3.4 Construction of the Hurst Index Estimator for Arbitrary Partition of the Interval ... 116
3.4.1 Asymptotic Behavior of Quadratic Variations of the Solution of SDE Driven by a fBm ... 116
3.4.2 On the Estimation of the Hurst Index for Arbitrary Partition ... 120

4 Parameter Estimation in the Mixed Models via Power Variations ... 125
4.1 Description of the Mixed Model and Mixed Power Variations ... 126
4.2 Exact Calculation and Asymptotic Behavior of the Moments of Higher Order of Mixed Power Variations ... 127
4.3 Weak and Strong Limit Theorems for the Centered and Normalized Mixed Power Variations ... 134
4.4 Statistical Estimation in Mixed Model ... 139
4.4.1 Statistical Estimation Based on Quadratic Variation ... 140
4.4.2 Statistical Estimation Using 4th Power Variation ... 152

4.4.3 Estimation of Hurst Parameter for Known Scale Coefficients ... 153
4.5 Simulations ... 155
4.5.1 Estimators Based on Quadratic Variation ... 155
4.5.2 Estimators Based on Quartic Variation ... 157
4.5.3 Estimation When *a* and *b* Are Known ... 159
4.5.4 Comparison of Estimators and Some Recommendations ... 159

5 Drift Parameter Estimation in Diffusion and Fractional Diffusion Models ... 161
5.1 Drift Parameter Estimation in the Homogeneous Diffusion Model: Standard MLE Is Always Strongly Consistent ... 162
5.1.1 Strong Consistency of the Drift Parameter Maximum-Likelihood Estimator Constructed for Continuous Observations ... 163
5.1.2 Discretization and Strong Consistency ... 167
5.1.3 Some Simulation Results ... 173
5.2 Estimation in Fractional Diffusion Model by Continuous Observations ... 175
5.2.1 Model Description: The Standard MLE ... 176
5.2.2 A Non-standard Transferred MLE ... 179
5.2.3 Examples of the Remainder Terms with the Estimation of the Rate of Convergence to Zero ... 181
5.2.4 Simulations ... 189
5.2.5 Sequential Drift Parameter Estimators ... 190
5.2.6 Linear Fractional Models and Strong Consistency ... 191
5.2.7 Estimation in the Mixed Model by Continuous Observations: Transferred MLE ... 192
5.3 Estimation in Homogeneous Fractional Diffusion Model by Discrete Observations ... 195
5.3.1 Bounds for Solution of SDE Driven by fBm ... 196
5.3.2 Drift Parameter Estimation by Discretized Observations ... 198
5.3.3 Simulations ... 206
5.4 Statistical Inference for the Fractional Ornstein–Uhlenbeck Model ... 208
5.4.1 Maximum Likelihood Estimation ... 208
5.4.2 Least Square Estimator for the Case $\theta > 0$ ... 210
5.4.3 Ergodic-Type Estimator for the Case $\theta < 0$ ... 219
5.4.4 Hypothesis Testing of the Drift Parameter Sign for Fractional Ornstein–Uhlenbeck Process ... 221
5.4.5 Maximum Likelihood Estimation in the Generalized Fractional Ornstein–Uhlenbeck Model with Time-Dependent Coefficients ... 230
5.5 Maximum Likelihood Drift Estimation in the Linear Model Containing Two fBms ... 233

5.5.1 Preliminaries: The Reduction of the Original Problem to the Integral Equation ... 235
5.5.2 Compactness of the Integral Operator: Existence-Uniqueness Result for the Fredholm Integral Equation ... 240
5.5.3 Statistical Results: The Form of the MLE, Its Consistency and Asymptotic Normality ... 252
5.6 Drift Parameter Estimation in Models with mfBm ... 260
5.6.1 Linear Multifractional Model ... 261
5.6.2 Multifractional Ornstein–Uhlenbeck Process ... 262

6 The Extended Orey Index for Gaussian Processes ... 269
6.1 Gaussian Processes with the Orey Index ... 270
6.1.1 The Extended Definition of the Orey Index ... 271
6.1.2 Examples of Processes with the Orey Index ... 279
6.2 The Convergence of the Quadratic Variation of Gaussian Process with the Orey Index ... 285
6.2.1 Examples ... 296
6.3 On the Estimation of the Orey Index for Arbitrary Partition ... 303
6.4 Exact Confidence Intervals of the Extended Orey Index for Gaussian Processes ... 304
6.4.1 Confidence Intervals ... 305
6.4.2 Examples ... 308
6.4.3 Simulations ... 318

A Selected Facts from Mathematical and Functional Analysis ... 321
A.1 Some Properties of the Hypergeometric Function ... 321
A.2 Regularly Varying Functions ... 322
A.3 p-Variation Calculus ... 323
A.4 The Wasserstein Distance and Kantorovich–Rubinstein Duality Theorem ... 324

B Selected Facts from Probability, Stochastic Processes and Stochastic Calculus ... 327
B.1 Weak and Strong Limit Theorems ... 327
B.1.1 Ergodic Theorem ... 327
B.1.2 The Strong Law of Large Numbers for Martingales ... 327
B.1.3 Classical and Non-classical Limit Theorems ... 328
B.1.4 Central and Non-central Limit Theorems for Stationary Gaussian Sequences ... 329
B.2 Wiener Chaos, Multiple Integrals, and Rosenblatt Distribution ... 333
B.2.1 Rosenblatt Random Variable ... 337
B.3 Exponential Maximal Bounds and Asymptotic Growth of Trajectories of Gaussian Processes ... 338

B.3.1 Exponential Maximal Upper Bound for Gaussian Process in Terms of Metric Massiveness 338
B.3.2 Exponential Maximal Upper Bound for the Weighted Gaussian Process Defined on the Half-Axis.................. 345
B.3.3 Exponential Maximal Upper Bound for Gaussian Process in the Bounded Strip on the Plane 348
B.3.4 Exponential Maximal Upper Bound for Gaussian Process Defined in the Angle of the Plane.................... 353
B.3.5 Asymptotic Growth with Probability 1 of Fractional Derivative of fBm... 357
B.4 Asymptotic Growth with Probability 1 of mfBm and Its Increments 368
B.4.1 Upper Bounds for the Incremental Variances of mfBm...... 368
B.4.2 Asymptotic Growth of the Trajectories of mfBm with Probability 1 371
B.4.3 Asymptotic Growth with Probability 1 of the Increments of mfBm................................... 375
B.5 Boxplot.. 379

References... 381

Index... 389

About the Authors

Prof. Kęstutis Kubilius received his Ph.D. in mathematics at Vilnius University in 1981. Currently he is a professor of mathematics at the same university. His research work mainly focuses on limit theorems for semimartingales, theory of stochastic differential equations, and parameter estimation for fractional diffusion processes. He is the author of more than 50 published papers.

Prof. Yuliya Mishura received her Ph.D. in probability and statistics at Kyiv University in 1978 and received her postdoctoral degree in probability and statistics (habilitation) in 1990. She is currently a professor at Taras Shevchenko National University of Kyiv. She is the author of more than 250 research papers and six books. Her research interests include theory and statistics of stochastic processes, fractional processes, stochastic analysis and financial mathematics.

Dr. Kostiantyn Ralchenko is a postdoctoral researcher at the Department of Probability Theory, Statistics and Actuarial Mathematics, Taras Shevchenko National University of Kyiv, where he also completed his Ph.D. in mathematics in 2012. His research interests include stochastic differential equations, fractional and multifractional processes, and statistics of stochastic processes. He is the author of 24 papers.

Acronyms

Abbreviations

a. s.	Almost surely
bifBm	Bifractional Brownian motion
CI	Confidence interval
CLT	Central limit theorem
fBm	Fractional Brownian motion
fO-U	Fractional Ornstein-Uhlenbeck process
i.i.d	Independent identically distributed
IQR	Inter-quartile range
mfBm	Multifractional Brownian motion
r. v.	Random variable
r. vs.	Random variables
SDE	Stochastic differential equation
sfBm	Sub-fractional Brownian motion
sssi	Self-similar with stationary increments

Notation

$(\Omega, \mathfrak{F}, \mathbb{F}, \mathsf{P})$	Complete probability space with filtration $\mathbb{F} = \{\mathfrak{F}_t\}_{t\geq 0}$
$\mathcal{B}(\mathbb{R})$	Borel σ-algebra on $\mathbb{R}$
$\mathbb{R}^+$	$[0, \infty)$
B^H	Fractional Brownian motion with Hurst parameter H
$\overset{d}{=}$	Equality in distribution
$\overset{d}{\longrightarrow}$	Convergence in distribution
$\overset{a.s.}{\longrightarrow}$	Almost sure convergence
$\overset{\mathsf{P}}{\rightarrow}$	Convergence in probability

$f(t) \asymp g(t)$	$f(t) = O\big(g(t)\big)$ and $g(t) = O\big(f(t)\big)$, as $t \to \infty$
$O_\omega(a_n), o_\omega(a_n)$	Let (Y_n) be a sequence of r. vs., ς is an a. s. non-negative r. v. and $(a_n) \subset (0, \infty)$ vanishes. $Y_n = O_\omega(a_n)$ means that $\lvert Y_n \rvert \le \varsigma \cdot a_n$; $Y_n = o_\omega(a_n)$ means that $\lvert Y_n \rvert \le \varsigma \cdot b_n$ with $b_n = o(a_n)$
$\mathbf{1}_A$	Indicator function of a set A
$\lfloor a \rfloor$	An integer part of a real number a
$\Gamma(\alpha)$	The gamma function
$\mathrm{B}(\alpha, \beta)$	The beta function
$F(a, b, c; x)$	The Gauss hypergeometric function
$\log_{2+} x$	$= \begin{cases} \log_2 x, & x > 0, \\ 0, & x \le 0. \end{cases}$
$\log^+ x$	$= \begin{cases} \log x, & x > 1, \\ 0, & x \le 1. \end{cases}$
$f_{a+}(x)$	$= (f(x) - f(a+))\mathbf{1}_{(a,b)}(x)$
$g_{b-}(x)$	$= (g(b-) - g(x))\mathbf{1}_{(a,b)}(x)$
$x \vee y$	$\max\{x, y\}$
$x \wedge y$	$\min\{x, y\}$
$\mathcal{N}(m, \sigma^2)$	Gaussian random variable with mean m and variance σ^2
$\mathcal{N}(\mathbf{m}; \Sigma)$	Gaussian random vector with mean $\mathbf{m}$ and covariance matrix Σ
A^T	The transpose of the matrix A
$\lambda_{\max}(A)$	The maximal eigenvalue of the matrix A
$\lambda_{\min}(A)$	The minimal eigenvalue of the matrix A
$\mathcal{I}^\alpha_{a+} f$	Riemann–Liouville left-sided fractional integral of order α
$\mathcal{I}^\alpha_{b-} f$	Riemann–Liouville right-sided fractional integral of order α
$\mathcal{D}^\alpha_{a+} f$	Riemann–Liouville left-sided fractional derivative of order α
$\mathcal{D}^\alpha_{b-} f$	Riemann–Liouville right-sided fractional derivative of order α
$L_p(a, b)$	Space of measurable p-integrable functions $f\colon [a, b] \to \mathbb{R}$
$C^\lambda([a, b])$	Space of Hölder continuous functions $f\colon [a, b] \to \mathbb{R}$ with Hölder exponent $\lambda \in (0, 1]$ equipped with a norm $$\lVert f \rVert_\lambda := \lVert f \rVert_\infty + \sup_{\substack{s,t \in [a,b] \\ s \ne t}} \frac{\lvert f(t) - f(s) \rvert}{\lvert s - t \rvert^\lambda}, \qquad \lVert f \rVert_\infty = \sup_{t \in [a,b]} \lvert f(t) \rvert.$$
$W^\alpha_\infty([0, T])$	Space of measurable functions $f\colon [0, T] \to \mathbb{R}$ such that $$\lVert f \rVert_{\infty,\alpha;T} = \sup_{s \in [0,T]} \left(\lvert f(s) \rvert + \int_0^s \lvert f(s) - f(u) \rvert \, (s - u)^{-1-\alpha} du \right)$$ is finite.

$\mathcal{W}_p([a,b])$ Class of functions on $[a,b]$ with bounded p-variation and norm

$$V_{p,\infty}(f;[a,b]) = V_p(f;[a,b]) + \sup_{x\in[a,b]} |f(x)|,\ V_p(f;[a,b])$$

$$= v_p^{1/p}(f;[a,b]),\ v_p(f;[a,b]) = \sup_{\varkappa} \sum_{k=1}^{n} |f(x_k) - f(x_{k-1})|^p,$$

where sup is taken over all partitions of $[a,b]$.

$C\mathcal{W}_p([a,b])$ Class of continuous functions on $[a,b]$ with bounded p-variation.

Chapter 1
Description and Properties of the Basic Stochastic Models

In this chapter we give definitions of the main stochastic processes and models used in the book and describe their properties. The statistical aspects of these models will be studied in the subsequent chapters.

Section 1.1 is devoted to the fractional Brownian motion (fBm), which plays a key role in the book. We review basic properties of this process, and describe Wiener and path-wise integration with respect to it.

Section 1.2 deals with the homogeneous diffusion model. We give sufficient conditions for existence and uniqueness of a solution to the corresponding SDE. We also analyze the distribution of the local time and its asymptotic behavior together with the moments of the integral functionals of transient diffusion processes.

In Sect. 1.3 we study the fractional diffusion models. We start with general conditions for existence and uniqueness of a solution to SDEs driven by fBm with $H > 1/2$. Then we study in more detail three special cases: the SDE with polynomial drift, the fractional Gompertz diffusion model and the fractional Ornstein–Uhlenbeck process. For these equations we give the explicit solutions. Also we study covariance structure of the fractional Ornstein–Uhlenbeck process and the ergodic properties of related stationary process.

Mixed SDEs with long-range dependence, involving both a Wiener process and a fBm with $H > 1/2$ are considered in Sect. 1.4.

Section 1.5 is devoted to some extensions of the fBm. In particular, we consider multifractional Brownian motion, bifractional Brownian motion, and sub-fractional Brownian motion.

References to primary sources are included in the relevant sections.

© Springer International Publishing AG 2017

K. Kubilius et al., *Parameter Estimation in Fractional Diffusion Models*,
B&SS – Bocconi & Springer Series 8,
https://doi.org/10.1007/978-3-319-71030-3_1

1.1 Fractional Brownian Motion

A.N. Kolmogorov was the first who considered in 1940, in the paper [91] continuous Gaussian processes with stationary increments and with the self-similarity property; the latter means that for any $a > 0$ there exists $b > 0$ such that

$$\{X_{at};\ t \geq 0\} \stackrel{d}{=} \{bX_t;\ t \geq 0\},$$

where $\stackrel{d}{=}$ denotes the distributional equivalence. Such processes B^H with zero mean and defined on the parameter set $\mathbb{R}^+ = [0, \infty)$ have a covariance function of the following form:

$$\mathsf{E} B_t^H B_s^H = \frac{1}{2}\left(s^{2H} + t^{2H} - |t-s|^{2H}\right),$$

where $0 < H < 1$. A.N. Kolmogorov called such Gaussian processes "Wiener Spirals" ("Wiener screw-lines"). Its modern name is the fractional Brownian motion (fBm). So, the fBm is the unique continuous self-similar zero–mean Gaussian process (with self-similarity coefficient $b = a^H$, or, as people often say, with self-similarity index H), and with stationary increments. Note that, on the one hand, the presence of these properties makes this process extremely convenient for the application, on the other hand it limits the set of appropriate models. The stochastic calculus of the fBm originated with the pioneering work of Mandelbrot and van Ness [115] who considered the integral moving average representation of B^H via the Wiener process on an infinite interval. They came up with the name fractional Brownian motion. Since then, many properties and possible applications of fBm have been studied. The wide application capability naturally leads to the need for statistical estimation of the parameters of the model used. In turn, statistical estimation of parameters is impossible without studying the properties of models involving fBm. Therefore, in this chapter we collected necessary preliminary information on the models in question. Those who are interested in further properties of the fBm, we can recommend the books [25, 120, 136] and references therein.

1.1.1 *Definition, Basic Properties and Integral Representations*

Let $(\Omega, \mathfrak{F}, \mathbb{F}, \mathsf{P})$ be a complete probability space with filtration $\mathbb{F} = \{\mathfrak{F}_t\}_{t\geq 0}$ satisfying the standard assumptions. It is assumed that all processes under consideration are adapted to filtration $\mathbb{F}$.

Definition 1.1 *Fractional Brownian motion (fBm)* with Hurst index $H \in (0, 1)$ is a Gaussian process $B^H = \{B_t^H, t \in \mathbb{R}^+\}$ on $(\Omega, \mathfrak{F}, \mathsf{P})$ featuring the properties

(i) $B_0^H = 0$;
(ii) $\mathsf{E}B_t^H = 0, t \in \mathbb{R}^+$;
(iii) $\mathsf{E}B_t^H B_s^H = \frac{1}{2}\left(t^{2H} + s^{2H} - |t-s|^{2H}\right), s, t \in \mathbb{R}^+$.

It is not hard to see that for $H = 1/2$ fBm is Brownian motion. For $H \neq 1/2$ an fBm is neither a semimartingale nor a Markov process.

Remark 1.2 From time to time, especially for introducing a fractional Ornstein–Uhlenbeck process, we will need also the fBm defined on the whole $\mathbb{R}$. The difference is that in this case $\mathsf{E}B_t^H B_s^H = \frac{1}{2}\left(|t|^{2H} + |s|^{2H} - |t-s|^{2H}\right), s, t \in \mathbb{R}$.

Definition 1.1 implies that an fBm is self-similar with the self-similarity parameter H, that is, $\{B_{ct}^H\} \overset{d}{=} \{c^H B_t^H\}$ for any $c > 0$. An fBm has stationary increments in the sense that $\mathsf{E}\left(B_t^H - B_s^H\right)^2 = |t-s|^{2H}$. Moreover, sample paths of fBm are a. s. Hölder continuous in the sense of the next definition.

Definition 1.3 A function $f : [a, b] \to \mathbb{R}$ is a *Hölder continuous function* with exponent $\lambda \in (0, 1]$, if there exists a constant $C > 0$ such that for all $s, t \in [a, b]$

$$|f(s) - f(t)| \le C\,|s-t|^{\lambda}.$$

In order to show the Hölder continuity of fBm, we recall the following Kolmogorov continuity theorem.

Theorem 1.4 (Kolmogorov)

1) Let $X = \{X_t, t \in [0, T]\}$ be a separable stochastic process satisfying the assumption: there exist constants $K > 0$, $\alpha > 0$ and $\beta > 0$ such that

$$\mathsf{E}|X_t - X_s|^{\alpha} \le K|t-s|^{1+\beta}, \quad 0 \le s < t \le T. \tag{1.1}$$

Then X is a continuous process.

2) Moreover, let the stochastic process $X = \{X_t, t \in [0, T]\}$ be separable and satisfy condition (1.1). *Then a.s. its trajectories are Hölder continuous up to order β/α.*

Taking into account that the process B^H is Gaussian, one can deduce from Theorem 1.4, 1) that its separable modification is continuous, therefore in what follows we consider its continuous modification. Moreover, it follows from Theorem 1.4, 2) that almost all sample paths of continuous modifications of B^H are locally Hölder continuous up to order H. A more precise result can be established by applying the following theorem from [7]. It gives necessary and sufficient conditions for Hölder continuity of a Gaussian process as well as the existence of exponential moments of its Hölder coefficient.

Theorem 1.5 ([7]) *The Gaussian process $X = \{X_t : t \in [0,T]\}$ is Hölder continuous up to order H, i. e. there exist r. vs. C_ε such that*

$$|X_t - X_s| \le C_\varepsilon |t-s|^{H-\varepsilon}, \qquad \text{for all } \varepsilon \in (0,H) \tag{1.2}$$

if and only if there exist constants c_ε such that

$$\mathsf{E}(X_t - X_s)^2 \le c_\varepsilon |t-s|^{2H-\varepsilon}, \qquad \text{for all } \varepsilon \in (0,H).$$

Moreover, the r. vs. C_ε in (1.2) satisfy

$$\mathsf{E}\exp\{\alpha C_\varepsilon^\kappa\} < \infty$$

for any constants $\alpha \in \mathbb{R}$ and $\kappa < 2$; and also for $\kappa = 2$ for small enough positive α. In particular, the moments of all orders of C_ε are finite.
Consequently, for all $0 < \varepsilon < H$ and $T > 0$, there exists a positive random variable $G_{\varepsilon,T}$ such that for all $s, t \in [0,T]$

$$\sup_{s,t\in[0,T]} |B_t^H - B_s^H| \le G_{\varepsilon,T}|t-s|^{H-\varepsilon} \quad \text{a. s.} \tag{1.3}$$

Moreover, according to Lemma 7.4 from [140], it is possible to choose $G_{\varepsilon,T} = C_{\varepsilon,T}\xi$, where $C_{\varepsilon,T}$ is a constant, and

$$\xi = \left(\int_0^T\int_0^T \frac{|B_r^H - B_u^H|^{\frac{2}{\varepsilon}}}{|r-u|^{\frac{2H}{\varepsilon}}}\, dr\, du\right)^{\frac{\varepsilon}{2}},$$

so that $\mathsf{E}(|G_{\varepsilon,T}|^p) < \infty$ for all $p \ge 1$.

The increments of an fBm are independent only in the case $H = 1/2$. They are negatively correlated for $H \in (0,1/2)$ and positively correlated for $H \in (1/2,1)$. Moreover, for $H \in (1/2,1)$ an fBm has the property of long-range dependence. The aforementioned means that $\sum_{n=1}^{\infty} r(n) = \infty$, where $r(n) = \mathsf{E}B_1^H\left(B_{n+1}^H - B_n^H\right) > 0$ is the autocovariance function.

Mandelbrot and van Ness [115] obtained the following integral representation of an fBm that is valid for an fBm defined both on $\mathbb{R}^+$ and on $\mathbb{R}$:

$$B_t^H = a_H\left\{\int_{-\infty}^0 \left[(t-s)^{H-\frac{1}{2}} - (-s)^{H-\frac{1}{2}}\right] dW_s + \int_0^t (t-s)^{H-\frac{1}{2}} dW_s\right\}, \tag{1.4}$$

where $W = \{W_t,\ t \in \mathbb{R}\}$ is a Wiener process, $a_H = \left(\frac{2H\Gamma(\frac{3}{2}-H)}{\Gamma(H+\frac{1}{2})\Gamma(2-2H)}\right)^{\frac{1}{2}}$, and $\Gamma(\cdot)$ denotes the Gamma function.

Another, compact-interval representation for an fBm on $\mathbb{R}^+$ was obtained in [131]. It has the form

$$B_t^H = \int_0^t K_H(t,s)\, d\widetilde{W}_s, \quad t \in [0,T], \tag{1.5}$$

where $\widetilde{W} = \{\widetilde{W}_t,\ t \geq 0\}$ is a Wiener process, and

$$K_H(t,s) = a_H \left[\left(\frac{t}{s}\right)^{H-\frac{1}{2}} (t-s)^{H-\frac{1}{2}} - \left(H - \tfrac{1}{2}\right) s^{\frac{1}{2}-H} \int_s^t (v-s)^{H-\frac{1}{2}} v^{H-\frac{3}{2}}\, dv \right]$$

is a square-integrable kernel. For $H > 1/2$ the form of kernel can be simplified:

$$K_H(t,s) = \beta_H s^{\frac{1}{2}-H} \int_s^t (v-s)^{H-\frac{3}{2}} v^{H-\frac{1}{2}}\, dv, \tag{1.6}$$

where $\beta_H = \left(H - \frac{1}{2}\right) a_H$.

1.1.2 Wiener Integration with Respect to fBm

Throughout this section, we assume that $H \in (1/2, 1)$. For an interval $[0, T]$, where $0 < T \leq \infty$ denote by $L_H^2[0,T]$ the completion of the space of simple functions $f\colon [0,T] \to \mathbb{R}$ with respect to the scalar product

$$\langle f, g\rangle_H^2 := H(2H-1) \int_0^T \int_0^T f(t) g(s)\, |t-s|^{2H-2}\, dsdt. \tag{1.7}$$

(It is worth mentioning that this completion contains not only classical functions, but also some distributions.) For a step function of the form

$$f(t) = \sum_{k=0}^{n-1} a_k \mathbf{1}_{[t_k, t_{k+1})},$$

where $\{t_0 < t_1 < \cdots < t_n\} \subset [0, T]$, the integral $I^H(f)$ of f with respect to B^H is defined by

$$I^H(f) = \int_0^T f(t)\, dB_t^H = \sum_{k=0}^{n-1} a_k \left(B_{t_{k+1}}^H - B_{t_k}^H\right).$$

It can be verified that I^H maps isomorphically the space of step functions on $[0, T]$ with the scalar product $\langle \cdot, \cdot \rangle_H$ into $L^2(\Omega)$, hence, I^H can be extended to $L_H^2[0, T]$. In particular, we have for $f, g \in L_H^2[0, T]$

$$\mathsf{E}\left(\int_0^T f(s)dB_s^H \int_0^T g(s)dB_s^H\right) = H(2H-1)\int_0^T \int_0^T f(t)g(s)\,|t-s|^{2H-2}\,ds\,dt.$$

Obviously, consider an fBm on $\mathbb{R}$, we can define in the same way the integral $\int_a^b f(s)dB_s^H$ for any $-\infty \leq a < b \leq \infty$, and

$$\mathsf{E}\left(\int_a^b f(s)dB_s^H\right)^2 = H(2H-1)\int_a^b \int_a^b f(t)f(s)\,|t-s|^{2H-2}\,ds\,dt. \tag{1.8}$$

If the right-hand side of (1.8) is finite, then the integral $\int_a^b f(s)dB_s^H$ is correctly defined. In what follows, we will often use the following moment bounds for Wiener integrals (for other possible moment inequalities see [120, Sec. 1.9–1.10]):

Proposition 1.6 (Inequalities for the Moments of Wiener Integrals with Respect to fBm [118]) *Let $H \in (\frac{1}{2}, 1)$, $f, g \in L_{\frac{1}{H}}[a, b]$ and $f = 0$ outside (a, b). Then for any $r > 0$, there exists a constant $C(H, r)$ such that*

$$\mathsf{E}\left|\int_a^b f(s)dB_s^H\right|^r \leq C(H,r)\|f\|_{L_{\frac{1}{H}}[a,b]}^r$$

and

$$\mathsf{E}\left|\int_a^b f(s)dB_s^H \int_a^b g(s)dB_s^H\right|^r \leq C(H,r)\|f\|_{L_{\frac{1}{H}}[a,b]}^r \|g\|_{L_{\frac{1}{H}}[a,b]}^r.$$

Proposition 1.7 ([120, Th. 1.10.3]) *Let $H \in (\frac{1}{2}, 1)$, $f \in L_{\frac{1}{H}}[0, T]$. Then for any $r > 0$, there exists a constant $C(H, r)$ such that*

$$\mathsf{E}\sup_{0\leq t\leq T}\left|\int_0^t f(s)dB_s^H\right|^r \leq C(H,r)\|f\|_{L_{\frac{1}{H}}[0,T]}^r.$$

Let $K(t, s)$ be the kernel (1.6). Then the map

$$(K_H^* f)(s) = \int_s^T f(t)\partial_t K(t,s)\,dt = \beta_H s^{1/2-H}\int_s^T f(t)t^{H-1/2}(t-s)^{H-3/2}\,dt \tag{1.9}$$

is an isometry between the space of step functions and can be extended to a Hilbert space isomorphism between $L_H^2[0, T]$ and $L^2[0, T]$. This implies that the process

$$W_t = I^H\left((K_H^*)^{-1}\mathbf{1}_{[0,t]}\right)$$

is a standard Wiener process on $[0, T]$, moreover, for any function $f \in L^2_H[0, T]$,

$$I^H(f) = \int_0^T (K_H^* f)(s)\, dW_s. \tag{1.10}$$

In particular, putting in the last formula $f = \mathbf{1}_{[0,t]}$, one gets the representation (1.5). For these transformations, see [82].

Finally, we define the so-called fundamental martingale, or Molchan martingale M^H, for B^H. To this end, introduce the kernel

$$l_H(t, s) = c_H s^{\frac{1}{2}-H}(t-s)^{\frac{1}{2}-H}\mathbf{1}_{(0,t)}(s), \tag{1.11}$$

with

$$c_H = \left(\frac{\Gamma(3-2H)}{2H\Gamma(\frac{3}{2}-H)^3\Gamma(H+\frac{1}{2})}\right)^{\frac{1}{2}}. \tag{1.12}$$

Then the process

$$M_t^H = \int_0^t l_H(t, s)\, dB_s^H = (2-2H)^{1/2}\int_0^t s^{1/2-H}\, dW_s. \tag{1.13}$$

is a square-integrable Gaussian martingale with quadratic variation $\langle M^H \rangle_t = t^{2-2H}$, see [131].

1.1.3 Elements of Fractional Calculus: Path-Wise Integration with Respect to fBm

In this section we describe a construction of the path-wise integral following the approach developed by Zähle [172–174]. We start by introducing fractional integrals and derivatives, see [155] for the details on the concept of fractional calculus. Since in the special case the path-wise integral introduced by Zähle coincides with the Riemann–Stieltjes integral, we give the Love–Young inequality for the Riemann–Stieltjes integral.

Definition 1.8 Let $f \in L_1(a, b)$. *The Riemann–Liouville left- and right-sided fractional integrals of order* $\alpha > 0$ are defined for almost all $x \in (a, b)$ by

$$\mathcal{I}^\alpha_{a+} f(x) := \frac{1}{\Gamma(\alpha)}\int_a^x (x-y)^{\alpha-1} f(y)\, dy, \tag{1.14}$$

$$\mathcal{I}^\alpha_{b-} f(x) := \frac{(-1)^{-\alpha}}{\Gamma(\alpha)}\int_x^b (y-x)^{\alpha-1} f(y)\, dy,$$

respectively, where $(-1)^{-\alpha} = e^{-i\pi\alpha}$.

Definition 1.9 For a function $f\colon [a,b] \to \mathbb{R}$ *the Riemann–Liouville left- and right-sided fractional derivatives of order* α $(0 < \alpha < 1)$ are defined by

$$\mathcal{D}^{\alpha}_{a+}f(x) := \mathbf{1}_{(a,b)}(x)\,\frac{1}{\Gamma(1-\alpha)}\,\frac{d}{dx}\int_a^x \frac{f(y)}{(x-y)^{\alpha}}\,dy,$$

$$\mathcal{D}^{\alpha}_{b-}f(x) := \mathbf{1}_{(a,b)}(x)\,\frac{(-1)^{1+\alpha}}{\Gamma(1-\alpha)}\,\frac{d}{dx}\int_x^b \frac{f(y)}{(y-x)^{\alpha}}\,dy.$$

The following proposition [155, Lemma 2.2] gives a simple sufficient condition for the existence of fractional derivatives.

Proposition 1.10 *Let f be an absolutely continuous function on $[a,b]$. Then $\mathcal{D}^{\alpha}_{a+}f$ and $\mathcal{D}^{\alpha}_{b-}f$ exist almost everywhere for $0 < \alpha < 1$. Moreover $\mathcal{D}^{\alpha}_{a+}f, \mathcal{D}^{\alpha}_{b-}f \in L_r(a,b)$, $1 \le r < 1/\alpha$, and*

$$\mathcal{D}^{\alpha}_{a+}f(x) = \frac{1}{\Gamma(1-\alpha)}\left(\frac{f(a)}{(x-a)^{\alpha}} + \int_a^x \frac{f'(y)}{(x-y)^{\alpha}}\,dy\right),$$

$$\mathcal{D}^{\alpha}_{b-}f(x) = \frac{(-1)^{\alpha}}{\Gamma(1-\alpha)}\left(\frac{f(b)}{(b-x)^{\alpha}} - \int_x^b \frac{f'(y)}{(y-x)^{\alpha}}\,dy\right).$$

Denote by $\mathcal{I}^{\alpha}_{a+}(L_p)$ (resp. $\mathcal{I}^{\alpha}_{b-}(L_p)$) the class of functions f that can be presented as $f = \mathcal{I}^{\alpha}_{a+}\varphi$ (resp. $f = \mathcal{I}^{\alpha}_{b-}\varphi$) for $\varphi \in L_p(a,b)$.

Proposition 1.11 *For $f \in \mathcal{I}^{\alpha}_{a+}(L_p)$ (resp. $f \in \mathcal{I}^{\alpha}_{b-}(L_p)$), $p \ge 1$, the corresponding Riemann–Liouville fractional derivatives admit the following* Weyl representation

$$\mathcal{D}^{\alpha}_{a+}f(x) = \frac{1}{\Gamma(1-\alpha)}\left(\frac{f(x)}{(x-a)^{\alpha}} + \alpha\int_a^x \frac{f(x)-f(y)}{(x-y)^{\alpha+1}}\,dy\right),$$

$$\mathcal{D}^{\alpha}_{b-}f(x) = \frac{(-1)^{\alpha}}{\Gamma(1-\alpha)}\left(\frac{f(x)}{(b-x)^{\alpha}} + \alpha\int_x^b \frac{f(x)-f(y)}{(y-x)^{\alpha+1}}\,dy\right),$$

where the convergence of the integrals holds pointwise for a. a. $x \in (a,b)$ for $p = 1$ and in $L_p(a,b)$ for $p > 1$.

Fractional differentiation and integration are inverse operators on the appropriate functional classes. Also, we shall use the following integration by parts formula for fractional derivatives [155, Sec. 2.6, Cor. 2].

Proposition 1.12 *The formula*

$$(-1)^{\alpha}\int_a^b (\mathcal{D}^{\alpha}_{a+}f)(x)g(x)dx = \int_a^b f(x)(\mathcal{D}^{\alpha}_{b-}g)(x)dx, \quad 0 < \alpha < 1, \tag{1.15}$$

is valid under the assumption that $f(x) \in \mathcal{I}^{\alpha}_{a+}(L_p)$, $g(x) \in \mathcal{I}^{\alpha}_{b-}(L_q)$, $\frac{1}{p} + \frac{1}{q} \le 1 + \alpha$.

Let $f, g\colon [a,b] \to \mathbb{R}$. Assume that the limits

$$f(u+) := \lim_{\delta \downarrow 0} f(u+\delta) \quad \text{and} \quad g(u-) := \lim_{\delta \downarrow 0} g(u-\delta)$$

exist for $a \le u \le b$. Denote

$$f_{a+}(x) = (f(x) - f(a+))\mathbf{1}_{(a,b)}(x),$$
$$g_{b-}(x) = (g(b-) - g(x))\mathbf{1}_{(a,b)}(x).$$

Definition 1.13 Assume that $f_{a+} \in \mathcal{I}^{\alpha}_{a+}(L_p)$, $g_{b-} \in \mathcal{I}^{1-\alpha}_{b-}(L_q)$ for some $\frac{1}{p} + \frac{1}{q} \le 1$, $0 < \alpha < 1$. *The generalized (fractional) Lebesgue–Stieltjes integral* of f with respect to g is defined by

$$\int_a^b f(x)\,dg(x) := (-1)^{\alpha} \int_a^b \mathcal{D}^{\alpha}_{a+} f_{a+}(x)\, \mathcal{D}^{1-\alpha}_{b-} g_{b-}(x)\,dx + f(a+)\big(g(b-) - g(a+)\big). \tag{1.16}$$

Note that this definition is correct, i. e. independent of the choice of α [172, Prop. 2.1]. If $\alpha p < 1$, then (1.16) can be simplified to

$$\int_a^b f(x)\,dg(x) := (-1)^{\alpha} \int_a^b \mathcal{D}^{\alpha}_{a+} f(x)\, \mathcal{D}^{1-\alpha}_{b-} g_{b-}(x)\,dx. \tag{1.17}$$

In particular, Definition 1.13 allows us to integrate Hölder continuous functions. Let $C^{\lambda}([a,b])$, $\lambda \in (0,1]$, denote the space of Hölder continuous functions equipped with the norm

$$\|f\|_{\lambda} := \|f\|_{\infty} + \sup_{\substack{s,t \in [a,b] \\ s \ne t}} \frac{|f(t) - f(s)|}{|s-t|^{\lambda}}, \qquad \|f\|_{\infty} = \sup_{t \in [a,b]} |f(t)|.$$

Proposition 1.14 ([172, Th 4.2.1]) *Let $f \in C^{\lambda}([a,b])$, $g \in C^{\mu}([a,b])$ with $\lambda + \mu > 1$. Then the assumptions of Definition 1.13 are satisfied with any $\alpha \in (1-\mu, \lambda)$ and $p = q = \infty$. Moreover, the generalized Lebesgue–Stieltjes integral $\int_a^b f(x)\,dg(x)$ defined by* (1.16) *coincides with the Riemann–Stieltjes integral*

$$\int_a^b f(x)\,dg(x) := \lim_{|\pi| \to 0} \sum_i f(x_i^*)(g(x_{i+1}) - g(x_i)),$$

where $\pi = \{a = x_0 \le x_0^* \le x_1 \le \ldots \le x_{n-1} \le x_{n-1}^* \le x_n = b\}$ and $|\pi| = \max_i |x_{i+1} - x_i|$.

Proposition 1.15 (Chain Rule [172, Comment to Th 4.3.1]) *Let $f \in C^{\lambda}([a,b])$ and $F \in C^{1}(\mathbb{R} \times [a,b])$ be real-valued functions such that $F'(f(\cdot),\cdot) \in C^{\mu}([a,b])$ with $\lambda + \mu > 1$. Then for any $y \in (a,b)$*

$$F(f(y),y) - F(f(a),a) = \int_a^y F_1'(f(x),x)\,df(x) + \int_a^y F_2'(f(x),x)\,dx,$$

where F_1' and F_2' are the partial derivatives of F with respect to the first and second variable, respectively.

The Riemann–Stieltjes integral can be defined for a wider class of functions than the Hölder functions. Denote by $\mathcal{W}_q([a,b])$ (resp. $C\mathcal{W}_q([a,b])$) the class of (resp. continuous) functions on $[a,b]$, with bounded p-variation, $p \in (0,\infty)$ (see Appendix A, Sect. A.3). The following statement is called Love–Young inequality.

Theorem 1.16 (see [52]) *Let $f \in \mathcal{W}_q([a,b])$ and $g \in \mathcal{W}_p([a,b])$ with $p,q \in (0,\infty)$ such that $1/p + 1/q > 1$. If f and g have no common discontinuities then the Riemann–Stieltjes integral $\int_a^b f(x)dg(x)$ exists and, for any $y \in [a,b]$, the following inequality holds:*

$$\left|\int_a^b f\,dg - f(y)\big[g(b) - g(a)\big]\right| \le C_{p,q} V_q\big(f;[a,b]\big) V_p\big(g;[a,b]\big), \tag{1.18}$$

where $C_{p,q} = \zeta(p^{-1} + q^{-1})$, $\zeta(s)$ denotes the Riemann zeta function, i.e., $\zeta(s) = \sum_{n\ge 1} n^{-s}$.

Remark 1.17 Let $f \in C^{\lambda}([a,b])$ and $g \in C^{\mu}([a,b])$ with $\lambda + \mu > 1$. Then $f \in \mathcal{W}_{1/\lambda}([a,b])$, $g \in \mathcal{W}_{1/\mu}([a,b])$, $V_{1/\lambda}\big(f;[a,b]\big) \le K_f^{\lambda}(b-a)^{\lambda}$, $V_{1/\mu}\big(g;[a,b]\big) \le K_g^{\mu}(b-a)^{\mu}$, where

$$K_f = \sup_{\substack{s,t\in[a,b]\\ s\ne t}} \frac{|f(t) - f(s)|}{|s-t|^{\lambda}}, \qquad K_g = \sup_{\substack{s,t\in[a,b]\\ s\ne t}} \frac{|g(t) - g(s)|}{|s-t|^{\mu}},$$

and the Love–Young inequality takes the form: for any $y \in [a,b]$

$$\left|\int_a^b f\,dg - f(y)\big[g(b) - g(a)\big]\right| \le C_{1/\mu,1/\lambda} K_f^{\lambda} K_g^{\mu} (b-a)^{\lambda+\mu}. \tag{1.19}$$

Theorem 1.18 (Chain Rule (See [52, 154])) *Let $p \in [1,2)$ and $f = (f_1,\dots,f_d) : [a,b] \to \mathbb{R}^d$ be a function such that for each $k = 1,\dots,d$, $f_k \in C\mathcal{W}_p([a,b])$. Let $g : \mathbb{R}^d \to \mathbb{R}$ be a differentiable function with locally Lipschitz partial derivatives g_k', $k = 1,\dots,d$. Then each $g_l' \circ f$ is Riemann–Stieltjes integrable with respect to f_k and*

$$(g\circ f)(b) - (g\circ f)(a) = \sum_{k=1}^{d} \int_a^b (g_k' \circ f)\,df_k.$$

Proposition 1.19 (Substitution Rule, See [52, 154]) *Let f, g, and h be functions in $CW_p([a,b])$. Then for the Riemann–Stieltjes integral the following equality holds:*

$$\int_a^b f(x)\, d\left(\int_a^x g(y)\, dh(y)\right) = \int_a^b f(x)g(x)\, dh(x).$$

Remark 1.20 Recall that for any $\mu \in (0, H)$ the trajectories of the fBm B^H are μ-Hölder continuous consequently have finite $1/\mu$-variation. Furthermore, Hölder continuity supplies for any $1 - H < \alpha < 1$ the existence of fractional derivatives $\mathcal{D}_{b-}^{1-\alpha} B_{b-}^H$ and $\mathcal{D}_{b-}^{1-\alpha} B_{b-}^H \in L_\infty[a,b]$ for any $0 \le a < b$. Therefore, for a function $f \in \mathcal{I}_{a+}^{\alpha}(L_1[a,b])$ we can define the integral w. r. t. fBm in the following way:

$$\int_a^b f\, dB^H := (-1)^\alpha \int_a^b (\mathcal{D}_{a+}^\alpha f)(x)(\mathcal{D}_{b-}^{1-\alpha} B_{b-}^H)(x) dx. \tag{1.20}$$

An evident upper bound follows immediately from (1.20):

$$\left|\int_a^b f dB^H\right| \le \sup_{a \le x \le b} |(\mathcal{D}_{b-}^{1-\alpha} B_{b-}^H)(x)| \int_a^b |(\mathcal{D}_{a+}^\alpha f)(x)| dx. \tag{1.21}$$

Moreover, if $Z = \{Z_t, t \ge 0\}$ is a stochastic process whose trajectories are λ-Hölder continuous with $\lambda > 1 - H$, then the path-wise integral $\int_0^T Z_t\, dB_t^H$ is well defined and coincides with the Riemann–Stieltjes integral.

1.2 Homogeneous Diffusion Model Involving a Wiener Process

Consider a family $\{X_t^x, t \ge 0, x \in \mathbb{R}\}$ of one-dimensional homogeneous diffusion processes defined on a complete filtered probability space $(\Omega, \mathfrak{F}, \mathbb{F} = \{\mathfrak{F}_t\}_{t \ge 0}, \mathsf{P})$ by an SDE

$$dX_t^x = a(X_t^x)dt + b(X_t^x)\, dW_t, \quad t \ge 0, \tag{1.22}$$

with the initial condition $X_0^x = x \in \mathbb{R}$; here $\{W_t, t \ge 0\}$ is the standard $\mathbb{F}$-Wiener process. If the initial condition is not important, we will denote the process simply by X.

For the reader's convenience, we mention the most well-known conditions of existence–uniqueness for both strong and weak solution in Sect. 1.2.1. In Sect. 1.2.2 we introduce hitting and exit moments, the notion of local time L_t^x and its limit L_∞^x, and describe the conditions under which the diffusion process is recurrent or transient. We produce the explicit distribution of local time L_∞^x for transient processes. In Sect. 1.2.3 we apply the explicit distribution of L_∞^x to study integral functionals of the form $J_\infty(f) = \int_0^\infty f(X_s^x)ds$, which can be interpreted as continuous perpetuities in the framework of financial or insurance mathematics. We follow the approach of Salminen and Yor, used in [153] to study integral functionals of a Wiener process with a positive drift, and generalize their results to homogeneous transient diffusion processes. Applying and adapting the results of Mijatović and Urusov from [119], we establish criteria of convergence, i.e., of almost sure finiteness, of the functionals $J_\infty(f)$, calculate their moments and potentials and bound their exponential moments.

1.2.1 Existence and Uniqueness of Weak and Strong Solutions

In this section we review some well-known results concerning various sufficient conditions for the existence and uniqueness of strong and weak solutions to Eq. (1.22). The first existence-uniqueness theorem for a strong solution was obtained by Itô [79]. For a one-dimensional homogeneous equation it can be formulated in the following form.

Proposition 1.21 (Itô) *Assume that the coefficients a and b of an SDE* (1.22) *satisfy the following conditions:*

(i) for every $N > 0$ there exists a constant $C_N > 0$ such that

$$|a(x) - a(y)| + |b(x) - b(y)| \le C_N|x - y|, \quad \text{for all } x, y \in [-N, N],$$

(ii) there exists a constant $C > 0$ such that

$$|a(x)| + |b(x)| \le C(1 + |x|), \quad \text{for all } x \in \mathbb{R}.$$

Then strong existence and pathwise uniqueness hold.

For the proof of Proposition 1.21 see, e. g., [151, p. 132]. Note that the condition (ii) is needed to guarantee the solution to be non-exploding. Without (ii) the equation still has a unique strong solution up to the eventual explosion time.

Proposition 1.22 (Zvonkin [175]) *Suppose that for an SDE* (1.22) *the coefficient a is measurable and bounded, the coefficient b is continuous and bounded, and there exist constants $C > 0$, $\varepsilon > 0$ such that*

$$|b(x) - b(y)| \le C\sqrt{|x - y|}, \quad \text{for all } x, y \in \mathbb{R},$$

$$|b(t, x)| \ge \varepsilon, \quad \text{for all } x \in \mathbb{R}.$$

Then strong existence and pathwise uniqueness hold.

A stronger result was obtained by H.J. Engelbert and W. Schmidt, see [54, Th. 5.53].

Proposition 1.23 (Engelbert, Schmidt) *Suppose that for an SDE* (1.22) $b \neq 0$ *at each point,* a/b^2 *is locally integrable on* $\mathbb{R}$*, and there exists a constant* $C > 0$ *such that*

$$|b(x) - b(y)| \leq C\sqrt{|x - y|}, \quad \textit{for all } x, y \in \mathbb{R},$$
$$|a(x)| + |b(x)| \leq C(1 + |x|), \quad \textit{for all } x \in \mathbb{R}.$$

Then strong existence and pathwise uniqueness hold.

The following proposition guarantees only pathwise uniqueness.

Proposition 1.24 (Yamada, Watanabe) *Suppose that for an SDE* (1.22) *there exist a constant* $C > 0$ *and a strictly increasing function* $h: \mathbb{R}_+ \to \mathbb{R}_+$ *with* $\int_{0+} h^{-2}(x)dx = +\infty$ *such that for all* $x, y \in \mathbb{R}$*,*

$$|a(x) - a(y)| \leq C|x - y|,$$
$$|b(x) - b(y)| \leq h(|x - y|).$$

Then pathwise uniqueness holds.
For the proof see [77, 171].

Now we consider the results for weak solutions. We start with weak existence.

Proposition 1.25 (Skorokhod [158]) *Suppose that for an SDE* (1.22) *the coefficients a and b are continuous and bounded. Then weak existence holds.*

The next weak existence-uniqueness result was established in [160].

Proposition 1.26 (Stroock, Varadhan) *Suppose that for an SDE* (1.22) *the coefficient a is measurable and bounded, the coefficient b is continuous and bounded, and* $b \neq 0$ *at each point. Then weak existence and uniqueness in law hold.*

In the following proposition from [93], the diffusion coefficient b need not be continuous.

Proposition 1.27 (Krylov) *Suppose that for a homogeneous SDE* (1.22) *the coefficient a is measurable and bounded, the coefficient b is measurable and bounded, and there exists a constant* $\varepsilon > 0$ *such that for all* $x \in \mathbb{R}$

$$|b(x)| \geq \varepsilon.$$

Then weak existence and uniqueness in law hold.

More information on this topic can be found, e. g., in the books [39, 77, 112, 151].

1.2.2 The Distribution of the Local Time of the Recurrent and Transient Diffusion Processes

Let the coefficients a, b of Eq. (1.22) be continuous and satisfy any conditions of the existence of a non-explosive weak solution on $\mathbb{R}$.

Introduce the functions

$$c(x) = \frac{a(x)}{b^2(x)}, \quad \varphi(x_0, x) = \exp\left\{-2\int_{x_0}^{x} c(u)du\right\}, \quad \Phi(x_0, x) = \int_{x_0}^{x} \varphi(x_0, z)dz,$$
$$x_0, x \in \mathbb{R} \cup \{-\infty, +\infty\}. \tag{1.23}$$

For these functions to be correctly defined, from now on, we assume that the following conditions hold: $b(x) \neq 0$ for $x \in \mathbb{R}$, and the function $c = c(x)$ is locally integrable.

Now we introduce several objects related to the family $\{X_t^x, t \geq 0, x \in \mathbb{R}\}$. The generator of diffusion process X is defined for $f \in C^2(\mathbb{R})$ as

$$\mathcal{L}f(x) = \frac{b^2(x)}{2}f''(x) + a(x)f'(x).$$

It is easy to see that for a fixed $x_0 \in \mathbb{R}$ the function $\Phi(x_0, \cdot)$ solves the second order homogeneous differential equation $\mathcal{L}\Phi(x_0, \cdot) = 0$.

For $x, y \in \mathbb{R}$, let $\tau_y^x = \inf\{t \geq 0, X_t^x = y\}$ be the first moment of hitting point y. For any $(\alpha, \beta) \subset \mathbb{R}$ and $x \in (\alpha, \beta)$, let $\tau_{\alpha,\beta}^x = \inf\{t \geq 0, X_t^x \notin (\alpha, \beta)\} = \tau_\alpha^x \wedge \tau_\beta^x$ be the first moment of exiting interval (α, β). (We use the convention $\inf \varnothing = +\infty$.)

For any $t > 0$ and $y \in \mathbb{R}$, define the local time of the process X^x at the point y on the interval $[0, t]$ by

$$L_t^x(y) = b^2(y) \lim_{\varepsilon \downarrow 0} \frac{1}{2\varepsilon} \int_0^t \mathbf{1}_{\{|X_s^x - y| \leq \varepsilon\}} ds. \tag{1.24}$$

(The factor $b^2(y)$ is included to agree with the general Meyer–Tanaka definition of a local time of a semimartingales formulated e. g. in [144].) The limit in (1.24) exists almost surely and defines a continuous non-decreasing process $\{L_t^x(y), t \geq 0\}$ for any $x, y \in \mathbb{R}$. The local time on the whole interval $[0, +\infty)$ will be denoted by $L_\infty^x(y) = \lim_{t \to +\infty} L_t^x(y)$.

Definition 1.28 A diffusion process X is called *recurrent* if

$$\mathsf{P}\left(\sup_{t>0} X_t = \limsup_{t \to +\infty} X_t^x = +\infty, \inf_{t>0} X_t = \liminf_{t \to +\infty} X_t^x = -\infty\right) = 1,$$

and $L_\infty^x(y) = \infty$ for any $x, y \in \mathbb{R}$ P-a. s., and is called *transient* otherwise.

According to Theorem 3.1, [77, p. 447], see also [80], in the case when

$$\Phi(x_0, \infty) = -\Phi(x_0, -\infty) = +\infty$$

for some $x_0 \in \mathbb{R}$ and consequently

$$\Phi(x, \infty) = -\Phi(x, -\infty) = +\infty$$

for any $x \in \mathbb{R}$, the diffusion process X is recurrent. For recurrent diffusion processes $L^x_\infty(y) = \infty$ for any $x, y \in \mathbb{R}$ P-a. s.

Therefore from now on we focus on the transient diffusion processes, for which one of the integrals $\Phi(x_0, \infty)$ or $\Phi(x_0, -\infty)$ is finite. Such processes converge to $+\infty$ or $-\infty$ as $t \to \infty$, and in more detail their asymptotic behavior is described in Theorem 3.1, [77, p. 447]. We formulate it in the following form.

Theorem 1.29 *If $\Phi(x, \infty)$ and $\Phi(x, -\infty)$ are finite for some and therefore for any $x \in \mathbb{R}$, then*

$$\mathsf{P}\left\{\sup_{t>0} X^x_t = +\infty\right\} = \mathsf{P}\left\{\lim_{t\to\infty} X^x_t = +\infty\right\} = \frac{-\Phi(x, -\infty)}{-\Phi(x, -\infty) + \Phi(x, +\infty)},$$
$$\mathsf{P}\left\{\inf_{t>0} X^x_t = -\infty\right\} = \mathsf{P}\left\{\lim_{t\to\infty} X^x_t = -\infty\right\} = \frac{\Phi(x, +\infty)}{-\Phi(x, -\infty) + \Phi(x, +\infty)}. \tag{1.25}$$

If $\Phi(x, -\infty) > -\infty$, $\Phi(x, +\infty) = +\infty$, then

$$\mathsf{P}\left\{\sup_{t>0} X^x_t < \infty\right\} = \mathsf{P}\left\{\inf_{t>0} X^x_t = -\infty\right\} = \mathsf{P}\left\{\lim_{t\to\infty} X^x_t = -\infty\right\} = 1.$$

If $-\Phi(x, -\infty) = +\infty$, $\Phi(x, +\infty) < \infty$, then

$$\mathsf{P}\left\{\sup_{t>0} X^x_t = +\infty\right\} = \mathsf{P}\left\{\inf_{t>0} X^x_t > -\infty\right\} = \mathsf{P}\left\{\lim_{t\to\infty} X^x_t = +\infty\right\} = 1.$$

Now we concentrate on the explicit distribution of $L^x_\infty(y)$, considering only the case of a transient process X. We formulate the following statement concerning the distribution of $L^x_\infty(y)$. Generally speaking, it can be easily deduced from the results of [28, 59]; corresponding results are summarized also in Theorem 1 [144]. However, we give a brief proof, for the reader's convenience.

Proposition 1.30

1. *In each of the cases: $x = y$; $x < y$ and $-\Phi(0, -\infty) = +\infty$; $x > y$ and $\Phi(0, +\infty) = +\infty$, the local time $L^x_\infty(y)$ is exponentially distributed with*

parameter $\psi_y(0)$ given by the equation

$$\psi_y(0) = \frac{1}{2}\left(\frac{1}{\Phi(y,+\infty)} - \frac{1}{\Phi(y,-\infty)}\right), \tag{1.26}$$

with $\frac{1}{\infty} := 0$.

2. *If* $x < y$ *and* $-\Phi(0,-\infty) < +\infty$, *then the local time* $L^x_\infty(y)$ *is distributed as* $\kappa\xi$, *where* ξ *is exponentially distributed with parameter* $\psi_y(0)$, κ *is a Bernoulli random variable independent of* ξ *with*

$$\mathsf{P}(\kappa = 0) = 1 - \mathsf{P}(\kappa = 1) = \frac{\Phi(y,x)}{\Phi(y,-\infty)}.$$

3. *If* $x > y$ *and* $\Phi(0,+\infty) < +\infty$, *then the local time* $L^x_\infty(y)$ *is distributed as* $\kappa\xi$, *where* ξ *is exponentially distributed with parameter* $\psi_y(0)$, κ *is a Bernoulli random variable independent of* ξ *with*

$$\mathsf{P}(\kappa = 0) = 1 - \mathsf{P}(\kappa = 1) = \frac{\Phi(y,x)}{\Phi(y,+\infty)}.$$

Proof By the strong Markov property of the process X, for any $l \geq 0$ and $x, y \in \mathbb{R}$

$$\mathsf{P}(L^x_\infty(y) > l) = \mathsf{P}(L^y_\infty(y) > l)\mathsf{P}(\tau^x_y < +\infty).$$

The probability $\mathsf{P}(\tau^x_y < +\infty) = 1 - \mathsf{P}(\tau^x_y = +\infty)$ can be found with the help of the well-known formula (see e.g. [59, Section VIII.6, (18)]): for $x \in (\alpha, \beta)$

$$\mathsf{P}(X^x_{\tau_{\alpha,\beta}} = \beta) = \frac{\Phi(\alpha,x)}{\Phi(\alpha,\beta)}.$$

Then the value of probability in question depends on x, y, and integrals $\Phi(x,+\infty)$, $\Phi(-\infty,x)$. Specifically, if $x > y$, then

$$\mathsf{P}(\tau^x_y = \infty) = \lim_{\alpha\to+\infty}\mathsf{P}(X^x_{\tau_{y,\alpha}} = \alpha) = \lim_{\alpha\to+\infty}\frac{\Phi(y,x)}{\Phi(y,\alpha)},$$

whence

$$\mathsf{P}(\tau^x_y = +\infty) = \begin{cases} \frac{\Phi(y,x)}{\Phi(y,+\infty)}, & \Phi(x,+\infty) < +\infty, \\ 0, & \Phi(x,+\infty) = +\infty. \end{cases} \tag{1.27}$$

For $x < y$

$$\mathsf{P}(\tau^x_y = \infty) = \lim_{\alpha\to-\infty}\left(1 - \mathsf{P}(X^x_{\tau_{\alpha,y}} = y)\right) = \lim_{\alpha\to-\infty}\frac{\Phi(\alpha,y) - \Phi(\alpha,x)}{\Phi(\alpha,y)} =$$

$$= \lim_{\alpha\to-\infty}\frac{\varphi(\alpha,x)\Phi(x,y)}{-\varphi(\alpha,y)\Phi(y,\alpha)} = \lim_{\alpha\to-\infty}\frac{-\varphi(\alpha,x)\varphi(x,y)\Phi(y,x)}{-\varphi(\alpha,y)\Phi(y,\alpha)} = \lim_{\alpha\to-\infty}\frac{\Phi(x,y)}{\Phi(\alpha,y)};$$

therefore

$$\mathsf{P}(\tau_y^x = +\infty) = \begin{cases} \frac{\Phi(y,x)}{\Phi(y,-\infty)}, & \Phi(x,-\infty) > -\infty, \\ 0, & \Phi(x,-\infty) = -\infty. \end{cases} \tag{1.28}$$

Thus it is sufficient to determine the distribution of the variables $L_\infty^x(x)$. But it was proved in [28, II.13,II.27] that $\mathsf{P}(L_\infty^x(x) > l) = \exp(-l\psi_x(0))$, where

$$\psi_x(0) = \frac{1}{2}\left(\frac{1}{\Phi(x,+\infty)} - \frac{1}{\Phi(x,-\infty)}\right),$$

with $\frac{1}{\infty} := 0$. Hence the proof follows. □

Consider two examples when the parameters of the distribution of the local time can be calculated explicitly.

Example 1.31 Let $a(x) \equiv a$ be constant and $b(x) \equiv b \neq 0$. Then $\varphi(x,y) = e^{-\frac{2a(y-x)}{b^2}}$, $\Phi(x,y) = \frac{b^2}{2a}(1 - e^{-2a(y-x)/b^2})$ for $a \neq 0$ and $\Phi(x,y) = y - x$ for $a = 0$. In this case the diffusion process X is transient if and only if $a \neq 0$, moreover, $\Phi(0,-\infty) = -\infty$ and $\Phi(0,+\infty) = \frac{b^2}{2a} < +\infty$ for $a > 0$, and $\Phi(0,-\infty) = \frac{b^2}{2a} > -\infty$, $\Phi(0,+\infty) = +\infty$ for $a < 0$. The cases are symmetric, therefore we will consider only the case $a > 0$.

Now $\psi_x(0) = \frac{a}{b^2}$. Thus, for $x \leq y$ the local time $L_\infty^x(y)$ is exponentially distributed with parameter $\frac{a}{b^2}$. For $x > y$ the local time is distributed as $\kappa\xi$, where ξ has an exponential distribution with parameter $\frac{a}{b^2}$ and κ is Bernoulli random variable independent of ξ and distributed as $\mathsf{P}(\kappa = 1) = 1 - \mathsf{P}(\kappa = 0) = e^{-2a(x-y)/b^2}$. Using the properties of the exponential distribution, we see that these cases can be combined: $L_\infty^x(y) \overset{d}{=} (\xi - 2(x-y)_+)_+$, where $\gamma_+ = \gamma \vee 0$.

Example 1.32 Let $a(x) = x$ and $b(x) = \sqrt{x^2+1}$. Then $\varphi(x,y) = \frac{x^2+1}{y^2+1}$ and $\Phi(x,y) = (1+x^2)(\arctan y - \arctan x)$. We see that the process is transient, $-\Phi(0,-\infty) = \Phi(0,\infty) = \frac{\pi}{2}$.

Due to Proposition 1.30, for $x < y$ the local time $L_\infty^x(y)$ is distributed as $\kappa\xi$, where ξ has an exponential distribution with parameter

$$\begin{aligned} \psi_y(0) &= \frac{1}{2\Phi(y,+\infty)} - \frac{1}{2\Phi(y,-\infty)} \\ &= \frac{1}{(1+y^2)(\pi - 2\arctan y)} - \frac{1}{(1+y^2)(\pi + 2\arctan y)} \\ &= \frac{4\arctan y}{(1+y^2)(\pi^2 - 4\arctan^2 y)}, \end{aligned}$$

and κ is a Bernoulli random variable, which is independent of ξ and distributed as

$$\mathsf{P}(\kappa = 0) = 1 - \mathsf{P}(\kappa = 1) = \frac{\arctan y - \arctan x}{\frac{\pi}{2} + \arctan y}.$$

For $x > y$ the local time $L^x_\infty(y)$ is distributed as $\kappa\xi$, where ξ has an exponential distribution with parameter $\psi_y(0)$, and κ is a Bernoulli random variable, which is independent of ξ and distributed as

$$\mathsf{P}(\kappa = 0) = 1 - \mathsf{P}(\kappa = 1) = \frac{\arctan x - \arctan y}{\frac{\pi}{2} - \arctan y}.$$

For $x = y$ the local time $L^y_\infty(y)$ is exponentially distributed with parameter $\psi_y(0)$.

1.2.3 Integral Functionals of a Transient Diffusion Processes

For a measurable function $f\colon \mathbb{R} \to \mathbb{R}$ such that f and f/b^2 are locally integrable, define the integral functionals

$$J^x_t(f) = \int_0^t f(X^x_s)ds$$

and

$$J^x_\infty(f) = \int_0^\infty f(X^x_s)ds.$$

We will study the questions of finiteness and existence of moments of $J^x_\infty(f)$. We start with the well-known occupation density formula (see e. g. [144, formula (21)]). It states that for any locally integrable function f the following equalities hold: for any $t > 0$

$$J^x_t(f) = \int_{\mathbb{R}} \frac{f(y)}{b^2(y)} L^x_t(y)dx, \tag{1.29}$$

and

$$J^x_\infty(f) = \int_{\mathbb{R}} \frac{f(y)}{b^2(y)} L^x_\infty(y)dy. \tag{1.30}$$

As it was mentioned before, if the process X^x is recurrent, then $L^x_\infty(y) = \infty$ a. s. for all $y \in \mathbb{R}$, so $J^x_\infty(f)$ is undefined unless f is identically zero. Of course, it can equal $+\infty$ or $-\infty$ if f is positive or negative, correspondingly. Therefore, we will require that the process X is transient. We recall that this holds if and only if

$\Phi(0,+\infty)$ or $\Phi(0,-\infty)$ is finite, and for other details see Sect. 1.2.2. Moreover, if $\Phi(0,+\infty) < +\infty$, $\Phi(0,-\infty) = -\infty$, then $X_s^x \to +\infty$ a. s.; if $\Phi(0,+\infty) = +\infty$, $\Phi(0,-\infty) > -\infty$ then $X_s^x \to -\infty$ a. s.; if $\Phi(0,+\infty) < +\infty$, $\Phi(0,-\infty) > -\infty$ then $X_s^x \to +\infty$ on a set A_+ of positive probability, and $X_s^x \to -\infty$ on a set $A_- = \Omega \setminus A_+$ of positive probability. The values of these probabilities are represented by formulas (1.25).

We start with a criterion of almost sure finiteness of $J_\infty^x(f)$. It was obtained in [88] in the case where only one of the integrals $\Phi(0,+\infty)$ of $\Phi(0,-\infty)$ is finite; the complete analysis was made in [119]. To acquaint the reader with these results, we define

$$I_1(f) = \int_0^{+\infty} \frac{|f(y)|}{b^2(y)} \Phi(y,+\infty)dy, \quad I_2(f) = \int_{-\infty}^0 \frac{|f(y)|}{b^2(y)} \Phi(y,-\infty)dy.$$

Theorem 1.33 ([119]) *For arbitrary $x \in \mathbb{R}$, the following statements hold.*

- *Let $\Phi(0,+\infty) < +\infty$, $\Phi(0,-\infty) = -\infty$.*
  If the integral $I_1(f)$ is finite, i.e. $I_1(f) < +\infty$, then the integral functional $J_\infty^x(f)$ is finite a. s., i. e. $J_\infty^x(f) \in \mathbb{R}$ a. s.
  If the integral $I_1(f)$ is infinite, i.e. $I_1(f) = \infty$ then the integral functional $J_\infty^x(f)$ is infinite a. s., i. e. $J_\infty^x(f) = \infty$ a. s.
- *Let $\Phi(0,+\infty) = +\infty$, $\Phi(0,-\infty) > -\infty$.*
  If the integral $I_2(f)$ is finite, i. e. $I_2(f) > -\infty$, then the integral functional $J_\infty^x(f)$ is finite a. s., i. e. $J_\infty^x(f) \in \mathbb{R}$ a. s.
  If the integral $I_2(f)$ is infinite, i. e. $I_2(f) = -\infty$ then the integral functional $J_\infty^x(f)$ is infinite a. s., i. e. $J_\infty^x(f) = \infty$ a. s.
- *Let $\Phi(0,+\infty) < +\infty$, $\Phi(0,-\infty) > -\infty$.*
  If the integral $I_1(f)$ is finite, i. e. $I_1(f) < +\infty$, then $J_\infty^x(f) \in \mathbb{R}$ a. s. on the set A_+.
  If the integral $I_1(f)$ is infinite, i.e. $I_1(f) = \infty$ then $J_\infty^x(f) = \infty$ a. s. on the set A_+.
  If the integral $I_2(f)$ is finite, i. e. $I_2(f) > -\infty$, then $J_\infty^x(f) \in \mathbb{R}$ a. s. on the set A_-.
  If the integral $I_2(f)$ is infinite, i. e. $I_2(f) = -\infty$ then $J_\infty^x(f) = \infty$ a. s. on the set A_-.

In what follows consider the case when $\Phi(0,+\infty) < +\infty$, $\Phi(0,-\infty) = -\infty$, other cases being similar. The next result is a direct consequence of Proposition 1.30.

Lemma 1.34 *Let $\Phi(0,+\infty) < +\infty$, $\Phi(0,-\infty) = -\infty$. Then for any $k \geq 1$*

$$\mathsf{E}\left[L_\infty^x(y)^k\right] = k!(2\Phi(y,+\infty))^k \text{ for } x \leq y$$

and

$$\mathsf{E}\left[L_\infty^x(y)^k\right] = 2^k k!\Phi(y,+\infty)^{k-1}\big(\Phi(y,+\infty) - \Phi(y,x)\big)$$
$$= 2^k k!\Phi(y,+\infty)^{k-1}\varphi(y,x)\Phi(x,+\infty)$$

for $x > y$.

Example 1.35 Return to Example 1.31 and even simplify it. More precisely, let $a(x) = a > 0$ for some constant a, $b(x) = 1$, so that X is a Brownian motion with constant positive drift. In this case $\varphi(x,y) = \exp\{-2a(y-x)\}$, $\Phi(x,+\infty) = 1/2a$, $\Phi(x,-\infty) = -\infty$ for any $x \in \mathbb{R}$. Therefore, the criterion for $J_\infty(f)$ to be finite is $\int_0^\infty |f(x)|dx < \infty$, which coincides with that of [153]. As to the moments of local times, in this case $\mathsf{E}\left[L_\infty^x(y)\right] = 1/a$ for $x \le y$ and $\mathsf{E}\left[L_\infty^x(y)\right] = \frac{1}{a}\exp\{-2a(x-y)\}$ for $x > y$.
Further we derive conditions for $\mathsf{E}\left[J_\infty^x(f)\right]$ to be finite.

Theorem 1.36 *Let* $\Phi(0,+\infty) < +\infty, \Phi(0,-\infty) = -\infty$ *and* $I_1(f) < +\infty$. *Assume additionally that*

$$\int_{-\infty}^x \frac{|f(u)|}{b^2(u)}\varphi(u,x)du < +\infty.$$

Then

$$\mathsf{E}\left[J_\infty^x(f)\right] = 2\int_x^{+\infty} \frac{f(u)}{b^2(u)}\Phi(u,+\infty)du + 2\Phi(x,+\infty)\int_{-\infty}^x \frac{f(u)}{b^2(u)}\varphi(u,x)du.$$

Proof The statement immediately follows from Lemma 1.34 and the Fubini theorem. □

Remark 1.37 For a Brownian motion with positive drift a, a sufficient condition for $\mathsf{E}\left[J_\infty^x(f)\right]$ to be finite is

$$\int_x^{+\infty} |f(u)|du + e^{-2ax}\int_{-\infty}^x |f(u)|e^{2au}du < \infty;$$

and in that case the equality

$$\mathsf{E}\left[J_\infty^x(f)\right] = \frac{1}{a}\int_x^{+\infty} f(u)du + \frac{1}{a}e^{-2ax}\int_{-\infty}^x f(u)e^{2au}du$$

holds. Obviously, the requirement: f is non-negative, locally integrable, integrable at $+\infty$ and $f(x) = O\left(|x|^m\right)_{x\to-\infty}$, is also sufficient, which is stated in [153, Prop. 4.1].

Now we continue with the moments of $J_\infty(f)$ of higher order.

Theorem 1.38 *Let $\Phi(0,+\infty) < +\infty, \Phi(0,-\infty) = -\infty$. The moments of higher order admit the following bound: for any $k > 1$*

$$\left(\mathsf{E}\left[|J^x_\infty(f)|^k\right]\right)^{1/k} \le 2(k!)^{1/k}\left(\int_x^{+\infty}\frac{|f(u)|}{b^2(u)}\Phi(u,+\infty)du\right.$$
$$\left.+\,\Phi(x,+\infty)^{1/k}\int_{-\infty}^x \frac{|f(u)|}{b^2(u)}\Phi(u,+\infty)^{1-1/k}\varphi(u,x)^{1/k}du\right). \tag{1.31}$$

Proof We use representation (1.30) and the generalized Minkowski inequality to get the following equalities and bounds:

$$\left(\mathsf{E}\left[|J^x_\infty(f)|^k\right]\right)^{1/k} = \left(\mathsf{E}\left[\left|\int_{\mathbb{R}}\frac{f(y)}{b^2(u)}L^x_\infty(y)dy\right|^k\right]\right)^{1/k}$$
$$\le \int_{\mathbb{R}}\frac{|f(y)|}{b^2(u)}\left(\mathsf{E}\left[L^x_\infty(y)^k\right]\right)^{1/k}dy. \tag{1.32}$$

Now (1.31) follows immediately from (1.32) and Lemma 1.34. □

We conclude with the existence of potential and exponential moments for diffusion processes. Related results were obtained in [88], only, we are in doubt why in some formulas of [88] the factor b^2 is placed in the numerator instead of denominator.

Definition 1.39 The integral functional $J_\infty(f)$ has a bounded potential P if

$$P := \sup_{x\in\mathbb{R}}\mathsf{E}\left[J^x_\infty(f)\right] < \infty.$$

The following result is an immediate corollary of Theorem 1.36.

Theorem 1.40 *Let $\Phi(0,+\infty) < +\infty, \Phi(0,-\infty) = -\infty$ and*

$$P_0 = 2\sup_{x\in\mathbb{R}}\left(\int_x^{+\infty}\frac{|f(u)|}{b^2(u)}\Phi(u,+\infty)du + \Phi(x,+\infty)\int_{-\infty}^x\frac{|f(u)|}{b^2(u)}\varphi(u,x)du\right) < \infty.$$

Then the integral functional $J_\infty(f)$ has a bounded potential $P \le P_0$.

Theorem 1.41 *Let $\Phi(0,+\infty) < +\infty, \Phi(0,-\infty) = -\infty$ and $P_0 < \infty$. Then*

$$\mathsf{E}\left[\exp(\lambda J^x_\infty(f))\right] \le \frac{1}{1-\lambda P_0}$$

for $\lambda P_0 < 1$.

Proof We apply the following result of Dellacherie and Meyer [43], see also [153, Lemma 5.2]. Let A be a continuous adapted non-decreasing process starting at zero such that there exists a constant $C > 0$ satisfying $\mathsf{E}[A_\infty - A_t \mid \mathfrak{F}_t] \le C$ for any $t \ge 0$. Then

$$\mathsf{E}[\exp(\lambda A_\infty)] \le \frac{1}{1-\lambda C}$$

for $\lambda < C^{-1}$.

Set $A_t = \int_0^t |f(X_s^x)|ds$. Then it follows from the Markov property of X and Theorems 1.36 and 1.40 that $\mathsf{E}[A_\infty - A_t \mid \mathfrak{F}_t] \le P_0$ for any $t \ge 0$ whence the proof follows. □

1.3 Stochastic Differential Equations Involving fBm

Consider an SDE driven by an fBm $B^H = \{B_t^H,\ t \in [0,T]\}$, $H \in (1/2, 1)$ on a complete probability space $(\Omega, \mathfrak{F}, \mathsf{P})$:

$$X_t = x_0 + \int_0^t a(s, X_s)ds + \int_0^t b(s, X_s)dB_s^H, \quad t \in [0,T]. \tag{1.33}$$

Here $x_0 \in \mathbb{R}$, $a, b : [0,T] \times \mathbb{R} \to \mathbb{R}$ are measurable functions. Such equations are very frequently met in various applications. The list, though being far from complete, includes the following fractional versions of well known models (see [65, 76, 120, 139, 157] and references therein) with corresponding fields of applications given in the brackets:

- Fractional Brownian Traffic equation $X_t = at + \sigma B_t^H$ (networking);
- Black–Scholes equation $X_t = x_0 + \lambda \int_0^t X_s ds + \sigma \int_0^t X_s dB_s^H$ (finance);
- Langevin equation $X_t = x_0 - \lambda \int_0^t X_s ds + \sigma B_t^H$, or, in more general form with drift, $X_t = x_0 + \int_0^t (\alpha - \beta X_s)\, ds + \sigma B_t^H$, whose solution is called the Ornstein–Uhlenbeck process or fractional Vasicek model (physics, finance, networking);
- Verhulst equation $X_t = x_0 + \int_0^t (\lambda X_s - X_s^2) ds + \sigma \int_0^t X_s dB_s^H$ (demography, biology);
- Landau–Ginzburg equation $X_t = x_0 + \int_0^t (\lambda X_s - X_s^3) ds + \sigma \int_0^t X_s dB_s^H$ (physics);
- Gompertz diffusion process $X_t = x_0 + \int_0^t (\alpha X_s - \beta X_s \log X_s)\, ds + \sigma \int_0^t X_s\, dB_s^H$.

It is therefore clear that an area of applications of such equations is very wide and there are many results devoted to estimation problems in models of this type.

This subsection is devoted to the existence-uniqueness theorems for SDEs driven by fBm. We start with the fractional diffusion model of general form and formulate the sufficient conditions for unique solvability of Eq. (1.33). Then we study in detail two particular cases: the SDE with polynomial drift and the Gompertz diffusion process. We get their explicit solutions. Note that both types of equations are transformable, or reducible, in the sense that they can be reduced to linear equations by some smooth transformations which do not change the initial flows of sigma-fields.

1.3.1 Existence and Uniqueness of Solution

The conditions ensuring existence and uniqueness of $X = \{X_t, t \in [0,T]\}$, which satisfies (1.33) were established in [140]. We formulate the following version of this result. Let us consider the following assumptions on the coefficients.

(A$_1$) $b(t,x)$ is differentiable in x, and there exist some constants $0 < \beta, \delta \le 1$ and for every $N \ge 0$ there exists $M_N > 0$ such that the following properties hold:

(i) Lipschitz continuity in x

$$|b(t,x) - b(t,y)| \le M_0 |x-y|, \quad \forall x,y \in \mathbb{R},\ t \in [0,T];$$

(ii) local uniform Hölder continuity of the derivative in x

$$\left|b'_x(t,x) - b'_x(t,y)\right| \le M_N |x-y|^\delta, \qquad \forall x,y \in [-N,N],\ \forall t \in [0,T];$$

(iii) Hölder continuity in t

$$|b(s,x) - b(t,x)| + \left|b'_x(s,x) - b'_x(t,x)\right| \le M_0 |t-s|^\beta, \qquad \forall x,\ \forall t,s \in [0,T];$$

(A$_2$) For every $N \ge 0$ there exists $L_N > 0$ such that the following properties hold:

(i) local uniform Lipschitz continuity in x

$$|a(t,x) - a(t,y)| \le L_N |x-y|, \qquad \forall x,y \in [-N,N],\ \forall t \in [0,T];$$

(ii) linear growth

$$|a(t,x)| \le L_0(|x|+1), \qquad \forall x \in \mathbb{R},\ \forall t \in [0,T].$$

Fix a parameter $\alpha \in (0,1/2)$. Let $W_\infty^\alpha([0,T])$ be the space of real-valued measurable functions $f\colon [0,T] \to \mathbb{R}$ such that

$$\|f\|_{\infty,\alpha;T} = \sup_{s\in[0,T]} \left(|f(s)| + \int_0^s |f(s) - f(u)|\,(s-u)^{-1-\alpha} du \right) < \infty.$$

Theorem 1.42 ([140, Theorem 2.1]) *Suppose that $x_0 \in \mathbb{R}$, and coefficients a and b satisfy assumptions* (A$_1$) *and* (A$_2$) *with $\beta > 1 - H$, $\delta > \frac{1}{H} - 1$. Then there exists a unique solution $X = \{X_t, t \in [0,T]\}$ of Eq.* (1.33) *such that $X \in W_\infty^\alpha([0,T]) \cap C^{H-\varepsilon}([0,T])$ a. s. for any $0 < \varepsilon < H$.*

1.3.2 Fractional Ornstein–Uhlenbeck Process

The present subsection deals with the Langevin equation that will be written as

$$X_t = x_0 + \theta \int_0^t X_s \, ds + \sigma B_t^H, \quad t \geq 0, \tag{1.34}$$

where $x_0 \in \mathbb{R}$, $\theta \in \mathbb{R}$, $\sigma > 0$ are real parameters, and B^H is an fBm with the Hurst index H. Since the model does not contain stochastic integrals with respect to fBm, it is possible to consider an fBm with arbitrary $H \in (0, 1)$. For all $H \in (0, 1)$ the Eq. (1.34) has a unique solution

$$X_t = x_0 e^{\theta t} + \theta \sigma e^{\theta t} \int_0^t e^{-\theta s} B_s^H \, ds + \sigma B_t^H, \quad t \geq 0, \tag{1.35}$$

which is called a *fractional Ornstein–Uhlenbeck process* (fO-U) [38].

Note that for $H \in (1/2, 1)$ the process (1.35) can be written in the following form

$$X_t = x_0 e^{\theta t} + \sigma \int_0^t e^{\theta(t-s)} dB_s^H,$$

where $\int_0^t e^{\theta(t-s)} dB_s^H$ is a Wiener integral w.r.t. the fBm. Evidently, as it was also emphasized in [38], the integral $\int_0^t e^{\theta(t-s)} dB_s^H$ exists for all $H \in (0, 1)$, if we understand it via integration by parts: $\int_0^t e^{-\theta s} dB_s^H = \theta \int_0^t e^{-\theta s} B_s^H \, ds + e^{-\theta t} B_t^H$.

Let us calculate the covariance function of the fO-U and describe the properties of a stationary Gaussian process related to the model.

1.3.2.1 Covariance Structure of the Fractional Ornstein–Uhlenbeck Process

Theorem 1.43 *Let $H \in (0, 1)$, $t \geq s \geq 0$. Then the covariance function of the fO-U* (1.35) *is given by*

$$\begin{aligned} \operatorname{cov}(X_t, X_s) = \frac{H\sigma^2}{2} \Bigg(& -e^{\theta t - \theta s} \int_0^{t-s} e^{-\theta z} z^{2H-1} dz + e^{-\theta t + \theta s} \int_{t-s}^t e^{\theta z} z^{2H-1} dz \\ & - e^{\theta t + \theta s} \int_s^t e^{-\theta z} z^{2H-1} dz + e^{\theta t - \theta s} \int_0^s e^{\theta z} z^{2H-1} dz \\ & + 2 e^{\theta t + \theta s} \int_0^t e^{-\theta z} z^{2H-1} dz \Bigg). \end{aligned} \tag{1.36}$$

Proof Using (1.35) and the formula for covariance of an fBm, we can write

$$\begin{aligned}
\operatorname{cov}(X_t, X_s) &= \mathsf{E}\left[\left(X_t - x_0 e^{\theta t}\right)\left(X_s - x_0 e^{\theta s}\right)\right] \\
&= \mathsf{E}\left[\left(\theta\sigma e^{\theta t}\int_0^t e^{-\theta u} B_u^H du + \sigma B_t^H\right)\left(\theta\sigma e^{\theta s}\int_0^s e^{-\theta v} B_v^H dv + \sigma B_s^H\right)\right] \\
&= \frac{\theta\sigma^2}{2} e^{\theta t}\int_0^t e^{-\theta u}\left(u^{2H} + s^{2H} - |u-s|^{2H}\right) du \\
&\quad + \frac{\theta\sigma^2}{2} e^{\theta s}\int_0^s e^{-\theta v}\left(v^{2H} + t^{2H} - |v-t|^{2H}\right) dv \\
&\quad + \frac{\sigma^2}{2}\left(t^{2H} + s^{2H} - |t-s|^{2H}\right) \\
&\quad + \frac{\theta^2\sigma^2}{2} e^{\theta t + \theta s}\int_0^t\int_0^s e^{-\theta u - \theta v}\left(u^{2H} + v^{2H} - |u-v|^{2H}\right) du\, dv \\
&= \frac{\sigma^2}{2}\sum_{n=1}^{10} I_n,
\end{aligned}$$

where

$$I_1 = \theta e^{\theta t}\int_0^t e^{-\theta u} s^{2H} du, \quad I_2 = \theta e^{\theta t}\int_0^t e^{-\theta u} u^{2H} du, \quad I_3 = -\theta e^{\theta t}\int_0^t e^{-\theta u}|u-s|^{2H} du,$$

$$I_4 = \theta e^{\theta s}\int_0^s e^{-\theta v} t^{2H} dv, \quad I_5 = \theta e^{\theta s}\int_0^s e^{-\theta v} v^{2H} dv, \quad I_6 = -\theta e^{\theta s}\int_0^s e^{-\theta v}(t-v)^{2H} dv,$$

$$I_7 = t^{2H} + s^{2H} - (t-s)^{2H}, \quad I_8 = \theta^2 e^{\theta t + \theta s}\int_0^t e^{-\theta v} dv \int_0^s e^{-\theta u} u^{2H} du,$$

$$I_9 = \theta^2 e^{\theta t + \theta s}\int_0^s e^{-\theta u} du \int_0^t e^{-\theta v} v^{2H} dv,$$

$$I_{10} = -\theta^2 e^{\theta t + \theta s}\int_0^t\int_0^s e^{-\theta u - \theta v}|u-v|^{2H}\, du\, dv.$$

The first two integrals are equal to

$$I_1 = s^{2H}\left(e^{\theta t} - 1\right) \quad \text{and} \quad I_2 = -e^{\theta t}\int_0^t u^{2H} de^{-\theta u} = -t^{2H} + 2He^{\theta t}\int_0^t e^{-\theta u} u^{2H-1} du.$$

By changes of variables and integration by parts, we obtain

$$\begin{aligned}
I_3 &= -\theta e^{\theta t}\int_0^s e^{-\theta u}(s-u)^{2H} du - \theta e^{\theta t}\int_s^t e^{-\theta u}(u-s)^{2H} du \\
&= -\theta e^{\theta t - \theta s}\int_0^s e^{\theta z} z^{2H} dz - \theta e^{\theta t - \theta s}\int_0^{t-s} e^{-\theta z} z^{2H} dz
\end{aligned}$$

$$= -e^{\theta t-\theta s}\left(e^{\theta s}s^{2H} - 2H\int_0^s e^{\theta z}z^{2H-1}dz - e^{-\theta(t-s)}(t-s)^{2H}\right.$$
$$\left. + 2H\int_0^{t-s} e^{-\theta z}z^{2H-1}dz\right)$$
$$= -e^{\theta t}s^{2H} + (t-s)^{2H} + 2He^{\theta t-\theta s}\int_0^s e^{\theta z}z^{2H-1}dz - 2He^{\theta t-\theta s}\int_0^{t-s} e^{-\theta z}z^{2H-1}dz.$$

Similarly to I_1, I_2, I_3, we transform I_4, I_5, I_6:

$$I_4 = t^{2H}\left(e^{\theta s} - 1\right), \qquad I_5 = -s^{2H} + 2He^{\theta s}\int_0^s e^{-\theta v}v^{2H-1}dv,$$
$$I_6 = -\theta e^{\theta s-\theta t}\int_{t-s}^t e^{\theta z}z^{2H}dz = -e^{\theta s-\theta t}\int_{t-s}^t z^{2H}de^{\theta z}$$
$$= -e^{\theta s}t^{2H} + (t-s)^{2H} + 2He^{\theta s-\theta t}\int_{t-s}^t e^{\theta z}z^{2H-1}dz.$$

Further,

$$I_8 = e^{\theta t+\theta s}\left(e^{-\theta t} - 1\right)\int_0^s u^{2H}de^{-\theta u}$$
$$= (1-e^{\theta t})s^{2H} - 2He^{\theta s}(1-e^{\theta t})\int_0^s e^{-\theta u}u^{2H-1}du,$$

and similarly

$$I_9 = (1-e^{\theta s})t^{2H} - 2He^{\theta t}(1-e^{\theta s})\int_0^t e^{-\theta v}v^{2H-1}dv.$$

Finally, we consider the term I_{10}. It can be represented as the sum of the following integrals:

$$I_{10} = -\theta^2 e^{\theta t+\theta s}\int_0^s\int_0^v e^{-\theta u-\theta v}(v-u)^{2H}\,du\,dv$$
$$-\theta^2 e^{\theta t+\theta s}\int_0^s\int_v^s e^{-\theta u-\theta v}(u-v)^{2H}\,du\,dv$$
$$-\theta^2 e^{\theta t+\theta s}\int_s^t\int_0^s e^{-\theta u-\theta v}(v-u)^{2H}\,du\,dv$$
$$= -2\theta^2 e^{\theta t+\theta s}\int_0^s\int_0^v e^{-\theta u-\theta v}(v-u)^{2H}\,du\,dv$$
$$-\theta^2 e^{\theta t+\theta s}\int_s^t\int_0^s e^{-\theta u-\theta v}(v-u)^{2H}\,du\,dv =: I'_{10} + I''_{10}.$$

Using the change of variables $v - u = z$, the change of order of integration, and integration by parts, we get

$$I'_{10} = -2\theta^2 e^{\theta t+\theta s} \int_0^s e^{-2\theta v} \int_0^v e^{\theta z} z^{2H} dz dv = -2\theta^2 e^{\theta t+\theta s} \int_0^s e^{\theta z} z^{2H} \int_z^s e^{-2\theta v} dv dz$$

$$= -2\theta^2 e^{\theta t+\theta s} \int_0^s e^{\theta z} z^{2H} \frac{e^{-2\theta s} - e^{-2\theta z}}{-2\theta} dz$$

$$= \theta e^{\theta t-\theta s} \left(\int_0^s e^{\theta z} z^{2H} dz - \int_0^s e^{-\theta u} u^{2H} du \right)$$

$$= e^{\theta t} s^{2H} - 2He^{\theta t-\theta s} \int_0^s e^{\theta z} z^{2H-1} dz + e^{\theta t} s^{2H} - 2He^{\theta t+\theta s} \int_0^s e^{-\theta u} u^{2H-1} du.$$

In order to simplify I''_{10}, we need to consider two cases.

If $t > 2s$, then using the change of variables $v - u = z$ in the inner integral, changing the order of integration, and integrating with respect to v, we obtain

$$I''_{10} = -\theta^2 e^{\theta t+\theta s} \int_s^t \int_{v-s}^v e^{\theta z-2\theta v} z^{2H} dz dv = -\theta^2 e^{\theta t+\theta s} \left(\int_0^s \int_s^{z+s} e^{\theta z-2\theta v} z^{2H} dv dz \right.$$

$$\left. + \int_s^{t-s} \int_z^{z+s} e^{\theta z-2\theta v} z^{2H} dv dz + \int_{t-s}^t \int_z^t e^{\theta z-2\theta v} z^{2H} dv dz \right)$$

$$= -\theta^2 e^{\theta t+\theta s} \left(\int_0^s e^{\theta z} z^{2H} \frac{e^{-2\theta(z+s)} - e^{-2\theta s}}{-2\theta} dz \right.$$

$$\left. + \int_s^{t-s} e^{\theta z} z^{2H} \frac{e^{-2\theta(z+s)} - e^{-2\theta z}}{-2\theta} dz + \int_{t-s}^t e^{\theta z} z^{2H} \frac{e^{-2\theta t} - e^{-2\theta z}}{-2\theta} dz \right)$$

$$= \frac{\theta}{2} e^{\theta t-\theta s} \int_0^{t-s} e^{-\theta z} z^{2H} dz - \frac{\theta}{2} e^{\theta t+\theta s} \int_s^t e^{-\theta z} z^{2H} dz - \frac{\theta}{2} e^{\theta t-\theta s} \int_0^s e^{\theta z} z^{2H} dz$$

$$+ \frac{\theta}{2} e^{\theta s-\theta t} \int_{t-s}^t e^{\theta z} z^{2H} dz.$$

Integrating by parts in each integral, we arrive at the following equality

$$I''_{10} = e^{\theta s} t^{2H} - e^{\theta t} s^{2H} - He^{\theta s-\theta t} \int_{t-s}^t e^{\theta z} z^{2H-1} dz + He^{\theta t-\theta s} \int_0^s e^{\theta z} z^{2H-1} dz$$

$$- He^{\theta t+\theta s} \int_s^t e^{-\theta z} z^{2H-1} dz + He^{\theta t-\theta s} \int_0^{t-s} e^{-\theta z} z^{2H-1} dz - (t-s)^{2H}.$$

Similarly one can verify the last formula for the case $s < t < 2s$.

Thus, summing up all the terms, we get (1.36). □

Corollary 1.44 *The r. v. X_t has Gaussian distribution $\mathcal{N}\left(x_0 e^{\theta t}, v(\theta,t)\right)$, with variance*

$$v(\theta,t) := \operatorname{Var} X_t = H\sigma^2 \int_0^t z^{2H-1}\left(e^{\theta z} + e^{\theta(2t-z)}\right) dz. \tag{1.37}$$

Let us investigate the asymptotic behavior of the variance $v(\theta,t)$ as $t \to \infty$.

Lemma 1.45

(i) If $\theta > 0$, then $v(\theta,t) \sim \frac{\sigma^2 H\Gamma(2H)}{\theta^{2H}} e^{2\theta t}$, as $t \to \infty$.
(ii) If $\theta < 0$, then $v(\theta,t) \to \frac{\sigma^2 H\Gamma(2H)}{(-\theta)^{2H}}$, as $t \to \infty$.
(iii) $v(0,t) = \sigma^2 t^{2H}$, $t \geq 0$.

Proof

(*i*) If $\theta > 0$, then by formula (1.37),

$$\frac{v(\theta,t)}{e^{2\theta t}} = H\sigma^2 e^{-2\theta t} \int_0^t z^{2H-1} e^{\theta z} dz + H\sigma^2 \int_0^t z^{2H-1} e^{-\theta z} dz \to \frac{\sigma^2 H\Gamma(2H)}{\theta^{2H}},$$

as $t \to \infty$.

(*ii*) Note that $v(\theta,t) = e^{2\theta t} v(-\theta,t)$, by (1.37). Then the convergence follows from (*i*).

(*iii*) The statement follows directly from (1.37). □

1.3.2.2 Stationary Fractional Ornstein–Uhlenbeck Process

Let $\theta < 0$. Consider an fBm defined on $\mathbb{R}$. In this case the process

$$Y_t = \sigma \int_{-\infty}^t e^{\theta(t-s)}\, dB_s^H := \theta\sigma e^{\theta t} \int_{-\infty}^t e^{-\theta s} B_s^H\, ds + \sigma B_t^H, \quad t \geq 0. \tag{1.38}$$

is well-defined. Indeed,

$$\int_{-\infty}^t \int_{-\infty}^t e^{-\theta(s+u)} |s-u|^{2H-2} du\, ds < \infty,$$

and the integral $\int_{-\infty}^t e^{-\theta s} B_s^H\, ds$ is well defined by (1.8).

Then the fO-U (1.35) can be expressed as

$$X_t = x_0 e^{\theta t} + Y_t - e^{\theta t} Y_0, \quad t \geq 0. \tag{1.39}$$

In view of the next result, the process Y is called a *stationary fractional Ornstein–Uhlenbeck process*.

Theorem 1.46 *Let $\theta < 0$, $H \in (0,1)$. Then the centered Gaussian process $Y = \{Y_t, t \geq 0\}$ defined by* (1.38) *is stationary and ergodic.*

The stationarity of Y follows immediately from the stationarity of the increments of an fBm. However, we can compute the covariance function explicitly.

Lemma 1.47 *Let $\theta < 0$, $H \in (0,1)$. For $t \geq s \geq 0$, the covariance function of Y is given by*

$$\operatorname{cov}(Y_t, Y_s) = \frac{H\sigma^2}{2}\left(\frac{\Gamma(2H)}{(-\theta)^{2H}}e^{\theta(t-s)} - e^{\theta(t-s)}\int_0^{t-s} e^{-\theta z}z^{2H-1}dz\right.$$
$$\left. + e^{-\theta(t-s)}\int_{t-s}^{+\infty} e^{\theta z}z^{2H-1}dz\right) =: r(t-s).$$

Proof By (1.39),

$$\operatorname{cov}(Y_t, Y_s) = \operatorname{cov}(X_t, X_s) + e^{\theta s}\operatorname{cov}(X_t, Y_0) + e^{\theta t}\operatorname{cov}(X_s, Y_0) + e^{\theta t+\theta s}\mathsf{E}Y_0^2. \tag{1.40}$$

The first term was calculated in Theorem 1.43. Similarly one can compute

$$\operatorname{cov}(X_t, Y_0) = \frac{H\sigma^2}{2}\left(-\frac{\Gamma(2H)}{(-\theta)^{2H}}e^{\theta t} - e^{\theta t}\int_0^t e^{-\theta z}z^{2H-1}dz + e^{-\theta t}\int_t^\infty e^{\theta z}z^{2H-1}dz\right),$$
$$\operatorname{cov}(X_s, Y_0) = \frac{H\sigma^2}{2}\left(-\frac{\Gamma(2H)}{(-\theta)^{2H}}e^{\theta s} - e^{\theta s}\int_0^s e^{-\theta z}z^{2H-1}dz + e^{-\theta s}\int_s^\infty e^{\theta z}z^{2H-1}dz\right),$$
$$\mathsf{E}Y_0^2 = \frac{\sigma^2 H\Gamma(2H)}{(-\theta)^{2H}}. \tag{1.41}$$

The proof can be concluded now by inserting these expressions together with (1.36) into (1.40). Here we take into account the equality $\int_0^\infty e^{\theta z}z^{2H-1}dz = \frac{\Gamma(2H)}{(-\theta)^{2H}}$. □

Remark 1.48 The second moment of Y_0 can be also calculated by applying Corollary 1.44 and Lemma 1.45. Indeed,

$$\begin{aligned}
\mathsf{E}Y_0^2 &= \sigma^2 \lim_{t\to-\infty} \mathsf{E}\left(-e^{-\theta t}B_t^H + \theta\int_t^0 e^{-\theta s}B_s^H\, ds\right)^2 \\
&= \sigma^2 \lim_{t\to\infty} \mathsf{E}\left(-e^{\theta t}B_{-t}^H + \theta\int_0^t e^{\theta s}B_{-s}^H\, ds\right)^2 \\
&= \sigma^2 \lim_{t\to\infty} e^{2\theta t}\left(\mathsf{E}\left(B_{-t}^H\right)^2 - 2\theta e^{-\theta t}\int_0^t e^{\theta s}\mathsf{E}B_{-t}^H B_{-s}^H\, ds\right. \\
&\quad \left. + \theta^2 e^{-2\theta t}\int_0^t\int_0^t e^{\theta(s+u)}\mathsf{E}B_{-s}^H B_{-u}^H\, ds\, du\right) \\
&= \sigma^2 \lim_{t\to\infty} e^{2\theta t}v(-\theta, t) = \frac{\sigma^2 H\Gamma(2H)}{(-\theta)^{2H}}.
\end{aligned}$$

In order to prove the ergodicity of the Gaussian stationary process Y it suffices to show that its autocovariance function $r(t) = \mathrm{cov}(Y_t, Y_0)$ vanishes at infinity.

Lemma 1.49 *Let $\theta < 0$, $H \in (0,1)$. Then $r(t) \to 0$, as $t \to \infty$.*

Proof By the change of variables, we get

$$r(t) = \frac{H\sigma^2}{2(-\theta)^{2H}} \left(\Gamma(2H)e^{\theta t} - e^{\theta t} \int_0^{-\theta t} e^y y^{2H-1} dy + e^{-\theta t} \int_{-\theta t}^{+\infty} e^{-y} y^{2H-1} dy \right) \tag{1.42}$$

If $H = 1/2$, then $r(t) = -\frac{\sigma^2}{2\theta} e^{\theta t} \to 0$, as $t \to \infty$. If $H \neq 1/2$, then using integration by parts, we obtain

$$\begin{aligned} r(t) &= \frac{H\sigma^2}{2(-\theta)^{2H}} \left(-e^{\theta t} \int_1^{-\theta t} e^y y^{2H-1} dy + e^{-\theta t} \int_{-\theta t}^{+\infty} e^{-y} y^{2H-1} dy \right) + O\left(e^{\theta t}\right) \\ &= \frac{H(2H-1)\sigma^2}{2(-\theta)^{2H}} \left(e^{\theta t} \int_1^{-\theta t} e^y y^{2H-2} dy + e^{-\theta t} \int_{-\theta t}^{+\infty} e^{-y} y^{2H-2} dy \right) + O\left(e^{\theta t}\right), \end{aligned} \tag{1.43}$$

as $t \to \infty$. We have

$$e^{-\theta t} \int_{-\theta t}^{+\infty} e^{-y} y^{2H-2}\, dy \le e^{-\theta t} \int_{-\theta t}^{+\infty} e^{-y} (-\theta t)^{2H-2}\, dy = (-\theta t)^{2H-2}, \tag{1.44}$$

and

$$\begin{aligned} e^{\theta t} \int_1^{-\theta t} e^y y^{2H-2}\, dy &\le e^{\theta t} \left(\int_1^{-\frac{\theta t}{2}} e^y\, dy + \int_{-\frac{\theta t}{2}}^{-\theta t} e^y \left(-\frac{\theta t}{2}\right)^{2H-2} dy \right) \\ &\le e^{\frac{\theta t}{2}} + \left(-\frac{\theta t}{2}\right)^{2H-2}. \end{aligned} \tag{1.45}$$

Thus, both terms in (1.43) converge to zero as $t \to \infty$. This concludes the proof. □

Remark 1.50 The stationary fO-U was introduced and investigated in [38]. Moreover, the following asymptotic relation for the autocovariance function $r(t)$ was established for $H \neq \frac{1}{2}$ and $N = 1, 2 \ldots$:

$$r(t) = \frac{\sigma^2}{2} \sum_{n=1}^{N} (-\theta)^{-2n} \left(\prod_{k=0}^{2n-1} (2H-k) \right) t^{2H-2n} + O\left(t^{2H-2N-2}\right), \tag{1.46}$$

as $t \to \infty$. This implies that for $H \in (\frac{1}{2}, 1]$, the process Y exhibits long-range dependence, that is, $\sum_{n=0}^{\infty} r(n) = \infty$. The expansion (1.46) can be deduced from (1.43). In order to obtain this result, one should use integration by parts N times, and then bound the remaining terms similarly to (1.44)–(1.45).

A similar relation holds for the fO-U X, namely for $H \neq \frac{1}{2}$, $N = 1, 2 \ldots$, and $s \geq 0$

$$\operatorname{cov}(X_s, X_{s+t}) = \frac{\sigma^2}{2} \sum_{n=1}^{N} (-\theta)^{-2n} \left(\prod_{k=0}^{2n-1} (2H - k) \right) \left(t^{2H-2n} - e^{\theta s}(s+t)^{2H-2n} \right) + O\left(t^{2H-2N-2}\right), \quad \text{as } t \to \infty.$$

It follows immediately from (1.46) if we note that, by (1.39),

$$\begin{aligned} \operatorname{cov}(X_s, X_{s+t}) &= \operatorname{cov}(Y_s, Y_{s+t}) - e^{\theta s} \operatorname{cov}(Y_0, Y_{s+t}) - e^{\theta(s+t)} \operatorname{cov}(Y_s, Y_0) + e^{\theta(2s+t)} \mathsf{E} Y_0^2 \\ &= r(t) - e^{\theta s} r(s+t) + O\left(e^{\theta t}\right), \quad t \to \infty. \end{aligned}$$

1.3.3 SDE with Polynomial Drift

Let $(\Omega, \mathfrak{F}, \mathbb{F}, \mathsf{P})$ be a filtered probability space generated by a fixed fBm $B^H = \{B_t^H, t \in [0, T]\}$. Let us consider the SDE with polynomial drift given by

$$X_t = x_0 + \int_0^t (aX_s^m + bX_s) \mathrm{d}s + c \int_0^t X_s \, \mathrm{d}B_s^H, \quad x_0 > 0, \quad a \leq 0, \quad t \in [0, T], \tag{1.47}$$

where $m \in \mathbb{N}$, $m \geq 2$, and B^H is an fBm with Hurst index $\frac{1}{2} < H < 1$. We shall prove that there always exists a unique solution $X = \{X_t, t \in [0, T]\}$ of (1.47) having the following properties:

- X is adapted to the filtration $\mathbb{F}$;
- $X \in C^{\gamma}([0, T])$ a.s. for all $1/2 < \gamma < H$ and

$$\mathsf{P}\left(\int_0^t \left(|aX_s^m| + |bX_s| \right) \mathrm{d}s + \left| \int_0^t X_s \mathrm{d}B_s^H \right| < \infty \right) = 1, \quad t \in [0, T];$$

- the explicit solution is given by

$$X_t = e^{bt + cB_t^H} \left(x_0^{1-m} + (1-m)a \int_0^t e^{(m-1)(bs + cB_s^H)} ds \right)^{1/(1-m)}, \quad t \in [0, T]. \tag{1.48}$$

Note that by the change of variable $Y_t = X_t^{1-m}$ Eq. (1.47) can be reduced to the linear equation

$$Y_t = Y_0 + (1-m)\int_0^t (a + bY_s)\mathrm{d}s + c(1-m)\int_0^t Y_s\,\mathrm{d}B_s^H, \quad Y_0 \geq 0, \quad t \in [0, T].$$

First we consider the differential equation, rewritten in the integral form,

$$x_t = x_0 + \int_0^t (ax_s^m + bx_s)\,ds + c\int_0^t x_s\,dh_s, \qquad x_0 > 0, \quad a \leq 0,\ t \in [0, T], \tag{1.49}$$

where $m \in \mathbb{N}$, $m \geq 2$, $h \in C^\gamma([0, T])$, for some $1/2 < \gamma < 1$, and prove two auxiliary theorems used in the sequel.

Theorem 1.51 *The function*

$$x_t = e^{bt+ch_t}\left(x_0^{1-m} + (1-m)a\int_0^t e^{(m-1)(bs+ch_s)}ds\right)^{1/(1-m)}, \qquad t \in [0, T], \tag{1.50}$$

is an element of $C^\gamma([0, T])$, for the same γ, and satisfies Eq. (1.49).

Proof First we show that $x \in C^\gamma([0, T])$. Let $z_t = \exp\{bt + ch_t\}$. By the mean value theorem for e^x, $x \in \mathbb{R}$,

$$\begin{aligned}|z_t - z_s| &\leq \exp\left\{2|b|T + |c| \cdot \|h\|_\gamma\left(1 + (t-s)^\gamma\right)\right\}\left[|b|(t-s) + |c| \cdot \|h\|_\gamma (t-s)^\gamma\right]\\ &\leq \exp\left\{2|b|T + |c| \cdot \|h\|_\gamma\left(1 + T^\gamma\right)\right\}\left[|b|T^{1-\gamma} + |c| \cdot \|h\|_\gamma\right](t-s)^\gamma.\end{aligned}$$

Define

$$f_t = x_0^{1-m} + (1-m)a\int_0^t e^{(m-1)(bs+ch_s)}ds.$$

Application of the mean value theorem for $u \mapsto u^{1/(1-m)}$, $u > 0$, $m \geq 2$, gives the bound

$$|f_t^{1/(1-m)} - f_s^{1/(1-m)}| \leq \frac{1}{m-1}\,|f_s + \theta(f_t - f_s)|^{m/(1-m)}|f_t - f_s|$$

where $\theta \in (0, 1)$. Since

$$(1-m)a\int_0^t e^{(m-1)(bs+ch_s)}ds > 0,$$

it follows that

$$f_s + \theta(f_t - f_s) > x_0^{1-m}.$$

Thus,

$$|f_t^{1/(1-m)} - f_s^{1/(1-m)}| \le x_0^m |f_t - f_s| \le |a| \exp\{(m-1)(|b|T + |c|\,|h|_\infty)\}(t-s).$$

Since $z \in C^\gamma([0,T]), f \in C^1([0,T])$, we have $x \in C^\gamma([0,T])$. Indeed,

$$\begin{aligned}|z_t f_t^{1/(1-m)} - z_s f_s^{1/(1-m)}| \le & |z_t f_t^{1/(1-m)} - z_s f_t^{1/(1-m)}| + |z_s f_t^{1/(1-m)} - z_s f_s^{1/(1-m)}| \\ \le & |f|_\infty^{1/(1-m)} |z_t - z_s| + |z|_\infty x_0^m |f_t - f_s| .\end{aligned}$$

Now we verify that the function (1.50) satisfies (1.49). Let $F(t,u,v) = e^{bt+cu} v^{\frac{1}{1-m}}$. Then by Theorem 1.18 and Proposition 1.19

$$\begin{aligned}x_t =& F(t, h_t, f_t) = F(0, h_0, f_0) + b \int_0^t x_s\, ds + c \int_0^t x_s\, dh_s \\ &+ \frac{1}{1-m} \int_0^t e^{bs+ch_s} f_s^{1/(1-m)-1}\, df_s \\ =& x_0 + b \int_0^t x_s\, ds + c \int_0^t x_s\, dh_s + \frac{1}{1-m} \int_0^t e^{bs+ch_s} f_s^{m/(1-m)}\, df_s \\ =& x_0 + b \int_0^t x_s\, ds + c \int_0^t x_s\, dh_s + \frac{1}{1-m} \int_0^t (1-m) a e^{m(bs+ch_s)} f_s^{m/(1-m)}\, ds \\ =& x_0 + b \int_0^t x_s\, ds + c \int_0^t x_s\, dh_s + a \int_0^t x_s^m\, ds.\end{aligned}$$

□

Now we shall use the notations of the spaces $\mathcal{W}_p([a,b])$, $C\mathcal{W}_p([a,b])$ and the respective norms. In particular, the norm on $\mathcal{W}_p([a,b])$ is defined by

$$V_{p,\infty}(f;[a,b]) = V_p(f;[a,b]) + \sup_{x\in[a,b]} |f(x)|,$$

$$V_p(f;[a,b]) = v_p^{1/p}(f;[a,b]), \quad v_p(f;[a,b]) = \sup_{\varkappa} \sum_{k=1}^{n} |f(x_k) - f(x_{k-1})|^p ,$$

where sup is taken over all partitions of $[a,b]$. For more details see Appendix A and Acronyms.

Theorem 1.52 *Let $h \in C^\gamma([0,T])$ for some $1/2 < \gamma < 1$. Then the differential equation (1.49) has a unique solution in the class $C\mathcal{W}_{1/\gamma}([0,T])$.*

Proof We have already shown that at least one solution $x \in C^\gamma([0,T])$ exists. Assume it is not unique and $y \in C^\gamma([0,T])$, is a different one. It should be noted that x and y have a bounded $1/\gamma$-variation. Let us denote $p = 1/\gamma$.

Since x, y are continuous the equality

$$a^m - b^m = (a-b)\sum_{k=0}^{m-1} a^k b^{m-1-k}$$

implies

$$|x_t^m - y_t^m| \le m L_{x,y,m,T} |x_t - y_t|,$$

where

$$L_{x,y,m,T} = \left(\max\left\{\max_{0\le t\le T}|x_t|, \max_{0\le t\le T}|y_t|\right\}\right)^{m-1}.$$

Further, for a fixed constant $(4|c|C_{p,p})^{-1}$ one can find a set of points $0 = \tau_0 < \tau_1 < \tau_2 < \cdots < \tau_n = T$ such that (see Appendix A.3)

$$V_p\big(h; [\tau_{k-1}, \tau_k]\big) \le \big(4|c|C_{p,p}\big)^{-1}$$

for all k, where $C_{p,p}$ is a constant from the Love–Young inequality (1.18). Assume we have proved that $x_{\tau_{k-1}} = y_{\tau_{k-1}}$. Then by the Love–Young inequality

$$\begin{aligned}
V_{p,\infty}(x-y; [\tau_{k-1}, \tau_k]) =& V_{p,\infty}(x - y - (x_{\tau_{k-1}} - y_{\tau_{k-1}}); [\tau_{k-1}, \tau_k]) \\
\le& 2|a| \int_{\tau_{k-1}}^{\tau_k} |x_t^m - y_t^m|\, dt + 2|b| \int_{\tau_{k-1}}^{\tau_k} |x_t - y_t|\, dt \\
&+ 2|c| C_{p,p} V_{p,\infty}(x-y; [\tau_{k-1}, \tau_k]) V_p(h; [\tau_{k-1}, \tau_k]) \\
\le& 2\big(L_{x,y,m,T}|a| + |b|\big) \int_{\tau_{k-1}}^{\tau_k} |x_t - y_t|\, dt \\
&+ 2|c| C_{p,p} V_{p,\infty}(x-y; [\tau_{k-1}, \tau_k]) V_p(h; [\tau_{k-1}, \tau_k])
\end{aligned}$$

and

$$\begin{aligned}
V_{p,\infty}(x-y; [\tau_{k-1}, \tau_k]) \le& 4\big(L_{x,y,m,T}|a| + |b|\big) \int_{\tau_{k-1}}^{\tau_k} |x_t - y_t|\, dt \\
\le& 4\big(L_{x,y,m,T}|a| + |b|\big) \int_{\tau_{k-1}}^{\tau_k} V_{p,\infty}(x-y; [\tau_{k-1}, t])\, dt.
\end{aligned}$$

Therefore by Gronwall's inequality $V_{p,\infty}(x-y; [\tau_{k-1}, \tau_k]) = 0$ and we can conclude that $x = y$ on $[\tau_{k-1}, \tau_k]$. Since $x_{\tau_0} = x_0 = y_{\tau_0}$ the claim of the theorem follows from the repetitive application of the above reasoning. □

Remark 1.53 It follows from Theorem 1.52 that $\sup_{t\le T}|x_t - y_t| = 0$, i.e. the solution of equation (1.49) is unique.

According to Remark 1.20, if $X \in C^{\lambda}([0,T])$ with $\gamma + \lambda > 1$, then the path-wise integral $\int_0^t X_s\,dB_s^H$, $t \in [0,T]$, exists in the sense of Definition 1.13 and coincides with the Riemann–Stieltjes integral. Therefore, the SDE (1.47) is well defined for almost all ω and the obtained result for the deterministic differential equation (1.49) can be applied to the corresponding SDE driven by an fBm.

Theorem 1.54 *Suppose that $x_0 > 0$ and $m \ge 2$. The stochastic process* (1.48) *for almost all ω belongs to $C^{\gamma}([0,T])$ for any $1/2 < \gamma < H$, and is the unique solution of* (1.47).

1.3.4 Fractional Gompertz Diffusion Process

Let us consider the following fractional Gompertz diffusion process

$$X_t = x_0 + \int_0^t (\alpha X_s - \beta X_s \log X_s)\,ds + \sigma \int_0^t X_s\,dB_s^H, \quad x_0 > 0, \quad 0 \le t \le T, \tag{1.51}$$

where α, $\beta \neq 0$, and $\sigma > 0$ are real parameters and B^H is an fBm with the Hurst index $H \in (1/2, 1)$. We shall prove that this equation has a unique explicit solution presented as

$$X_t = \exp\left\{ e^{-\beta t}\log x_0 + \alpha \int_0^t e^{-\beta(t-s)}ds + \sigma \int_0^t e^{-\beta(t-s)}\,dB_s^H \right\}.$$

Note that using the change of variable $Z_t = \log X_t$ the Eq. (1.51) can be reduced to the fractional Vasicek model

$$Z_t = Z_0 + \int_0^t (\alpha - \beta Z_s)\,ds + \sigma B_t^H, \qquad 0 \le t \le T.$$

In order to prove the existence and uniqueness of the solution of the Gompertz equation (1.51) it is enough to consider a differential equation as in the case of SDEs with polynomial drift. Namely, let us consider the differential equation

$$x_t = x_0 + \int_0^t (\alpha x_s - \beta x_s \log x_s)\,ds + \sigma \int_0^t x_s\,dh_s, \qquad x_0 \ge 0, \quad \beta \neq 0, \quad 0 \le t \le T, \tag{1.52}$$

where $h \in C^{\gamma}([0,T])$ for some $1/2 < \gamma < 1$.

Lemma 1.55 *The function*

$$x_t = \exp\left\{e^{-\beta t}\log x_0 + \frac{\alpha}{\beta}\left(1-e^{-\beta t}\right) + \sigma\int_0^t e^{-\beta(t-s)}dh_s\right\}, \qquad t \in [0, T], \qquad (1.53)$$

is an element of $C^\gamma([0,T])$ *and satisfies Eq.* (1.52).

Proof Let ϕ be a Lipschitz function and let $f \in C^\gamma([0,T])$. Then $\phi \circ f \in C^\gamma([0,T])$.
First, let us show that $x \in C^\gamma([0,T])$. Let

$$z_t = e^{-\beta t}\log x_0 + \frac{\alpha}{\beta}\left(1-e^{-\beta t}\right) + \sigma\int_0^t e^{-\beta(t-s)}dh_s =: e^{-\beta t}y_t.$$

We can now easily show that by the Love–Young inequality

$$\begin{aligned}|y_t - y_s| \leq& |\alpha| e^{|\beta|T}(t-s) + C_{1,1/\gamma}\sigma V_{1,\infty}\left(e^{\beta \cdot}; [s,t]\right)V_{1/\gamma}(h; [s,t])\\ \leq& |\alpha| e^{|\beta|T}(t-s) + C_{1,1/\gamma}\sigma e^{|\beta|T}\left(1+(t-s)\right)(t-s)^\gamma.\end{aligned}$$

Thus $z \in C^\gamma([0,T])$. Since e^u is a locally Lipschitz function, it follows that $x \in C^\gamma[0,T]$.

Now we verify that the function (1.53) satisfies (1.52). This statement can be checked by the application of the Chain rule (Theorem 1.18) and the Substitution rule (Proposition 1.19). Namely, let $F(t,x,y) = \exp\{e^{-\beta t}(\log x_0 + \alpha x + \sigma y)\}$ and denote

$$a_t = \int_0^t e^{\beta s}ds, \qquad c_t = \int_0^t e^{\beta s}dh_s.$$

Then

$$\begin{aligned}F(t,a_t,c_t) =& x_0 - \beta\int_0^t x_s\log x_s\,ds + \alpha\int_0^t x_s e^{-\beta s}da_s + \sigma\int_0^t x_s e^{-\beta s}dc_s\\ =& x_0 - \beta\int_0^t x_s\log x_s\,ds + \alpha\int_0^t x_s\,ds + \sigma\int_0^t x_s\,dh_s.\end{aligned}$$

□

Theorem 1.56 *The differential equation* (1.52) *has a unique solution in the class* $C\mathcal{W}_{1/\gamma}([0,T])$.

Proof We have already shown that at least one solution $x \in C^\gamma([0,T])$ exists. Assume it is not unique and $y \in C^\gamma([0,T])$ is a different one. It should be noted that x and y have a bounded $1/\gamma$-variation. Let us denote $p = 1/\gamma$.

Further, for a fixed constant $(4\sigma C_{p,p})^{-1}$ one can find a set of points $0 = \tau_0 < \tau_1 < \tau_2 < \cdots < \tau_n = T$ such that (see Appendix A.3)

$$V_p\big(h; [\tau_{k-1}, \tau_k]\big) \le \big(4\sigma C_{p,p}\big)^{-1}$$

for all k, where $C_{p,p}$ is the constant from the Love–Young inequality. Assume we have proved that $x_{\tau_{k-1}} = y_{\tau_{k-1}}$.

By the well known inequality $\log(1+x) \le x$, $x > 0$, we get

$$|x_s \log y_s - x_s \log x_s| \le x_s \left| \log\left(1 + \frac{|y_s - x_s|}{x_s}\right)\right| \le |y_s - x_s|$$

and

$$|\log y_s| \le \left|\log\left(\max_{0\le t\le T} y_t\right)\right| =: L_{y,T}.$$

Recall that the norm on $\mathcal{W}_p$ is defined by

$$V_{p,\infty}(f; [a,b]) = V_p(f; [a,b]) + \sup_{x\in[a,b]} |f(x)|.$$

Then

$$\begin{aligned}
V_{p,\infty}(x-y; [\tau_{k-1}, \tau_k]) =& V_{p,\infty}(x - y - (x_{\tau_{k-1}} - y_{\tau_{k-1}}); [\tau_{k-1}, \tau_k]) \\
\le& 2|\alpha| \int_{\tau_{k-1}}^{\tau_k} |x_t - y_t|\, dt + 2|\beta| \int_{\tau_{k-1}}^{\tau_k} |x_t \log x_t - y_t \log y_t|\, dt \\
&+ 2\sigma C_{p,p} V_{p,\infty}(x-y; [\tau_{k-1}, \tau_k]) V_p(h; [\tau_{k-1}, \tau_k]) \\
\le& 2\big(|\alpha| + |\beta| + |\beta| L_{y,T}\big) \int_{\tau_{k-1}}^{\tau_k} |x_t - y_t|\, dt \\
&+ 2\sigma C_{p,p} V_{p,\infty}(x-y; [\tau_{k-1}, \tau_k]) V_p(h; [\tau_{k-1}, \tau_k])
\end{aligned}$$

and

$$\begin{aligned}
V_{p,\infty}(x-y; [\tau_{k-1}, \tau_k]) \le& 4\big(|\alpha| + |\beta| + |\beta| L_{y,T}\big) \int_{\tau_{k-1}}^{\tau_k} |x_t - y_t|\, dt \\
\le& 4\big(|\alpha| + |\beta| + |\beta| L_{y,T}\big) \int_{\tau_{k-1}}^{\tau_k} V_{p,\infty}(x-y; [\tau_{k-1}, t])\, dt.
\end{aligned}$$

Therefore by Gronwall's inequality $V_{p,\infty}(x-y; [\tau_{k-1}, \tau_k]) = 0$ and we can conclude that $x = y$ on $[\tau_{k-1}, \tau_k]$. Since $x_{\tau_0} = x_0 = y_{\tau_0}$ the claim of the theorem follows from the repetitive application of the above reasoning. □

Remark 1.57 It follows from Theorem 1.56 that $\sup_{t\le T}|x_t - y_t| = 0$, i. e. the solution of equation (1.52) is unique.

1.4 Mixed SDE with Wiener Process and Fractional Brownian Motion

Let $(\Omega, \mathfrak{F}, \mathbb{F}, \mathsf{P})$ be a complete probability space equipped with a filtration satisfying standard assumptions, and $W = \{W_t, t \in [0,T]\}$ be a standard $\mathbb{F}$-Wiener process. We consider an $\mathbb{F}$-adapted stochastic process $Z = \{Z_t, t \in [0,T]\}$, which is Hölder continuous with Hölder exponent $\gamma > 1/2$. For example, one can take $Z = B^H$, an fBm, with $H > 1/2$. The processes W and Z can be dependent. We study a mixed SDE

$$X_t = x_0 + \int_0^t a(s, X_s)\,ds + \int_0^t b(s, X_s)\,dZ_s + \int_0^t c(s, X_s)\,dW_s,\ t \in [0,T]\,. \tag{1.54}$$

Here $x_0 \in \mathbb{R}$, and $a, b, c\colon [0,T] \times \mathbb{R} \to \mathbb{R}$ are measurable functions. The integral w. r. t. Wiener process W is the standard Itô integral, and the integral w. r. t. Z is pathwise generalized Lebesgue–Stieltjes integral, see Definition 1.13.

Assume that the coefficients satisfy the following conditions.

(B$_1$) There exists $C > 0$ such that

$$|a(t,x)| + |b(t,x)| + |c(t,x)| \le C(1 + |x|),$$

for all $t \in [0,T]$, $x \in \mathbb{R}$.

(B$_2$) $b(t,x)$ is differentiable in x and there exists $C > 0$ such that

$$|\partial_x b(t,x)| \le C,$$

for all $t \in [0,T]$, $x \in \mathbb{R}$.

(B$_3$) For every $N > 0$ there exists $C_N > 0$ such that

$$|a(t,x) - a(t,y)| + |c(t,x) - c(t,y)| + |\partial_x b(t,x) - \partial_x b(t,y)| \le C_N|x - y|,$$

for all $t \in [0,T]$, $x, y \in [-N, N]$.

(B$_4$) There exist constants $C > 0$ and $\beta \in (1-\gamma, 1/2)$ such that

$$|b(t,x) - b(s,x)| \le C|t-s|^\beta(1 + |x|),$$

$$|\partial_x b(t,x) - \partial_x b(s,x)| \le C|t-s|^\beta,$$

for all $t, s \in [0,T]$, $x \in \mathbb{R}$.

Theorem 1.58 ([156]) *Under the assumptions* (B_1)–(B_4)*, Eq.* (1.54) *has a unique solution.*

Remark 1.59 Theorem 1.58 is a particular case of the main result of [156], where a more general SDE (namely, multidimensional mixed stochastic delay equation) was considered. It is worth noting that the existence and uniqueness theorem for the SDE (1.54) was earlier proved in [125] and [126] under stronger assumptions on the coefficients. Mention also that Eq. (1.54) with $Z = B^H$, a fBm, was first considered in [94], where existence and uniqueness of a solution was proved for time-independent coefficients and zero drift. For inhomogeneous coefficients, unique solvability was established in [120] for $H \in (3/4, 1)$ and bounded coefficients, in [66] for any $H > 1/2$, but under the assumption that W and B^H are independent.

1.5 Sub-fractional, Bifractional and Multifractional Brownian Motions

The stationarity of increments of fBm means that the behavior of them is the same at each point, and this substantially restricts the area of their application. Therefore in this section we introduce two Gaussian self-similar processes, which also preserve many properties of fBm, but have no stationary increments except for particular parameter values. One of them is a sub-fractional Brownian motion (it is considered in Sect. 1.5.1), and another one is a bifractional Brownian motion (it is defined in Sect. 1.5.2). Furthermore, stochastic processes whose path regularity and "memory depth" evolve in time do not fit the model of fBm. Thus several authors recently provided different *multifractional* generalizations of fBm by assuming that the Hurst parameter H depends on the time t (see [16, 29, 142, 149]). One of such generalizations, a so called harmonizable multifractional Brownian motion, is considered in Sect. 1.5.3, where we also give the existence–uniqueness theorem for a corresponding SDE.

1.5.1 Sub-fractional Brownian Motion

Bojdecki et al. [27] introduced and studied the following class of self-similar Gaussian processes.

Definition 1.60 A *sub-fractional Brownian motion* (sfBm) with index $H \in (0, 1)$ is a zero-mean Gaussian stochastic process $S^H = \{S_t^H, t \geq 0\}$ with covariance function

$$G_H(s,t) := s^{2H} + t^{2H} - \frac{1}{2}\left[(s+t)^{2H} + |s-t|^{2H}\right].$$

The case $H = 1/2$ corresponds to Brownian motion. The sfBm preserves many properties of fBm, see [27, 165]. In particular, it is self-similar with self-similarity parameter H and its sample paths are γ-Hölder continuous for $\gamma \in (0, H)$. The incremental variance function has the following form

$$\mathsf{E}\left(S_t^H - S_s^H\right)^2 = |t-s|^{2H} + (s+t)^{2H} - 2^{2H-1}(t^{2H} + s^{2H}).$$

Consequently, the increments of S^H are not stationary for $H \neq 1/2$. This property is replaced by the following inequalities: for any $0 \le s \le t$

$$(t-s)^{2H} \le \mathsf{E}\left(S_t^H - S_s^H\right)^2 \le \left(2 - 2^{2H-1}\right)(t-s)^{2H}, \qquad \text{if} \quad 0 < H < 1/2, \tag{1.55}$$

$$\left(2 - 2^{2H-1}\right)(t-s)^{2H} \le \mathsf{E}\left(S_t^H - S_s^H\right)^2 \le (t-s)^{2H}, \qquad \text{if} \quad 1/2 < H < 1. \tag{1.56}$$

Hence, by Theorems 1.4–1.5, S^H has γ-Hölder continuous paths for any $\gamma \in (0, H)$.

Note also that if $B^H = \{B_t^H,\ t \in \mathbb{R}\}$ is a two-sided fBm, then the process $\frac{B_t^H + B_{-t}^H}{\sqrt{2}}$ is an sfBm.

1.5.2 *Bifractional Brownian Motion*

Houdré and Villa [70] introduced the following extension of fBm.

Definition 1.61 A *bifractional Brownian motion* (bifBm) $B^{H,K} = \left\{B_t^{H,K}, t \ge 0\right\}$ with parameters $H \in (0,1)$ and $K \in (0,1]$ is a centered Gaussian process with covariance function

$$R_{HK}(t,s) = 2^{-K}\left((t^{2H} + s^{2H})^K - |t-s|^{2HK}\right), \qquad s, t \ge 0.$$

Note that, if $K = 1$ then $B^{H,1}$ is an fBm with Hurst parameter $H \in (0,1)$. Since the covariance function $R_{HK}(t,s)$ is homogeneous of order $2HK$, we see that a bifBm $B^{H,K}$ is self-similar with self-similarity parameter HK. Its incremental variance function is of the following form

$$\mathsf{E}\left(B_t^{H,K} - B_s^{H,K}\right)^2 = 2^{1-K}\left[|t-s|^{2HK} - \left(t^{2H} + s^{2H}\right)^K\right] + t^{2HK} + s^{2HK}.$$

In the case $K \neq 1$ a bifBm has no stationary increments, but it has the following quasi-helix property (in the sense of Kahane [83]),

$$2^{-K}|t-s|^{2HK} \le \mathsf{E}\left(B_t^{H,K} - B_s^{H,K}\right)^2 \le 2^{1-K}|t-s|^{2HK} \tag{1.57}$$

for all $s, t \in [0, \infty)$. Hence, by Theorems 1.4–1.5, $B^{H,K}$ has γ-Hölder continuous paths for any $\gamma \in (0, HK)$. Note also that the definition of bifBm can be extended for $1 < K < 2$ with $H \in (0, 1)$ and $HK \in (0, 1)$ (see [9, 109]). For more details on bifBm we refer to [70, 108, 152].

1.5.3 Multifractional Brownian Motion

Let $H: \mathbb{R}_+ \to (0, 1)$ be a continuous function satisfying the following conditions:

($\mathbf{C}_1$) There exist constants $0 < h_1 < h_2 \leq 1$ such that for any $t \geq 0$

$$h_1 \leq H_t \leq h_2.$$

($\mathbf{C}_2$) There exist constants $D > 0$ and $\kappa \in (0, 1]$ such that for all $t \geq s > 0$

$$|H_t - H_s| \leq D\,|t - s|^{\kappa}.$$

A *multifractional Brownian motion* (mfBm) with functional parameter H was introduced in [16]. It is defined by

$$Y_t = \int_{\mathbb{R}} \frac{e^{itu} - 1}{|u|^{H_t + 1/2}} \widetilde{W}(du), \quad t \geq 0, \tag{1.58}$$

where $\widetilde{W}(du)$ is the "Fourier transform" of the white noise $W(du)$, that is a unique complex-valued random measure such that for all $f \in L^2(\mathbb{R})$

$$\int_{\mathbb{R}} f(u) W(du) = \int_{\mathbb{R}} \widehat{f}(u) \widetilde{W}(du) \quad \text{a. s.,}$$

see [16, 159].

The covariance function of mfBm is given by

$$\mathsf{E} Y_s Y_t = D(H_s, H_t) \left(s^{H_s + H_t} + t^{H_s + H_t} - |s - t|^{H_s + H_t} \right),$$

where $D(x, y) = \frac{\pi}{\Gamma(x+y+1)\sin(\pi(x+y)/2)}$, see [6] or [159, p. 213]. Note that if H_t is a constant, then the process Y is an fBm (up to multiplicative constant).

In particular,

$$\left(\mathsf{E} \,|\, Y_t|^2 \right)^{1/2} = C(H_t) t^{H_t}, \tag{1.59}$$

where $C(H) = \left(\frac{\pi}{H\Gamma(2H)\sin(\pi H)}\right)^{1/2}$. Since the function $C(H)$ is bounded on $[h_1, h_2]$, we have under assumptions (C_1)–(C_2)

$$\left(\mathsf{E}\,|Y_t|^2\right)^{1/2} \leq K_1 t^{h_2}, \quad t \geq 1, \tag{1.60}$$

for some $K_1 > 0$.

Denote

$$h_3 = \min\{h_1, \kappa\}. \tag{1.61}$$

Remark 1.62 It follows from the bound (B.60) in the Appendix, from the fact that a multifractional process is Gaussian, and from Theorems 1.4–1.5 that under conditions (C_1) and (C_2) the process Y with probability 1 has Hölder trajectories up to order h_3 on any finite interval.

Remark 1.63 There exist different definitions of mfBm in the literature. In particular, Peltier and Lévy Véhel [142] introduced mfBm based on the Mandelbrot–van Ness representation of fBm (1.4). This version of mfBm is often termed a *moving-average mfBm*. It has a different covariance structure, see [46]. Another version of mfBm, a *Volterra-type mfBm*, is based on the representation (1.5). It was studied in [29, 149]. The process introduced in (1.58) is sometimes called a *harmonizable mfBm* to distinguish it from other types of mfBm.

Assume that the Hurst function satisfies the conditions (C_1)–(C_2) and, additionally, $h_3 > 1/2$. In this case, according to Remark 1.62, the process Y has with probability 1 Hölder trajectories up to order h_3 on any finite interval $[0, T]$. Therefore, the path-wise integrals w. r. t. mfBm can be defined using the approach described in Sect. 1.1.3 for fBm. Assume that we have another process, say $Z = \{Z_t, t \in [0, T]\}$, also having Hölder trajectories up to some order h with $h + h_3 > 1$. In particular, it can be $h = h_3$. Then there exists an integral $\int_a^b Z_s dY_s$, which is the limit a.s. of the Riemann sums and has the standard properties (so called path-wise integral). This integral is defined as

$$\int_a^b Z dY := e^{i\pi\alpha} \int_a^b (\mathcal{D}_{a+}^{\alpha} Z)(x)(\mathcal{D}_{b-}^{1-\alpha} Y_{b-})(x)dx. \tag{1.62}$$

This enables us to consider an SDE driven by mfBm:

$$X_t = x_0 + \int_0^t a(s, X_s)ds + \int_0^t b(s, X_s)dY_s, \quad t \in [0, T], \tag{1.63}$$

where $x_0 \in \mathbb{R}$, $a, b\colon [0, T] \times \mathbb{R} \to \mathbb{R}$ are measurable functions. The sufficient conditions for the existence and uniqueness of a solution are similar to Theorem 1.42.

Theorem 1.64 ([148, Theorem 4.1]) *Let the Hurst function H_t satisfy the conditions* (C_1)–(C_2) *and, additionally,* $h_3 = \min\{h_1, \kappa\} > 1/2$. *Assume that the coefficients a and b satisfy assumptions* (A_1) *and* (A_2) *of Sect. 1.3.1 with* $1 - h_3 < \beta \leq 1$ *and* $\frac{1}{h_3} - 1 < \delta \leq 1$. *Then a solution* $X = \{X_t, t \in [0, T]\}$ *of Eq.* (1.63) *exists and is unique. Moreover, its trajectories belong to the space* $C^{h_3-\varepsilon}([0, T])$ *a. s. for any* $0 < \varepsilon < h_3$.

Chapter 2
The Hurst Index Estimators for a Fractional Brownian Motion

We state again that the phenomenon of long range dependence is observed in various fields, such as hydrology, biology, finance, economy, chemistry, physics, telecommunication networks and others. Since a long memory parameter (or the Hurst index/exponent, usually denoted H) determines the mathematical properties of the model, its estimation is of great importance. One of the main methods of the estimation of the Hurst index involves the convergent sequence of the quadratic variations of the underlying process.

The problem of the almost sure convergence of the quadratic variations has been solved for a wide class of Gaussian processes by Baxter [10] and Gladyshev [63]. These authors employed dyadic partitions. Klein and Giné [89] used more general partitions and proved that particular functions of the mesh of the partition must be at most $o(1/\log n)$ for the almost sure convergence of the quadratic variations to hold.

Since in this chapter we confine ourselves to the study of a fBm, we apply the known results for a class of Gaussian processes to an fBm. This greatly simplifies the presentation of the results.

Now we will only briefly mention some facts from the theory of the estimation of the Hurst index. We'll discuss them in more detail later in the chapter.

Using his result of convergence of the quadratic variations, Gladyshev [63] constructed a strongly consistent estimator of Guyon and León [78] proposed another estimator, which also involved the quadratic variations and was asymptotically normal only when $H \in (0, 3/4)$. To avoid this drawback second order quadratic variations have been considered by Istas and Lang [78] and Benassi et al. [18], to name a few.

The present chapter is devoted to the consideration of asymptotic behavior of the quadratic variations, as well as to review and consideration of the different estimators of the Hurst index. In Sect. 2.1 we consider convergence rate of quadratic and weighted quadratic variations. Moreover, we recall some asymptotic normality results for the quadratic variations of an fBm. In Sect. 2.2 besides the well-known

© Springer International Publishing AG 2017

K. Kubilius et al., *Parameter Estimation in Fractional Diffusion Models*,
B&SS – Bocconi & Springer Series 8,
https://doi.org/10.1007/978-3-319-71030-3_2

estimators of the Hurst index we give the new class of statistics which serve the same purpose.

2.1 Quadratic Variations of a Fractional Brownian Motion

Definition 2.1 For a real-valued zero-mean process $X = \{X_t, t \in [0, T]\}$, we define the first and second order *quadratic variations* along uniform partitions as

$$V_{n,T}^{(i)X} = \sum_{k=1}^{n-i+1} \left(\Delta_{n,k}^{(i)} X\right)^2, \qquad i = 1, 2, \tag{2.1}$$

where

$$\Delta_{n,k}^{(1)} X = X(t_k^n) - X(t_{k-1}^n), \qquad \Delta_{n,k}^{(2)} X = X(t_{k+1}^n) - 2X(t_k^n) + X(t_{k-1}^n),$$

and $t_k^n = kT/n$. We call quadratic variation *normalized* if X in (2.1) is replaced by $\widehat{X}$, where

$$\Delta_{n,k}^{(i)} \widehat{X} = \frac{\Delta_{n,k}^{(i)} X}{\sqrt{\mathsf{E}\left(\Delta_{n,k}^{(i)} X\right)^2}}, \qquad i = 1, 2.$$

Let B^H be an fBm. Random variables $\left(\Delta_{n,k}^{(i)} \widehat{B}^H\right)$ are centered stationary Gaussian r. vs. with variance 1 and covariance function $\rho_H^{(i)}(j-k)$, where for any real number x, $\rho_H^{(i)}(x)$ has the form

$$\rho_H^{(1)}(x) = \frac{1}{2}\left[|x+1|^{2H} + |x-1|^{2H} - 2|x|^{2H}\right],$$

$$\rho_H^{(2)}(x) = \frac{1}{2(4-2^{2H})}\left[-6|x|^{2H} + 4|x+1|^{2H} - |x+2|^{2H} - |x-2|^{2H} + 4|x-1|^{2H}\right].$$

Theorem 2.2 *For all $H \in (0, 1)$*

$$(n-1)^{-1} V_{n,T}^{(i)\widehat{B}^H} \xrightarrow{a.s.} 1, \quad n \to \infty, \quad i = 1, 2.$$

Proof Let $i = 1$. Then Theorem 2.2 is a particular case of Theorem 4.5, point 2, if one puts $p = 0$. For $i = 2$ the proof is carried out using the same methods. □

Remark 2.3 If to replace convergence a.s. with convergence in probability, then Theorem 2.2 is an immediate consequence of the ergodic theorem for the stationary Gaussian sequences and the self-similarity property of an fBm. See Theorem B.1

and Corollary B.2. Indeed, consider, for example, $i = 1$ (the case $i = 2$ is considered similarly). Note that $\mathsf{E}\left(\Delta_{n,k}^{(1)} B^H\right)^2 = \left(\frac{T}{n}\right)^{2H}$, and

$$\mathsf{E}(B_n^H - B_{n-1}^H)B_1^H = \frac{1}{2}\left(n^{2H} - 2(n-1)^{2H} + (n-2)^{2H}\right)$$
$$= \frac{1}{2}n^{2H}\left(1 - 2\left(1 - \frac{1}{n}\right)^{2H} + \left(1 - \frac{2}{n}\right)^{2H}\right)$$
$$= \frac{1}{2}n^{2H}\left(1 - 2 + 2\cdot\frac{2H}{n} - 2\frac{2H(2H-1)}{2n^2} + 1 - 2\cdot\frac{2H}{n} + 4\frac{2H(2H-1)}{2n^2} + o(n^{-2})\right)$$
$$= O(n^{2H-2}),$$

whence

$$n^{-1}V_{n,T}^{(1)\widehat{B}^H} = n^{-1}\left(\frac{n}{T}\right)^{2H}\sum_{k=1}^{n}(\Delta_{n,k}^{(1)}B^H)^2 \stackrel{d}{=} n^{-1}\sum_{k=1}^{n}(B_k^H - B_{k-1}^H)^2 \xrightarrow{a.s.} \mathsf{E}(B_1^H)^2 = 1.$$

For the non-uniform partitions the situation is much more involved, and there is a variety of approaches. For more detailed information see introduction to this chapter.

2.1.1 *The Rate of Convergence of Quadratic Variations of an fBm*

In this subsection, we show how concentration inequalities for Gaussian quadratic form can be used to obtain convergence rate of quadratic variation of an fBm. First we formulate the concentration inequality proposed by Breton et al. [31], whose proof relies on the Malliavin calculus techniques developed by Nourdin and Viens [137]. Later on we apply it to an fBm.

One more way to obtain convergence rate of quadratic variation of an fBm is to apply the Lévy–Octaviani inequality (see [100, Theorem 3]). But we prefer the first method, since it is more organic in the context of this book.

Consider a finite centered Gaussian family $X = \{X_k : k = 1, \dots, M\}$, and write $d_{kj} = \mathsf{E}X_kX_j$. In what follows, we shall consider two quadratic forms related to X, and involved some real coefficient c. Denote

$$Q_1(c, X) = c\sum_{k=1}^{M}\left(X_k^2 - d_{kk}\right)$$

and

$$Q_2(c, Y) = 2c^2 \sum_{k,j=1}^{m} X_k X_j \, d_{kj}.$$

Note that $Q_2(c, X) \geq 0$.

Theorem 2.4 ([31]) *Suppose that $Q_1(c, X)$ is not a. s. zero and fix $\alpha \geq 0$ and $\beta > 0$. Assume that $Q_2(c, X) \leq \alpha Q_1(c, X) + \beta$ a. s. Then, for all $z > 0$*

$$\mathsf{P}\big(|Q_1(c, X)| \geq z\big) \leq 2 \exp\left\{ -\frac{z^2}{2\alpha z + 2\beta} \right\}.$$

Now we prove the modified version of the concentration inequality from [31] by applying Theorem 2.4.

Let $(X_{n,k})_{k=1,\dots,n-1}$, $n \in \mathbb{N}$, be a triangle array of Gaussian random variables. Denote $X_n = (X_{n,1}, \dots, X_{n,n-1})$ and $d_{jk}^{X,n} = \mathsf{E} X_{n,j} X_{n,k}$.

Lemma 2.5 *Assume that*

$$\varepsilon_n := \left| \frac{1}{n-1} \sum_{k=1}^{n-1} d_{kk}^{X,n} - 1 \right| \longrightarrow 0 \quad \text{as } n \to \infty. \tag{2.2}$$

Then for all $z > 0$ we have

$$\mathsf{P}\left(\frac{1}{\sqrt{n-1}} \sum_{k=1}^{m} \left[(X_{n,k})^2 - d_{kk}^{X,n} \right] \geq z \right) \leq 2 \, \exp\left(-\frac{z^2}{4 \nu_n \left(\frac{z}{\sqrt{n-1}} + \varepsilon_n + 1 \right)} \right),$$

where $1 \leq m \leq n-1$, $\nu_n = \max_{1 \leq k \leq n-1} \sum_{j=1}^{n-1} |d_{kj}^{X,n}|$.

Proof Denote

$$Q_1\big((n-1)^{-1/2}, X_n\big) = \frac{1}{\sqrt{n-1}} \sum_{k=1}^{m} \left[(X_{n,k})^2 - d_{kk}^{X,n} \right]$$

and

$$Q_2\big((n-1)^{-1/2}, X_n\big) = \frac{2}{n-1} \sum_{k,j=1}^{m} X_{n,k} X_{n,j} \, d_{kj}^{X,n}.$$

Then

$$
\begin{aligned}
&Q_2\big((n-1)^{-1/2}, X_n\big) \\
&\le \frac{2}{n-1}\sum_{k,j=1}^{m}|X_{n,k}|\cdot|X_{n,j}|\cdot\left|d_{kj}^{X,n}\right| \le \frac{1}{n-1}\sum_{k,j=1}^{m}[X_{n,k}^2+X_{n,j}^2]\cdot\left|d_{kj}^{X,n}\right| \\
&= \frac{2}{n-1}\sum_{k,j=1}^{m}X_{n,k}^2\left|d_{kj}^{X,n}\right| = \frac{2}{n-1}\sum_{k=1}^{m}X_{n,k}^2\sum_{j=1}^{m}\left|d_{kj}^{X,n}\right| \\
&\le \frac{2}{n-1}\sum_{k=1}^{m}X_{n,k}^2\left(\max_{1\le k\le n-1}\sum_{j=1}^{n-1}\left|d_{kj}^{X,n}\right|\right) \\
&\le \frac{2\nu_n}{\sqrt{n-1}}\left(Q_1\big((n-1)^{-1/2}, X_n\big) + \frac{1}{\sqrt{n-1}}\sum_{k=1}^{n-1}d_{kk}^{X,n}\right) \\
&\le \frac{2\nu_n}{\sqrt{n-1}}\left(Q_1\big((n-1)^{-1/2}, X_n\big) + \frac{1}{\sqrt{n-1}}\sum_{k=1}^{n-1}\left|d_{kk}^{X,n}-1\right| + \sqrt{n-1}\right) \\
&\le \alpha_n Q_1\big((n-1)^{-1/2}, X_n\big) + \beta_n,
\end{aligned}
$$

where

$$\alpha_n = \frac{2\nu_n}{\sqrt{n-1}}, \qquad \beta_n = 2\nu_n(\varepsilon_n+1).$$

From Theorem 2.4 we obtain the required assertion. □

We apply obtained concentration inequality to a fBm. Before presenting this result, we prove two auxiliary lemmas.

Lemma 2.6 (see [30]) *The following equality*

$$2\sum_{j=2}^{\infty}|\widehat{\rho}_H^{(2)}(j)| = -\operatorname{sgn}(2H-1)\left(3-3\cdot 2^{2H}+3^{2H}\right), \qquad H\in(0,1), \tag{2.3}$$

holds, where

$$\widehat{\rho}_H^{(2)}(j) = \frac{1}{2}\big(-|j-2|^{2H}+4|j-1|^{2H}-6|j|^{2H}+4|j+1|^{2H}-|j+2|^{2H}\big),\ j\in\mathbb{N}\cup\{0\}.$$

Proof Let $\theta(f)(x) = f(x+1)-2f(x)+f(x-1)$. Observe that if f is a convex (resp. concave) function, then $\theta(f)(x)\ge 0$ (resp. $\theta(f)(x)\le 0$). The function $|x|^{2H}$, $H\in(0,1)$, is convex if $H\ge 1/2$, concave if $H\le 1/2$. The same holds true for $\theta(|x|^{2H})$ for $x\ge 1$ and $(\theta\circ\theta)(|x|^{2H})$ for $x\ge 2$.

Note that $-\frac{1}{2}(\theta \circ \theta)(|j|^{2H}) = \widehat{\rho}_H^{(2)}(j)$ and $\widehat{\rho}_H^{(2)}(j)$ has the same sign as $(1-2H)$ for $j \geq 2$.

Now we prove (2.3). Denote $\alpha_0 = 6, \alpha_1 = \alpha_{-1} = -4, \alpha_2 = \alpha_{-2} = 1$. First, for all $N \geq 2$, we have

$$
\begin{aligned}
-2\sum_{j=2}^{N}\widehat{\rho}_H^{(2)}(j) = &\sum_{j=0}^{N-2}|j|^{2H} - 4\sum_{j=1}^{N-1}|j|^{2H} + 6\sum_{j=2}^{N}|j|^{2H} - 4\sum_{j=3}^{N+1}|j|^{2H} + \sum_{j=4}^{N+2}|j|^{2H} \\
=&(1-4+6-4+1)\sum_{j=0}^{N-2}|j|^{2H} \\
&-4(N-1)^{2H} + 6\big(N^{2H} + (N-1)^{2H} - 1^{2H}\big) \\
&-4\big((N+1)^{2H} + N^{2H} + (N-1)^{2H} - 1^{2H} - 2^{2H}\big) \\
&+\big((N+2)^{2H} + (N+1)^{2H} + N^{2H} + (N-1)^{2H} - 1^{2H} - 2^{2H} - 3^{2H}\big) \\
=&(N+2)^{2H} + (1-4)(N+1)^{2H} + (1-4+6)N^{2H} \\
&+(1-4+6-4)(N-1)^{2H} + 6(-1^{2H}) - 4\big(-1^{2H} - 2^{2H}\big) \\
&+\big(-1^{2H} - 2^{2H} - 3^{2H}\big) \\
=&(N+2)^{2H} - 3(N+1)^{2H} + 3N^{2H} - (N-1)^{2H} - (3 - 3\cdot 2^{2H} + 3^{2H}) \\
=&(N+2)^{2H}\sum_{i=0}^{3}\left[\left(1-\frac{i}{N+2}\right)^{2H}\sum_{k=2-i}^{2}\alpha_k\right] - (3 - 3\cdot 2^{2H} + 3^{2H}) \\
=&(N+2)^{2H}\sum_{i=0}^{3}\left[\left(1-\frac{2Hi}{N+2} + \frac{2H(2H-1)i^2}{2(N+2)^2}\right.\right. \\
&\left.\left.+O\left(\frac{1}{(N+2)^3}\right)\right)\sum_{k=2-i}^{2}\alpha_k\right] - (3 - 3\cdot 2^{2H} + 3^{2H}).
\end{aligned}
$$

But

$$
\sum_{i=0}^{3}\sum_{k=2-i}^{2}\alpha_k = \sum_{i=0}^{3}\sum_{k=2-i}^{2}i\alpha_k = \sum_{i=0}^{3}\sum_{k=2-i}^{2}i^2\alpha_k = 0, \qquad \sum_{i=0}^{3}\sum_{k=2-i}^{2}i^3\alpha_k = -6.
$$

Thus, we conclude that

$$
-2\sum_{j=2}^{N}\widehat{\rho}_H^{(2)}(j) = O\left((N+2)^{2H-3}\right) - (3 - 3\cdot 2^{2H} + 3^{2H}).
$$

Note that the function $3 - 3 \cdot 2^{2H} + 3^{2H}$ is positive for $0 < H < 1/2$ and negative for $1/2 < H < 1$. Since

$$\left|\widehat{\rho}_H^{(2)}(j)\right| = -\operatorname{sgn}(2H-1)\widehat{\rho}_H^{(2)}(j), \tag{2.4}$$

we have

$$\begin{aligned} 2\sum_{j=2}^{N}\left|\widehat{\rho}_H^{(2)}(j)\right| &= -2\operatorname{sgn}(2H-1)\sum_{j=2}^{N}\widehat{\rho}_H^{(2)}(j) \\ &= -\operatorname{sgn}(2H-1)(3-3\cdot 2^{2H}+3^{2H}) + O\left((N+2)^{2H-3}\right) \end{aligned}$$

and we obtain the required result. □

Lemma 2.7

$$\sup_{H\in(0,1)}\sum_{j\in\mathbb{Z}}\left|\rho_H^{(2)}(j)\right| = \frac{8}{3}, \tag{2.5}$$

where $\mathbb{Z}$ is the set of integers.

Proof Note that $\widehat{\rho}_H^{(2)}(j) = \widehat{\rho}_H^{(2)}(-j)$ and

$$\begin{aligned} \sum_{j\in\mathbb{Z}}\left|\rho_H^{(2)}(j)\right| &= 1 + 2\left|\rho_H^{(2)}(1)\right| + 2\sum_{j=2}^{\infty}\left|\rho_H^{(2)}(j)\right| \\ &= 1 + \frac{7-4\cdot 2^{2H}+3^{2H}}{4-2^{2H}} - \operatorname{sgn}(2H-1)\frac{3-3\cdot 2^{2H}+3^{2H}}{4-2^{2H}} \\ &= \begin{cases} 1 + \frac{10-7\cdot 2^{2H}+2\cdot 3^{2H}}{4-2^{2H}} & \text{for } H \le 1/2, \\ 1 + \frac{4-2^{2H}}{4-2^{2H}} = 2 & \text{for } H \ge 1/2. \end{cases} \end{aligned}$$

Since the function $\frac{10-7\cdot 2^{2H}+2\cdot 3^{2H}}{4-2^{2H}}$, $H \in (0, 1/2)$, is decreasing, then

$$\lim_{H\to 0+}\frac{10-7\cdot 2^{2H}+2\cdot 3^{2H}}{4-2^{2H}} = \frac{5}{3}.$$

□

With the aid of the auxiliary lemmas and Lemma 2.5 we easily obtain the concentration inequality for quadratic variation of an fBm.

Lemma 2.8 *Let B^H be an fBm. For all $z > 0$ and any $H \in (0, 1)$,*

$$\mathsf{P}\left(\left|\frac{1}{\sqrt{n-1}}\sum_{k=1}^{m}\left[\left(\Delta_{n,k}^{(2)}\widehat{B}^H\right)^2 - 1\right]\right| > z\right) \le 2\exp\left(-\frac{z^2}{\frac{32}{3}\left(\frac{z}{\sqrt{n-1}}+1\right)}\right), \tag{2.6}$$

where $1 \le m \le n-1$.

Proof Set $X_{n,k} = \Delta^{(2)}_{n,k}\widehat{B}^H$, $1 \le k \le n-1$. Then $d_{kk}^{\widehat{B}^H,n} = 1$ and $d_{kj}^{\widehat{B}^H,n} = \rho_H^{(2)}(k-j)$. Thus, $\varepsilon_n = 0$ and

$$\max_{1\le k\le n-1} \sum_{j=1}^{n-1} \left| d_{kj}^{\widehat{B}^H,n} \right| \le \sum_{r\in\mathbb{Z}} \left| \rho_H^{(2)}(r) \right|.$$

From (2.5) it follows that

$$\nu_n \le \nu = \sup_{H\in(0,1)} \sum_{j\in\mathbb{Z}} \left| \rho_H^{(2)}(j) \right| = \frac{8}{3}.$$

This yields the required inequality. □

Now we apply the concentration inequality (2.6) to obtain the convergence rate of quadratic variation of an fBm.

To avoid cumbersome expressions, we introduce symbols O_ω and o_ω. Let (Y_n) be a sequence of r. vs., ς is an a.s. non-negative r.v. and $(a_n) \subset (0,\infty)$ vanishes. $Y_n = O_\omega(a_n)$ means that $|Y_n| \le \varsigma \cdot a_n$; $Y_n = o_\omega(a_n)$ means that $|Y_n| \le \varsigma \cdot b_n$ with $b_n = o(a_n)$. In particular, $Y_n = o_\omega(1)$ corresponds to the sequence (Y_n) which tends to 0 a.s. as $n \to \infty$ and $Y_n = O_\omega(1)$ corresponds to the sequence (Y_n) which is a.s. bounded.

Theorem 2.9 *For any $t \in [0,T]$ define $r_{nt} = \left[\frac{t}{T}n\right]$, $\rho_{nt} = \frac{r_{nt}}{n}T$. Then a. s.*

$$\sup_{t\in[0,T]} \left| Tn^{-1}V_{n,t}^{(2)\widehat{B}^H} - \rho_{nt} \right| = O_\omega\left(n^{-1/2}\log^{1/2} n\right), \tag{2.7}$$

where

$$V_{n,t}^{(2)\widehat{B}^H} = \sum_{k=1}^{r_{nt}} \left(\Delta^{(2)}_{n,k}\widehat{B}^H\right)^2 \quad \text{for } t < T \quad \text{and} \quad V_{n,T}^{(2)\widehat{B}^H} = \sum_{k=1}^{n-1} \left(\Delta^{(2)}_{n,k}\widehat{B}^H\right)^2.$$

Proof Note that

$$\mathsf{P}\left(\max_{1\le k\le n-1} \left| \frac{T}{n} \sum_{j=1}^{k} \left(\Delta^{(2)}_{n,k}\widehat{B}^H\right)^2 - \frac{k}{n}T \right| > \delta \right)$$
$$\le \sum_{k=1}^{n-1} \mathsf{P}\left(\left| \sum_{j=1}^{k} \left[\left(\Delta^{(2)}_{n,k}\widehat{B}^H\right)^2 - 1 \right] \right| > \frac{n\delta}{T} \right)$$

for all $\delta > 0$. From Lemma 2.8 we get

$$\mathsf{P}\Big(\Big|\sum_{j=1}^{k}\Big[\big(\Delta_{n,k}^{(2)}\widehat{B}^H\big)^2 - 1\Big]\Big| > \frac{n\delta}{T}\Big)$$

$$= \mathsf{P}\Big(\Big|\frac{1}{\sqrt{n-1}}\sum_{j=1}^{k}\Big[\big(\Delta_{n,k}^{(2)}\widehat{B}^H\big)^2 - 1\Big]\Big| \ge \frac{n\delta}{T\sqrt{n-1}}\Big)$$

$$\le \mathsf{P}\Big(\Big|\frac{1}{\sqrt{n-1}}\sum_{j=1}^{k}\Big[\big(\Delta_{n,k}^{(2)}\widehat{B}^H\big)^2 - 1\Big]\Big| \ge \frac{\delta\sqrt{n-1}}{T}\Big) \le 2\exp\Big\{-\frac{3\delta^2(n-1)}{32T(\delta+T)}\Big\}.$$

Thus,

$$\mathsf{P}\Big(\max_{1\le k\le n-1}\Big|\frac{T}{n}\sum_{j=1}^{k}\big(\Delta_{n,k}^{(2)}\widehat{B}^H\big)^2 - \frac{k}{n}T\Big| > \delta_n\Big) \le 2(n-1)\exp\Big\{-\frac{3\delta_n^2(n-1)}{32T(\delta_n+T)}\Big\}.$$

Set $\delta_n = T\sqrt{a\frac{\log n}{n}}$ and $n \ge 3$. Then

$$\mathsf{P}\Big(\max_{1\le k\le n-1}\Big|\frac{T}{n}\sum_{j=1}^{k}\big(\Delta_{n,k}^{(2)}\widehat{B}^H\big)^2 - \frac{k}{n}T\Big| > \delta_n\Big)$$

$$\le 2(n-1)\exp\left\{-\frac{3a(1-\frac{1}{n})\log n}{32\Big(\sqrt{a\frac{\log n}{n}}+1\Big)}\right\} \le 2n\exp\left\{-\frac{a\log n}{16\Big(\sqrt{a\frac{\log n}{n}}+1\Big)}\right\}.$$

For $a > 961$ we get

$$\mathsf{P}\Big(\max_{1\le k\le n-1}\Big|\frac{T}{n}\sum_{j=1}^{k}\big(\Delta_{n,k}^{(2)}\widehat{B}^H\big)^2 - \frac{k}{n}T\Big| > \delta_n\Big) \le 2n^{-2}.$$

It means that

$$\sum_{n=2}^{\infty}\mathsf{P}\Big(\max_{1\le k\le n-1}\Big|\frac{T}{n}\sum_{j=1}^{k}\big(\Delta_{n,k}^{(2)}\widehat{B}^H\big)^2 - \frac{k}{n}T\Big| > \delta_n\Big) < \infty.$$

Finally, we get the statement of the theorem from the Borel–Cantelli lemma and the evident equality

$$\sup_{t\le T}\Big|Tn^{-1}V_{nt}^{(2)\widehat{B}^H} - \rho_{nt}\Big| = \max_{1\le k\le n-1}\Big|\frac{T}{n}\sum_{j=1}^{k}\big(\Delta_{n,k}^{(2)}\widehat{B}^H\big)^2 - \frac{k}{n}T\Big|.$$

□

2.1.2 Asymptotic Behavior of Weighted Quadratic Variations

To study the asymptotic behavior of estimates of the Hurst index, we need to know the asymptotics of weighted quadratic variations. The purpose of this subsection is to find the rate of convergence of weighted quadratic variations to the limit process.

We prove the following result:

Lemma 2.10 ([104]) *Let $\alpha \in (0,1]$ and let $h : \Omega \times [0,T] \to \mathbb{R}$ be a random function which is Hölder continuous of order α, i.e. for almost each $\omega \in \Omega$*

$$|h_s(\omega) - h_t(\omega)| \le K_h(\omega)\,|s-t|^{\alpha}$$

with some a.s. finite and positive r.v. K_h. Then

$$\left(\frac{n}{T}\right)^{2H-1} \sum_{k=1}^{n-1} h_{\frac{k}{n}T}\left(\Delta_{n,k}^{(2)} B^H\right)^2 = (4-2^{2H}) \int_0^T h_t \mathrm{d}t + O_{\omega}(n^{-\alpha/2}\log^{\alpha/2} n). \qquad (2.8)$$

Proof For the sake of clarity we split the proof into three steps.

Step 1. Let $\widetilde{\Omega} = [0,1] = I_1$, $\mathcal{B}_1 = \{A \cap [0,1] \mid A \in \mathcal{B}(\mathbb{R})\}$, $\widetilde{\mathsf{P}} = \lambda_1$ and L_1 denotes a set of r.vs. on $(\widetilde{\Omega}, \mathcal{B}_1, \widetilde{\mathsf{P}})$ supported on I_1, i.e.

$$L_1 = \{Z : \widetilde{\Omega} \to I_1 \mid Z \text{ is } (\mathcal{B}_1, \mathcal{B}_1)\text{-measurable}\}.$$

For each $\tau \in (0,1]$ define a metric d_τ on I_1 as follows: $d_\tau(x,y) = |x-y|^{\tau}$. Then any d_τ induces the same topology on I_1 and corresponding Borel σ-fields coincide with $\mathcal{B}_1$. Therefore it does not matter whether we treat I_1 as a metric space (I_1, d_α) or as a metric space (I_1, d_1). In each case the set L_1 remains the same.

Let M_1 denote the set of probability measures on $\mathcal{B}(\mathbb{R})$ corresponding to r.vs. of L_1, i.e.

$$M_1 = \{\mu : \mathcal{B}(\mathbb{R}) \to [0,1] \mid \exists Z \in L_1 : \mathsf{P}_Z = \widetilde{\mathsf{P}}(Z \in \cdot) = \mu\}.$$

Define on M_1 two Wasserstein metrics:

$$d_{W_\tau}(\mu,\nu) = \inf_{Y \sim \mu, Z \sim \nu} \mathsf{E} d_\tau(Y,Z) = \inf_{Y \sim \mu, Z \sim \nu} \mathsf{E}\,|Y-Z|^{\tau}, \qquad \tau \in \{\alpha, 1\},$$

where the infimum is taken over all joint distributions (Y,Z) of random variables Y, Z with given marginals. Take arbitrary $Y, Z \in L_1 : Y \sim \mu, Z \sim \nu$. By Jensen's inequality, $\mathsf{E}\,|Y-Z|^{\alpha} \le (\mathsf{E}\,|Y-Z|)^{\alpha}$. Thus,

$$d_{W_\alpha}(\mu,\nu) = \inf_{\widetilde{Y} \sim \mu, \widetilde{Z} \sim \nu} \mathsf{E}\left|\widetilde{Y} - \widetilde{Z}\right|^{\alpha} \le (\mathsf{E}\,|Y-Z|)^{\alpha}.$$

Consequently, $d_{W_\alpha} \le (d_{W_1})^{\alpha}$.

Step 2. Let $V_{nt}^{(2)\widehat{B}^H}$, $t \in [0, T]$, be the same as in Theorem 2.9. Denote

$$p_{nk} = \left(\frac{n}{T}\right)^{2H} \frac{\left(\Delta_{n,k}^{(2)} B^H\right)^2}{(4 - 2^{2H}) V_{nT}^{(2)\widehat{B}^H}}, \qquad P_{nk} = \sum_{j=1}^{k} p_{nj} = \frac{V_{kT}^{(2)\widehat{B}^H}}{V_{nT}^{(2)\widehat{B}^H}}, \qquad k = 1, \ldots, n-1,$$

and

$$\mu_n(A) = \sum_{k=1}^{n-1} p_{nk}\, \delta_{\frac{k}{n}}(A) \quad \text{for } A \in \mathcal{B}(\mathbb{R}), \tag{2.9}$$

where δ_a denotes the Dirac measure, i.e. for each measurable set A $\delta_a(A) = \mathbf{1}_A(a)$. Then a.s. μ_n is a discrete measure from M_1. Let F_n, F be distribution functions corresponding to the measures μ_n, λ_1 accordingly. For definiteness, we use right-continuous versions here and further on. By (2.9),

$$F_n(x) = \begin{cases} 0, & x < \frac{1}{n}; \\ P_{nk}, & x \in [\frac{k}{n}; \frac{k+1}{n}), \quad k = 1, \ldots, n-1; \\ 1, & x \geq 1. \end{cases}$$

Denote $F(x) = (x \wedge 1)\mathbf{1}_{(0,\infty)}(x)$. Then

$$|F_n(x) - F(x)| = \begin{cases} x, & x \in [0, 1/n); \\ 0, & x \notin [0, 1); \\ |x - P_{nk}|, & x \in [\frac{k}{n}, \frac{k+1}{n}), k = 1, \ldots, n-1. \end{cases}$$

Theorem 2.9 implies relationship $Tn^{-1} V_{nt}^{(2)\widehat{B}^H} = t + O_\omega(n^{-1/2} \log^{1/2} n)$, since denoting by $\{x\} \in [0, 1)$ a fractional part of $x \in \mathbb{R}^+$ one has

$$\rho_{nt} = \frac{\frac{t}{T}n - \left\{\frac{t}{T}n\right\}}{n} T = t + O\left(\frac{1}{n}\right).$$

Consequently, $\forall x \in \left[\frac{k}{n}, \frac{k+1}{n}\right)$, $k = 1, \ldots, n-1$,

$$|F(x) - F_n(x)| = |x - P_{nk}| = \left| x - \frac{\frac{k}{n}T + O_\omega\left(\left(\frac{\log n}{n}\right)^{\frac{1}{2}}\right)}{T + O_\omega\left(\left(\frac{\log n}{n}\right)^{\frac{1}{2}}\right)} \right|$$

$$= \left| x - \frac{\frac{k}{n} + O_\omega\left(\left(\frac{\log n}{n}\right)^{\frac{1}{2}}\right)}{1 + O_\omega\left(\left(\frac{\log n}{n}\right)^{\frac{1}{2}}\right)} \right| = \left| \frac{x - \frac{k}{n} + O_\omega\left(\left(\frac{\log n}{n}\right)^{\frac{1}{2}}\right)}{1 + O_\omega\left(\left(\frac{\log n}{n}\right)^{\frac{1}{2}}\right)} \right|$$

$$= O_\omega\left(\left(\frac{\log n}{n}\right)^{\frac{1}{2}}\right),$$

and $d_K(\mu_n, \lambda_1) = \sup_x |F_n(x) - F(x)| = O_\omega\left(n^{-1/2}\log^{1/2} n\right)$, where d_K denotes the Kolmogorov metric on the set of probability measures on $\mathcal{B}(\mathbb{R})$.

Step 3. Let $\varphi(t) = h_{tT}/(T^\alpha K_h)$, $t \in [0, 1]$. Retaining notations introduced in the previous steps,

$$\begin{aligned}
&\left(\frac{n}{T}\right)^{2H-1}\sum_{k=1}^{n-1} h_{\frac{k}{n}T}\left(\Delta_{n,k}^{(2)}B^H\right)^2 \\
&= \frac{(4-2^{2H})T}{n} V_{nT}^{(2)\widehat{B}^H} T^\alpha K_h \sum_{k=1}^{n-1}\varphi\left(\frac{k}{n}\right)p_{nk} \\
&= \left((4-2^{2H})T + O_\omega(n^{-1/2}\log^{1/2} n)\right)T^\alpha K_h \int_{I_1}\varphi\,\mathrm{d}\mu_n \\
&= (4-2^{2H})T^{1+\alpha}K_h\int_{I_1}\varphi\,\mathrm{d}\mu_n + O_\omega(n^{-1/2}\log^{1/2} n).
\end{aligned} \tag{2.10}$$

By the Kantorovich duality theorem (see Appendix A.4 or [50, p. 421]),

$$\forall \mu, \nu \in M_1 \quad d_{W_\alpha}(\mu, \nu) = \sup_{\psi \in C_1^\alpha}\left|\int_{I_1}\psi\,\mathrm{d}\mu - \int_{I_1}\psi\,\mathrm{d}\nu\right|,$$

where $C_1^\alpha = \{\psi : I_1 \to \mathbb{R} \mid |\psi(s) - \psi(t)| \le d_\alpha(s,t) = |s-t|^\alpha\}$. Since $\varphi \in C_1^\alpha$, we have

$$\left|\int_{I_1}\varphi\,\mathrm{d}\mu_n - \int_{I_1}\varphi\,\mathrm{d}\lambda_1\right| \le d_{W_\alpha}(\mu_n, \lambda_1). \tag{2.11}$$

Now, recall that there is another explicit formula for d_{W_1} on M_1 (see Appendix A.4 or [44, p. 271]):

$$d_{W_1}(\mu, \nu) = \int_0^1 \left|F_\mu(x) - F_\nu(x)\right|\,\mathrm{d}x.$$

Thus, results of the previous steps yield

$$\begin{aligned}
d_{W_\alpha}(\mu_n, \lambda_1) \le & d_{W_1}^\alpha(\mu_n, \lambda_1) = \left(\int_0^1 \left|F_\mu(x) - F_\nu(x)\right|\,\mathrm{d}x\right)^\alpha \\
\le & d_K^\alpha(\mu_n, \lambda_1) = O_\omega\left(n^{-\alpha/2}\log^{\alpha/2} n\right),
\end{aligned} \tag{2.12}$$

and from (2.10)–(2.12) it follows

$$\left(\frac{n}{T}\right)^{2H-1}\sum_{k=2}^{n} h_{\frac{k}{n}T}\left(\Delta^{(2)}_{n,k}B^H\right)^2 = \left(4-2^{2H}\right)T^{1+\alpha}K_h\int_{I_1}\varphi\,\mathrm{d}\lambda_1 + O_\omega\left(n^{-\alpha/2}\log^{\alpha/2}n\right)$$

$$= \left(4-2^{2H}\right)T\int_0^1 h_{tT}\,\mathrm{d}t + O_\omega\left(n^{-\alpha/2}\log^{\alpha/2}n\right)$$

$$= \left(4-2^{2H}\right)\int_0^T h_t\,\mathrm{d}t + O\left(n^{-\alpha/2}\log^{\alpha/2}n\right). \qquad \square$$

2.1.3 CLT for Quadratic Variation of an fBm

The following limit results plays an important role in consideration of an asymptotic normality of the studied estimates.

At first we will prove bivariate CLT for quadratic variations of an fBm. This result is a particular case of a theorem proved by Begyn [13]. Our proof is completely different.

Let $B^H = \{B^H(t) : t \in [0,T]\}$, $T > 0$, be a fBm. Denote

$$V^{\widehat{B}^H}_{in,T} = \sum_{k=1}^{in-1}\left(\Delta^{(2)}_{in,k}\widehat{B}^H\right)^2, \qquad \Delta^{(2)}_{in,k}\widehat{B}^H = \frac{\Delta^{(2)}_{in,k}B^H}{\sqrt{\mathsf{E}(\Delta^{(2)}_{in,k}B^H)^2}} \qquad i = 1,2,$$

$$d^{\widehat{B}^H,in}_{k,j} = \mathsf{E}\Delta^{(2)}_{in,k}\widehat{B}^H\,\Delta^{(2)}_{in,j}\widehat{B}^H, \quad 1 \le j,k \le in-1, \quad i = 1,2,$$

$$c^{\widehat{B}^H,2n}_{j,k} = \mathsf{E}\Delta^{(2)}_{n,j}\widehat{B}^H\,\Delta^{(2)}_{2n,k}\widehat{B}^H, \qquad 1 \le j \le n-1, \quad 1 \le k \le 2n-1,$$

$$\Delta^{(2)}_{in,k}\widehat{B}^H = \widehat{B}^H_{\frac{k+1}{in}T} - 2\widehat{B}^H_{\frac{k}{in}T} + \widehat{B}^H_{\frac{k-1}{in}T}, \qquad 1 \le k \le in-1, \quad i = 1,2.$$

Note that

$$d^{\widehat{B}^H,n}_{k,j} = d^{\widehat{B}^H,2n}_{k,j} = \rho^{(2)}_H(j-k),$$

where function $\rho^{(2)}_H(\cdot)$ was defined in Sect. 2.1. Now we calculate covariance $c^{\widehat{B}^H,2n}_{j,k}$.

Set $t, h > 0$, $t-h \ge 0$ and $\Delta^{(2)}B^H_{t,h} = B^H_{t+h} - 2B^H_t + B^H_{t-h}$. It is easy to calculate that

$$\mathsf{E}\Delta^{(2)}B^H_{t_1,h_1}\Delta^{(2)}B^H_{t_2,h_2}$$
$$= \frac{1}{2}\Big[-|t_1+h_1-t_2-h_2|^{2H} + 2|t_1+h_1-t_2|^{2H} - |t_1+h_1-t_2+h_2|^{2H}$$
$$+ 2|t_1-t_2-h_2|^{2H} - 4|t_1-t_2|^{2H} + 2|t_1-t_2+h_2|^{2H}$$
$$- |t_1-h_1-t_2-h_2|^{2H} + 2|t_1-h_1-t_2|^{2H} - |t_1-h_1-t_2+h_2|^{2H}\Big].$$

Then

$$
\begin{aligned}
&\mathsf{E}\Delta^{(2)}B^H_{\frac{j}{n}T,\frac{T}{n}}\Delta^{(2)}B^H_{\frac{k}{2n}T,\frac{T}{2n}}\\
&=\frac{T^{2H}}{2}\Big[-\Big|\frac{j}{n}+\frac{1}{n}-\frac{k}{2n}-\frac{1}{2n}\Big|^{2H}+2\Big|\frac{j}{n}+\frac{1}{n}-\frac{k}{2n}\Big|^{2H}-\Big|\frac{j}{n}+\frac{1}{n}-\frac{k}{2n}+\frac{1}{2n}\Big|^{2H}\\
&\quad+2\Big|\frac{j}{n}-\frac{k}{2n}-\frac{1}{2n}\Big|^{2H}-4\Big|\frac{j}{n}-\frac{k}{2n}\Big|^{2H}+2\Big|\frac{j}{n}-\frac{k}{2n}+\frac{1}{2n}\Big|^{2H}\\
&\quad-\Big|\frac{j}{n}-\frac{1}{n}-\frac{k}{2n}-\frac{1}{2n}\Big|^{2H}+2\Big|\frac{j}{n}-\frac{1}{n}-\frac{k}{2n}\Big|^{2H}-\Big|\frac{j}{n}-\frac{1}{n}-\frac{k}{2n}+\frac{1}{2n}\Big|^{2H}\Big]\\
&=\frac{T^{2H}}{2^{2H+1}n^{2H}}\Big[\,|2j-k+1|^{2H}+2\,|2j-k+2|^{2H}-|2j-k+3|^{2H}\\
&\quad+|2j-k-1|^{2H}-4\,|2j-k|^{2H}-|2j-k-3|^{2H}+2\,|2j-k-2|^{2H}\Big]\\
&=:\frac{T^{2H}}{n^{2H}}\widetilde{\rho}_H(j,k).
\end{aligned}
\tag{2.13}
$$

Thus, $c^{\widehat{B}^H,2n}_{j,k}=\frac{\widetilde{\rho}_H(j,k)}{4-2^{2H}}$.

Theorem 2.11 *Let $B^H=\{B^H(t):t\in[0,T]\}$, $T>0$, be a fBm. Then*

$$
\mathbf{X}_n=\sqrt{n}\begin{pmatrix} n^{-1}V^{\widehat{B}^H}_{n,T}-1\\ (2n)^{-1}V^{\widehat{B}^H}_{2n,T}-1\end{pmatrix}\xrightarrow{d}\mathcal{N}(0;\Sigma_H),\qquad \Sigma_H=\begin{pmatrix}\Sigma_{11}&\Sigma_{12}\\ \Sigma_{12}&\Sigma_{22}\end{pmatrix},\tag{2.14}
$$

where $\mathcal{N}(0;\Sigma_H)$ is a Gaussian vector with

$$
\begin{aligned}
\Sigma_{11}&=2\Big(1+\frac{2}{(4-2^{2H})^2}\sum_{j=1}^{\infty}\widehat{\rho}^2_H(j)\Big),\qquad \Sigma_{22}=\frac{1}{2}\,\Sigma_{11},\\
\Sigma_{12}&=\Sigma_{21}=\frac{1}{(4-2^{2H})^2}\sum_{j\in\mathbb{Z}}\tilde{\rho}^2_H(j),\\
\widehat{\rho}_H(j)&=\frac{1}{2}\big[-6\,|j|^{2H}-|j-2|^{2H}-|j+2|^{2H}+4\,|j-1|^{2H}+4\,|j+1|^{2H}\big],\\
\widetilde{\rho}_H(j)&=\frac{1}{2^{2H+1}}\big[\,|j+1|^{2H}+2\,|j+2|^{2H}-|j+3|^{2H}+|j-1|^{2H}-4\,|j|^{2H}\\
&\qquad-|j-3|^{2H}+2\,|j-2|^{2H}\big].
\end{aligned}
$$

Proof To find out limiting distribution of $\mathbf{X}_n$, we compute a limiting moment generating function $\lim M_{\mathbf{X}_n}(\lambda) = M(\lambda)$.

Consider a centered Gaussian vector $\mathbf{G}_n = (G_n^{(i)}, 1 \le i \le 3n-2)$

$$\begin{aligned} G_n^{(i)} &= \Delta_{n,i}^{(2)} \widehat{B}^H, \quad 1 \le i \le n-1, \\ G_n^{(i)} &= \sqrt{2^{-1}}\, \Delta_{2n,i+1-n}^{(2)} \widehat{B}^H, \quad n \le i \le 3n-2 \end{aligned}$$

and a diagonal matrix

$$D_n = \operatorname{diag}(\underbrace{\lambda_1, \dots, \lambda_1}_{n-1}, \underbrace{\lambda_2, \dots, \lambda_2}_{2n-1}).$$

It is evident that $\mathsf{E} G_n^{(i)} = 0$ and $\mathsf{E}(G_n^{(i)})^2 = 1$ for all $1 \le i \le n-1$, $\mathsf{E}(G_n^{(i)})^2 = 2^{-1}$ for all $n \le i \le 3n-2$.

We denote the covariance matrix of the vector $\mathbf{G}_n$ by $\Sigma_{\mathbf{G}_n}$. Set

$$\widetilde{D}_n = \left(\Sigma_{\mathbf{G}_n}^{1/2}\right)^T D_n \Sigma_{\mathbf{G}_n}^{1/2}.$$

We give bound on eigenvalues of $\widetilde{D}_n$. It is obvious that $\widetilde{D}_n$ is symmetric. Denote by $\|A_n\| = \sup_{\|x\|=1} \|A_n x\|$ matrix A norm. For symmetric matrix $\widetilde{D}_n$ its norm is equal to its spectral norm, i.e. $\left\|\widetilde{D}_n\right\| = \rho(\widetilde{D}_n) := \max_k |\lambda_k(\widetilde{D}_n)|$. Since norm $\|\cdot\|$ is submultiplicative norm then

$$\begin{aligned} \max_k |\lambda_k(\widetilde{D}_n)| &= \left\|\widetilde{D}_n\right\| \le \left\|\Sigma_{\mathbf{G}_n}^{1/2}\right\| \cdot \|D_n\| \cdot \left\|\Sigma_{\mathbf{G}_n}^{1/2}\right\| = \left\|\Sigma_{\mathbf{G}_n}^{1/2}\right\|^2 \cdot \|D_n\| \\ &= \rho\big((\Sigma_{\mathbf{G}_n}^{1/2})^2\big) \cdot \rho(D_n) = \rho(\Sigma_{\mathbf{G}_n}) \cdot \rho(D_n) = \lambda_{\max}(\Sigma_{\mathbf{G}_n}) \cdot \max\{|\text{ł}_1|, |\text{ł}_2|\}. \end{aligned}$$

Now consider $\lambda_{\max}(\Sigma_{\mathbf{G}_n})$. In order to bound the maximal eigenvalue we again make use of the fact that the latter does not exceed maximal row sum of absolute values. Thus,

$$\lambda_{\max}(\Sigma_{\mathbf{G}_n}) \le \max_j \sum_{i=1}^{3n-2} \left|(\Sigma_{\mathbf{G}_n})_{ij}\right|. \tag{2.15}$$

Note that

$$
\begin{aligned}
(\Sigma_{\mathbf{G}_n})_{i,j+1-n} =& \frac{1}{\sqrt{2}} \mathsf{E}\big[\Delta^{(2)}_{n,i}\widehat{B}^H \Delta^{(2)}_{2n,j+1-n}\widehat{B}^H\big] \\
=& \frac{1}{\sqrt{2}} \mathsf{E}\big[\big(\Delta^{(2)}_{2n,2i+1}\widehat{B}^H + \Delta^{(2)}_{2n,2i-1}\widehat{B}^H + 2\Delta^{(2)}_{2n,2i}\widehat{B}^H\big)\Delta^{(2)}_{2n,j+1-n}\widehat{B}^H\big] \\
=& \frac{1}{\sqrt{2}} \big[d^{\widehat{B}^H,2n}_{2i+1,j+1-n} + d^{\widehat{B}^H,2n}_{2i-1,j+1-n} + 2d^{\widehat{B}^H,2n}_{2i,j+1-n}\big]
\end{aligned}
$$

for $1 \le i \le n-1$ and $1 \le j \le 2n-1$. By (2.15) and Lemma 2.8 it follows that

$$
\begin{aligned}
\lambda_{\max}(\Sigma_{\mathbf{G}_n}) \le& \max_{1\le j\le n-1} \sum_{i=1}^{n-1} |d^{\widehat{B}^H,n}_{i,j}| + \frac{1}{2} \max_{1\le j\le 2n-1} \sum_{i=1}^{2n-1} |d^{\widehat{B}^H,2n}_{i,j}| \\
&+ \frac{1}{\sqrt{2}} \max_{1\le j\le n-1} \sum_{i=1}^{2n-1} \big(|d^{\widehat{B}^H,2n}_{i,2j+1}| + |d^{\widehat{B}^H,2n}_{i,2j-1}| + 2|d^{\widehat{B}^H,2n}_{i,2j}|\big) \\
&+ \frac{1}{\sqrt{2}} \max_{1\le j\le 2n-1} \sum_{i=1}^{n-1} \big(|d^{\widehat{B}^H,2n}_{2i+1,j}| + |d^{\widehat{B}^H,2n}_{2i-1,j}| + 2|d^{\widehat{B}^H,2n}_{2i,j}|\big)\Big] \\
\le& \max_{1\le j\le n-1} \sum_{i=1}^{n-1} |d^{\widehat{B}^H,n}_{i,j}| + \Big(\frac{1}{2} + \frac{8}{\sqrt{2}}\Big) \max_{1\le j\le 2n-1} \sum_{i=1}^{2n-1} |d^{\widehat{B}^H,2n}_{i,j}| \le 20.
\end{aligned}
$$

Summing up, we finally come to conclusion that $\max_k |\lambda_k(\widetilde{D}_n)|$ is uniformly (in n, k) bounded by finite constant depending only on λ_1, λ_2.

Note that

$$
\lambda^T \mathbf{Y}_n := \sqrt{n}(\lambda_1, \lambda_2) \begin{pmatrix} n^{-1}\big(V^{\widehat{B}^H}_{n,T} - \mathsf{E}V^{\widehat{B}^H}_{n,T}\big) \\ (2n)^{-1}\big(V^{\widehat{B}^H}_{2n,T} - \mathsf{E}V^{\widehat{B}^H}_{2n,T}\big) \end{pmatrix} = \frac{1}{\sqrt{n}}\big(\mathbf{G}_n^T D_n \mathbf{G}_n - \mathsf{E}\mathbf{G}_n^T D_n \mathbf{G}_n\big).
$$

Recall that $\mathbf{G}_n \overset{d}{=} \sqrt{\Sigma_{\mathbf{G}_n}}\mathbf{Z}_n$ with $\mathbf{Z}_n \sim \mathcal{N}(0; I_{3n-2})$, where I_{3n-2} denotes an identity $(3n-2)$ matrix. So, one can write down the following equality

$$
\mathbf{G}_n^T D_n \mathbf{G}_n \overset{d}{=} (\Sigma^{1/2}_{\mathbf{G}_n}\mathbf{Z}_n)^T D_n \Sigma^{1/2}_{\mathbf{G}_n}\mathbf{Z}_n = \mathbf{Z}_n^T (\Sigma^{1/2}_{\mathbf{G}_n})^T D_n \Sigma^{1/2}_{\mathbf{G}_n}\mathbf{Z}_n = \mathbf{Z}_n^T \widetilde{D}_n \mathbf{Z}_n.
$$

Let $\widetilde{D}_n = Q_n^T \Lambda(\widetilde{D}_n) Q_n$ be canonical representation of $\widetilde{D}_n$ via diagonal matrix of eigenvalues and corresponding orthogonal matrix of eigenvectors. Since orthogonal transform does not change the distribution of Z_n, we have

$$
\mathbf{Z}_n^T \widetilde{D}_n \mathbf{Z}_n = \mathbf{Z}_n^T Q_n^T \Lambda(\widetilde{D}_n) Q_n \mathbf{Z}_n \overset{d}{=} \mathbf{Z}_n^T \Lambda(\widetilde{D}_n)\mathbf{Z}_n = \sum_{j=1}^{3n-2} Z^2_{n,j}\lambda_{n,j}.
$$

The estimation of eigenvalues of $\widetilde{D}_n$ show that we can choose n_0 such that $n^{-1/2}\max_k|\lambda_k(\widetilde{D}_n)| < 1/2$ for all $n \geq n_0$. To have $M_{\mathbf{Y}_n}(\lambda)$ well defined, we assume that all n in the sequel satisfy this condition.

Now moment generating function $M_{\mathbf{Y}_n}(\lambda)$ we can rewrite as

$$M_{\mathbf{Y}_n}(\lambda) = \exp\left\{-\frac{\mathsf{E}\mathbf{G}_n^T D_n \mathbf{G}_n}{\sqrt{n}}\right\}\mathsf{E}\exp\left\{\sum_{j=1}^{3n-2} Z_{n,j}^2 \frac{\lambda_{n,j}}{\sqrt{n}}\right\}.$$

Note that $(Z_{n,j}^2)$ are i.i.d. and each $Z_{n,j}^2$ has chi-square distribution $\chi^2(1)$ with

$$M_{Z_{n,j}^2}(x) = \frac{1}{\sqrt{1-2x}} \qquad \text{for } x \in (-1/2, 1/2).$$

Thus,

$$\begin{aligned} M_{\mathbf{Y}_n}(\lambda) &= \exp\left\{-\frac{\mathsf{E}\mathbf{G}_n^T D_n \mathbf{G}_n}{\sqrt{n}}\right\}\prod_{j=1}^{3n-2} M_{\chi^2(1)}\left(\frac{\lambda_{n,j}}{\sqrt{n}}\right) \\ &= \exp\left\{-\frac{\mathsf{E}\mathbf{G}_n^T D_n \mathbf{G}_n}{\sqrt{n}}\right\}\left(\prod_{j=1}^{3n-2}\frac{1}{1-2\frac{\lambda_{n,j}}{\sqrt{n}}}\right)^{\frac{1}{2}} \\ &= \exp\left\{-\frac{\mathsf{E}\mathbf{G}_n^T D_n \mathbf{G}_n}{\sqrt{n}} - \frac{1}{2}\sum_{j=1}^{3n-2}\log\left(1-2\frac{\lambda_{n,j}}{\sqrt{n}}\right)\right\}. \end{aligned}$$

By Maclaurin's expansion,

$$\log(1-x) = -x - \frac{x^2}{2} - \frac{x^3}{3} + o(x^3), \qquad x \to 0.$$

Since $\max_{n,j}\left|\lambda_{n,j}\right|$ is uniformly bounded and $\sum_{j=1}^{3n-2}\lambda_{n,j} = \operatorname{tr}(\widetilde{D}_n) = \mathsf{E}\mathbf{Z}_n^T\widetilde{D}_n\mathbf{Z}_n = \mathsf{E}\mathbf{G}_n^T D_n \mathbf{G}_n$, we can rewrite the expression for $M_{\mathbf{Y}_n}(t)$ as follows

$$\begin{aligned} M_{\mathbf{Y}_n}(\lambda) &= \exp\left\{-\frac{\mathsf{E}\mathbf{G}_n^T D_n \mathbf{G}_n}{\sqrt{n}} + \frac{1}{2}\sum_{j=1}^{3n-2}\left(2\frac{\lambda_{n,j}}{\sqrt{n}} + 4\frac{\lambda_{n,j}^2}{2n}\right) + o\left(\frac{1}{\sqrt{n}}\right)\right\} \\ &= \exp\left\{\frac{1}{n}\sum_{j=1}^{3n-2}\lambda_{n,j}^2\right\}\exp\left\{o\left(\frac{1}{\sqrt{n}}\right)\right\}. \end{aligned}$$

Therefore, it remains to compute limiting value of the first multiplier. By $\widetilde{D}_n$ definition,

$$\sum_{j=1}^{3n-2} \lambda_{n,j}^2 = \mathrm{tr}(\widetilde{D}_n^2) = \mathrm{tr}\big(((\sqrt{\Sigma_{\mathbf{G}_n}})^T D_n \sqrt{\Sigma_{\mathbf{G}_n}})^2\big) = \mathrm{tr}\big((D_n \Sigma_{\mathbf{G}_n})^2\big)$$

$$= \sum_{i=1}^{3n-2} \sum_{j=1}^{3n-2} (D_n \Sigma_{\mathbf{G}_n})_{ij} (D_n \Sigma_{\mathbf{G}_n})_{ji}. \tag{2.16}$$

Note that (2.16) may be rearranged in the following way

$$\sum_{i=1}^{3n-2} \sum_{j=1}^{3n-2} (D_n \Sigma_{\mathbf{G}_n})_{ij} (D_n \Sigma_{\mathbf{G}_n})_{ji}$$

$$= \frac{\lambda_1^2}{(4-2^{2H})^2} \sum_{i=1}^{n-1} \sum_{j=1}^{n-1} \widehat{\rho}_H^2(i-j) + \frac{\lambda_1 \lambda_2}{(4-2^{2H})^2} \sum_{i=1}^{n-1} \sum_{j=n}^{2n-1} \widetilde{\rho}_H(j,k)$$

$$+ \frac{\lambda_2^2}{4(4-2^{2H})^2} \sum_{i=1}^{2n-1} \sum_{j=1}^{2n-1} \widehat{\rho}_H^2(i-j) = \sum_{k=1}^{3} I_n^{(k)}.$$

Therefore, to obtain limiting expression for $M_{\mathbf{Y}_n}(\lambda)$ it suffices to divide each sum by n and to calculate the corresponding limits.

We treat the first sum $I_n^{(1)}$. Since $\widehat{\rho}_H(\cdot)$ is symmetric, we have

$$\sum_{i=1}^{n-1} \sum_{j=1}^{n-1} \widehat{\rho}_H^2(i-j) = (n-1)\widehat{\rho}_H^2(0) + 2\sum_{k=1}^{n-1} (n-1-k)\widehat{\rho}_H^2(k).$$

Thus,

$$\frac{1}{n(4-2^{2H})^2} \sum_{i=1}^{n-1} \sum_{j=1}^{n-1} \widehat{\rho}_H^2(i-j) = \left(1-\frac{1}{n}\right) + \frac{2}{(4-2^{2H})^2} \sum_{i=1}^{n-1} \widehat{\rho}_H^2(k)$$

$$- \frac{2}{n(4-2^{2H})^2} \sum_{k=1}^{n-1} (k+1)\widehat{\rho}_H^2(k). \tag{2.17}$$

For further consideration, we find the bound on $\rho_H^{(2)}(k)$. Fix $k \geq 3$. Then, using the formula

$$(1+u)^\alpha = 1 + \sum_{k=1}^{\infty} \frac{\alpha(\alpha-1)\dots(\alpha-k+1)}{k!} u^k \quad \text{for } -1 < u < 1, \tag{2.18}$$

we obtain

$$
\begin{aligned}
\widehat{\rho}_H^{(2)}(k) =& \frac{k^{2H}}{2}\left(-\left(1-\frac{2}{k}\right)^{2H}+4\left(1-\frac{1}{k}\right)^{2H}-6+4\left(1+\frac{1}{k}\right)^{2H}-\left(1+\frac{2}{k}\right)^{2H}\right)\\
=& \frac{k^{2H}}{2}\sum_{m=1}^{\infty}\frac{2H(2H-1)\cdots(2H-m+1)}{m!}\big(-(-2)^m+4(-1)^m+4-2^m\big)k^{-m}\\
=& k^{2H}\sum_{m=1}^{\infty}\frac{2H(2H-1)\cdots(2H-2m+1)}{(2m)!}(4-4^m)k^{-2m}.
\end{aligned}
$$

Note that the sign of $2H(2H-1)(2H-2)\cdots(2H-2m+1)$ is the same as that of $2H-1$ and

$$
\begin{aligned}
|2H(2H-1)(2H-2)\cdots(2H-2m+1)| =& 2H|2H-1|(2-2H)\cdots(2m-1-2H)\\
\le& 2(2m-1)!.
\end{aligned}
$$

Moreover, using estimates obtained in Breton et al. [31] and the inequality $(1-H)\log 4 < 4-2^{2H}$ we have

$$
\begin{aligned}
\left|\rho_H^{(2)}(k)\right| \le& \frac{k^{2H}}{4-2^{2H}}\sum_{m=1}^{\infty}\frac{2H|2H-1|(2-2H)\cdots(2m-1-2H)}{(2m)!}(4^m-4)k^{-2m}\\
\le& \frac{k^{2H}}{\log 4}\sum_{m=2}^{\infty}\frac{4^m-4}{m}\left(\frac{1}{k^2}\right)^m = \frac{k^{2H}}{\log 4}\left(4\log\left(1-\frac{1}{k^2}\right)-\log\left(1-\frac{4}{k^2}\right)\right)\\
\le& \frac{243}{20\log 4}k^{2H-4} \quad \text{for any } k\ge 3,
\end{aligned}
\tag{2.19}
$$

since

$$
\log(1-u) = -\sum_{k=1}^{\infty}\frac{u^k}{k} \quad \text{if } 0\le u<1,
$$

$$
4\log(1-u)-\log(1-4u) \le \frac{243}{20}u^2 \quad \text{if } 0\le u\le\frac{1}{6}.
$$

Using estimate (2.19), we obtain

$$
\begin{aligned}
\frac{1}{(4-2^{2H})^2}\sum_{k=3}^{n-1}\frac{k+1}{n}\widehat{\rho}_H^2(k) \le& \left(\frac{243}{20\log 4}\right)^2 n^{-1}\sum_{k=3}^{n-1}\frac{k+1}{k^4}\\
\le& \left(\frac{243}{20\log 4}\right)^2 n^{-1}\sum_{k=1}^{\infty}\frac{k+1}{k^4} \xrightarrow[n\to\infty]{} 0.
\end{aligned}
\tag{2.20}
$$

It follows from (2.17) and (2.20) that

$$\frac{1}{n(4-2^{2H})^2}\sum_{i=1}^{n-1}\sum_{j=1}^{n-1}\widetilde{\rho}^2(i-j) \longrightarrow 1+\frac{2}{(4-2^{2H})^2}\sum_{i=1}^{\infty}\widetilde{\rho}^2(k).$$

Now we consider the second sum $I_n^{(2)}$. Denote by $\lfloor a \rfloor$ an integer part of a real number a. Note that

$$\begin{aligned}
\sum_{j=1}^{n-1}\sum_{k=1}^{2n-1}\widetilde{\rho}_H^2(j,k) &= \sum_{j=1}^{n-1}\sum_{k=1}^{2j-1}\widetilde{\rho}_H^2(j,k) + \sum_{j=1}^{n-1}\widetilde{\rho}_H^2(j,2j) + \sum_{j=1}^{n-1}\sum_{k=2j+1}^{2n-1}\widetilde{\rho}_H^2(j,k) \\
&= \sum_{m=1}^{2n-3}\left(n-1-\left\lfloor \frac{m}{2}\right\rfloor\right)\widetilde{\rho}_H^2(m) + (n-1)\widetilde{\rho}_H^2(0) \\
&\quad + \sum_{m=1}^{2n-3}\left(n-1-\left\lfloor \frac{m}{2}\right\rfloor\right)\widetilde{\rho}_H^2(-m) \\
&=(n-1)\sum_{m=-2n+3}^{2n-3}\widetilde{\rho}_H^2(m) - \sum_{m=1}^{2n-3}\left\lfloor \frac{m}{2}\right\rfloor\widetilde{\rho}_H^2(m) \\
&\quad - \sum_{m=1}^{2n-3}\left\lfloor \frac{m}{2}\right\rfloor\widetilde{\rho}_H^2(-m),
\end{aligned} \tag{2.21}$$

since $\widetilde{\rho}_H(m)=\widetilde{\rho}_H(-m)$. Thus,

$$\begin{aligned}
&\frac{1}{n}\frac{1}{(4-2^{2H})^2}\sum_{j=1}^{n-1}\sum_{k=1}^{2n-1}\widetilde{\rho}_H^2(j,k) \\
&= \left(1-\frac{1}{n}\right)\frac{1}{(4-2^{2H})^2}\sum_{m=-2n+3}^{2n-3}\widetilde{\rho}_H^2(m) \\
&\quad - \frac{1}{n}\frac{1}{(4-2^{2H})^2}\left(\sum_{m=1}^{2n-3}\left\lfloor \frac{m}{2}\right\rfloor\widetilde{\rho}_H^2(m) + \sum_{m=1}^{2n-3}\left\lfloor \frac{m}{2}\right\rfloor\widetilde{\rho}_H^2(-m)\right).
\end{aligned} \tag{2.22}$$

Using the formula (2.18), we obtain

$$\begin{aligned}
\widetilde{\rho}_H(k) =&k^{2H}\left[\left(1-\frac{1}{k}\right)^{2H} + 2\left(1-\frac{2}{k}\right)^{2H} - \left(1-\frac{3}{k}\right)^{2H} - 4 + \left(1+\frac{1}{k}\right)^{2H}\right. \\
&\left. + 2\left(1+\frac{2}{k}\right)^{2H} - \left(1+\frac{3}{k}\right)^{2H}\right]
\end{aligned}$$

$$
\begin{aligned}
=&k^{2H}\Bigg[\sum_{m=1}^{\infty}\frac{2H(2H-1)\cdots(2H-m+1)}{m!}\left(\frac{1}{k}\right)^{m}[(-1)^m+2\cdot(-2)^m \\
&-(-3)^m+1+2\cdot 2^m-3^m]\Bigg] \\
=&2k^{2H}\Bigg[\sum_{m=1}^{\infty}\frac{2H(2H-1)\cdots(2H-2m+1)}{(2m)!}\left(\frac{1}{k}\right)^{2m}\left[1+2^{2m+1}-3^{2m}\right]\Bigg]
\end{aligned}
\tag{2.23}
$$

for $k \geq 4$, where function $1+2^{2m+1}-3^{2m}$ is decreasing for $m \geq 1$. Similarly to the proof of (2.19), we have

$$
\begin{aligned}
\frac{|\widetilde{\rho}_H(k)|}{4-2^{2H}} \leq& \frac{2k^{2H}}{\log 4}\sum_{m=2}^{\infty}\frac{2(2m-1)!}{(2m)!}\left(\frac{1}{k}\right)^{2m}|1+2^{2m+1}-3^{2m}| \leq \frac{2k^{2H}}{\log 4}\sum_{m=2}^{\infty}\frac{1}{m}\left(\frac{9}{k^2}\right)^{m} \\
\leq& -\frac{18k^{2H-2}}{\log 4}\log\left(1-\frac{9}{k^2}\right) \leq \frac{72}{\log 4}k^{2H-4} \leq \frac{72}{\log 4}k^{-2}
\end{aligned}
\tag{2.24}
$$

for any $k \geq 4$. We have used the inequality

$$
-\log(1-u) \leq 4u \quad \text{if } 0 \leq u \leq \frac{3}{4}.
$$

It follows from (2.24) that

$$
\begin{aligned}
\frac{1}{n(4-2^{2H})^2}\sum_{k=1}^{2n-3}\left[\frac{k}{2}\right]\widetilde{\rho}^2(k) \leq& \frac{1}{n(4-2^{2H})^2}\sum_{k=1}^{2n-3}k\widetilde{\rho}^2(k) \\
\leq& \frac{1}{n(4-2^{2H})^2}\sum_{k=1}^{3}k\widetilde{\rho}^2(k)+\left(\frac{72}{\log 4}\right)^2\frac{1}{n}\sum_{k=1}^{\infty}k^{-3} \longrightarrow 0.
\end{aligned}
$$

Thus,

$$
\frac{1}{n}\frac{1}{(4-2^{2H})^2}\sum_{j=1}^{n-1}\sum_{k=1}^{2n-1}\widetilde{\rho}_H^2(j,k) \longrightarrow \frac{1}{(4-2^{2H})^2}\sum_{m\in\mathbb{Z}}\widetilde{\rho}_H^2(m).
$$

Results obtained above imply that $M(\lambda)=\exp\{\frac{1}{2}\lambda^T\Sigma_H\lambda\}$. Thus,

$$
\sqrt{n}\begin{pmatrix} n^{-1}\left(V_{n,T}^{\widehat{B}^H}-\mathsf{E}V_{n,T}^{\widehat{B}^H}\right) \\ (2n)^{-1}\left(V_{2n,T}^{\widehat{B}^H}-\mathsf{E}V_{2n,T}^{\widehat{B}^H}\right)\end{pmatrix} \xrightarrow{d} \mathcal{N}(0;\Sigma_H).
$$

Application of Slutsky's theorem gives the required result. □

Remark 2.12 Note that application of the Delta method to (2.14) implies

$$\sqrt{n}\left(\frac{(2n)^{-1}V_{2n,T}^{\widehat{B^H}}}{n^{-1}V_{n,T}^{\widehat{B^H}}}-1\right)\xrightarrow{d}\mathcal{N}(0,\sigma_H^2)$$

and

$$\sqrt{n}\left(\log\frac{(2n)^{-1}V_{2n,T}^{\widehat{B^H}}}{n^{-1}V_{n,T}^{\widehat{B^H}}}-\log 1\right)=\sqrt{n}\log\frac{(2n)^{-1}V_{2n,T}^{\widehat{B^H}}}{n^{-1}V_{n,T}^{\widehat{B^H}}}\xrightarrow{d}\mathcal{N}(0,\sigma_H^2)$$

with $\sigma_H^2=\frac{3}{2}\Sigma_{11}-2\Sigma_{12}$.

The next theorem is a particular case of Theorem 3.4 in [24]. Let us introduce the following notations:

$$\sigma_{2,\ell}^2(\mathbf{k},\mathbf{d})=\sum_{i=1}^{\ell}\sum_{j=1}^{\ell}d_id_j\rho_2(k_i,k_j),\quad \mathbf{k}=(k_1,\dots,k_\ell)\in\mathbb{N}^\ell,\quad \mathbf{d}=(d_1,\dots,d_\ell)\in\mathbb{R}^\ell$$

$$\rho_2(k_i,k_j)=\frac{1}{\sqrt{k_ik_j}}\sum_{p=1}^{+\infty}c_{2p,2}^2\cdot(2p)!\left(\sum_{s=0}^{k_i-1}\sum_{r=-\infty}^{+\infty}\rho_{k_i,k_j}^{2p}(k_ir+k_js)\right),$$

$$c_{2p,2}=\frac{1}{(2p)!}\prod_{i=0}^{p-1}(2-2i),$$

$$\begin{aligned}\rho_{b,c}(x)=&\frac{1}{2(4-2^{2H})}(bc)^{-H}\big[-|x|^{2H}+2|x-b|^{2H}-|x-2b|^{2H}\\&+2|x+c|^{2H}-4|x+c-b|^{2H}+2|x+c-2b|^{2H}-|x+2c|^{2H}\\&+2|x+2c-b|^{2H}-|x+2c-2b|^{2H}\big].\end{aligned}$$

If $k_i=k_j=1$ then $\rho_{1,1}(x)=\rho_H^{(2)}(x)$ and

$$\rho_2(1,1)=2\sum_{r=-\infty}^{+\infty}\left(\rho_H^{(2)}(r)\right)^2=2\left(1+\frac{2}{(4-2^{2\gamma})^2}\sum_{j=1}^{\infty}\widehat{\rho}_\gamma^2(j)\right).$$

Theorem 2.13 *Let $n_i=r_in$, $i=1,\dots,\ell$, where $r_i,n\in\mathbb{N}$. Denote*

$$z_i=\frac{y_i}{\sum_{i=1}^{\ell}y_i^2}\quad\text{and}\quad y_i=\log r_i-\frac{1}{\ell}\sum_{i=1}^{\ell}\log r_i.$$

Then

$$\frac{1}{2}\sum_{i=1}^{\ell}\frac{z_i}{\sqrt{r_i}}\left(\sqrt{n_i}\,V_{n_i,T}^{(2)\widehat{B}^H}-1\right)\xrightarrow{d}\mathcal{N}\left(0,\sigma_{2,\ell}^2\left(\mathbf{r},\frac{1}{2}(\mathbf{z}/\sqrt{\mathbf{r}})\right)\right),$$

where $\mathbf{r}=(r_1,\dots,r_\ell)$, $\mathbf{z}=(z_1,\dots,z_\ell)$.

2.2 The Hurst Index Estimators

In this section we present several statistics suitable for estimation of the Hurst index of an fBm based on the second order increments of an observed discrete trajectory. Such type estimators are also applicable for other classes of processes such as solutions of SDE driven by fBm or Gaussian processes. In Sect. 2.2.1 we review the statistics that are classical, to some extent, and in Sect. 2.2.2 present a new one.

2.2.1 *Some Known Statistics*

In 1961, Gladyshev [63] derived a limit theorem for a statistic based on the first order quadratic variations of an fBm. This yielded an estimator

$$\widehat{H}_n^{(1)}=\frac{1}{2}-\frac{\log V_{n,T}^{(1)B^H}}{2\log n}$$

of H, which was strongly consistent but not asymptotically normal. X. Guyon and J. León in 1989 and Istas and Lang in 1997 [78] proved that if $H<3/4$ then

$$\sqrt{n}\left(n^{2H-1}V_{n,T}^{(1)B^H}-1\right)\xrightarrow{d}\mathcal{N}(0,\sigma_H^2),\qquad \text{as } n\to\infty,$$

where σ_H is a positive constant.

To obtain a CLT valid for all $H\in(0,1)$, one has to consider generalized variations. Istas and Lang [78] in 1997 and Benassi et al. [18] in 1998 have shown that it is enough to consider second order quadratic variations. In 2001, Coeurjolly [40] has shown that second order quadratic variations minimize the asymptotic variance of the variations. Therefore, it is natural to consider second order quadratic variations for the construction of consistent and asymptotically normal estimators. The estimator introduced in [18] and [78] has the form

$$\widehat{H}_n^{(2)}=\frac{1}{2}-\frac{1}{2\log 2}\log\frac{V_{2n,T}^{(2)B^H}}{V_{n,T}^{(2)B^H}},$$

and is strongly consistent and asymptotically normal.

In 2001, Coeurjolly [40] developed a class of consistent estimators of the Hurst parameter of the fractional Brownian motion based on the asymptotic behavior of the k-th absolute moment of discrete variations of its sample paths over a discrete grid of the interval [0, 1]. As a special case of estimators suggested by Coeurjolly is the estimator

$$\widehat{H}_n^{(3)} = \varphi_{n,T}^{-1}\left(\frac{1}{n}V_{n,T}^{(2)B^H}\right), \qquad \text{for } n > T,$$

$$\varphi_{n,T}(x) = \left(\frac{T}{n}\right)^{2x}(4 - 2^{2x}) \quad \text{and } \varphi_{n,T}^{-1} \quad \text{is the inverse of } \varphi_{n,T}, \quad x \in (0,1),$$

which is a strongly consistent and asymptotically normal.

In 2014 in the book [24] Berzin et al., via a regression model, considered a strongly consistent and asymptotically normal estimator

$$\widehat{H}_n^{(4)} = -\frac{1}{2}\sum_{j=1}^{\ell} z_j \log\left(\frac{V_{n_j,T}^{(2)B^H}}{n_j - 1}\right), \qquad n_j = r_j n,\ r_j \in \mathbb{N},\ j = 1, \ldots, \ell,$$

$$z_i = \frac{y_i}{\sum_{i=1}^{\ell} y_i^2} \quad \text{and} \quad y_i = \log r_i - \frac{1}{\ell}\sum_{i=1}^{\ell}\log r_i.$$

2.2.2 A New Class of Statistics for Estimation of the Hurst Index

The aim of this subsection is to present a new class of statistics suitable for estimation of the Hurst index from discretely observed trajectory of an fBm. The idea of construction is rather simple and makes use of self-similarity and stationarity of the second order increments. Employment of these properties immediately enables us to prove the usual asymptotic results, namely strong consistency and normality. We draw the attention of the reader to the fact that our class of statistics includes the statistics introduced by Bardet and Surgailis [8]. More precisely, in the case of fBm increment ratio (IR) statistic of [8] belongs to the class of statistics considered in this subsection.

We assume that an observed discrete sample corresponds to the uniform partition of a time interval of a trajectory $(B_t^H)_{t\in[0,T]}$ with fixed $T > 0$. Since B^H is self-similar, without loss of generality in the rest of this section we concentrate on samples $B_{\frac{i}{n}}^H, i = 0, \ldots, n$, corresponding to a trajectory $(B_t^H)_{t\in[0,1]}$.

Let $d_{n,i} = d_i = B_{\frac{i+1}{n}}^H - 2B_{\frac{i}{n}}^H + B_{\frac{i-1}{n}}^H$, $i = 1, \ldots, n-1$, be an array of second order differences obtained from the sample and $r_i = \frac{d_{i+1}}{d_i}$, $i = 1, \ldots, n-1$. Let K denote a standard Cauchy r.v., i.e. K is absolutely continuous and has a density with respect to Lebesgue measure on $\mathcal{B}(\mathbb{R})$ given by $f_K(x) = \frac{1}{\pi(1+x^2)}$, $x \in \mathbb{R}$.

The main result of this subsection is contained in the theorem given below.

Theorem 2.14 *Let $h : \mathbb{R} \to \mathbb{R}$ be a measurable function. Assume that:*

(i) $\mathsf{E}h^2\left(K + \rho_H^{(2)}(1)\right) < \infty$;

(ii) $x \mapsto \mathsf{E}h\left(K + \rho_x^{(2)}(1)\right)$ *possesses a non-zero derivative of constant sign in a neighborhood of H,*

where $\rho_x^{(2)}(1) = \frac{-7-9^x+4^{x+1}}{2(4-4^x)}$ *and consequently* $\rho_H^{(2)}(1) = \mathrm{Corr}(d_i, d_{i+1})$. *Then*

$$\widehat{H}_{n,h} := \varphi\left(\bar{h}_n\right) \to H \text{ a.s.} \quad \text{and} \quad \sqrt{n}\left(\widehat{H}_{n,h} - H\right) \xrightarrow{d} \mathcal{N}(0, \sigma_h^2),$$

where $\bar{h}_n = \frac{1}{n-1}\sum_{i=1}^{n-1} h(r_i)$, *$\varphi$ denotes the inverse of* $H \mapsto \mathsf{E}h\left(K + \rho_H^{(2)}(1)\right)$ *and σ_h^2 is precisely defined in the proof of the theorem.*

Remark 2.15 Note that choosing h one has to inspect the domain of its inverse φ. It may happen that the domain of φ does not include all values of $\bar{h}_n$ and formally for such values $\varphi(\bar{h}_n)$ is undefined. If this is the case, for these values one is free to modify the definition of $\widehat{H}_{n,h}$ anyhow. This does not change asymptotic result provided above since an inspection of the proof reveals that $\varphi(\bar{h}_n)$ is always well defined asymptotically because the limiting value of $\bar{h}_n$ belongs to the domain of φ. To make the remark more transparent, we give a note in Example 1 of Sect. 2.2.3.

Providing the proof we do not make any additional remarks regarding this issue and by default assume that the sample size n is large enough to ensure that all quantities are well defined.

Proof Retain the notions introduced in the previous subsection and consider a bivariate Gaussian sequence

$$Z_i = (Z_{i,1}, Z_{i,2}) = \frac{1}{\sqrt{4 - 2^{2H}}}(X_{i+1}, X_i), \quad i \geq 1,$$

where $X_i = B_{i+1}^H - 2B_i^H + B_{i-1}^H$, $i \geq 1$. The sequence $(X_i)_{i\geq 1}$ is stationary, $X_i \sim \mathcal{N}(0, 4 - 2^{2H})$ for all $i \geq 1$, and $\mathrm{Corr}(X_i, X_{i+k}) = \rho_H^{(2)}(k)$. Moreover, (see (2.19))

$$\left|\rho_H^{(2)}(k)\right| \leq \frac{243}{20\log 4} k^{2H-4} \quad \text{for any } k \geq 3.$$

Now we apply the result of Arcones (see Theorem B.11) and ergodic theorem (see Theorem B.1) for $(Z_i)_{i\geq 1}$ in order to obtain asymptotic results for the statistics of $\widehat{H}_{n,h}$. We use the notations given in Theorem B.11.

It follows from the properties listed above that $(Z_i)_{i\geq 1}$ is stationary and such that

$$\left|r^{(p,q)}(k)\right| := \left|\mathsf{E}Z_{1,p}Z_{1+k,q}\right| \leq \frac{243}{20\log 4}\,k^{2H-4} \tag{2.25}$$

for all $p, q = 1, 2$, $k \geq 3$.

The bound (2.25) implies convergence of $\sum_{k=1}^{\infty} r^{(p,q)}(k)$ and $\sum_{k=1}^{\infty}\left(r^{(p,q)}(k)\right)^2$. By (B.3)–(B.4)

$$\frac{1}{n}\sum_{j,k=1}^{n}\left(r^{(p,q)}(j-k)\right)^i = \left(r^{(p,q)}(0)\right)^i + \frac{2}{n}\sum_{k=1}^{n-1}(n-k)\left(r^{(q,p)}(k)\right)^i, \quad i = 1, 2.$$

Thus, it suffices to show that for any $p, q = 1, 2$, there exist limits

$$\lim_{n\to\infty} n^{-1}\sum_{k=1}^{n-1}(n-k)\left(r^{(q,p)}(k)\right)^i, \quad i = 1, 2.$$

Fix p, q and note that

$$\frac{1}{n}\sum_{k=1}^{n-1}(n-k)\left(r^{(q,p)}(k)\right)^i = \sum_{k=1}^{n-1}\left(r^{(q,p)}(k)\right)^i - \frac{1}{n}\sum_{k=1}^{n-1}k\left(r^{(q,p)}(k)\right)^i.$$

By (2.25),

$$\sum_{k=1}^{n-1}k\left|r^{(q,p)}(k)\right|^i \leq \left(\frac{243}{20\log 4}\right)^i\sum_{k=1}^{n-1}\frac{k}{k^{(4-2H)i}} < \infty.$$

Thus,

$$\lim_{n\to\infty} n^{-1}\sum_{k=1}^{n-1}(n-k)\left(r^{(q,p)}(k)\right)^i = \sum_{k=1}^{\infty}\left(r^{(p,q)}(k)\right)^i.$$

Therefore Theorem B.11 can be applied to $(Z_i)_{i\geq 1}$ provided function f is suitably chosen. Let

$$f(x,y) = \mathbf{1}_{\mathbb{R}\times(\mathbb{R}\setminus\{0\})}(x,y)h\left(\frac{x}{y}\right), \qquad R_i = \frac{Z_{i,1}}{Z_{i,2}}, \quad i \geq 1.$$

Then

- $R_i = \frac{Z_{i,1}}{Z_{i,2}} \stackrel{d}{=} K + \rho(H)$;
- by self-similarity $(R_i) \stackrel{d}{=} (r_i)$;

- because of absolute continuity of Gaussian r.v. and the previous facts $\mathsf{E}f^j(Z_1) = \mathsf{E}h^j(R_1) = \mathsf{E}h^j(r_1) = \mathsf{E}h^j(K+\rho(H)), j = 1, 2$.

Also note that in addition to stationarity $(X_i)_{i\geq 1}$ has vanishing correlation (see (2.19)). Therefore it is ergodic and conclusions listed above together with the ergodic theorem yield relationships

$$\begin{aligned} 1 =& \mathsf{P}\left(\lim_{n\to\infty} \frac{f(Z_1)+\cdots+f(Z_n)}{n} = \mathsf{E}f(Z_1)\right) \\ =& \mathsf{P}\left(\lim_{n\to\infty} \frac{h(r_1)+\cdots+h(r_n)}{n} = \mathsf{E}h(r_1) = \mathsf{E}h(R_1)\right). \end{aligned}$$

That is, $\bar{h}_n \to \mathsf{E}h(R_1)$ a.s. and by the continuous mapping theorem

$$\varphi(\bar{h}_n) \to \varphi(\mathsf{E}h(R_1)) = \varphi(\mathsf{E}h(K+\rho(H))) = H \quad \text{a. s.}$$

Next, note that from previously stated equality $\mathsf{E}f^2(Z_1) = \mathsf{E}h^2(K+\rho(H))$ and condition $\mathsf{E}h^2(K+\rho(H)) < \infty$ it follows that Theorem B.11 applies to the chosen f and gives (B.6) which may be rewritten as

$$\frac{1}{\sqrt{n}}\sum_{i=1}^{n}(h(r_i) - \mathsf{E}h(r_1)) = \sqrt{n}(\bar{h}_n - \mathsf{E}h(r_1)) \xrightarrow{d} \mathcal{N}(0,\sigma_f^2), \tag{2.26}$$

$$\sigma_f^2 = \mathrm{Var}(h(R_1)) + 2\sum_{k=1}^{\infty}\mathrm{cov}(h(R_1), h(R_{1+k})).$$

In order to complete the proof, one has to apply the Delta method, which also yields asymptotic variance $\sigma_h^2 = (\varphi'(\mathsf{E}h(K+\rho(H))))^2\sigma_f^2$. □

2.2.3 Examples

In this subsection we give three examples of functions which satisfy conditions (i)–(ii) stated in Theorem 2.14. The first function is considered because it usually happens that arcsin transform symmetrizes distribution and improves normal approximation. The second one demonstrates a connection with [8] discussed in the introduction. The third one is interesting itself.

For simplicity let us denote $\rho(x) = \rho_x^{(2)}(1)$, $x \in (0,1)$. Before proceeding to the mentioned examples note that $x \mapsto \rho(x)$ is increasing with a range equal to $\left(-\frac{2}{3}, -2+\frac{9}{8}\frac{\log 9}{\log 4}\right)$ and derivative

$$\rho'(x) = \frac{1}{2}\left(\frac{-9^x\cdot 4\cdot\log 9 + 4^x\cdot 9\cdot\log 4 + 36^x\log(9/4)}{(4-4^x)^2}\right).$$

In the rest of this subsection we omit an argument for $\rho(H)$ when it appears unnecessary and write ρ instead.

Example 1 Let $h(x) = \sin x$. Then h is bounded and (i) holds. Next, note that K is symmetric r.v. Hence, its characteristic function $\psi_K(t) = \mathsf{E}\cos(tK) = \mathrm{e}^{-|t|}$ and for any odd g it holds that $\mathsf{E}g(K) = 0$. Therefore

$$\mathsf{E}\sin(K+\rho) = \cos\rho\,\mathsf{E}\sin K + \sin\rho\,\mathsf{E}\cos K = \sin\rho\cdot\psi_K(1) = \frac{\sin\rho}{\mathrm{e}}.$$

Properties of $H \mapsto \rho(H)$ imply that $H \mapsto \mathrm{e}^{-1}\sin\rho(H)$ is increasing on $(0, 1)$ with the inverse

$$\varphi(y) = \rho^{-1}(\arcsin(\mathrm{e}y)), \qquad y \in \left(\frac{\sin(-2/3)}{\mathrm{e}}, \frac{\sin\left(-2+\frac{9}{8}\frac{\log 9}{\log 4}\right)}{\mathrm{e}}\right)$$

and derivative $\varphi'(\mathsf{E}\sin(K+\rho(H))) = \mathrm{e}\left(\rho'(H)\cos\rho(H)\right)^{-1}$.

Now we clarify Remark 2.15. It follows from the proof of the theorem that

$$\bar{h}_n \to \mathsf{E}\sin(K+\rho) = \frac{\sin\rho}{\mathrm{e}} \quad \text{a.s.}$$

Consequently, for n large enough

$$\bar{h}_n \in \operatorname{dom}(\varphi) = \left(\frac{\sin(-2/3)}{\mathrm{e}}, \frac{\sin\left(-2+\frac{9}{8}\frac{\log 9}{\log 4}\right)}{\mathrm{e}}\right)$$

and $\varphi(\bar{h}_n)$ is well defined. However, for small sample sizes it may happen that this is not the case since

$$\mathsf{P}\left(\bar{h}_n = \frac{1}{n}\sum_{i=1}^{n}\sin r_i \notin \operatorname{dom}(\varphi)\right) > 0.$$

For the estimator

$$\widehat{H}_n^{(1)} = \varphi\left(\frac{1}{n-1}\sum_{i=1}^{n-1} h(r_i)\right) = \varphi\left(\frac{1}{n-1}\sum_{i=1}^{n-1}\sin\frac{d_{i+1}}{d_i}\right)$$

we can apply the results of Theorem 2.14

Example 2 Let $h(x) = \frac{|1+x|}{1+|x|}$. It is evident that condition (i) of the theorem is satisfied. Note that $\mathsf{E}h(K + \rho(H)) = \lambda(\rho(H))$, where

$$\lambda(r) := \frac{1}{\pi} \arccos(-r) + \frac{1}{\pi} \sqrt{\frac{1+r}{1-r}} \log\left(\frac{2}{1+r}\right), \quad |r| < 1.$$

Really, denote $Y_i = X_i/\sqrt{4 - 2^{2H}}$. Recall that a ratio of two independent standard Gaussian r.vs. has the standard Cauchy distribution. We denote this ratio by K. Since $Y_{i+1} - \rho Y_i$ and $Y_i \sim \mathcal{N}(0, 1)$ are uncorrelated, then

$$\begin{aligned}\mathsf{E}h(K + \rho) &= \mathsf{E}h\left(\frac{Y_{i+1} - \rho Y_i}{Y_i} + \rho\right) = \mathsf{E}h\left(\frac{Y_{i+1}}{Y_i}\right) = \mathsf{E}\frac{|Y_i + Y_{i+1}|}{|Y_i| + |Y_{i+1}|} \\ &= \frac{1}{2\pi\sqrt{1-\rho^2}} \int_{\mathbb{R}^2} \frac{|x_1 + x_2|}{|x_1| + |x_2|} \exp\left\{-\frac{1}{2(1-\rho^2)}\left[x_1^2 - 2\rho x_1 x_2 + x_2^2\right]\right\} dx_1 dx_2.\end{aligned}$$

By the change of variables $x_1 = a\cos\phi$, $x_2 = a\sin\phi$, with $|r| < 1$, (see [8, p. 777]), we get that

$$\begin{aligned}\lambda(r) :=& \frac{1}{2\pi\sqrt{1-r}} \int_{\mathbb{R}^2} \frac{|x_1 + x_2|}{|x_1| + |x_2|} \exp\left\{-\frac{1}{2(1-r)}\left[x_1^2 - 2r x_1 x_2 + x_2^2\right]\right\} dx_1 dx_2 \\ =& \frac{\sqrt{1-r^2}}{\pi} \int_0^{\pi} \frac{|\cos\phi + \sin\phi|}{(|\cos\phi| + |\sin\phi|)(1 - r\sin(2\phi))} d\phi =: I_1 + I_2,\end{aligned}$$

where

$$\begin{aligned}I_1 =& \frac{\sqrt{1-r^2}}{\pi} \int_0^{\pi/2} \frac{1}{1 - r\sin(2\phi)} d\phi \\ =& \frac{\sqrt{1-r^2}}{\pi} \int_0^{\infty} \frac{1}{1 + t^2 - 2rt} dt = \frac{1}{2} + \frac{1}{\pi} \arctan\left(\frac{r}{\sqrt{1-r^2}}\right) = \frac{1}{\pi} \arccos(-r), \\ I_2 =& \frac{2\sqrt{1-r^2}}{\pi} \int_0^{\pi/4} \frac{\cos\phi - \sin\phi}{(\cos\phi + \sin\phi)(1 + r\sin(2\phi))} d\phi \\ =& \frac{\sqrt{1-r^2}}{\pi} \int_0^{1} \frac{1-t}{(1+t)(1 + 2rt + t^2)} dt = \frac{\sqrt{1-r^2}}{\pi(1-r)} \log\left(\frac{2}{1+r}\right).\end{aligned}$$

The function $\lambda(r)$ is monotone increasing on $[-1, 1]$; $\lambda(1) = 1$, $\lambda(-1) = 0$. It is easy to check that $\rho(H)$ is monotone increasing function; $\rho(1) = 0$ so that $\lambda(\rho(H))$ is also monotone for $H \in (0, 1)$. Thus function $H \mapsto \lambda(\rho(H))$ has the inverse function φ.

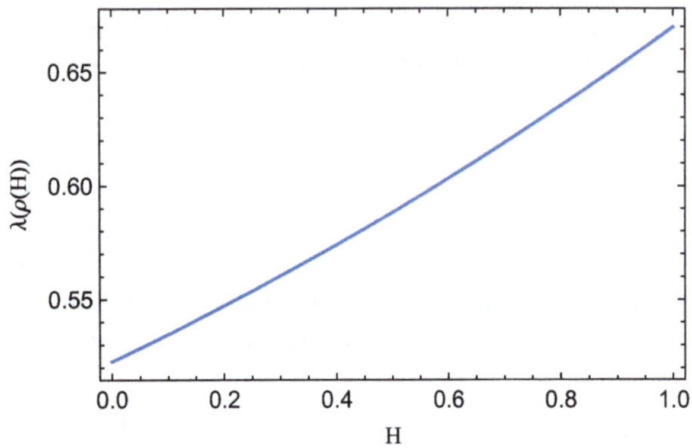

Fig. 2.1 The graph of $\lambda(\rho(H))$

For the estimator

$$\widehat{H}_n^{(2)} = \varphi\left(\frac{1}{n-1}\sum_{i=1}^{n-1} h(r_i)\right) = \varphi\left(\frac{1}{n-1}\sum_{i=1}^{n-1} \frac{|d_i + d_{i+1}|}{|d_i| + |d_{i+1}|}\right)$$

we can apply the results of Theorem 2.14.

Figure 2.1 shows that function $H \mapsto \lambda(\rho(H))$ is nearly linear and is well approximated by $0.1468H + 0.5174$. Thus, an estimator $\widehat{H}_n^{(2)}$ can be computed

$$\widehat{H}_n^{(2)} \simeq \frac{1}{0.1468}\left(\frac{1}{n-1}\sum_{i=1}^{n-1} \frac{|d_i + d_{i+1}|}{|d_i| + |d_{i+1}|} - 0.5174\right). \tag{2.27}$$

Example 3 Fix a. Set $h(x) = \mathbf{1}_{\{x<a\}}$. Then

$$\mathsf{E}h(K + \rho(H)) = \mathsf{P}(K + \rho(H) < a) = \frac{1}{2} + \frac{1}{\pi}\arctan\big(a - \rho(H)\big).$$

Note that $H \mapsto \mathsf{E}h(K + \rho(H))$ is decreasing. Thus, the function $H \mapsto \frac{1}{2} + \frac{1}{\pi}\arctan(a - \rho(H))$ has the inverse function φ and we can apply the results of Theorem 2.14 for the estimator

$$\widehat{H}_n^{(3)} = \varphi\left(\frac{1}{n-1}\sum_{i=1}^{n-1} \mathbf{1}_{\left\{\frac{d_{i+1}}{d_i} < a\right\}}\right).$$

Chapter 3
Estimation of the Hurst Index from the Solution of a Stochastic Differential Equation

The present chapter is devoted to the Hurst parameter estimation constructed from the solutions to SDE driven by a fBm. Our goal is to construct strongly consistent and asymptotically normal estimators of the Hurst parameter H based on discrete observations of the underlying processes. In Sect. 3.1 we propose strongly consistent estimator of the Hurst index and specify the rate of convergence of this estimator to the true value of H when the diameter of partition of observation interval tends to zero. This estimator preserves its properties, if we replace the solution with its Euler-Peano approximation. In Sect. 3.2 under some restriction on diffusion coefficient we propose two strongly consistent and asymptotically normal estimators of the Hurst index. In Sect. 3.3 we consider two reducible SDE. One is SDE driven by a fBm with polynomial drift and another one is fractional Gompertz diffusion process. The concrete type of the SDE's gives the chance to refuse restriction for the diffusion coefficient used in the previous section and receive strongly consistent and asymptotically normal estimators of the Hurst index and diffusion coefficient. Moreover, we compare asymptotic behavior of some estimators of the Hurst index and the diffusion coefficient of the fractional Gompertz diffusion process with the aid of computer simulations. In Sect. 3.4 we consider the limit behavior of the second-order quadratic variations of the solution of an SDE driven by a fBm along arbitrary partition of the interval and derive strongly consistent Hurst index estimator.

3.1 Strong Consistency of the Hurst Index Estimators Constructed from a Solution of SDE

In this section we give a strongly consistent Hurst index estimator constructed from the solution of SDE driven by a fBm. This estimator is based on the

© Springer International Publishing AG 2017

K. Kubilius et al., *Parameter Estimation in Fractional Diffusion Models*,
B&SS – Bocconi & Springer Series 8,
https://doi.org/10.1007/978-3-319-71030-3_3

second order increments of a trajectory observed in discrete moments of time (discretely observed). Also we will specify the rate of convergence of this estimator to the true value of H when the diameter of partition of observation interval tends to zero. This estimator preserves its properties, if we replace the solution with its Euler-Peano approximation.

3.1.1 Stochastic Differential Equation Involving a fBm: Bounds for the Increments of the Solution

Consider a stochastic differential equation

$$X_t = X_0 + \int_0^t a(s, X_s)\,\mathrm{d}s + \int_0^t b(s, X_s)\,\mathrm{d}B_s^H, \quad t \in [0, T], \tag{3.1}$$

where $T > 0$ is fixed, $(B_t^H)_{t\in[0,T]}$ is a fBm with the Hurst index $1/2 < H < 1$ defined on a complete probability space $(\Omega, \mathfrak{F}, \mathsf{P})$, $X_0 \in \mathbb{R}$, $a, b : [0, T] \times \mathbb{R} \to \mathbb{R}$ are measurable functions. Assume that conditions of Theorem 1.42 are satisfied and (3.1) has a unique solution. In what follows we provide additional constraint to the set of those imposed by Theorem 1.42. We assume that a is uniformly Hölder continuous in t with the same $\beta \in (1 - H, 1]$ as in (A_1) condition (*iii*). To be more precise, we assume

$$|a(s, x) - a(t, x)| \le M_0\, |t - s|^{\beta}, \qquad \forall t, s \in [0, T],\ \forall x \in \mathbb{R}. \tag{3.2}$$

Further, we assume that condition (3.2) is always fulfilled, if we use the increments of the solution of SDE.

The estimation of increments of the solution of SDE (3.1) is based on the following lemma.

Lemma 3.1 *Let $-\infty < a < b < \infty$, $h, r : [a, b] \to \mathbb{R}$ be measurable functions, $\varepsilon_1, \varepsilon_2 > 0$, $x \in (a, b)$ be such that $x + \varepsilon_1, x - \varepsilon_2 \in [a, b]$. Also, let*

- $\lambda + \mu > 1$;
- $h \in C^{\lambda}([a, b])$, $r \in C^{\mu}([a, b])$.

Then

$$\int_x^{x+\varepsilon_1} r\mathrm{d}h - \int_{x-\varepsilon_2}^{x} r\mathrm{d}h = r(x)\Delta^{(2)} h_{x,\varepsilon_1,\varepsilon_2} + \theta_{x,\varepsilon}\psi(\varepsilon), \tag{3.3}$$

where $\Delta^{(2)} h_{x,\varepsilon_1,\varepsilon_2} = h(x + \varepsilon_1) - 2h(x) + h(x - \varepsilon_2)$, $\varepsilon = \varepsilon_1 \vee \varepsilon_2$, $\theta_{x,\varepsilon} \in [-1, 1]$ and $\psi(\varepsilon) = O(\varepsilon^{\mu+\lambda})$ as $\varepsilon \to 0 + 0$.

Proof By the Love-Young inequality (1.18) and Hölder continuity of h, r,

$$\left|\int_x^{x+\varepsilon_1} r dh - r(x)(h(x+\varepsilon_1) - h(x))\right| \le C_{1/\lambda,1/\mu} K_r^\mu K_h^\lambda \varepsilon^{\mu+\lambda},$$

where

$$K_r = \sup_{\substack{s,t\in[a,b],\\ s\neq t}} \frac{|r(t)-r(s)|}{|t-s|^\mu}$$

and

$$K_h = \sup_{\substack{s,t\in[a,b],\\ s\neq t}} \frac{|h(t)-h(s)|}{|t-s|^\lambda}.$$

Therefore

$$\int_x^{x+\varepsilon_1} r dh = r(x)(h(x+\varepsilon_1) - h(x)) + \theta^+_{x,\varepsilon} C_{1/\mu,1/\lambda} K_r^\mu K_h^\lambda \varepsilon^{\mu+\lambda} \tag{3.4}$$

with some $\theta^+_{x,\varepsilon} \in [-1,1]$. Using the same argument,

$$\int_{x-\varepsilon_2}^{x} r dh = r(x)(h(x) - h(x-\varepsilon_2)) + \theta^-_{x,\varepsilon} C_{1/\mu,1/\lambda} K_r^\mu K_h^\lambda \varepsilon^{\mu+\lambda}, \quad \theta^-_{x,\varepsilon} \in [-1,1]. \tag{3.5}$$

Setting $\theta_{x,\varepsilon} = \frac{\theta^+_{x,\varepsilon} - \theta^-_{x,\varepsilon}}{2}$, $\psi(\varepsilon) = 2C_{1/\mu,1/\lambda} K_r^\mu K_h^\lambda \varepsilon^{\mu+\lambda}$ one obtains (3.3). □

Let $\pi_n = \{\tau_k^n,\ k = 0,\dots,i_n\}$, $n \ge 1$, $\mathbb{N} \ni i_n \uparrow \infty$, be a sequence of partitions of the interval $[0,T]$ such that $d_n \to 0$, where $d_n = \max_{1\le k\le i_n}(\tau_k^n - \tau_{k-1}^n)$. Moreover, assume that $d_n < 1$. If partition π_n is uniform then $\tau_k^n = \frac{kT}{i_n}$ for all $k \in \{0,\dots,i_n\}$. If $i_n \equiv n$, we write t_k^n instead of τ_k^n. Recall that

$$\Delta_{n,k} X = X_{\tau_k^n} - X_{\tau_{k-1}^n}, \qquad \Delta_{n,k}^{(2)} X = X_{\tau_{k+1}^n} - 2X_{\tau_k^n} + X_{\tau_{k-1}^n}.$$

Recall also that the notion of O_ω was introduced at Sect. 2.1.1.

Lemma 3.2 *For any fixed* $\gamma \in (\frac{1}{2} \vee (1-\beta), H)$

$$\Delta^{(2)} X_{\tau_k^n} = \begin{cases} b\left(\tau_k^n, X_{\tau_k^n}\right) \Delta^{(2)} B^H_{\tau_k^n} + O_\omega\left(d_n^{\gamma+\theta}\right) & \textit{for uniform partition},\\ b\left(\tau_k^n, X_{\tau_k^n}\right) \Delta^{(2)} B^H_{\tau_k^n} + O_\omega(d_n) & \textit{for arbitrary partition}, \end{cases} \tag{3.6}$$

where $\theta = \beta \wedge \gamma$.

Proof Fix $\gamma \in (\frac{1}{2} \vee (1-\beta), H)$ and let

$$A = \{\omega \mid X_\cdot(\omega) \in C^\gamma([0,T])\} \cap \{\omega \mid B^H_\cdot(\omega) \in C^\gamma([0,T])\}, \quad P(A) = 1.$$

For $\omega \in A$ let $N = N(\omega) = \sup_{t\in[0,T]} |X_t|$. Note that on A functions $a(\cdot, X_\cdot)$, $b(\cdot, X_\cdot) \in C^\theta([0,T])$. Indeed, by ($A_2$) condition (i) and (3.2),

$$\begin{aligned} |a(s,X_s) - a(t,X_t)| &\le |a(s,X_s) - a(t,X_s)| + |a(t,X_s) - a(t,X_t)| \\ &\le M_0 |s-t|^\beta + L_N |X_s - X_t| \le M_0 |s-t|^\beta + L_N K_{X_\cdot} |s-t|^\gamma \\ &\le |s-t|^\theta \left(M_0 T^{\beta-\theta} + L_N K_{X,\omega} T^{\gamma-\theta}\right), \end{aligned} \tag{3.7}$$

for $s,t \in [0,T]$ and $K_{X,\omega} = \sup_{s,t\in[0,T]; s\neq t} \frac{|X_t - X_s|}{|t-s|^\gamma}$. Hence, the claim holds for a. The case of b is handled in the same way.

Next, note that

$$\begin{aligned} \Delta^{(2)} X_{\tau_k^n} &= \left(\int_{\tau_k^n}^{\tau_{k+1}^n} a(t,X_t)\,\mathrm{d}t - \int_{\tau_{k-1}^n}^{\tau_k^n} a(t,X_t)\,\mathrm{d}t\right) \\ &+ \left(\int_{\tau_k^n}^{\tau_{k+1}^n} b(t,X_t)\,\mathrm{d}B_t^H - \int_{\tau_{k-1}^n}^{\tau_k^n} b(t,X_t)\,\mathrm{d}B_t^H\right), \quad k = 1,\dots,i_n - 1. \end{aligned} \tag{3.8}$$

Almost sure continuity of $t \mapsto X_t$ together with continuity of a and b implies a.s. boundedness on A of $\sup_{t\in[0,T]} |a(t,X_t)|$ and $\sup_{t\in[0,T]} |b(t,X_t)|$. Take $x = \tau_k^n$, $\varepsilon_1 = \tau_{k+1}^n - \tau_k^n$, $\varepsilon_2 = \tau_k^n - \tau_{k-1}^n$ and apply Lemma 3.1 to differences in the brackets to conclude that

$$\begin{aligned} \Delta^{(2)}_{n,k} X =& a(\tau_k^n, X_{\tau_k^n})\left[(\tau_{k+1}^n - \tau_k^n) - (\tau_k^n - \tau_{k-1}^n)\right] + O_\omega\left(d_n^{1+\theta}\right) \\ &+ b\left(\tau_k^n, X_{\tau_k^n}\right) \Delta^{(2)} B^H_{\tau_k^n} + O_\omega\left(d_n^{\theta+\gamma}\right), \end{aligned}$$

because $\theta + \gamma = (2\gamma) \wedge (\beta + \gamma) > 1$. Now the proof immediately follows. □

3.1.2 *Strongly Consistent Estimators of the Hurst Index*

First, we prove the following auxiliary result which can be widely used in applications.

Proposition 3.3 *Suppose that all conditions of Theorem 1.42 hold and the partitions of the interval* $[0,T]$ *are uniform. Then for any* $0 < \theta < H \wedge \beta$

$$\left(\frac{n}{T}\right)^{2H-1} V_{n,T}^{(2)X} = (4-2^{2H}) \int_0^T b^2(t,X_t)\,dt + O_\omega\left(\left(\frac{\log n}{n}\right)^{\theta/2}\right),$$

where X *is a solution of equation (3.1).*

Proof Fix $\gamma \in (\frac{1}{2} \vee (1-\beta), H)$, sufficiently close to H so that $\gamma + \theta/4 > H$ for $\theta = \beta \wedge \gamma$. It was already shown in Lemma 3.2 that $b(\cdot, X_\cdot) \in C^\theta([0,T])$. Therefore $b^2(\cdot, X_\cdot) \in C^\theta([0,T])$. Since $B_\cdot^H \in C^\gamma([0,T])$ a.s., Lemmas 3.2 and 2.10 yield

$$\left(\frac{n}{T}\right)^{2H-1} \sum_{k=1}^{n-1} \left(\Delta_{n,k}^{(2)} X\right)^2$$

$$= \left(\frac{n}{T}\right)^{2H-1} \sum_{k=1}^{n-1} b^2\left(t_k^n, X_{t_k^n}\right) \left(\Delta_{n,k}^{(2)} B^H\right)^2 + O_\omega\left(\left(\frac{1}{n}\right)^{\theta-2(H-\gamma)}\right)$$

$$= (4-2^{2H}) \int_0^T b^2(t,X_t)\mathrm{d}t + O_\omega\left(\left(\frac{\log n}{n}\right)^{\theta/2}\right) + O_\omega\left(\left(\frac{1}{n}\right)^{\theta-2(H-\gamma)}\right),$$

where $\pi_n = \{t_k^n, k = 0, \ldots, n\}$, $n \geq 1$, is a sequence of uniform partitions of the interval $[0,T]$. If $\gamma + \theta/4 > H$ then $\theta - 2(H-\gamma) > \theta/2$. Thus we get the statement of the proposition. □

Theorem 3.4 *Suppose that all conditions of Theorem 1.42 and non-degeneracy condition*

$$\int_0^T b^2(t,X_t)\mathrm{d}t > 0 \quad a.s. \tag{ND1}$$

hold. Let

$$\widehat{H}_n = \frac{1}{2} - \frac{1}{2\log 2} \log\left(\frac{\sum_{k=1}^{2n-1}\left(\Delta_{2n,k}^{(2)} X\right)^2}{\sum_{k=1}^{n-1}\left(\Delta_{n,k}^{(2)} X\right)^2}\right).$$

Then

$$\widehat{H}_n \xrightarrow{a.s.} H \qquad as\ n \to \infty.$$

Remark 3.5 Evidently, if $b(0, X_0) > 0$ and conditions of Theorem 1.42 hold, then it follows from continuity of b that (ND1) holds. Moreover, it follows from Proposition 3.3 that condition (ND1) supplies the non-degeneracy of $\sum_{k=1}^{n-1}\left(\Delta_{n,k}^{(2)} X\right)^2$

starting from some number $n = n(\omega)$ which means that the estimator $\widehat{H}_n$ is correctly defined a.s. starting from some number $n = n(\omega)$. Therefore everywhere below mentioning $\widehat{H}_n$ or similar estimators, we consider such numbers n for which it is correctly defined.

Proof The estimator $\widehat{H}_n$ can be rewritten as

$$\begin{aligned}\widehat{H}_n =&\frac{1}{2} - \frac{1}{2\log 2}\left[(2H-1)\log\frac{1}{2} + \log\frac{\left(\frac{2n}{T}\right)^{2H-1} V_{2n,T}^{(2)X}}{\left(\frac{n}{T}\right)^{2H-1} V_{n,T}^{(2)X}}\right]\\ =&H - \frac{1}{2\log 2}\log\frac{\left(\frac{2n}{T}\right)^{2H-1} V_{2n,T}^{(2)X}}{\left(\frac{n}{T}\right)^{2H-1} V_{n,T}^{(2)X}},\end{aligned}$$

which, coupled with the result of Proposition 3.3 implies the convergence

$$\widehat{H}_n \xrightarrow{a.s.} H \qquad \text{as } n \to \infty.$$

□

3.1.3 *The Rate of Convergence of the Hurst Index Estimator Constructed from a SDE*

From now on, we assume that the diffusion coefficient b satisfies additional conditions of non-degeneracy. The first one is the condition (ND1) introduced in Theorem 3.4.

Theorem 3.6 *Suppose that all conditions of Theorem 1.42 and condition (ND1) hold. Let γ be any number from the interval $(1/2 \vee (1-\beta), H)$ and $\theta = \beta \wedge \gamma$. Then the estimator $\widehat{H}_n$ from Theorem 3.4 has at least the following rate of convergence to the true value of H:*

$$\widehat{H}_n = H + O_\omega\left(\left(\frac{\log n}{n}\right)^{\theta/2}\right).$$

Proof Fix $\gamma \in (\frac{1}{2} \vee (1-\beta), H)$, which satisfies $\gamma + \theta/4 > H$. Using the fact that $\int_0^T b^2(t, X_t)\mathrm{d}t > 0$ a.s. and the arguments given in Proposition 3.3, we obtain

$$\begin{aligned}\left(\frac{n}{T}\right)^{2H-1}\sum_{k=1}^{n-1}\left(\Delta_{n,k}^{(2)}X\right)^2 =&(4-2^{2H})\int_0^T b^2(t, X_t)\mathrm{d}t + O_\omega\left(\left(\frac{\log n}{n}\right)^{\theta/2}\right)\\ =&(4-2^{2H})\int_0^T b^2(t, X_t)\mathrm{d}t\left(1 + O_\omega\left(\left(\frac{\log n}{n}\right)^{\theta/2}\right)\right).\end{aligned}$$

Consequently,

$$\left(\frac{2n}{T}\right)^{2H-1}\sum_{k=1}^{2n-1}\left(\Delta_{2n,k}^{(2)}X\right)^2 = (4-2^{2H})\int_0^T b^2(t,X_t)\,\mathrm{d}t\left(1+O_\omega\left(\left(\frac{\log n}{n}\right)^{\theta/2}\right)\right),$$

and by Maclaurin's expansion,

$$\log\left(\frac{\sum_{k=1}^{2n-1}\left(\Delta_{2n,k}^{(2)}X\right)^2}{\sum_{k=1}^{n-1}\left(\Delta_{n,k}^{(2)}X\right)^2}\right)$$

$$= \log\left(\frac{2^{-(2H-1)}(4-2^{2H})\int_0^T b^2(t,X_t)\mathrm{d}t\left[1+O_\omega\left(\left(\frac{\log n}{n}\right)^{\theta/2}\right)\right]}{(4-2^{2H})\int_0^T b^2(t,X_t)\mathrm{d}t\left[1+O_\omega\left(\left(\frac{\log n}{n}\right)^{\theta/2}\right)\right]}\right)$$

$$= (2H-1)\log 2^{-1} + \log\left(\frac{1+O_\omega\left(\left(\frac{\log n}{n}\right)^{\theta/2}\right)}{1+O_\omega\left(\left(\frac{\log n}{n}\right)^{\theta/2}\right)}\right)$$

$$= (2H-1)\log 2^{-1} + \log\left(1+O_\omega\left(\left(\frac{\log n}{n}\right)^{\theta/2}\right)\right)$$

$$= (2H-1)\log 2^{-1} + O_\omega\left(\left(\frac{\log n}{n}\right)^{\theta/2}\right).$$

□

Remark 3.7 Let X be a solution of SDE with coefficients which do not depend on time. In Theorem 8 of [100] it was proved under some additional restriction that

$$\left|\widehat{H}_n \mathbf{1}_{\widetilde{C}_n} - H\right| = O\left(n^{-1/4}(\log n)^{1/4+\alpha}\right)$$

for $\alpha > 0$, where

$$\widetilde{C}_n = \left\{2^{-1}\left(1-2n^{-\frac{1}{4}}(\log n)^{\frac{1}{4}+\alpha}\right) \le \frac{\sum_{k=1}^{2n-1}\left(\Delta_{2n,k}^{(2)}X\right)^2}{\sum_{k=1}^{n-1}\left(\Delta_{n,k}^{(2)}X\right)^2} \le 1+2n^{-\frac{1}{4}}(\log n)^{\frac{1}{4}+\alpha}\right\}.$$

The result obtained in Theorem 3.6 is sharper and for more general equation. Let X be a solution of SDE with coefficients which do not depend on time. If $\gamma = \frac{1}{2}+\delta$ with $0<\delta<H-\frac{1}{2}$, we get $O_\omega((\log n/n)^{1/4+\delta/2})$.

3.1.4 On Euler–Peano Approximation of SDE and Consistency of Respective Estimator

In practice, it is very interesting to compare different estimators. However, we can do it if we explicitly know the solution of the considered SDE, which is often not the case. In such situations, we replace these "exact" sample paths with their approximations (for example, the Euler or Milstein ones).

Let π_n be a sequence of uniform partitions of the interval $[0, T]$. Define the Euler–Peano approximation

$$Y_t^n = X_0 + \int_0^t a\big(\tau_s^n, Y^n(\tau_s^n)\big)ds + \int_0^t b\big(\tau_s^n, Y^n(\tau_s^n)\big)dB_s^H, \quad t \in [0, T],\ n \in \mathbb{N}, \tag{3.9}$$

where $\tau_s^n = t_k^n$ for $s \in [t_k^n, t_{k+1}^n)$, $0 \le k \le n-1$, $n \in \mathbb{N}$.

The next result allows us to assert that, if we replace the solution by its Euler-Peano approximation, the estimators of H are strongly consistent.

Theorem 3.8 *Suppose that all conditions of Theorem 1.42 and condition (ND1) hold, and the partitions of the interval* $[0, T]$ *are uniform. Define*

$$\widehat{H}_n^E = \frac{1}{2} - \frac{1}{2\log 2}\log\left(\frac{\sum_{k=1}^{2n-1}\left(\Delta_{2n,k}^{(2)}Y^n\right)^2}{\sum_{k=1}^{n-1}\left(\Delta_{n,k}^{(2)}Y^n\right)^2}\right).$$

Then $\widehat{H}_n^E - H \to 0$ *a.s. as* $n \to \infty$.

Proof Note that, at the subdivision points $\{t_k^n\}$, the second order increments of the Euler-Peano approximation can be written in the form

$$\begin{aligned}
\Delta_{n,k}^{(2)}Y^n =& \big[a\big(t_k^n, Y^n(t_k^n)\big) - a\big(t_{k-1}^n, Y^n(t_{k-1}^n)\big)\big]\frac{T}{n} \\
&+ \big[b\big(t_k^n, Y^n(t_k^n)\big)\Delta_{n,k+1}^{(1)}B^H - b\big(t_{k-1}^n, Y^n(t_{k-1}^n)\big)\Delta_{n,k}^{(1)}B^H\big] \\
=& \big[a\big(t_k^n, Y^n(t_k^n)\big) - a\big(t_{k-1}^n, Y^n(t_{k-1}^n)\big)\big]\frac{T}{n} + b\big(t_k^n, Y^n(t_k^n)\big)\Delta_{n,k}^{(2)}B^H \\
&+ \big[b\big(t_k^n, Y^n(t_k^n)\big) - b\big(t_{k-1}^n, Y^n(t_{k-1}^n)\big)\big]\Delta_{n,k}^{(1)}B^H, \qquad k = 1, \ldots, n-1.
\end{aligned}$$

Also,

$$\begin{aligned}
\max_{1\le k\le n}\left|Y^n(t_k^n) - Y^n(t_{k-1}^n)\right| \le& \max_{1\le k\le n}\left|a\big(t_{k-1}^n, Y^n(t_{k-1}^n)\big)\right|\frac{T}{n} \\
&+ \max_{1\le k\le n}\left|b\big(t_{k-1}^n, Y^n(t_{k-1}^n)\big)\Delta_{n,k}^{(1)}B^H\right|.
\end{aligned}$$

Almost sure continuity of $t \mapsto Y_t^n$ together with continuity of a and b implies the a.s. boundedness of $\sup_{t\in[0,T]} \left|a(t, Y_t^n)\right|$ and $\sup_{t\in[0,T]} \left|b(t, Y_t^n)\right|$. Thus,

$$\max_{1\le k\le n} \left|Y^n(t_k^n) - Y^n(t_{k-1}^n)\right| = O_\omega\left(n^{-\gamma}\right), \tag{3.10}$$

where $\gamma \in (1/2 \vee (1-\beta), H)$. Fix $\gamma \in (1/2 \vee (1-\beta), H)$ and $\theta = \beta \wedge \gamma$ satisfying $H - \gamma < \theta/2$. Similarly as in Lemma 3.2 we get

$$\begin{aligned}
&\left|a(t_k^n, Y^n(t_k^n)) - a(t_{k-1}^n, Y^n(t_{k-1}^n))\right| \\
&\le \left|a(t_k^n, Y^n(t_k^n)) - a(t_{k-1}^n, Y^n(t_k^n))\right| + \left|a(t_{k-1}^n, Y^n(t_k^n)) - a(t_{k-1}^n, Y^n(t_{k-1}^n))\right| \\
&\le M_0 \left|t_k^n - t_{k-1}^n\right|^\beta + L_N \left|Y^n(t_k^n) - Y^n(t_{k-1}^n)\right| = O_\omega(n^{-\theta}).
\end{aligned}$$

The case of b is handled in the same way. Since

$$\Delta_{n,k}^{(2)} Y^n = b(t_k^n, Y^n(t_k^n))\Delta_{n,k}^{(2)} B^H + O_\omega\left(n^{-(\theta+\gamma)}\right),$$

we have

$$\left(\frac{n}{T}\right)^{2H-1} V_{n,T}^{(2)Y^n} - \left(\frac{n}{T}\right)^{2H-1} \sum_{k=1}^{n-1} b^2(t_k^n, Y^n(t_k^n))(\Delta_{n,k}^{(2)} B^H)^2 \xrightarrow{a.s.} 0 \tag{3.11}$$

as $n \to \infty$. Further, we observe that

$$\begin{aligned}
&\left|\left(\frac{n}{T}\right)^{2H-1} \sum_{k=1}^{n-1} [b^2(t_k^n, X(t_k^n)) - b^2(t_k^n, Y^n(t_k^n))](\Delta_{n,k}^{(2)} B^H)^2\right| \\
&\le \max_{1\le k\le n-1} \left|b^2(t_k^n, X(t_k^n)) - b^2(t_k^n, Y^n(t_k^n))\right| \left(\frac{n}{T}\right)^{2H-1} \sum_{k=1}^{n-1} (\Delta_{n,k}^{(2)} B^H)^2 \\
&\le M_0 \max_{1\le k\le n-1} \left|X(t_k^n) - Y^n(t_k^n)\right| \max_{1\le k\le n-1} \left[\left|b(t_k^n, X(t_k^n))\right| + \left|b(t_k^n, Y^n(t_k^n))\right|\right] \\
&\times \left(\frac{n}{T}\right)^{2H-1} \sum_{k=1}^{n-1} (\Delta_{n,k}^{(2)} B^H)^2
\end{aligned} \tag{3.12}$$

The uniform distance between the solution and its Euler-Peano approximation was estimated in the book of Mishura [120] (see also [124]). It follows from this that $\widehat{H}_n^E$ is correctly defined a.s. starting from some number $n = n(\omega)$ and

$$\sup_{t\le T} |Y_t^n - X_t| \xrightarrow{a.s.} 0 \quad \text{as} \quad n \to \infty. \tag{3.13}$$

From (3.13) and convergence

$$\left(\frac{n}{T}\right)^{2H-1}\sum_{k=1}^{n-1}(\Delta_{n,k}^{(2)}B^H)^2 \xrightarrow{a.s.} (4-2^{2H})T$$

we have that the last term of (3.12) tends to 0 as $n \to \infty$. From Lemma 2.10 and (3.11)–(3.12) it follows that

$$\left(\frac{n}{T}\right)^{2H-1} V_{n,T}^{(2)Y^n} \xrightarrow{a.s.} (4-2^{2H})\int_0^T b^2\big(t,X(t)\big)dt.$$

Together with non-degeneracy condition (ND1) it means that the limit behavior of the second order increments of the Euler-Peano approximation is the same as the limit behavior of the solution of SDE (see Proposition 3.3). Moreover, this result allows us to assert that, if we replace the solution by its Euler-Peano approximation, the estimator $\widehat{H}_n^E$ of H is strongly consistent. □

3.2 Strongly Consistent and Asymptotically Normal Estimators of the Hurst Index Constructed from a SDE

Now we introduce additional restriction on the diffusion coefficient b which allows to get the strongly consistent and asymptotically normal estimators of the Hurst parameter constructed from the solutions of stochastic differential equations. The estimators are based on discrete observations of the underlying processes.

3.2.1 Estimators of the Hurst Index of the Solution of SDE Driven by a fBm Under Additional Restrictions on Diffusion Coefficient: Asymptotic Normality

We consider two possibilities. First, we consider the case when the coefficient b is explicitly included in the definition of statistics. Next, we give the statistics where the coefficient b is not necessary in an explicit form. In both cases, boundedness of $1/b$ plays an important role and is assumed to hold.

Theorem 3.9 *Suppose that all conditions of Theorem 1.42 hold, and moreover,* $\beta > 1/2$ *in assumption* (A_1). *Additionally, let the following non-degeneracy condition hold:*

(ND2) *There exists a random variable* ς *such that* $\mathsf{P}(\varsigma < \infty) = 1$ *and*

$$\sup_{t\in[0,T]}\frac{1}{|b(t,X_t)|} \le \varsigma \quad a.s. \tag{3.14}$$

Then

$$\widehat{H}_n^{(1)} \xrightarrow{a.s.} H,$$

$$2\sqrt{n}\log\left(\frac{n}{T}\right)(\widehat{H}_n^{(1)} - H) \xrightarrow{d} \mathcal{N}(0, \Sigma_{11}) \qquad \text{for } H \in (1/2, 1), \qquad (3.15)$$

where Σ_{11} *is a known variance defined in Theorem 2.11,* $t_k^n = \frac{k}{n}T$, $1 \le k \le n$,

$$\widehat{H}_n^{(1)} = \varphi_{n,T}^{-1}\left(\frac{1}{n}\sum_{k=1}^{n-1}\left(\frac{\Delta_{n,k}^{(2)}X}{b(t_k^n, X(t_k^n))}\right)^2\right) \quad \text{for } n > T,$$

$\varphi_{n,T}(x) = \left(\frac{T}{n}\right)^{2x}(4 - 2^{2x})$ *and* $\varphi_{n,T}^{-1}$ *denotes the inverse of* $\varphi_{n,T}$, $x \in (0, 1)$, $n > T$.

Remark 3.10 On one hand, let $b(t, x) > b > 0$. Then (ND2) holds. On the other hand, (ND2) is evidently stronger than (ND1).

Remark 3.11 Since $(\frac{n}{T})^{2H-1}\sum_{k=1}^{n-1}\left(\Delta_{n,k}^{(2)}B^H\right)^2 \to (4 - 2^{2H})T$ (see Theorem 2.2), we can state that

$$\frac{1}{n}\sum_{k=1}^{n-1}\left(\frac{\Delta_{n,k}^{(2)}X}{b\big(t_k^n, X(t_k^n)\big)}\right)^2 = \frac{1}{n}\sum_{k=1}^{n-1}\left(\frac{b\big(t_k^n, X_{t_k^n}\big)\Delta_{n,k}^{(2)}B^H + O_\omega(n^{-(\gamma+\theta)})}{b\big(t_k^n, X_{t_k^n}\big)}\right)^2$$

$$= \frac{1}{n}\sum_{k=1}^{n-1}\left(\Delta_{n,k}^{(2)}B^H\right)^2 + O_\omega\left(\frac{1}{n^{2\gamma+\theta}}\right) \xrightarrow{a.s.} 0.$$

It means that asymptotically the random variables $\frac{1}{n}\sum_{k=1}^{n-1}\left(\frac{\Delta_{n,k}^{(2)}X}{b\left(t_k^n, X(t_k^n)\right)}\right)^2$ belong to the domain of the function $\varphi_{n,T}^{-1}$ which is interval [0, 3] (more details are in the proof of Theorem 3.9).

Proof Observe first that

$$\varphi_{n,T}(x) = \left(\frac{T}{n}\right)^{2x}(4 - 2^{2x}), \qquad x \in (0, 1),$$

is continuous and strictly decreasing for $n > T$. Thus, it has an inverse $\varphi_{n,T}^{-1}$ for $n > T$.

From Lemma 3.2 for any fixed $\gamma \in (\frac{1}{2} \vee (1 - \beta), H)$ we have

$$\Delta_{n,k}^{(2)}X = b\left(t_k^n, X_{t_k^n}\right)\Delta_{n,k}^{(2)}B^H + O_\omega\left(n^{-(\gamma+\theta)}\right), \qquad (3.16)$$

where $\theta = \beta \wedge \gamma$.

Fix γ satisfying $\theta/2 > H - \gamma$. By (3.14) and Theorem 2.2, we get

$$\begin{aligned}
\beta_n := \frac{\varphi_{n,T}(\widehat{H}_n^{(1)})}{\varphi_{n,T}(H)} &= \left[\left(\frac{T}{n}\right)^{2H}(4-2^{2H})\right]^{-1}\varphi_{n,T}\left(\varphi_{n,T}^{-1}\left(\frac{1}{n}\sum_{k=1}^{n-1}\left(\frac{\Delta_{n,k}^{(2)}X}{b(t_k^n, X_{t_k^n})}\right)^2\right)\right)\\
&= \left[\left(\frac{T}{n}\right)^{2H}(4-2^{2H})\right]^{-1}\left(\frac{1}{n}\sum_{k=1}^{n-1}\left(\frac{b(t_k^n, X_{t_k^n})\Delta_{n,k}^{(2)}B^H + O_\omega(n^{-(\gamma+\theta)})}{b(t_k^n, X_{t_k^n})}\right)^2\right)\\
&= \frac{n^{2H-1}}{T^{2H}(4-2^{2H})}\left(\sum_{k=1}^{n-1}\left(\Delta_{n,k}^{(2)}B^H\right)^2 + O_\omega\left(\frac{1}{n^{2\gamma+\theta-1}}\right)\right)\\
&= \frac{n^{2H-1}}{T^{2H}(4-2^{2H})}\sum_{k=1}^{n-1}\left(\Delta_{n,k}^{(2)}B^H\right)^2 + O_\omega\left(\frac{1}{n^{\theta-2(H-\gamma)}}\right)\\
&= n^{-1}V_{n,T}^{\widehat{B^H}} + O_\omega\left(\frac{1}{n^{\theta-2(H-\gamma)}}\right) \xrightarrow{a.s.} 1.
\end{aligned}$$

Fix $0 < \delta < \min\{H - 1/2, 1 - H\}$. Note that $\varphi_{n,T}(H)$ is not only decreasing in H under n fixed, but also

$$\frac{\varphi_{n,T}(H_2)}{\varphi_{n,T}(H_1)} = \frac{\left(\frac{T}{n}\right)^{2H_2}(4-2^{2H_2})}{\left(\frac{T}{n}\right)^{2H_1}(4-2^{2H_1})} \longrightarrow 0 \qquad \text{as } n \to \infty,$$

for $H_1 < H_2$. Therefore there exists $n_0(\omega)$ such that

$$\beta_n \in \left(\frac{\varphi_{n,T}(H+\delta)}{\varphi_{n,T}(H)}, \frac{\varphi_{n,T}(H-\delta)}{\varphi_{n,T}(H)}\right),$$

or, in other words,

$$\varphi_{n,T}(\widehat{H}_n^{(1)}) \in \left(\varphi_{n,T}(H+\delta), \varphi_{n,T}(H-\delta)\right),$$

$n \geq n_0$. Since $\varphi_{n,T}$ is strictly decreasing in H for any fixed $n > T$, we conclude that $\widehat{H}_n^{(1)} \in (H-\delta, H+\delta)$ for all $n \geq \max(n_0, T)$. Thus, $\widehat{H}_n^{(1)}$ is strongly consistent.

Now we prove the asymptotic normality of $\widehat{H}_n^{(1)}$. Note that

$$\begin{aligned}
\log\left(\frac{\varphi_{n,T}(\widehat{H}_n^{(1)})}{\varphi_{n,T}(H)}\right) &= \log\left(\frac{n}{T}\right)^{-2(\widehat{H}_n^{(1)}-H)} + \log\frac{4-2^{2\widehat{H}_n^{(1)}}}{4-2^{2H}}\\
&= -2(\widehat{H}_n^{(1)}-H)\log\left(\frac{n}{T}\right) + \log\left(4-2^{2\widehat{H}_n^{(1)}}\right) - \log\left(4-2^{2H}\right).
\end{aligned} \tag{3.17}$$

Let $h(x) = \log(4 - 2^{2x})$, $x \in (0, 1)$. By the mean value theorem

$$\log\left(4 - 2^{2\widehat{H}_n^{(1)}}\right) - \log\left(4 - 2^{2H}\right) = (\widehat{H}_n^{(1)} - H)h'(H + \lambda_n(\widehat{H}_n^{(1)} - H)),$$

for some $\lambda_n \in (0, 1)$. Since $h'(H + \lambda_n(\widehat{H}_n^{(1)} - H)) \to h'(H)$ as $\widehat{H}_n^{(1)} \to H$, the sequence $(h'(H + \lambda_n(\widehat{H}_n^{(1)} - H)))_{n \ge 1}$ is bounded.

The equality (3.17) can be rewritten in the following way:

$$-2(\widehat{H}_n^{(1)} - H)\log\left(\frac{n}{T}\right) = \log\left(\frac{\varphi_{n,T}(\widehat{H}_n^{(1)})}{\varphi_{n,T}(H)}\right) - (\widehat{H}_n^{(1)} - H)h'(H + \lambda_n(\widehat{H}_n^{(1)} - H)).$$

Thus,

$$\left(-2\log\left(\frac{n}{T}\right) + h'(H + \lambda_n(\widehat{H}_n^{(1)} - H))\right)(\widehat{H}_n^{(1)} - H) = \log\left(\frac{\varphi_{n,T}(\widehat{H}_n^{(1)})}{\varphi_{n,T}(H)}\right),$$

and

$$-2\log\left(\frac{n}{T}\right)(\widehat{H}_n^{(1)} - H)\left(1 - \frac{h'(H + \lambda_n(\widehat{H}_n^{(1)} - H))}{2\log(\frac{n}{T})}\right) = \log\left(\frac{\varphi_{n,T}(\widehat{H}_n^{(1)})}{\varphi_{n,T}(H)}\right).$$

Consequently,

$$-2\sqrt{n}\log\left(\frac{n}{T}\right)(\widehat{H}_n^{(1)} - H) = \frac{\sqrt{n}\left(\log\left(\frac{\varphi_{n,T}(\widehat{H}_n^{(1)})}{\varphi_{n,T}(H)}\right) - \log 1\right)}{\left(1 - \frac{h'\left(H+\lambda_n(\widehat{H}_n^{(1)}-H)\right)}{2\log\left(\frac{n}{T}\right)}\right)}. \tag{3.18}$$

Since $h'(H + \lambda_n(\widehat{H}_n^{(1)} - H)$ is bounded, the denominator is equal to $1 + o_\omega(1)$.

Fix $\gamma \in (\frac{1}{2} \vee (1 - \beta), H)$, which satisfies $H - \gamma < (\theta - 1/2)/2$. It is possible due to additional assumption that $\beta > 1/2$. From Theorem 2.11 we have that

$$\sqrt{n}\left(n^{-1}V_{n,T}^{\widehat{B}^H} - 1\right) \xrightarrow{d} \mathcal{N}(0, \Sigma_{11}).$$

Thus, by Slutsky's Theorem (see Appendix B.5) it follows that

$$\sqrt{n}\left(\frac{\varphi_{n,T}(\widehat{H}_n^{(1)})}{\varphi_{n,T}(H)} - 1\right) = \sqrt{n}\left(n^{-1}V_{n,T}^{\widehat{B}^H} - 1\right) + O_\omega\left(\frac{1}{n^{\theta-2(H-\gamma)-1/2}}\right) \xrightarrow{d} \mathcal{N}(0, \Sigma_{11}).$$

The Delta method (see Appendix B.6) implies that

$$\sqrt{n}\log\left(\frac{\varphi_{n,T}(\widehat{H}_n^{(1)})}{\varphi_{n,T}(H)}\right) = \sqrt{n}\left(\log\left(\frac{\varphi_{n,T}(\widehat{H}_n^{(1)})}{\varphi_{n,T}(H)}\right) - \log 1\right) \xrightarrow{d} \mathcal{N}(0, \Sigma_{11}). \tag{3.19}$$

It remains to observe that assertion (3.15) immediately follows from (3.18), (3.19) and Slutsky's theorem. □

It is natural to try to remove the coefficient b from the definition of statistics. It is enough to know that b is separated from zero, for example, we can know b up to a constant multiplier. In this connection, we construct a new statistics not involving b explicitly.

For this purpose we need several additional notations. Assume that the process X is observed at time points $\frac{i}{m_n}T$, $i = 1, \ldots, m_n$, where $m_n = nk_n$, and k_n grows faster than $n \log n$, but the growth does not exceed polynomial, e.g. $k_n = n \log^{\mu} n$, $\mu > 1$, or $k_n = n^2$.

Denote

$$W_{n,k} = \sum_{j=-k_n+1}^{k_n+1} \left(\Delta^{(2)} X_{s_j^n + t_k^n}\right)^2 = \sum_{j=-k_n+1}^{k_n-1} \left(X_{s_{j+1}^n + t_k^n} - 2X_{s_j^n + t_k^n} + X_{s_{j-1}^n + t_k^n}\right)^2,$$

where $1 \le k \le n-1$ and $\{s_j^n = \frac{j}{nk_n}T : j = -k_n+1, \ldots, k_n-1\}$ is a sequence of uniform partitions of the interval $[t_{k-1}^n, t_{k+1}^n]$, $n \ge 1$.

Theorem 3.12 *Assume that conditions of Theorem 3.9 hold. Then*

$$\widehat{H}_n^{(2)} \xrightarrow{a.s.} H,$$

$$2\sqrt{n}\log(k_n)(\widehat{H}_n^{(2)} - H) \xrightarrow{d} \mathcal{N}(0, \Sigma_{11}) \qquad \textit{for } H \in (1/2, 1),$$

where

$$\widehat{H}_n^{(2)} = \frac{1}{2} + \frac{1}{2\log k_n} \log\left(\frac{2}{n}\sum_{k=1}^{n-1} \frac{\left(\Delta_{n,k}^{(2)} X\right)^2}{W_{n,k-1}}\right),$$

and Σ_{11} is a known variance defined in Theorem 2.11.

Before presenting the proof of this theorem, we give two auxiliary lemmas.

Lemma 3.13 *Let*

$$V_{n,T}(k) = \frac{m_n^{2H}}{2k_n T^{2H}(4 - 2^{2H})} \sum_{j=-k_n+1}^{k_n-1} \left(B^H_{s_{j+1}^n + t_k^n} - 2B^H_{s_j^n + t_k^n} + B^H_{s_{j-1}^n + t_k^n}\right)^2, 1 \le k \le n-1.$$

Then the following relation holds:

$$\max_{1 \le k \le n-1} \left|V_{n,T}(k) - 1\right| = O_{\omega}\left(\sqrt{\frac{\log n}{k_n}}\right).$$

Proof By self similarity and stationarity of increments of fBm,

$$V_{n,T}(k) \stackrel{d}{=} \frac{2^{2H-1} m_n^{2H}}{2^{2H} k_n (4-2^{2H})} \sum_{j=-k_n+1}^{k_n-1} \left(B^H_{\frac{j+k_n+1}{m_n}} - 2B^H_{\frac{j+k_n}{m_n}} + B^H_{\frac{j+k_n-1}{m_n}} \right)^2$$

$$\stackrel{d}{=} \frac{(2m_n)^{2H-1} n}{(2n)^{2H}(4-2^{2H})} \sum_{j=-k_n+1}^{k_n-1} \left(B^H_{\frac{j+1}{k_n}+1} - 2B^H_{\frac{j}{k_n}+1} + B^H_{\frac{j-1}{k_n}+1} \right)^2$$

$$\stackrel{d}{=} \frac{(2k_n)^{2H-1}}{2^{2H}(4-2^{2H})} \sum_{j=1}^{2k_n-1} \left(B^H_{\frac{j+1}{k_n}} - 2B^H_{\frac{j}{k_n}} + B^H_{\frac{j-1}{k_n}} \right)^2$$

$$\stackrel{d}{=} \frac{(2k_n)^{2H-1}}{4-2^{2H}} \sum_{j=1}^{2k_n-1} \left(B^H_{\frac{j+1}{2k_n}} - 2B^H_{\frac{j}{2k_n}} + B^H_{\frac{j-1}{2k_n}} \right)^2 \stackrel{d}{=} \frac{1}{2k_n} V^{(2)\widehat{B}^H}_{2k_n,1}.$$

Therefore,

$$\mathsf{P}\left(\max_{1\le k\le n-1} |V_{n,T}(k) - 1| > \delta \right) \le \sum_{k=1}^{n-1} \mathsf{P}\left(|V_{n,T}(k) - 1| > \delta \right)$$

$$\le n\mathsf{P}\left(\left| (2k_n)^{-1} V^{(2)\widehat{B}^H}_{2k_n,1} - 1 \right| > \delta \right)$$

for all $\delta > 0$. Put $\widehat{V}^{(2)\widehat{B}^H}_{2k_n,1} = \frac{1}{2k_n-1} V^{(2)\widehat{B}^H}_{2k_n,1}$. Note that

$$\left| (2k_n)^{-1} V^{(2)\widehat{B}^H}_{2k_n,1} - 1 \right| \le \frac{2k_n - 1}{2k_n} \left| \widehat{V}^{(2)\widehat{B}^H}_{2k_n,1} - 1 \right| + \frac{1}{2k_n} \le \left| \widehat{V}^{(2)\widehat{B}^H}_{2k_n,1} - 1 \right| + \frac{1}{2k_n}.$$

Let (δ_n) be a sequence of positive numbers such that $\delta_n \downarrow 0$ as $n \to \infty$ and $k_n^{-1} < \delta_n$. By Lemma 2.8,

$$\mathsf{P}\left(\left| (2k_n)^{-1} V^{(2)\widehat{B}^H}_{2k_n,1} - 1 \right| > 2\delta_n \right) \le \mathsf{P}\left(\left| \widehat{V}^{(2)\widehat{B}^H}_{2k_n,1} - 1 \right| + \frac{1}{2k_n} > 2\delta_n \right)$$

$$\le \mathsf{P}\left(\left| \widehat{V}^{(2)\widehat{B}^H}_{2k_n,1} - 1 \right| > \delta_n \right) \le 2 \exp\left\{ - \frac{\delta_n^2 (2k_n - 1)}{\frac{32}{3}(\delta_n + 1)} \right\}.$$

Set $\delta_n = \sqrt{a\frac{\log n}{2k_n-1}}$, where $a > 0$ is a constant that will be specified later. Then $\delta_n \to 0$. If $a \geq 3$ and $n \geq 2$, then $\delta_n > k_n^{-1}$ and

$$\mathsf{P}\left(\left|(2k_n)^{-1}V_{2k_n,1}^{(2)\widehat{B}^H} - 1\right| > 2\delta_n\right) \leq 2\,\exp\left\{-\frac{3a\log n}{32\left(\sqrt{a\frac{\log n}{2k_n-1}}+1\right)}\right\}.$$

Moreover, if we put $a = 64$ and consider n such that $\sqrt{a\frac{\log n}{2k_n-1}} < 1$, then

$$\mathsf{P}\left(\left|(2k_n)^{-1}V_{2k_n,1}^{(2)\widehat{B}^H} - 1\right| > 2\delta_n\right) \leq 2n^{-3}.$$

Therefore series $\sum_n \mathsf{P}(\max_{1\leq k\leq n-1}\left|V_{n,T}(k) - 1\right| > \delta_n)$ converges and by the Borel-Cantelli lemma

$$\max_{1\leq k\leq n-1}\left|V_{n,T}(k) - 1\right| < \delta_n \qquad \text{a.s.}$$

for n sufficiently large, whence the proof follows. □

Lemma 3.14 *Let in condition* (A_1), *(iii) of Theorem 1.42 we have* $\beta > 1/2$. *Then for any* $\gamma \in (\frac{1}{2}\vee(1-\beta), H)$ *and for each* $k = 1,\ldots,n-1$

$$W_{n,k} = b^2\left(t_k^n, X_{t_k^n}\right)\frac{2k_nT^{2H}(4-2^{2H})}{m_n^{2H}} + O_\omega\left(\frac{\sqrt{k_n}\log n}{m_n^{2H}}\right) + O_\omega\left(\frac{k_n}{n^\theta m_n^{2\gamma}}\right),$$

where $\theta = \gamma\wedge\beta$.

Proof

Step 1. By Lemma 3.2,

$$\Delta^{(2)}X_{s_j^n+t_k^n} = b(s_j^n+t_k^n, X_{s_j^n+t_k^n})\Delta^{(2)}B^H_{s_j^n+t_k^n} + O_\omega\left(\frac{1}{m_n^{\gamma+\theta}}\right), \quad j = -k_n+1,\ldots,k_n-1.$$

From (3.7) in the proof of Lemma 3.2 we get

$$\begin{aligned}\left|b^2(t,X_t) - b^2(s,X_s)\right| &= \left|b(t,X_t) - b(s,X_s)\right|\left|b(t,X_t) + b(s,X_s)\right| \\ &\leq 2\left|s-t\right|^\theta\left(M_0T^{\beta-\theta} + L_NK_{X.}T^{\gamma-\theta}\right)\sup_{t\in[0,T]}\left|b(t,X_t)\right|.\end{aligned}$$

Thus, almost sure continuity of $t\mapsto b(t,X_t)$ lead to

$$b^2\left(s_j^n+t_k^n, X_{s_j^n+t_k^n}\right) - b^2\left(t_k^n, X_{t_k^n}\right) = O_\omega\left(\frac{1}{n^\theta}\right).$$

Step 2. Fix $\gamma \in (\frac{1}{2} \vee (1-\beta), H)$, which satisfies $H-\gamma < (\theta - \frac{1}{2})/2$. *Step 1* and Lemma 3.13 yield

$$W_{n,k} = \sum_{j=-k_n+1}^{k_n-1} \left(\Delta^{(2)} X_{s_j^n+t_k^n}\right)^2 = \sum_{j=-k_n+1}^{k_n-1} b^2\left(s_j^n + t_k^n, X_{s_j^n+t_k^n}\right)\left(\Delta^{(2)} B^H_{s_j^n+t_k^n}\right)^2$$

$$+ O_\omega\left(\frac{k_n}{m_n^{2\gamma+\theta}}\right)$$

$$= b^2\left(t_k^n, X_{t_k^n}\right) \sum_{j=-k_n+1}^{k_n-1} \left(\Delta^{(2)} B^H_{s_j^n+t_k^n}\right)^2$$

$$+ \sum_{j=-k_n+1}^{k_n-1} \left(b^2\left(s_j^n + t_k^n, X_{s_j^n+t_k^n}\right) - b^2\left(t_k^n, X_{t_k^n}\right)\right)\left(\Delta^{(2)} B^H_{s_j^n+t_k^n}\right)^2 + O_\omega\left(\frac{k_n}{m_n^{2\gamma+\theta}}\right)$$

$$= b^2\left(t_k^n, X_{t_k^n}\right) \frac{2k_n T^{2H}(4-2^{2H})}{m_n^{2H}} V_{n,T}(k) + O_\omega\left(\frac{k_n}{n^\theta m_n^{2\gamma}}\right) + O_\omega\left(\frac{k_n}{m_n^{2\gamma+\theta}}\right)$$

$$= b^2\left(t_k^n, X_{t_k^n}\right) \frac{2k_n T^{2H}(4-2^{2H})}{m_n^{2H}} + O_\omega\left(\frac{\sqrt{k_n \log n}}{m_n^{2H}}\right) + O_\omega\left(\frac{k_n}{n^\theta m_n^{2\gamma}}\right).$$

Consequently, the proof of lemma is completed. □

Proof of Theorem 3.12 Fix $\gamma \in (\frac{1}{2} \vee (1-\beta), H)$, $\theta = \gamma \wedge \beta$, satisfying $H - \gamma < (\theta - \frac{1}{2})/2$. Put

$$S_{n,T} := \frac{2}{n k_n^{2H-1}} \sum_{k=1}^{n-1} \frac{\left(\Delta^{(2)} X_{t_k^n}\right)^2}{W_{n,k-1}}.$$

It follows from (3.16), (3.14) and Lemma 3.14 that

$$S_{n,T} = \frac{2}{n k_n^{2H-1}} \sum_{k=1}^{n-1} \frac{b^2\left(t_k^n, X_{t_k^n}\right)\left(\Delta^{(2)}_{n,k} B^H\right)^2 + O_\omega(n^{-(2\gamma+\theta)})}{b^2\left(t_k^n, X_{t_k^n}\right) \frac{2k_n T^{2H}(4-2^{2H})}{m_n^{2H}} + O_\omega\left(\frac{\sqrt{k_n \log n}}{m_n^{2H}}\right) + O_\omega\left(\frac{k_n}{n^\theta m_n^{2\gamma}}\right)}$$

$$= \frac{2}{n k_n^{2H-1}} \sum_{k=1}^{n-1} \frac{\left(\Delta^{(2)}_{n,k} B^H\right)^2 + O_\omega\left(n^{-(2\gamma+\theta)}\right)}{\frac{2k_n T^{2H}(4-2^{2H})}{m_n^{2H}} + O_\omega\left(\frac{\sqrt{k_n \log n}}{m_n^{2H}}\right) + O_\omega\left(\frac{k_n}{n^\theta m_n^{2\gamma}}\right)}$$

$$= \frac{m_n^{2H}}{n k_n^{2H} T^{2H}(4-2^{2H})} \sum_{k=1}^{n-1} \frac{\left(\Delta^{(2)}_{n,k} B^H\right)^2 + O_\omega\left(n^{-(2\gamma+\theta)}\right)}{1 + O_\omega\left(\sqrt{\frac{\log n}{k_n}}\right) + O_\omega\left(\frac{m_n^{2(H-\gamma)}}{n^\theta}\right)}$$

$$= \frac{n^{-1} V_{n,T}^{(2)\widehat{B}^H} + O_\omega\left(n^{-2(\gamma-H)-\theta}\right)}{1 + O_\omega\left(\sqrt{\frac{\log n}{k_n}}\right) + O_\omega\left(\frac{m_n^{2(H-\gamma)}}{n^\theta}\right)}.$$

The terms $O_\omega(n^{-2(\gamma-H)-\theta})$ and $O_\omega\left(\frac{m_n^{2(H-\gamma)}}{n^\theta}\right)$ vanish provided $\theta/2 > H-\gamma$ and $H-\gamma$ is small enough. Convergence $n^{-1}\widehat{V_{n,T}^{B^H}} \xrightarrow{a.s.} 1$ implies that $S_{n,T} \xrightarrow{a.s.} 1$. Hence

$$\widehat{H}_n^{(2)} = H + \frac{\log S_{n,T}}{2\log k_n} \xrightarrow{a.s.} H.$$

In order to prove asymptotic normality of the estimator $H_n^{(2)}$, observe that

$$\begin{aligned}
&\sqrt{n}\left(S_{n,T}-1\right)\\
&= \sqrt{n}\left(\frac{n^{-1}\widehat{V_{n,T}^{B^H}} - 1 + O_\omega\left(\sqrt{\frac{\log n}{k_n}}\right) + O_\omega\left(\frac{m_n^{2(H-\gamma)}}{n^\theta}\right) + O_\omega\left(n^{-2(\gamma-H)-\theta}\right)}{1 + O_\omega\left(\sqrt{\frac{\log n}{k_n}}\right) + O_\omega\left(\frac{m_n^{2(H-\gamma)}}{n^\theta}\right)}\right)\\
&= \frac{\sqrt{n}(n^{-1}\widehat{V_{n,T}^{B^H}} - 1)}{1 + O_\omega\left(\sqrt{\frac{\log n}{k_n}}\right) + O_\omega\left(\frac{m_n^{2(H-\gamma)}}{n^\theta}\right)}\\
&\quad + \frac{O_\omega\left(\sqrt{\frac{n\log n}{k_n}}\right) + O_\omega\left(\frac{m_n^{2(H-\gamma)}}{n^{\theta-1/2}}\right) + O_\omega\left(n^{-2(\gamma-H)-\theta-1/2}\right)}{1 + O_\omega\left(\sqrt{\frac{\log n}{k_n}}\right) + O_\omega\left(\frac{m_n^{2(H-\gamma)}}{n^\theta}\right)} \xrightarrow{d} \mathcal{N}(0,\Sigma_{11})
\end{aligned}$$

for $H-\gamma < (\theta-1/2)/2$. Now apply Slutsky's theorem and Theorem 2.11. □

3.3 Estimation of the Hurst Index and of Diffusion Coefficient for Transformable SDEs

In this section we consider two types of nonlinear stochastic differential equations involving a fractional Brownian motion, namely, we study the SDE with polynomial drift and Gompertz diffusion processes. We apply the previously considered Hurst estimators to the solutions of these equations. Additionally, we consider some estimators of diffusion coefficient. Moreover, we compare the asymptotic behavior of these estimators with the help of computer simulations. Note that both types of equations are transformable, or reducible, in the sense that they can be reduced to a linear equations by some smooth transformations which do not change the initial flows of sigma-fields. More precisely, a SDE with polynomial drift can be transformed into a linear SDE while Gompertz diffusion process can be transformed into fractional Ornstein-Uhlenbeck process (fractional Vasicek model). More information about this is given in Chap. 1, Sects. 1.3.3 and 1.3.4.

3.3.1 Estimators of the Hurst Index and the Diffusion Coefficient for Fractional Diffusion Processes with Polynomial Drift

We start with the structure of increments of a solution of fractional stochastic differential equation with polynomial drift (that is the same, of fractional diffusion process with polynomial drift). Throughout this subsection we assume that all partitions are uniform.

3.3.1.1 Expansions and Asymptotical Behavior of the Increments on a Small Interval

Let $(\Omega, \mathfrak{F}, \mathbb{F}, \mathsf{P})$ be a filtered probability space generated by a fixed fBm $(B_t^H)_{t\in[0,T]}$. We consider the SDE with polynomial drift given by the equation

$$X_t = X_0 + \int_0^t (aX_s^m + bX_s)\mathrm{d}s + c\int_0^t X_s\,\mathrm{d}B_s^H, \quad X_0 > 0, \quad a \le 0, \quad t \in [0,T], \tag{3.20}$$

where $1/2 < H < 1$, $m \in \mathbb{N}$, $m \ge 2$, and B^H is a fBm with Hurst index $1/2 < H < 1$. This equation has a unique explicit solution of the form

$$X_t = e^{bt+cB_t^H}\left(X_0^{1-m} + (1-m)a\int_0^t e^{(m-1)(bs+cB_s^H)}ds\right)^{1/(1-m)}, \quad t \in [0,T]. \tag{3.21}$$

The direct proof of uniqueness with validation of (3.21) is given in Sect. 1.3.3.

First of all we will find an expression for the increments of solution of Eq. (3.20) as it has been done for the general form of the equation.

Lemma 3.15 *Assume that $H \in (1/2, 1), a \le 0, m \ge 2$ and X is the solution of Eq. (3.20). Then for every $\varepsilon \in (0, H - 1/2)$*

$$\Delta^{(1)}_{t+h,h}X := X_{t+h} - X_t = X_t\big(bh + c\Delta^{(1)}_{t+h,h}B^H + O_\omega(h)\big), \qquad h \to 0+, \quad t \in [0,T),$$

and

$$\Delta^{(2)}_{t,h}X := \Delta^{(1)}_{t+h,h}X - \Delta^{(1)}_{t,h}X = X_t\big(c\Delta^{(2)}_{t,h}B^H + O_\omega(h^{2(H-\varepsilon)})\big), \quad h \to 0+, \quad t \in [0,T).$$

Proof We rewrite the solution of Eq. (3.20) in the following form:

$$X_t = X_0 Z_t A_t,$$

where

$$Z_t = \exp\{bt + cB_t^H\} \quad \text{and} \quad A_t = \left(1 + (1-m)aX_0^{m-1}\int_0^t Z_s^{m-1}ds\right)^{1/(1-m)}.$$

Let $h \to 0+$. By (1.3) and the Maclaurin's expansion of e^x we get that

$$\begin{aligned} Z_{t+h} =& e^{b(t+h)}e^{cB_{t+h}^H} = e^{bt}(1 + bh + o(h^{2(H-\varepsilon)}))e^{cB_t^H + c\Delta_{t+h,h}^{(1)}B^H} \\ =& Z_t\left(1 + bh + o(h^{2(H-\varepsilon)})\right)\left(1 + c\Delta_{t+h,h}^{(1)}B^H + O_\omega(h^{2(H-\varepsilon)})\right) \\ =& Z_t\left(1 + bh + c\Delta_{t+h,h}^{(1)}B^H + O_\omega(h^{2(H-\varepsilon)})\right). \end{aligned} \tag{3.22}$$

Respectively, applying the Maclaurin's expansion of e^{-x} we get that

$$\begin{aligned} Z_{t-h} =& Z_t e^{-bh}e^{-c\Delta_{t,h}^{(1)}B^H} = Z_t(1 - bh + o(h^{2(H-\varepsilon)}))e^{-c\Delta_{t,h}^{(1)}B^H} \\ =& Z_t\left(1 - bh + o(h^{2(H-\varepsilon)})\right)\left(1 - c\Delta_{t,h}^{(1)}B^H + O_\omega(h^{2(H-\varepsilon)})\right) \\ =& Z_t\left(1 - bh - c\Delta_{t,h}^{(1)}B^H + O_\omega(h^{2(H-\varepsilon)})\right). \end{aligned} \tag{3.23}$$

Next note that $t \mapsto A_t^{1-m} \geq 1$ is non-decreasing whereas

$$\int_t^{t+h} Z_s^{m-1}ds = hZ_{t+\theta h}^{m-1} = O_\omega(h) \quad \text{and} \quad \int_{t-h}^t Z_s^{m-1}ds = hZ_{t-h+\theta h}^{m-1} = O_\omega(h), \tag{3.24}$$

where $\theta \in (0,1)$, vanishes as $h \to 0$. Therefore the Maclaurin's expansion of $x \mapsto (1+x)^\alpha$ gives

$$\begin{aligned} A_{t+h} =& \left(A_t^{1-m} + (1-m)aX_0^{m-1}\int_t^{t+h} Z_s^{m-1}ds\right)^{1/(1-m)} \\ =& A_t\left(1 + I_{t,t+h} + o_\omega(h^{2(H-\varepsilon)})\right) \end{aligned} \tag{3.25}$$

and

$$\begin{aligned} A_{t-h} =& \left(A_t^{1-m} - (1-m)aX_0^{m-1}\int_{t-h}^{t} Z_s^{m-1}ds\right)^{1/(1-m)} \\ =& A_t\left(1 - I_{t-h,t} + o_\omega(h^{2(H-\varepsilon)})\right), \end{aligned} \tag{3.26}$$

where

$$I_{t,t+h} = \frac{aX_0^{m-1}}{A_t^{1-m}}\int_t^{t+h} Z_s^{m-1}ds \quad \text{and} \quad I_{t-h,t} = \frac{aX_0^{m-1}}{A_t^{1-m}}\int_{t-h}^t Z_s^{m-1}ds.$$

Hence

$$X_{t+h} = X_t\left(1 + bh + c\Delta^{(1)}_{t+h,h}B^H + O_\omega(h^{2(H-\varepsilon)})\right)\left(1 + I_{t,t+h} + o_\omega(h^{2(H-\varepsilon)})\right)$$
$$= X_t\left(1 + bh + c\Delta^{(1)}_{t+h,h}B^H + I_{t,t+h} + O_\omega(h^{2(H-\varepsilon)})\right), \tag{3.27}$$

and

$$X_{t-h} = X_t\left(1 - bh - c\Delta^{(1)}_{t,h}B^H + O_\omega(h^{2(H-\varepsilon)})\right)\left(1 - I_{t-h,t} + o_\omega(h^{2(H-\varepsilon)})\right)$$
$$= X_t\left(1 - bh - c\Delta^{(1)}_{t,h}B^H - I_{t-h,t} + O_\omega(h^{2(H-\varepsilon)})\right). \tag{3.28}$$

Consequently, $\Delta^{(1)}_{t+h,h}X = X_t(bh + c\Delta B^H_{t,t+h} + I_{t,t+h} + O_\omega(h^{2(H-\varepsilon)}))$ and $-\Delta^{(1)}_{t,h}X = X_t(-bh - c\Delta B^H_{t,h} - I_{t-h,t} + O_\omega(h^{2(H-\varepsilon)}))$. Taking into account (3.27) and (3.28), we get

$$\Delta^{(2)}_{t,h}X = X_t\left(bh + c\Delta^{(1)}_{t+h,h}B^H + I_{t,t+h} + O_\omega(h^{2(H-\varepsilon)})\right)$$
$$+ X_t\big(-bh - c\Delta^{(1)}_{t,h}B^H - I_{t-h,t} + O_\omega(h^{2(H-\varepsilon)})\big)$$
$$= X_t\left(c\Delta^{(2)}_{t,h}B^H + I_{t,t+h} - I_{t-h,t} + O_\omega(h^{2(H-\varepsilon)})\right).$$

It remains to show that $I_{t,t+h} - I_{t-h,t} = O_\omega(h^{2(H-\varepsilon)})$. In order to see this observe that by (3.22)–(3.26) and the mean value theorem, for some $\theta_i \in (0,1), i = 1,2$,

$$I_{t,t+h} - I_{t-h,t} = \frac{aX_0^{m-1}}{A_t^{1-m}}\left(\int_t^{t+h} Z_s^{m-1}ds - \int_{t-h}^t Z_s^{m-1}ds\right)$$
$$= \frac{aX_0^{m-1}}{A_t^{m-1}}\left(\int_t^{t+h} Z_s^{m-1}ds - \int_{t-h}^t Z_s^{m-1}ds\right)$$
$$= \frac{aX_0^{m-1}}{A_t^{1-m}}\left(hZ^{m-1}_{t+\theta_1 h} - hZ^{m-1}_{t-h+\theta_2 h}\right)$$
$$= \frac{aX_0^{m-1}}{A_t^{1-m}}\left(hZ_s^{m-1}\left(\left(\frac{Z_{s+\tau}}{Z_s}\right)^{m-1} - 1\right)\right)$$
$$= \frac{aX_0^{m-1}}{A_t^{1-m}}\left(hZ_s^{m-1}\left((1 + O_\omega(h^{H-\varepsilon}))^{m-1} - 1\right)\right)$$
$$= \frac{aX_0^{m-1}}{A_t^{1-m}}O_\omega(h^{1+H-\varepsilon}) = O_\omega(h^{2(H-\varepsilon)}),$$

where $s = t - h + \theta_2 h$ and $\tau = h(1 + \theta_1 - \theta_2)$. □

3.3.1.2 Construction of the Estimators of Hurst Index and of Diffusion Coefficient

Now our goal is to construct strongly consistent and asymptotically normal estimators of the Hurst parameter H from discrete observations of a single sample path. The properties of the estimators depend significantly on whether the diffusion coefficient c is known or unknown. Therefore we consider the estimation when c is known and when it is unknown.

Assume that X is a solution of SDE (3.20) with $a \le 0$. Let $X_0 > 0$, $c > 0$, and $1/2 < H < 1$. The explicit form of X_t implies that $X_t \ge e^{bt+cB_t^H} X_0$, therefore, there exists an a.s. finite r.v. ς such that $\sup_{t\in[0,T]} \frac{1}{|X_t|} \le \varsigma$ a.s. The structure of the increments of the solution of SDE with polynomial drift allows us to apply the theorems obtained in Sect. 3.2.1. Thus, one can use Theorem 3.9 if constant c is known and Theorem 3.12 in the general situation. We reformulate Theorem 3.9 for the solution X of Eq. (3.20) under the assumption that c is known.

Theorem 3.16 *Assume that X is a solution of SDE (3.20) with $a \le 0$. Let $X_0 > 0$, $c > 0$, and $1/2 < H < 1$. Then*

$$\widehat{H}_n^{(1)} \xrightarrow{a.s.} H,$$

$$2\sqrt{n}\,\log\frac{n}{T}\,(\widehat{H}_n^{(1)} - H) \xrightarrow{d} \mathcal{N}(0, \Sigma_{11}),$$

where variance Σ_{11} is defined in Theorem 2.11,

$$\widehat{H}_n^{(1)} = \varphi_{n,T}^{-1}\left(\frac{1}{n}\sum_{k=1}^{n-1}\left(\frac{\Delta_{n,k}^{(2)}X}{cX(t_k^n)}\right)^2\right) \quad \text{for } n > T,$$

$$\varphi_{n,T}(x) = \left(\frac{T}{n}\right)^{2x}(4 - 2^{2x}) \quad \text{and} \quad \varphi_{n,T}^{-1} \text{ is the inverse of } \varphi_{n,T}, \quad x \in (0, 1).$$

As we have noted above, for an unknown constant c, we can use Theorem 3.12. Moreover, we present another estimator which is suitable when c is unknown.

Theorem 3.17 *Let the assumptions of Theorem 3.16 hold and*

$$\widehat{H}_n^{(2)} = \frac{1}{2} - \frac{1}{2\log 2}\log\left(\frac{\widetilde{V}_{2n,T}^{(2)X}}{\widetilde{V}_{n,T}^{(2)X}}\right),$$

where

$$\widetilde{V}_{n,T}^{(2)X} = \sum_{k=1}^{n-1}\left(\frac{\Delta_{n,k}^{(2)}X}{X(t_k^n)}\right)^2.$$

Then

$$\widehat{H}_n^{(2)} \xrightarrow{a.s.} H$$

$$2 \log 2 \sqrt{n}(\widehat{H}_n^{(2)} - H) \xrightarrow{d} \mathcal{N}(0, \sigma_H^2)$$

with a known variance σ_H^2 *defined in Remark 2.12.*

In order to estimate c one can use the theorem below. In this theorem $\widehat{H}_n$ stands for a sequence of random variables.

Theorem 3.18 *Assume that* $1/2 < H < 1$ *and* $\widehat{H}_n = H + O_\omega(\phi(n))$. *If* $\phi(n) = o\left(\frac{1}{\log n}\right)$ *then*

$$\widehat{c}_n^2 = \frac{n^{2\widehat{H}_n - 1}}{T^{2\widehat{H}_n}(4 - 2^{2\widehat{H}_n})} \widetilde{V}_{n,T}^{(2)X} \xrightarrow{a.s.} c^2 \tag{3.29}$$

If $\phi(n) = o\left(\frac{1}{\sqrt{n}\log n}\right)$ *then additionally to* (3.29)

$$\sqrt{n}\left(\frac{\widehat{c}_n^2}{c^2} - 1\right) \xrightarrow{d} \mathcal{N}(0, \Sigma_{11})$$

with a known variance Σ_{11} *defined in Sect. 2.1.3.*

Hence the estimation of c and the construction of confidence intervals is possible under additional restrictions.

3.3.1.3 Proof of Theorem 3.17

Fix $\varepsilon \in \left(0, \frac{H-1/2}{3}\right)$. By Lemma 3.15

$$\begin{aligned}\widetilde{V}_{n,T}^{(2)X} &= \sum_{k=1}^{n-1} \left(c\Delta_{n,k}^{(2)} B^H + O_\omega\left(n^{-2(H-\varepsilon)}\right)\right)^2 \\ &= c^2 \sum_{k=1}^{n-1} \left(\Delta_{n,k}^{(2)} B^H\right)^2 + O_\omega\left(n^{1-3(H-\varepsilon)}\right) \\ &= \frac{c^2 T^{2H}(4 - 2^{2H})}{n^{2H}} V_{n,T}^{(2)\widehat{B^H}} + O_\omega\left(n^{1-3(H-\varepsilon)}\right)\end{aligned}$$

and in the same way

$$\widetilde{V}_{2n,T}^{(2)X} = \frac{c^2 T^{2H}(4 - 2^{2H})}{(2n)^{2H}} V_{2n,T}^{(2)\widehat{B^H}} + O_\omega\left(n^{1-3(H-\varepsilon)}\right).$$

So,

$$\begin{aligned}\widehat{H}_n^{(2)} =& \frac{1}{2} - \frac{1}{2\log 2}\log\left(\frac{\frac{c^2T^{2H}(4-2^{2H})}{(2n)^{2H-1}}(2n)^{-1}V_{2n,T}^{(2)\widehat{B}^H} + O_\omega\left(n^{1-3(H-\varepsilon)}\right)}{\frac{c^2T^{2H}(4-2^{2H})}{n^{2H-1}}n^{-1}V_{n,T}^{(2)\widehat{B}^H} + O_\omega\left(n^{1-3(H-\varepsilon)}\right)}\right)\\ =& \frac{1}{2} - \frac{1}{2\log 2}\log\left(\frac{(2n)^{-1}V_{2n,T}^{(2)\widehat{B}^H}}{2^{2H-1}n^{-1}V_{n,T}^{(2)\widehat{B}^H}}\left(\frac{1+O_\omega\left(n^{-H+3\varepsilon}\right)}{1+O_\omega\left(n^{-H+3\varepsilon}\right)}\right)\right)\\ =& \widetilde{H}_n - \frac{1}{2\log 2}\log\left(1+O_\omega\left(n^{-H+3\varepsilon}\right)\right) = \widetilde{H}_n + O_\omega\left(n^{-H+3\varepsilon}\right),\end{aligned}$$

where

$$\widetilde{H}_n = \frac{1}{2} - \frac{1}{2\log 2}\log\left(\frac{(2n)^{-1}V_{2n,T}^{(2)\widehat{B}^H}}{2^{2H-1}n^{-1}V_{n,T}^{(2)\widehat{B}^H}}\right) = H - \frac{1}{2\log 2}\log\left(\frac{(2n)^{-1}V_{2n,T}^{(2)\widehat{B}^H}}{n^{-1}V_{n,T}^{(2)\widehat{B}^H}}\right).$$

Since $n^{-1}V_{n,T}^{(2)\widehat{B}^H} \xrightarrow{a.s.} 1$ and $(2n)^{-1}V_{2n,T}^{(2)\widehat{B}^H} \xrightarrow{a.s.} 1$ (see Theorem 2.2) then $\widetilde{H}_n \xrightarrow{a.s.} H$. By the limit results of Sect. 2.1.3 we get

$$2\log 2\sqrt{n}(\widetilde{H}_n - H) \xrightarrow{d} \mathcal{N}(0,\sigma_H^2).$$

Now to finish the proof we will apply the Slutsky's theorem and the results obtained above. Note that the limit variance σ_H^2 of $\widehat{H}_n^{(2)}$ equals that to $\widetilde{H}_n$.

3.3.1.4 Proof of Theorem 3.18

Let $\varepsilon \in \left(0, \frac{H-1/2}{3}\right)$ and

$$\widetilde{c}_n^2 = \frac{n^{2(H-\widehat{H}_n)}(4-2^{2\widehat{H}_n})}{T^{2(H-\widehat{H}_n)}(4-2^{2H})}\widehat{c}_n^2.$$

By the Maclaurin's expansion and the mean value theorem

$$\begin{aligned}\frac{n^{2(H-\widehat{H}_n)}(4-2^{2\widehat{H}_n})}{T^{2(H-\widehat{H}_n)}(4-2^{2H})} &= \exp\left\{O_\omega(\phi(n))\log\left(\frac{n}{T}\right)^2 + \log\left(4-2^{2\widehat{H}_n}\right) - \log(4-2^{2H})\right\}\\ &= (1+O_\omega(\phi(n)\log n))\exp\left\{O_\omega(\phi(n))h'(H+\theta_n(\widehat{H}_n-H))\right\}\\ &= (1+O_\omega(\phi(n)\log n))(1+O_\omega(\phi(n))) = (1+O_\omega(\phi(n)\log n))\\ &= (1+o_\omega(1)) \qquad (3.30)\end{aligned}$$

(since $h'(H + \theta_n(\widehat{H}_n - H))$ is bounded), where $\theta_n \in (0, 1)$ and $h(x) = \log(4 - 2^{2x})$, $x \in (0, 1)$. Hence it suffices to show that $\widetilde{c_n^2}$ satisfies

$$\widetilde{c_n^2} \xrightarrow{d} c^2, \qquad \sqrt{n}\left(\frac{\widetilde{c_n^2}}{c^2} - 1\right) \xrightarrow{d} \mathcal{N}(0, \Sigma_{11}) \quad \text{as} \quad n \to \infty,$$

and the claim will be proved. By Lemma 3.15

$$\widetilde{V}_{n,T}^{(2)X} = c^2 \sum_{k=1}^{n-1} (\Delta_{n,k}^{(2)} B^H)^2 + O_\omega\big(n^{1-3(H-\varepsilon)}\big).$$

Thus,

$$\begin{aligned}\widetilde{c_n^2} - c^2 =& c^2\left(\frac{n^{2H-1}}{T^{2H}(4 - 2^{2H})} \sum_{k=1}^{n-1} (\Delta_{n,k}^{(2)} B^H)^2 - 1\right) + O_\omega\big(n^{-H+3\varepsilon}\big) \\ =& c^2\big(n^{-1} V_{n,T}^{(2)\widehat{B^H}} - 1\big) + O_\omega\big(n^{-H+3\varepsilon}\big).\end{aligned}$$

It remains to apply Slutsky's theorem and the limit result stated in Theorem 2.11.

3.3.1.5 Examples of Fractional Diffusion Processes with Polynomial Drift: Parameter Estimation and Simulations

The purpose of this subsection is twofold. First of all we want to demonstrate that the Eq. (3.20) is rather flexible and can be used in several popular models frequently met in practical applications. Secondly we provide some simulations in order to illustrate various aspects of performance of the suggested statistics.

Table 3.1 lists three popular models covered by the Eq. (3.20). For each type of the model we have drawn 500 sample paths and calculated several characteristics depicted in the figures provided below. In each case, unless explicitly stated otherwise, the parameters of the models were $X_0 = 3$, $b = 0.5$ and $c = 0.7$. The

Table 3.1 Several models covered by Eq. (3.1)

Name of equation	Ranges of parameters	Explicit form of the solution	Common field of application
Black-Scholes	$a = 0, b > 0$ $c > 0, m = 2$	$X_t = X_0 \exp\{bt + cB_t^H\}$	Financial modeling
Verhulst	$a = -1, b > 0,$ $c > 0, m = 2$	$X_t = \frac{X_0 \exp\{bt + cB_t^H\}}{1 + X_0 \int_0^t \exp\{bs + cB_s^H\} ds}$	Demography
Landau-Ginzburg	$a = -1, b > 0$ $c > 0, m = 3$	$X_t = \frac{X_0 \exp\{bt + cB_t^H\}}{\sqrt{1 + 2X_0^2 \int_0^t \exp\{2(bs + cB_s^H)\} ds}}$	Physics

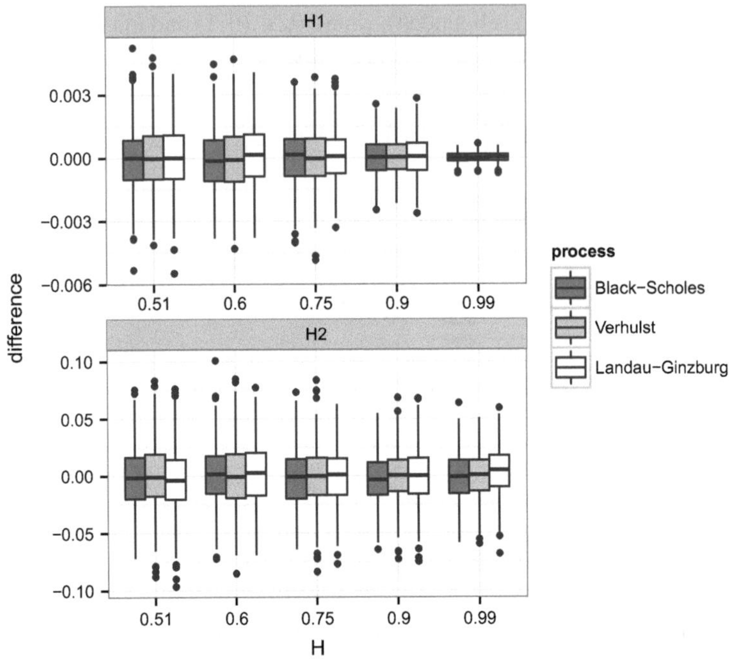

Fig. 3.1 The differences between the estimated and real values of H

simulations were performed using the R software environment [147]. We summarize our observations by giving a short discussion of the figures.

- In Fig. 3.1 we consider the dependence of numeric characteristics of the estimators $\widehat{H}_n^{(1)}$ (defined in Theorem 3.16 and denoted in the figures as H1) and $\widehat{H}_n^{(2)}$ (defined in Theorem 3.17 and denoted in the figures as H2) on the true value of the parameter H.
  Results imply that the accuracy of $\widehat{H}_n^{(1)}$ surpasses that of $\widehat{H}_n^{(2)}$ by at least an order of magnitude, and becomes even more precise for values of H close to 1. The practical use of $\widehat{H}_n^{(1)}$, however, requires the knowledge of the true value of volatility c which is not always available. Nonetheless if that is the case then the estimator $\widehat{H}_n^{(1)}$ performs well even for very short sample paths, as illustrated in Fig. 3.2. It is also worth noting that the estimators do not display notable dependence on the type of the underlying process.
- Figure 3.3 presents the boxplots of both estimators as the sample size, n, varies from 1024 to 8192 points. It can be seen that the estimator $\widehat{H}_n^{(1)}$ is considerably more precise than $\widehat{H}_n^{(2)}$, yet the rate at which their accuracy improves with increased sample sizes is roughly the same. For both of these estimators the interquartile ranges shrink by roughly 30% as the sample size is doubled.

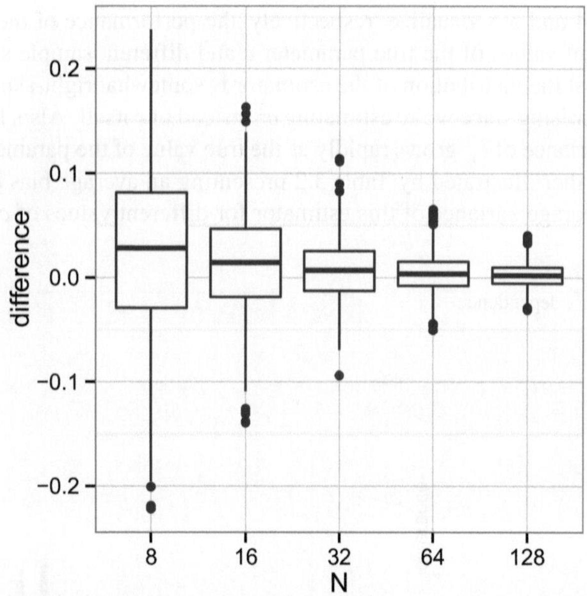

Fig. 3.2 Behavior of $\widehat{H}_n^{(1)}$ in case of short sample paths

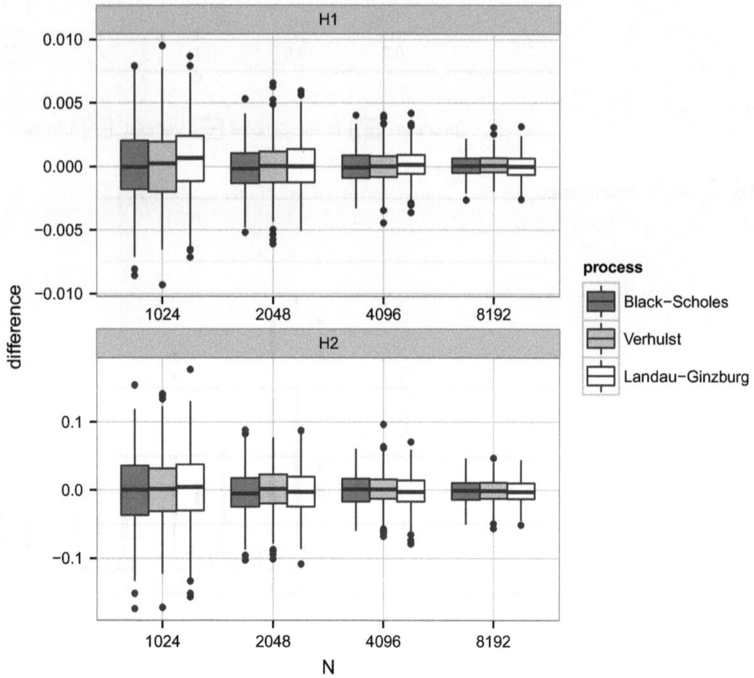

Fig. 3.3 The differences between the estimated and real values of H

- Figures 3.4 and 3.5 visualize, respectively, the performance of the estimator $\widehat{c}_n^2$ for different values of the true parameter c and different sample sizes. One can observe that the distribution of the estimator is somewhat right-skewed, which is hardly surprising since we're estimating c^2 instead of c itself. Also, Fig. 3.4 shows that the variance of $\widehat{c}_n^2$ grows rapidly as the true value of the parameter increases. This is further illustrated by Table 3.2 presenting an average bias $\widehat{c}_n^2 - c^2$ along with an average variance of this estimator for different values of c. An adjusted

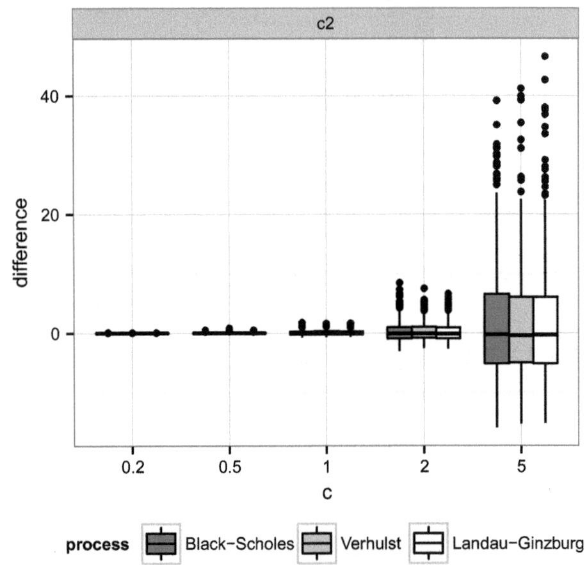

Fig. 3.4 $\widehat{c}_n^2 - c^2$, dependence on c

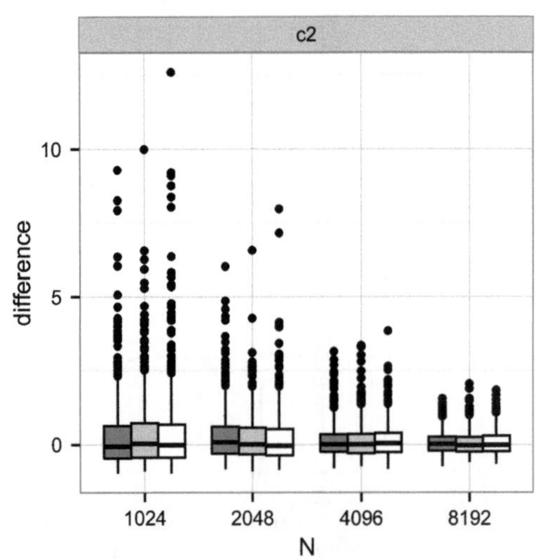

Fig. 3.5 $\widehat{c}_n^2 - c^2$, dependence on N

Table 3.2 Numeric characteristics of $\widehat{c_n^2}$

c	0.2	0.5	1	2	5
Bias	0.006	0.031	0.133	0.460	3.188
Variance	0.001	0.019	0.306	5.159	193.4

Table 3.3 Comparison of theoretical and empirical variances of $\widehat{H}_n^{(1)}$

Sample path length, n						
	128	512	2048	128	512	2048
Hurst index, H	Theoretical variance $\times 10^4$			Empirical variance $\times 10^6$		
0.6	234.8	35.51	5.944	175.6	26.16	4.842
0.75	215.2	32.55	5.447	119.8	20.94	3.399
0.9	198.0	29.95	5.012	51.81	10.73	1.936

Table 3.4 Comparison of theoretical and empirical variances of $\widehat{H}_n^{(2)}$

Sample path length, n						
	128	512	2048	128	512	2048
Hurst index, H	Theoretical variance $\times 10^4$			Empirical variance $\times 10^4$		
0.6	57.53	14.38	3.596	214.8	56.51	14.77
0.75	52.73	13.18	3.295	192.0	47.88	11.65
0.9	48.52	12.13	3.032	161.2	36.78	9.767

R-squared of the linear model $Var(\widehat{c_n^2}) = kc^4 + b$ equals to 0.9838 and is in agreement with asymptotical results of Theorem 3.16.

- The remaining tables characterize the convergence of $Var(\widehat{H}_n^{(i)})$ to their theoretical limits. One can see that the theoretical and empirical variances decrease at the same rate as the sample size is increased and differ by only a constant factor. The empirical variances were evaluated using sample paths of the Black-Scholes model (Tables 3.3 and 3.4). In case of the other models considered in this paper, the results are analogous.

3.3.2 Estimators of the Hurst Index and the Diffusion Coefficient for Fractional Gompertz Diffusion Process

In this subsection for the estimation of the Hurst index of the fractional Gompertz diffusion process we used three estimators and three estimators for estimation of the diffusion coefficient. We prove that these estimators for the fractional Gompertz diffusion process are strongly consistent and some of them are asymptotically normal. Later on we compare used estimators with the aid of computer simulation.

3.3.2.1 Expansions and Asymptotical Behavior of the Increments on a Small Interval

Let $(\Omega, \mathfrak{F}, \mathbb{F}, \mathsf{P})$ be a filtered probability space generated by a fixed fBm $(B_t^H)_{t\in[0,T]}$. We consider the fractional Gompertz diffusion process given by the stochastic differential equation

$$X_t = X_0 + \int_0^t (\alpha X_s - \beta X_s \log X_s)\, ds + \sigma \int_0^t X_s\, dB_s^H, \quad X_0 > 0, \quad 0 \le t \le T, \tag{3.31}$$

where α, $\beta \neq 0$, and $\sigma > 0$ are real parameters and B^H is a fBm with the Hurst index $H \in (1/2, 1)$. This equation has the unique explicit solution of the form (see Sect. 1.3.4)

$$X_t = \exp\left\{ e^{-\beta t} \log X_0 + \alpha \int_0^t e^{-\beta(t-s)} ds + \sigma \int_0^t e^{-\beta(t-s)}\, dB_s^H \right\}.$$

First of all we will find an expression for the increments of solution of Eq. (3.31) as it has been done for the general form of the equation.

Lemma 3.19 *Suppose that X satisfies (3.31), $\varepsilon \in (0, H - \frac{1}{2})$ and a partition $\pi_n = \{\tau_k^{m_n},\ k = 0, \ldots, m_n\}$, $n \ge 1$, $\mathbb{N} \ni m_n \uparrow \infty$, of the interval $[0, T]$ is uniform. Then the following relations hold:*

$$\Delta_{m_n,k}^{(1)} X = X_{\tau_{k-1}^{m_n}} \left[\sigma \Delta_{m_n,k}^{(1)} B^H + O_\omega(d_n) \right] = X_{\tau_{k-1}^{m_n}} O_\omega\left(d_n^{H-\varepsilon}\right), \quad k = 1, \ldots, m_n, \tag{3.32}$$

$$\Delta_{m_n,k}^{(2)} X = X_{\tau_k^{m_n}} \left[\sigma \Delta_{m_n,k}^{(2)} B^H + O_\omega\left(d_n^{2(H-\varepsilon)}\right) \right], \quad k = 1, \ldots, m_n, \tag{3.33}$$

where $d_n = \tau_k^{m_n} - \tau_{k-1}^{m_n} = \frac{T}{m_n}$ and $d_n \to 0$ as $n \to \infty$.

Proof For the sake of simplicity we will omit the index m_n for the points $\tau_k^{m_n}$. We first prove (3.32). Note that

$$\Delta_k^{(1)} X = X_{\tau_k} - X_{\tau_{k-1}} \quad \text{and} \quad X_{\tau_k} = X_{\tau_{k-1}} \exp\{\Delta_k^{(1)} Y\},$$

where

$$Y_t = e^{-\beta t} \log X_0 + \frac{\alpha}{\beta}(1 - e^{-\beta t}) + \sigma \int_0^t e^{-\beta(t-s)} dB_s^H.$$

It is clear that

$$\Delta_k^{(1)} Y = e^{-\beta\tau_{k-1}}\left(e^{-\beta(\tau_k-\tau_{k-1})}-1\right)\left(\log X_0 - \frac{\alpha}{\beta} + \sigma\int_0^{\tau_{k-1}} e^{\beta s} dB_s^H\right)$$
$$+\sigma e^{-\beta\tau_k}\int_{\tau_{k-1}}^{\tau_k}\left[e^{\beta s}-e^{\beta\tau_k}\right]dB_s^H + \sigma\int_{\tau_{k-1}}^{\tau_k} dB_s^H.$$

From the Chain rule it follows that

$$\int_0^t e^{\beta s} dB_s^H = e^{\beta t} B_t^H - \beta\int_0^t e^{\beta s} B_s^H ds.$$

Thus,

$$\left|\int_0^{\tau_{k-1}} e^{\beta s} dB_s^H\right| = \left|e^{\beta\tau_{k-1}} B_{\tau_{k-1}}^H - \beta\int_0^{\tau_{k-1}} e^{\beta s} B_s^H ds\right| \le e^{|\beta| T}(|\beta| T+1)\sup_{t\le T}\left|B_t^H\right|.$$

Provided

$$e^{-\beta(\tau_k-\tau_{k-1})} = 1 + O(d_n),$$

it follows that

$$Z_{k-1} := e^{-\beta\tau_{k-1}}\left(e^{-\beta(\tau_k-\tau_{k-1})}-1\right)\left[\log X_0 - \frac{\alpha}{\beta} + \sigma\int_0^{\tau_{k-1}} e^{\beta s} dB_s^H\right] = O_\omega(d_n).$$

Further, by the Love-Young inequality (1.18) applied with $p=1$ and $q=\frac{1}{H-\varepsilon}$ and inequality (1.3) it follows that

$$\left|\int_{\tau_{k-1}}^{\tau_k}\left[e^{\beta s}-e^{\beta\tau_k}\right]dB_s^H\right| \le C_{1,q} G_{\varepsilon,T} e^{|\beta| T}|\beta|(\tau_k-\tau_{k-1})^{1+H-\varepsilon} = O_\omega\left(d_n^{1+H-\varepsilon}\right).$$

Consequently,

$$X_{\tau_k} = X_{\tau_{k-1}}\exp\left\{Z_{k-1} + O_\omega\left(d_n^{1+H-\varepsilon}\right) + \sigma\Delta_k^{(1)} B^H\right\}$$
$$= X_{\tau_{k-1}}\left[1 + Z_{k-1} + O_\omega\left(d_n^{1+H-\varepsilon}\right) + \sigma\Delta_k^{(1)} B^H + O_\omega\left(d_n^{2(H-\varepsilon)}\right)\right] \tag{3.34}$$

and

$$\Delta_k^{(1)} X = X_{\tau_{k-1}}\left[Z_{k-1} + O_\omega\left(d_n^{2(H-\varepsilon)}\right) + \sigma\Delta_k^{(1)} B^H\right] = X_{\tau_{k-1}} O_\omega\left(d_n^{H-\varepsilon}\right). \tag{3.35}$$

Next we prove (3.33). In the same way as we proved (3.34) and (3.35), we obtain

$$X_{\tau_{k-1}} = X_{\tau_k}\exp\{-\Delta_k^{(1)} Y\} = X_{\tau_k}\exp\left\{-Z_{k-1} + O_\omega\left(d_n^{1+H-\varepsilon}\right) - \sigma\Delta_k^{(1)} B^H\right\}$$
$$= X_{\tau_k}\left[1 - Z_{k-1} + O_\omega\left(d_n^{1+H-\varepsilon}\right) - \sigma\Delta_k^{(1)} B^H + O_\omega\left(d_n^{2(H-\varepsilon)}\right)\right] \tag{3.36}$$

and

$$-\Delta_k^{(1)}X = X_{\tau_k}\Big[-Z_{k-1} + O_\omega\big(d_n^{2(H-\varepsilon)}\big) - \sigma\Delta_k^{(1)}B^H\Big] = X_{\tau_k}O_\omega\big(d_n^{H-\varepsilon}\big). \tag{3.37}$$

Taking into account (3.34)– (3.37), we get

$$\begin{aligned}\Delta_k^{(2)}X &= X_{\tau_k}\Big[Z_k + O_\omega\big(d_n^{2(H-\varepsilon)}\big) + \sigma\Delta_{k+1}^{(1)}B^H\Big]\\ &\quad - X_{\tau_k}\Big[Z_{k-1} + O_\omega\big(d_n^{2(H-\varepsilon)}\big) + \sigma\Delta_k^{(1)}B^H\Big]\\ &= X_{\tau_k}\Big[(Z_k - Z_{k-1}) + O_\omega\big(d_n^{2(H-\varepsilon)}\big) + \sigma\Delta_k^{(2)}B^H\Big].\end{aligned}$$

By the Love–Young inequality, (1.18) applied with $p = 1$ and $q = \frac{1}{H-\varepsilon}$,

$$\left|\int_{\tau_{k-1}}^{\tau_k} e^{\beta s}\, dB_s^H\right| \le (1+|\beta|\,T)C_{1,q}G_{\varepsilon,T}e^{|\beta|T}(\tau_k-\tau_{k-1})^{H-\varepsilon} = O_\omega\big(d_n^{H-\varepsilon}\big).$$

Thus,

$$\begin{aligned}Z_k - Z_{k-1} =& \big(e^{-\beta\tau_k} - e^{-\beta\tau_{k-1}}\big)\big(e^{-\beta d_n} - 1\big)\Big[\log X_0 - \frac{\alpha}{\beta} + \sigma\int_0^{\tau_{k-1}} e^{\beta s} dB_s^H\Big]\\ &+ \sigma e^{-\beta\tau_k}\big(e^{-\beta d_n} - 1\big)\int_{\tau_{k-1}}^{\tau_k} e^{\beta s} dB_s^H = O_\omega\big(d_n^{2(H-\varepsilon)}\big).\end{aligned}$$

Finally, we obtain

$$\Delta_k^{(2)}X = X_{\tau_k}\big[\sigma\Delta_k^{(2)}B^H + O_\omega\big(d_n^{2(H-\varepsilon)}\big)\big]. \qquad \square$$

3.3.2.2 Hurst Index Estimators

Let as before $\pi_n = \{\tau_k^{m_n},\ k = 0,\dots,m_n\}$, $n \ge 1$, $\mathbb{N} \ni m_n \uparrow \infty$, be a sequence of uniform partitions of the interval $[0, T]$. Further in this subsection the sequence (m_n) for different statistics takes different values. So for the statistics $\widehat{H}_n^{(1)}$ it takes two values n and $2n$; for the statistics $\widehat{H}_n^{(2)}$ it takes two values $m_n = n$ and $m_n = nk_n = n^3$; for the statistics $\widehat{H}_n^{(3)}$ it takes ℓ values $m_n = r_j n$, where $r_j \in \mathbb{N}, j = 1,\dots,\ell$.

Let $(X_t)_{t\in[0,T]}$ be a stochastic process and

$$\Delta_{m_n,k}^{(1)}X = X(\tau_k^{m_n}) - X(\tau_{k-1}^{m_n}), \qquad \Delta_{m_n,k}^{(2)}X = X(\tau_{k+1}^{m_n}) - 2X(\tau_k^{m_n}) + X(\tau_{k-1}^{m_n}),$$
$$k = 1,\dots,m_n - (i-1), \quad i = 1, 2.$$

Denote

$$\widetilde{V}^{(i)X}_{m_n,T} = \sum_{k=1}^{m_n-(i-1)} \left(\frac{\Delta^{(i)}_{m_n,k} X}{X(\tau_k^{m_n})} \right)^2, \qquad i = 1, 2,$$

and

$$W_{n,k} = \sum_{j=-k_n+1}^{k_n-1} \left(\Delta^{(2)} X_{s_j^n + t_k^n} \right) = \sum_{j=-k_n+1}^{k_n-1} \left(X_{s_{j+1}^n + t_k^n} - 2X_{s_j^n + t_k^n} + X_{s_{j-1}^n + t_k^n} \right)^2,$$

where $s_j^n = \frac{jT}{m_n}$, $m_n = n^3$, and $t_k^n = \frac{kT}{n}$.

Theorem 3.20 *Assume that X is a solution of the fractional Gompertz SDE and $1/2 < H < 1$. Then*

$$\widehat{H}_n^{(j)} \xrightarrow{a.s.} H, \qquad j = 1, 2, 3,$$

and

$$2 \log 2 \sqrt{n} (\widehat{H}_n^{(1)} - H) \xrightarrow{d} \mathcal{N}(0, \sigma_H^2),$$

$$2\sqrt{n} \log \frac{n}{T} (\widehat{H}_n^{(2)} - H) \xrightarrow{d} \mathcal{N}(0, \Sigma_{11}),$$

$$\sqrt{n} (\widehat{H}_n^{(3)} - H) \xrightarrow{d} \mathcal{N}\left(0, \sigma_\ell^2 \left(\mathbf{r}, \frac{1}{2} (\mathbf{z}/\sqrt{\mathbf{r}}) \right)\right)$$

with known variances σ_H^2, Σ_{11}, $\sigma_\ell^2(\mathbf{r}, \frac{1}{2}(\mathbf{z}/\sqrt{\mathbf{r}}))$ defined in Sect. 2.1.3, where

$$\widehat{H}_n^{(1)} = \frac{1}{2} - \frac{1}{2 \log 2} \log \left(\frac{\widetilde{V}^{(2)X}_{2n,T}}{\widetilde{V}^{(2)X}_{n,T}} \right),$$

$$\widehat{H}_n^{(2)} = \frac{1}{2} + \frac{1}{2 \log k_n} \log \left(\frac{2}{n} \sum_{k=1}^{n-1} \frac{(\Delta^{(2)}_{n,k} X)^2}{W_{n,k-1}} \right),$$

$$\widehat{H}_n^{(3)} = -\frac{1}{2} \sum_{j=1}^{\ell} z_j \log \left(\frac{\widetilde{V}^{(2)X}_{n_j,T}}{n_j - 1} \right), \qquad n_j = r_j n, \; r_j \in \mathbb{N}, \; j = 1, \ldots, \ell,$$

$$z_j = \frac{y_j}{\sum_{i=1}^{\ell} y_i^2} \quad \textit{and} \quad y_j = \log r_j - \frac{1}{\ell} \sum_{i=1}^{\ell} \log r_i.$$

Remark 3.21 The estimators $\widehat{H}_n^{(i)}$, $i = 1, 2, 3$, were considered in [101, 102], and [24], respectively.

3.3.2.3 Estimators of Diffusion Coefficient

In this subsection, we describe three estimators of the diffusion coefficient σ.

Theorem 3.22 *Assume that X is a solution of the fractional Gompertz SDE,* $1/2 < H < 1$, *and* $\widehat{H}_n = H + O_\omega(\phi(n))$. *If* $\phi(n) = o\left(\frac{1}{\log n}\right)$ *then*

$$\widehat{\sigma}^2_{1,n} = \frac{n^{2\widehat{H}_n - 1}}{T^{2\widehat{H}_n}} \widetilde{V}^{(1)X}_{n,T} \xrightarrow{a.s.} \sigma^2,$$

$$\widehat{\sigma}^2_{2,n} = \frac{n^{2\widehat{H}_n - 1}}{T^{2\widehat{H}_n}(4 - 2^{2\widehat{H}_n})} \widetilde{V}^{(2)X}_{n,T} \xrightarrow{a.s.} \sigma^2,$$

$$\widehat{\sigma}^2_{3,n} = \frac{\sum_{k=1}^{n} \left(\Delta^{(2)}_{n,k} X\right)^2}{(\frac{T}{n})^{2\widehat{H}_n}(4 - 2^{2\widehat{H}_n}) \sum_{k=1}^{n} X^2(t^n_{k-1})} \xrightarrow{a.s.} \sigma^2,$$

and

$$\sqrt{n}\left(\frac{\widehat{\sigma}^2_{i,n}}{\sigma^2} - 1\right) \xrightarrow{d} \mathcal{N}(0, \Sigma_{11}), \qquad i = 1, 2,$$

where variance Σ_{11} *is defined in Sect. 2.1.3.*

Remark 3.23 The estimators $\widehat{\sigma}^2_{i,n}$, $i = 1, 2$, are similar to the estimators used in the book [24] for the evaluation of the diffusion coefficient σ of the solutions of linear SDE when H is known.

3.3.2.4 Proof of Theorem 3.20

1. The convergence of the statistics $\widehat{H}^{(1)}_n$ and $\widehat{H}^{(2)}_n$ considered in Theorem 3.20 follows from Lemma 3.19. It suffices to repeat the proofs of theorems 3.17 and 3.12.
2. Now we prove the convergence of the statistic $\widehat{H}^{(3)}_n$. The proof follows the outline of the proof of Theorem 3.18 in [24]. Fix $\varepsilon \in (0, \frac{H-1/2}{3})$. By Lemma 3.19 we get

$$\begin{aligned}
&\left(\frac{n^H}{\sigma T^H \sqrt{4 - 2^{2H}}}\right)^2 \frac{\widetilde{V}^{(2)X}_{n,T}}{n-1} \\
&= \left(\frac{n^H}{T^H \sqrt{4 - 2^{2H}}}\right)^2 \frac{1}{n-1} \sum_{k=1}^{n-1} \left[\left(\Delta^{(2)}_{n,k} B^H\right)^2 + O_\omega\left(n^{-3(H-\varepsilon)}\right)\right] \\
&= \frac{n}{n-1} n^{-1} V^{(2)\widehat{B}^H}_{n,T} + O_\omega\left(n^{-H+3\varepsilon}\right) \xrightarrow{a.s.} 1. \qquad (3.38)
\end{aligned}$$

By (3.38) and Theorem 2.9 we get

$$\begin{aligned}\log \frac{\widetilde{V}^{(2)X}_{n,T}}{n-1} =& -2H\log\frac{n}{T} + 2\log\left(\sigma\sqrt{4-2^{2H}}\right) + \log\frac{n}{n-1} \\ &+ \log\left[\left(n^{-1}V^{(2)\widehat{B^H}}_{n,T} - 1\right) + 1 + O_\omega\left(n^{-H+3\varepsilon}\right)\right] \\ =& -2H\log\frac{n}{T} + 2\log\left(\sigma\sqrt{4-2^{2H}}\right) + \log\frac{n}{n-1} \\ &+ \log\left[O_\omega(n^{-1/2}\log^{1/2} n) + 1 + O_\omega\left(n^{-H+3\varepsilon}\right)\right] \\ =& -2H\log\frac{n}{T} + 2\log\left(\sigma\sqrt{4-2^{2H}}\right) + O_\omega(n^{-1/2}\log^{1/2} n).\end{aligned}$$

Thus,

$$\widehat{H}^{(3)}_n = -\frac{1}{2}\sum_{j=1}^{\ell} z_j\left[-2H\log\left(\frac{n_j}{T}\right) + 2\log\left(\sigma\sqrt{4-2^{2H}}\right)\right] + O_\omega(n^{-1/2}\log^{1/2} n).$$

We notice the following properties

$$\sum_{j=1}^{\ell} y_j = 0, \qquad \sum_{j=1}^{\ell} z_j y_j = 1, \qquad \sum_{j=1}^{\ell} z_j = \frac{\sum_{j=1}^{\ell} y_j}{\sum_{j=1}^{\ell} y_j^2} = 0.$$

Using those we get

$$\begin{aligned}\widehat{H}^{(3)}_n =& H\sum_{j=1}^{\ell} z_j\log(r_j n) - \log\left(\sigma\sqrt{4-2^{2H}}\right)\sum_{j=1}^{\ell} z_j + O_\omega(n^{-1/2}\log^{1/2} n) \\ =& H\sum_{j=1}^{\ell} z_j\log(r_j n) + O_\omega(n^{-1/2}\log^{1/2} n) \\ =& H\sum_{j=1}^{\ell} z_j\left[y_j - y_j + \log(r_j n)\right] + O_\omega(n^{-1/2}\log^{1/2} n) \\ =& H + H\sum_{j=1}^{\ell} z_j\left[\log n + \frac{1}{\ell}\sum_{j=1}^{\ell}\log r_j\right] + O_\omega(n^{-1/2}\log^{1/2} n) \\ =& H + O_\omega(n^{-1/2}\log^{1/2} n). \qquad (3.39)\end{aligned}$$

So the estimator $\widehat{H}^{(3)}_n$ is strongly consistent.

Now we prove the asymptotic normality of the estimator $\widehat{H}_n^{(3)}$. From (3.38) and (3.39) it follows that

$$\begin{aligned}\widehat{H}_n^{(3)} =& -\frac{1}{2}\sum_{j=1}^{\ell} z_j\Big[-2H\log\Big(\frac{n_j}{T}\Big)+2\log\big(\sigma\sqrt{4-2^{2H}}\big)\Big]\\ &-\frac{1}{2}\sum_{j=1}^{\ell} z_j\log\big[\big(n_j^{-1}V_{n_j,T}^{(2)\widehat{B}^H}-1\big)+1+O_\omega\big(n^{-H+3\varepsilon}\big)\big]\\ =&H-\frac{1}{2}\sum_{j=1}^{\ell} z_j\big(n_j^{-1}V_{n_j,T}^{(2)\widehat{B}^H}-1\big)+O_\omega\big(n^{-H+3\varepsilon}\big).\end{aligned}$$

Thus,

$$\sqrt{n}\big(\widehat{H}_n^{(3)}-H\big)=-\frac{1}{2}\sum_{j=1}^{\ell}\frac{z_j}{\sqrt{r_j}}\Big(\sqrt{n_j}\,V_{n_j,T}^{(2)\widehat{B}^H}-1\Big)+O_\omega\big(n^{1/2-H+3\varepsilon}\big)$$

and we obtain the asymptotic normality of the estimator $\widehat{H}_n^{(3)}$ by the application of Theorem 2.13.

3.3.2.5 Proof of Theorem 3.22

The proof of convergence of $\widehat{\sigma}_{2,n}^2$ is analogous to that of $\widehat{c}_n^2$ in Theorem 3.18. Let us prove that $\widehat{\sigma}_{1,n}^2 \xrightarrow{a.s.} \sigma^2$, as $n\to\infty$. Suppose that $d_n=\frac{T}{n}$. From Lemma 3.19 it follows that

$$\begin{aligned}d_n^{-2H}n^{-1}\widetilde{V}_{n,T}^{(1)X} =&\sigma^2 d_n^{-2H}n^{-1}\sum_{i=1}^{n}\big(\Delta_{n,k}B^H\big)^2+d_n^{-2H}O_\omega\big(d_n^{3(H-\varepsilon)}\big)\\ =&\sigma^2 n^{-1}V_{n,T}^{(1)\widehat{B}^H}+O_\omega\big(d_n^{H-3\varepsilon}\big).\end{aligned}$$

Since

$$n^{-1}V_{n,T}^{(1)\widehat{B}^H}\xrightarrow{a.s.}1 \quad\text{and}\quad \frac{n^{2(H-\widehat{H}_n)}}{T^{2(H-\widehat{H}_n)}}=\exp\Big\{O_\omega(\phi(n))\log\Big(\frac{n}{T}\Big)^2\Big\}\longrightarrow 1, \tag{3.40}$$

it can be concluded that

$$\widehat{\sigma}_{1,n}^2=\frac{n^{2(\widehat{H}_n-H)}}{T^{2(\widehat{H}_n-H)}}\frac{n^{2H-1}}{T^{2H}}\widetilde{V}_{n,T}^{(1)X}=\frac{n^{2(\widehat{H}_n-H)}}{T^{2(\widehat{H}_n-H)}}\Big[\frac{\sigma^2}{n}V_{n,T}^{(1)\widehat{B}^H}+O_\omega\big(d_n^{H-3\varepsilon}\big)\Big]\xrightarrow{a.s.}\sigma^2.$$

Let $\varepsilon \in (0, \frac{H-1/2}{3})$ and

$$\widetilde{\sigma}_{1,n}^2 = \frac{n^{2(H-\widehat{H}_n)}}{T^{2(H-\widehat{H}_n)}} \widehat{\sigma}_{1,n}^2.$$

We obtain the asymptotic normality of $\widehat{\sigma}_{1,n}^2$ from the equality

$$\widetilde{\sigma}_{1,n}^2 - \sigma^2 = \sigma^2 \left(n^{-1} V_{n,T}^{(1)\widehat{B}^H} - 1\right) + O_\omega\left(d_n^{H-3\varepsilon}\right),$$

the Slutsky theorem, and the limit results stated in Theorem 2.11.

Further, let us prove that $\widehat{\sigma}_{3,n}^2 \xrightarrow{a.s.} \sigma^2$. Denote $d_n = \frac{T}{n}$. By (3.30) it suffices to show that

$$\widetilde{\sigma}_n^2 = \frac{\sum_{k=1}^n \left(\Delta_{n,k}^{(2)} X\right)^2}{d_n^{2H}(4-2^{2H}) \sum_{k=1}^n X^2(t_{k-1}^n)} \xrightarrow{a.s.} \sigma^2.$$

Notice that

$$\frac{\sum_{k=1}^n \left(\Delta_{n,k}^{(2)} X\right)^2}{d_n^{2H}(4-2^{2H}) \sum_{k=1}^n X^2(t_{k-1}^n)} = \frac{d_n^{1-2H} \sum_{k=1}^n \left(\Delta_{n,k}^{(2)} X\right)^2}{d_n(4-2^{2H}) \sum_{k=1}^n X^2(t_{k-1}^n)}$$

and

$$d_n^{1-2H} \sum_{k=1}^n \left(\Delta_{n,k}^{(2)} X\right)^2 = \sigma^2 d_n^{1-2H} \sum_{k=1}^n X^2(t_{k-1}^n) \left(\Delta_{n,k}^{(2)} B^H\right)^2 + O_\omega\left(d_n^{H-3\varepsilon}\right). \tag{3.41}$$

In order to estimate (3.41), observe that (see Proposition 3.3)

$$\lim_{n\to\infty} d_n^{1-2H} \sum_{k=1}^n X^2(t_{k-1}^n) \left(\Delta_{n,k}^{(2)} B^H\right)^2 = (4-2^{2H}) \int_0^T X_t^2 \, ds$$

and

$$d_n \sum_{k=1}^n X^2(t_{k-1}^n) \xrightarrow{a.s.} \int_0^T X_t^2 \, dt.$$

Thus

$$\widetilde{\sigma}_n^2 \xrightarrow{a.s.} \sigma^2.$$

3.3.2.6 The Convergence Rate of $\widehat{H}_n^{(i)}$, $i = 1, 2, 3.$

Theorem 3.16 makes use of the conditions $\widehat{H}_n = H + O_\omega(\phi(n))$, $\phi(n) = o\left(\frac{1}{\log n}\right)$. Let us show that this indeed holds.
The convergence rate of $\widehat{H}_n^{(1)}$. Fix $\varepsilon \in \left(0, \frac{H-1/2}{3}\right)$. From Lemma 3.19 and the proof of Theorem 3.17 it follows that

$$\widehat{H}_n^{(1)} = \widetilde{H}_n + O_\omega\left(n^{-H+3\varepsilon}\right),$$

where

$$\widetilde{H}_n = \frac{1}{2} - \frac{1}{2\log 2}\log\left(\frac{V_{2n,T}^{(2)\widehat{B}^H}}{2^{2H}V_{n,T}^{(2)\widehat{B}^H}}\right) = H - \frac{1}{2\log 2}\log\left(\frac{(2n)^{-1}V_{2n,T}^{(2)\widehat{B}^H}}{n^{-1}V_{n,T}^{(2)\widehat{B}^H}}\right). \tag{3.42}$$

It suffices to consider the convergence rate of the logarithmic term in the Eq. (3.42). By Theorem 2.9 we get

$$\begin{aligned}\log\left(\frac{(2n)^{-1}V_{2n,T}^{(2)\widehat{B}^H}}{n^{-1}V_{n,T}^{(2)\widehat{B}^H}}\right) &= \log\left(\frac{1 + O_\omega((2n)^{-1/2}\log^{1/2}(2n))}{1 + O_\omega(n^{-1/2}\log^{1/2} n)}\right)\\ &= \log\left(1 + o_\omega(n^{-1/2}\log n)\right) = o_\omega\left(n^{-1/2}\log n\right).\end{aligned}$$

Then the statistic $\widetilde{H}_n$ has the convergence rate $o_\omega(n^{-1/2}\log n)$. Consequently, $\widehat{H}_n^{(1)}$ satisfies the required condition.
The convergence rate of $\widehat{H}_n^{(2)}$. Denote

$$S_{n,T} := \frac{2}{nk_n^{2H-1}}\sum_{k=1}^{n-1}\frac{\left(\Delta^{(2)}X_{t_k^n}\right)^2}{W_{n,k-1}}.$$

Then

$$\widehat{H}_n^{(2)} = H + \frac{\log S_{n,T}}{4\log n}.$$

Fix $\varepsilon \in \left(0, \frac{H-1/2}{7}\right)$. Proceeding along the lines of the proof of Theorem 3.12, it can be concluded that

$$S_{n,T} = \frac{n^{-1}\widetilde{V}_{n,T}^{(2)B^H} + O_\omega\left(n^{-(H-3\varepsilon)}\right)}{1 + O_\omega\left(\sqrt{\frac{\log n}{n^2}}\right) + O_\omega\left(\frac{1}{n^{H-7\varepsilon}}\right)}.$$

Then

$$S_{n,T} = \frac{1 + O_\omega\left(\sqrt{\frac{\log n}{n}}\right)}{1 + O_\omega\left(\frac{1}{n^{H-7\varepsilon}}\right)} = 1 + o_\omega\left(\frac{\log n}{\sqrt{n}}\right).$$

Consequently, $\widehat{H}_n^{(2)}$ satisfies the required condition.

The convergence rate of $\widehat{H}_n^{(3)}$ was obtained in the proof of Theorem 3.20.

3.3.2.7 Modeling of the Estimators and Numerical Simulations

In this subsection we describe the numerical simulations that were performed in order to compare the behavior of the estimators considered in Sects. 3.3.2.2 and 3.3.2.3.

The sample paths of a fractional Brownian motion, which were further used to construct the sample paths of the fractional Gompertz diffusion process, were simulated using the Wood-Chan circulant matrix embedding method [168]. The values of the constants involved in these simulations were, unless explicitly stated otherwise, $X_0 = 3, \alpha = 0.5, \beta = 2$, and $\sigma = 1.5$. We considered these sample paths on the unit interval, hence $T = 1$. In what follows we present the dependencies of the estimators both on the true parameter value and on the length of the sample path. We have also checked for possible dependencies of the estimators of the Hurst index and the variance on the values of the other parameters of the considered equation, namely the drift coefficients α and β and the initial condition X_0. No such dependencies of significant impact have been observed.

Modeling of the Hurst Index Estimators

Figures 3.6 and 3.7 display, respectively, the dependence of the four estimators of the Hurst index on the true parameter value and on the length of the sample path.

Modeling of the Variance Estimators

In order to calculate the estimators $\widehat{\sigma}^2_{1,n}$, $\widehat{\sigma}^2_{2,n}$ and $\widehat{\sigma}^2_{3,n}$ we need to supply them with the estimated values of the Hurst index. In the Figs. 3.8 and 3.9 presented below, the variance estimator $\widehat{\sigma}^2_{i,n}$, using the Hurst index estimator $\widehat{H}_n^{(j)}$, is denoted as 'si_hj', $i, j = 1, 2, 3$. The graphs present the relative differences, namely $(\widehat{\sigma}_{i,n} - \sigma)/\sigma$. It can be seen that the performance of all the considered estimators is roughly similar. Of the other estimators, it appears that using $\widehat{H}_n^{(3)}$ yields better numerical characteristics.

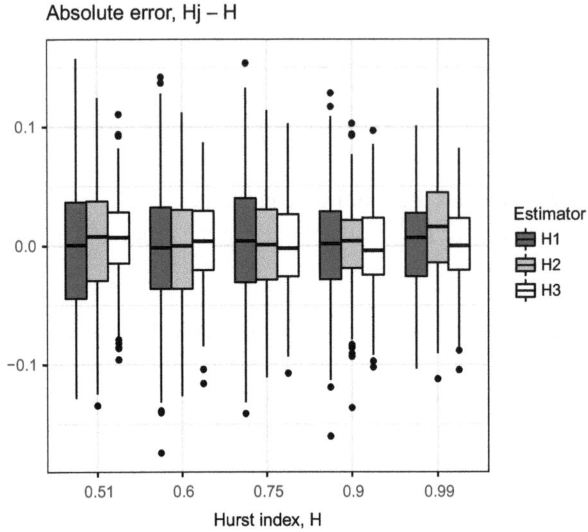

Fig. 3.6 Dependence of the absolute error on H

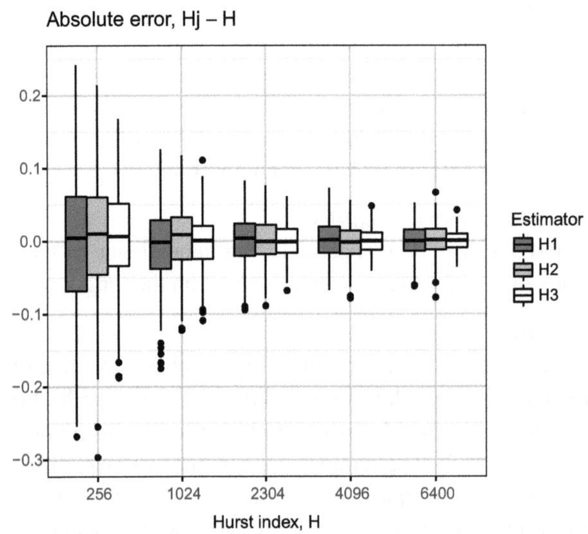

Fig. 3.7 Dependence of the absolute error on n

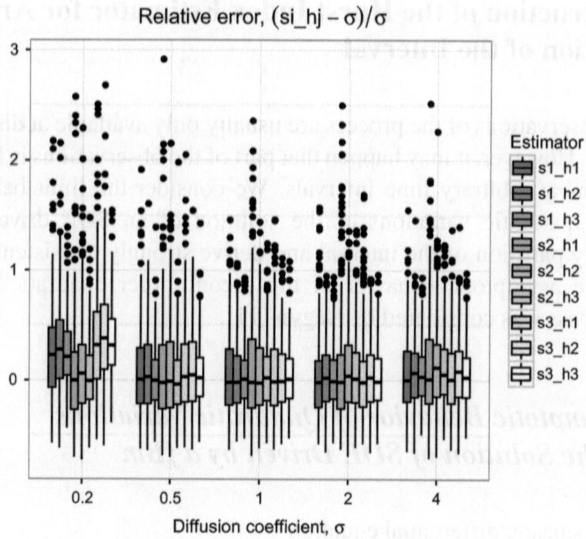

Fig. 3.8 Dependence of the relative error on σ

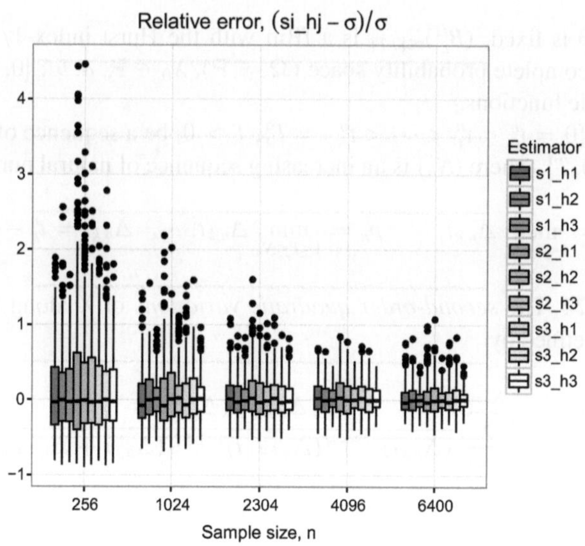

Fig. 3.9 Dependence of the relative error on n

3.4 Construction of the Hurst Index Estimator for Arbitrary Partition of the Interval

In practice, observations of the process are usually only available at discrete regular time intervals. However, it may happen that part of the observations is lost, resulting in observations at arbitrary time intervals. We consider the limit behavior of the second-order quadratic variations of the solution of an SDE driven by a fBm along arbitrary partition of the interval and derive strongly consistent Hurst index estimator. The asymptotic behavior of the second-order quadratic variations of Gaussian process was considered by Bégyn [11].

3.4.1 *Asymptotic Behavior of Quadratic Variations of the Solution of SDE Driven by a fBm*

Consider a stochastic differential equation

$$X_t = X_0 + \int_0^t a(s, X_s)\mathrm{d}s + \int_0^t b(s, X_s)\mathrm{d}B_s^H, \quad t \in [0, T], \tag{3.43}$$

where $T > 0$ is fixed, $(B_t^H)_{t\in[0,T]}$ is a fBm with the Hurst index $1/2 < H < 1$ defined on a complete probability space $(\Omega, \mathfrak{F}, \mathsf{P})$, $X_0 \in \mathbb{R}$, $a, b : [0, T] \times \mathbb{R} \to \mathbb{R}$ are measurable functions.

Let $\pi_n = \{0 = t_0^n < t_1^n < \cdots < t_{N_n}^n = T\}$, $T > 0$, be a sequence of partitions of the interval $[0, T]$, where (N_n) is an increasing sequence of natural numbers. Define

$$m_n = \max_{1\le k\le N_n} \Delta_{n,k}t, \qquad p_n = \min_{1\le k\le N_n} \Delta_{n,k}t, \qquad \Delta_{n,k}t = t_k^n - t_{k-1}^n.$$

Definition 3.24 The *second-order quadratic variations* of X along the partitions $(\pi_n)_{n\in\mathbb{N}}$ are defined by

$$V_{\pi_n,T}^{(2)X} = 2\sum_{k=1}^{N_n-1} \frac{\Delta_{n,k+1}t(\Delta_{\mathrm{ir},k}^{(2)n}X)^2}{(\Delta_{n,k}t)^{1/2+H}(\Delta_{n,k+1}t)^{1/2+H}(\Delta_{n,k}t + \Delta_{n,k+1}t)},$$

where

$$\Delta_{\mathrm{ir},k}^{(2)n}X = \Delta_{n,k}t\, X\big(t_{k+1}^n\big) + \Delta_{n,k+1}t\, X\big(t_{k-1}^n\big) - \big(\Delta_{n,k}t + \Delta_{n,k+1}t\big)X\big(t_k^n\big).$$

If $(\pi_n)_{n\in\mathbb{N}}$ is a sequence of uniform partitions then one has

$$V_{N_n,T}^{(2)X} = (T^{-1}N_n)^{2H-1}\sum_{k=1}^{N_n-1}\big(\Delta_{n,k}^{(2)}X\big)^2, \qquad \Delta_{n,k}^{(2)}X = X(t_{k+1}^n) - 2X(t_k^n) + X(t_{k-1}^n)\,.$$

To study the almost sure convergence of the second order quadratic variation of X we need additional assumptions on the sequence $(\pi_n)_{n\in\mathbb{N}}$.

Definition 3.25 (see [11]) Let $(\ell_k)_{k\geq 1}$ be a sequence of real numbers in the interval $(0,\infty)$. We say that $(\pi_n)_{n\in\mathbb{N}}$ is a *sequence of partitions* with asymptotic ratios $(\ell_k)_{k\geq 1}$ if it satisfies the following assumptions:

1. There exists $c \geq 1$ such that $m_n \leq cp_n$ for all n.
2. $\lim_{n\to\infty} \max_{1\leq k\leq N_n-1} \left|\frac{\Delta_k^n t}{\Delta_{k+1}^n t} - \ell_k\right| = 0$.

The set $\mathcal{L} = \{\ell_1; \ell_2; \ldots; \ell_k; \ldots\}$ will be called a *range of the asymptotic ratios* of the sequence $(\pi_n)_{n\in\mathbb{N}}$.

It is clear that if the sequence $(\pi_n)_{n\in\mathbb{N}}$ is regular, then it is a sequence with asymptotic ratios $\ell_k = 1$ for all $k \geq 1$.

Definition 3.26 (see [11]) A function $g : (0,\infty) \to \mathbb{R}$ is *invariant* on $\mathcal{L}$ if for all $\ell, \hat{\ell} \in \mathcal{L}$, $g(\ell) = g(\hat{\ell})$.

For example, let $\mathcal{L} = \{\alpha, \alpha^{-1}\}$ be a set containing two real positive numbers, and let

$$h(\lambda) = \frac{1 + \lambda^{2H-1} - (1+\lambda)^{2H-1}}{\lambda^{H-1/2}}.$$

The function h is invariant on $\mathcal{L}$.

Definition 3.27 A function $f : [a,b] \to \mathbb{R}$ is called a *step function* provided there are numbers $a = x_0 < x_1 < \cdots < x_n = b$ such that $f(x)$ is constant on each of the open intervals (x_{i-1}, x_i).

Definition 3.28 (see [45]) A function $f : [a,b] \to \mathbb{R}$ is called *regulated* provided there is a sequence $(f_n)_{n\geq 1}$ of step functions which converges uniformly to f.

Theorem 3.29 *Suppose that all conditions of Theorem 1.42 hold and let X be a solution of Eq.* (3.43)*. Let $(\pi_n)_{n\in\mathbb{N}}$ be a sequence of partitions with asymptotic ratios $(\ell_k)_{k\geq 1}$ and a range of the asymptotic ratios $\mathcal{L}$. Assume that the lower mesh of the subdivisions π_n satisfy $p_n = o(\log^{-1} n)$. If the function*

$$h(\lambda) = \frac{1 + \lambda^{2H-1} - (1+\lambda)^{2H-1}}{\lambda^{H-1/2}}$$

is invariant on $\mathcal{L}$ or if the sequence of step functions $\ell_n(t)$, i.e. $\ell_n(t) = \ell_k$ on (t_k^n, t_{k+1}^n), $0 \leq k \leq N_n - 1$ and $\ell_0 = \ell_1$, converges to a regulated function $\ell(t)$ on the interval $[0,T]$*, then*

$$V_{\pi_n,T}^{(2)X} \xrightarrow{a.s.} \begin{cases} 2h(\ell_1)\int_0^T b^2(t,X_t)\,\mathrm{d}t, & \textit{if } h \textit{ is invariant on } \mathcal{L}, \\ 2\int_0^s b^2(t,X_t)h(\ell(t))\,\mathrm{d}t, & \textit{if } \ell(s) \textit{ is a regulated function.} \end{cases}$$

Proof It is obvious that we can rewrite the square of the second-order difference in the following way:

$$\Delta^{(2)n}_{\mathrm{ir},k} X = \left(\Delta_{n,k} t \int_{t_k^n}^{t_{k+1}^n} a(t, X_t)\,\mathrm{d}t - \Delta_{n,k+1} t \int_{t_{k-1}^n}^{t_k^n} a(t, X_t)\,\mathrm{d}t \right) + \left(\Delta_{n,k} t \int_{t_k^n}^{t_{k+1}^n} b(t, X_t)\,\mathrm{d}B_t^H - \Delta_{n,k+1} t \int_{t_{k-1}^n}^{t_k^n} b(t, X_t)\,\mathrm{d}B_t^H \right) \quad (3.44)$$

Take $x = t_k^n$, $\varepsilon_1 = t_{k+1}^n - t_k^n$, $\varepsilon_2 = t_k^n - t_{k-1}^n$, $\gamma = H - \varepsilon$, $\varepsilon \in (0, (H - 1/2) \wedge \frac{\theta}{2})$, and apply Lemmas 3.1–3.2 to differences in the brackets, where the parameter θ was defined in Lemma 3.2. For the first term in (3.44) we get

$$\Delta_{n,k} t \int_{t_k^n}^{t_{k+1}^n} a(t, X_t)\,\mathrm{d}t - \Delta_{n,k+1} t \int_{t_{k-1}^n}^{t_k^n} a(t, X_t)\,\mathrm{d}t = O_\omega(m_n^{2+\theta}).$$

Similarly,

$$\begin{aligned} &\Delta_{n,k} t \int_{t_k^n}^{t_{k+1}^n} b(t, X_t)\,\mathrm{d}B_t^H - \Delta_{n,k+1} t \int_{t_{k-1}^n}^{t_k^n} b(t, X_t)\,\mathrm{d}B_t^H \\ &= \Delta_{n,k} t \big[b(t_k^n, X_{t_k^n}) \Delta^{(1)}_{n,k+1} B^H + O_\omega(m_n^{H-\varepsilon+\theta}) \big] \\ &\quad - \Delta_{n,k+1} t \big[b(t_k^n, X_{t_k^n}) \Delta^{(1)}_{n,k} B^H + O_\omega(m_n^{H-\varepsilon+\theta}) \big] \\ &= b(t_k^n, X_{t_k^n}) \Delta^{(2)n}_{\mathrm{ir},k} B^H + O_\omega(m_n^{1+H-\varepsilon+\theta}). \end{aligned}$$

Thus,

$$\Delta^{(2)n}_{\mathrm{ir},k} X = b(t_k^n, X_{t_k^n}) \Delta^{(2)n}_{\mathrm{ir},k} B^H + O_\omega(m_n^{1+H-\varepsilon+\theta}).$$

Set

$$\mu_k^n = (\Delta_{n,k} t)^{1/2+H} (\Delta_{n,k+1} t)^{1/2+H} (\Delta_{n,k} t + \Delta_{n,k+1} t).$$

Note that

$$\max_{1 \le k \le N_n - 1} \left| \Delta^{(2)n}_{\mathrm{ir},k} B^H \right| \le 2 m_n \max_{1 \le k \le N_n} \left| \Delta^{(1)}_{n,k} B^H \right| \le 2 m_n^{1+H-\varepsilon} G_{\varepsilon,T} \quad \text{and} \quad \frac{1}{\mu_k^n} \le \frac{1}{2 p_n^{2+2H}}.$$

Thus,

$$V^{(2)X}_{\pi_n,T} = \sum_{k=1}^{N_n - 1} \frac{\Delta_{n,k+1} t}{\mu_k^n} b^2(t_k^n, X_{t_k^n}) \big(\Delta^{(2)n}_{\mathrm{ir},k} B^H \big)^2 + O_\omega(m_n^{\theta - 2\varepsilon}).$$

Denote $r^n(t) = \max\{k: t_k^n < t\}$. Then

$$V_{\pi_n,t}^{(2)B^H} = \sum_{k=1}^{r^n(t)} \frac{\Delta_{n,k+1}t}{\mu_k^n} (\Delta_{\mathrm{ir},k}^{(2)n} B^H)^2$$

and

$$V_{\pi_n,T}^{(2)X} = \int_0^T b^2(t, X_t)\, \mathrm{d}V_{\pi_n,t}^{(2)B^H} + O_\omega(m_n^{\theta-2\varepsilon}). \tag{3.45}$$

Now we prove that

$$V_{\pi_n,t}^{(2)B^H} \xrightarrow{a.s.} \begin{cases} 2h(\ell_1)t, & \text{if } h \text{ is invariant on } \mathcal{L}, \\ 2\int_0^t h(\ell(u))\, du, & \text{if } \ell(t) \text{ is a regulated function.} \end{cases} \tag{3.46}$$

By definition we have

$$\begin{aligned} \mathsf{E}(\Delta_{\mathrm{ir},k}^{(2)n} B^H)^2 =& (\Delta_{n,k}t)^2(\Delta_{n,k+1}t)^{2H} + \Delta_{n,k}t\Delta_{n,k+1}t\big[(\Delta_{n,k+1}t)^{2H} \\ &- (\Delta_{n,k+1}t + \Delta_{n,k}t)^{2H} + (\Delta_{n,k}t)^{2H}\big] + (\Delta_{n,k+1}t)^2(\Delta_{n,k}t)^{2H} \\ =& \Delta_{n,k}t(\Delta_{n,k}t + \Delta_{n,k+1}t)(\Delta_{n,k+1}t)^{2H} + \Delta_{n,k+1}t(\Delta_{n,k}t \\ &+ \Delta_{n,k+1}t)(\Delta_{n,k}t)^{2H} - \Delta_{n,k}t\Delta_{n,k+1}t(\Delta_{n,k}t + \Delta_{n,k+1}t)^{2H}. \end{aligned}$$

Since

$$(\Delta_{n,k}t + \Delta_{n,k+1}t)(\Delta_{n,k}t)^{1/2+H}(\Delta_{n,k+1}t)^{1/2+H}\left(\frac{\Delta_{n,k}t(\Delta_{n,k+1}t)^{2H}}{(\Delta_{n,k}t)^{1/2+H}(\Delta_{n,k+1}t)^{1/2+H}}\right)$$

$$= \mu_k^n\left(\frac{\Delta_{n,k}t}{\Delta_{n,k+1}t}\right)^{1/2-H} = \mu_k^n(\ell_k^n)^{1/2-H},$$

$$(\Delta_{n,k}t + \Delta_{n,k+1}t)(\Delta_{n,k}t)^{1/2+H}(\Delta_{n,k+1}t)^{1/2+H}\left(\frac{(\Delta_{n,k}t)^{2H}\Delta_{n,k+1}t}{(\Delta_{n,k}t)^{1/2+H}(\Delta_{n,k+1}t)^{1/2+H}}\right)$$

$$= \mu_k^n\left(\frac{\Delta_{n,k}t}{\Delta_{n,k+1}t}\right)^{H-1/2} = \mu_k^n(\ell_k^n)^{H-1/2},$$

$$\begin{aligned} &(\Delta_{n,k}t + \Delta_{n,k+1}t)(\Delta_{n,k}t)^{1/2+H}(\Delta_{n,k+1}t)^{1/2+H} \\ &\times\left(\frac{\Delta_{n,k}t\Delta_{n,k+1}t(\Delta_{n,k}t + \Delta_{n,k+1}t)^{2H}}{(\Delta_{n,k}t)^{1/2+H}(\Delta_{n,k+1}t)^{1/2+H}(\Delta_{n,k}t + \Delta_{n,k+1}t)}\right) \\ &= \mu_k^n(\Delta_{n,k}t)^{1/2-H}(\Delta_{n,k+1}t)^{1/2-H}(\Delta_{n,k}t + \Delta_{n,k+1}t)^{2H-1} \\ &= \mu_k^n(\Delta_{n,k}t)^{1/2-H}(\Delta_{n,k+1}t)^{H-1/2}\left(\frac{\Delta_{n,k}t}{\Delta_{n,k+1}t} + 1\right)^{2H-1} \\ &= \mu_k^n(\ell_k^n)^{1/2-H}(\ell_k^n + 1)^{2H-1}, \end{aligned}$$

we have

$$\mathsf{E}(\Delta^{(2)n}_{\mathrm{ir},k}B^H)^2 = \mu^n_k\big[(\ell^n_k)^{1/2-\gamma} + (\ell^n_k)^{\gamma-1/2} - (\ell^n_k)^{1/2-\gamma}(1+\ell^n_k)^{2\gamma-1}\big] = \mu^n_k\, h(\ell^n_k),$$

where

$$\ell^n_k = \frac{\Delta_{n,k}t}{\Delta_{n,k+1}t}.$$

Thus,

$$\mathsf{E}V^{(2)}_{\pi_n}(B^H,2)_t = \sum_{k=1}^{r^n(t)} \Delta^n_{k+1}t \cdot h(\ell^n_k)$$

$$\longrightarrow \begin{cases} 2h(\ell_1)t, & \text{if } h \text{ is invariant on } \mathcal{L}, \\ 2\int_0^t h(\ell(u))\,du, & \text{if } \ell(t) \text{ is a regulated function.} \end{cases}$$

To complete the proof of (3.46), it suffices to apply simply modified proof of Theorem 6.12 and result from Sect. 6.2.1.1.

Since the function $b^2(t, X_t)$ is continuous, the statement of the theorem follows from the Helly–Bray theorem and (3.45)–(3.46). □

3.4.2 On the Estimation of the Hurst Index for Arbitrary Partition

Let $(\pi_n)_{n\geq 1}$ be a sequence of partitions of $[0,T]$ such that $0 = t^n_0 < t^n_1 < \cdots < t^n_{m(n)} = T$ for all $n \geq 1$. Assume that we have two sequences of partitions $(\pi_{i(n)})_{n\geq 1}$ and $(\pi_{j(n)})_{n\geq 1}$ of $[0,T]$ such that $\pi_{i(n)} \subset \pi_{j(n)} \subseteq \pi_n$ for $i(n) < j(n) \leq m(n)$ and for all $n \in \mathbb{N}$, where $\pi_{i(n)} = \{0 = t^n_0 < t^n_{i(1)} < t^n_{i(2)} < \cdots < t^n_{i(n)} = T\}$ and $\pi_{j(n)} = \{0 = t^n_0 < t^n_{j(1)} < t^n_{j(2)} < \cdots < t^n_{j(n)} = T\}$. Set

$$\Delta_{n,i(k)}t = t^n_{i(k)} - t^n_{i(k-1)}, \qquad m_{i(n)} = \max_{1\leq k\leq i(n)} \Delta_{n,i(k)}t, \qquad p_{i(n)} = \min_{1\leq k\leq i(n)} \Delta_{n,i(k)}t.$$

Moreover, assume that $p_{j(n)} \neq m_{i(n)}$ and $m_{i(n)} \leq c p_{i(n)}$, for all $i(n)$, $n \geq 1$, $c \geq 1$. Note that $p_{j(n)} \leq p_{i(n)}$.

Let X be a solution of Eq. (3.43). Set

$$V^{(2)X}_{\pi_{i(n)},T} = 2\sum_{k=1}^{i(n)-1} \frac{\Delta_{n,i(k+1)}t\big(\Delta^{(2)n}_{\mathrm{ir},i(k)}X\big)^2}{(\Delta_{n,i(k)}t)^{H+1/2}(\Delta_{n,i(k+1)}t)^{H+1/2}[\Delta_{n,i(k)}t + \Delta_{n_{i(k+1)}}t]},$$

where

$$\Delta^{(2)n}_{\mathrm{ir},i(k)}X = \Delta_{n,i(k)}t \cdot X(t^n_{i(k+1)}) + \Delta_{n,i(k+1)}t \cdot X(t^n_{i(k-1)}) - \left[\Delta_{n,i(k)}t + \Delta_{n,i(k+1)}t\right]X(t^n_{i(k)}).$$

Denote $\mu^n_{i(k)} = (\Delta_{n,i(k)}t)^{H+1/2}(\Delta_{n,i(k+1)}t)^{H+1/2}[\Delta_{n,i(k)}t + \Delta_{n_{i(k+1)}}t]$ and

$$V^{(2)X}_{i(n),T} = \sum_{k=1}^{i(n)-1} (\Delta^{(2)n}_{\mathrm{ir},i(k)}X)^2.$$

Define

$$\widehat{H}_n = -\frac{1}{2} + \frac{1}{2\log(p_{j(n)}/m_{i(n)})} \log \frac{V^{(2)X}_{j(n),T}}{V^{(2)X}_{i(n),T}}.$$

Theorem 3.30 *Assume that the conditions of Theorem 3.29 are satisfied for two sequences of partitions $(\pi_{i(n)})_{n\geq 1}$ and $(\pi_{j(n)})_{n\geq 1}$ of $[0,T]$ with the properties mentioned above. Then*

$$V^{(2)X}_{\pi_{k(n)},T} \xrightarrow{a.s.} \begin{cases} 2h(\ell_1)\int_0^T b^2(t,X_t)\,\mathrm{d}t, & \text{if } h \text{ is invariant on } \mathcal{L}, \\ 2\int_0^s b^2(t,X_t)h(\ell(t))\,\mathrm{d}t, & \text{if } \ell(s) \text{ is a regulated function.} \end{cases} \tag{3.47}$$

for $k(n) = i(n)$ and for $k(n) = j(n)$. If the sequences of partitions $\{\pi_{i(n)}\}$ and $\{\pi_{j(n)}\}$, $i(n) < j(n)$, are such that $p_{j(n)}/p_{i(n)} \to 0$ as $n \to \infty$, then

$$\widehat{H}_n \xrightarrow{a.s.} H.$$

Proof Theorem 3.29 yields the limit (3.47). It is evident that

$$\frac{1}{2m^{2H+1}_{i(n)}} \leq \frac{\Delta_{n,i(k+1)}t}{\mu^n_{i(k)}} \leq \frac{1}{2p^{2H+1}_{i(n)}}$$

and

$$\left(\frac{p_{i(n)}}{m_{j(n)}}\right)^{2H+1} \frac{V^{(2)X}_{j(n),T}}{V^{(2)X}_{i(n),T}} \leq \frac{V^{(2)X}_{\pi_{j(n)},T}}{V^{(2)X}_{\pi_{i(n)},T}} \leq \left(\frac{m_{i(n)}}{p_{j(n)}}\right)^{2H+1} \frac{V^{(2)X}_{j(n),T}}{V^{(2)X}_{i(n),T}}.$$

Next, since $\log(p_{j(n)}/m_{i(n)}) \leq 0$ and

$$\frac{m^{2H+1}_{i(n)} V^{(2)X}_{j(n),T}}{p^{2H+1}_{j(n),T} V^{(2)X}_{i(n),T}} \Big/ \frac{V^{(2)X}_{\pi_{j(n)},T}}{V^{(2)X}_{\pi_{i(n)},T}} \geq 1,$$

we have

$$\begin{aligned}
\widehat{H}_n =& -\frac{1}{2}+\frac{1}{2\log(p_{j(n)}/m_{i(n)})}\left((2H+1)\log(p_{j(n)}/m_{i(n)})+\log\frac{m_{i(n)}^{2H+1}V_{j(n),T}^{(2)X}}{p_{j(n)}^{2H+1}V_{i(n),T}^{(2)X}}\right)\\
=&H+\frac{1}{2\log(p_{j(n)}/m_{i(n)})}\log\frac{m_{i(n)}^{2H+1}V_{j(n),T}^{(2)X}}{p_{j(n)}^{2H+1}V_{i(n),T}^{(2)X}}\\
=&H+\frac{1}{2\log(p_{j(n)}/m_{i(n)})}\log\frac{V_{\pi_{j(n)},T}^{(2)X}}{V_{\pi_{i(n)},T}^{(2)X}}\\
&+\frac{1}{2\log(p_{j(n)}/m_{i(n)})}\log\left(\frac{m_{i(n)}^{2H+1}V_{j(n),T}^{(2)X}}{p_{j(n)}^{2H+1}V_{i(n),T}^{(2)X}}\Big/\frac{V_{\pi_{j(n)},T}^{(2)X}}{V_{\pi_{i(n)},T}^{(2)X}}\right)\\
\le&H+\frac{1}{2\log(p_{j(n)}/m_{i(n)})}\log\frac{V_{\pi_{j(n)},T}^{(2)X}}{V_{\pi_{i(n)},T}^{(2)X}}.
\end{aligned}$$

In the same way we get

$$\begin{aligned}
\widehat{H}_n =& -\frac{1}{2}+\frac{1}{2\log(p_{j(n)}/m_{i(n)})}\left((2H+1)\log(m_{j(n)}/p_{i(n)})+\log\frac{p_{i(n)}^{2H+1}V_{j(n),T}^{(2)X}}{m_{j(n)}^{2H+1}V_{i(n),T}^{(2)X}}\right)\\
=&-\frac{1}{2}+\left(H+\frac{1}{2}\right)\frac{\log(m_{j(n)}/p_{i(n)})}{\log(p_{j(n)}/m_{i(n)})}+\frac{1}{2\log(p_{j(n)}/m_{i(n)})}\log\frac{p_{i(n)}^{2H+1}V_{j(n),T}^{(2)X}}{m_{j(n)}^{2H+1}V_{i(n),T}^{(2)X}}\\
=&H+\left(H+\frac{1}{2}\right)\frac{\log(m_{j(n)}/p_{i(n)})-\log(p_{j(n)}/m_{i(n)})}{\log(p_{j(n)}/m_{i(n)})}\\
&+\frac{1}{2\log(p_{j(n)}/m_{i(n)})}\log\frac{V_{\pi_{j(n)},T}^{(2)X}}{V_{\pi_{i(n)},T}^{(2)X}}\\
&+\frac{1}{2\log(p_{j(n)}/m_{i(n)})}\log\left(\frac{p_{i(n)}^{2H+1}V_{j(n),T}^{(2)X}}{m_{j(n)}^{2H+1}V_{i(n),T}^{(2)X}}\Big/\frac{V_{\pi_{j(n)},T}^{(2)X}}{V_{\pi_{i(n)},T}^{(2)X}}\right)\\
\ge&H+\left(H+\frac{1}{2}\right)\frac{\log(m_{j(n)}/p_{i(n)})-\log(p_{j(n)}/m_{i(n)})}{\log(p_{j(n)}/m_{i(n)})}\\
&+\frac{1}{2\log(p_{j(n)}/m_{i(n)})}\log\frac{V_{\pi_{j(n)},T}^{(2)X}}{V_{\pi_{i(n)},T}^{(2)X}},
\end{aligned}\tag{3.48}$$

since

$$\frac{1}{2\log(p_{j(n)}/m_{i(n)})}\log\left(\frac{p_{i(n)}^{2H+1}V_{j(n),T}^{(2)X}}{m_{j(n)}^{2H+1}V_{i(n),T}^{(2)X}}\Big/\frac{V_{\pi_{j(n)},T}^{(2)X}}{V_{\pi_{i(n)},T}^{(2)X}}\right)\geq 0$$

and

$$\left(H+\frac{1}{2}\right)\frac{\log(m_{j(n)}/p_{i(n)})-\log(p_{j(n)}/m_{i(n)})}{\log(p_{j(n)}/m_{i(n)})}\leq 0.$$

If the sequences of partitions $\{\pi_{i(n)}\}$ and $\{\pi_{j(n)}\}$, $i(n) < j(n)$, are regular then the second term in the inequality (3.48) is equal to 0 and

$$|\widehat{H}_n - H| \leq \frac{1}{2\log(m_{i(n)}/p_{j(n)})}\left|\log\frac{V_{\pi_{j(n)},T}^{(2)X}}{V_{\pi_{i(n)},T}^{(2)X}}\right|.$$

Under conditions of the theorem in the regular case of partitions the statement of the theorem holds. For arbitrary partitions we obtain inequalities

$$\left|\widehat{H}_n - H - \frac{1}{2\log(p_{j(n)}/m_{i(n)})}\log\frac{V_{\pi_{j(n)},T}^{(2)X}}{V_{\pi_{i(n)},T}^{(2)X}}\right|$$
$$\leq\left(H+\frac{1}{2}\right)\frac{\log(m_{j(n)}/p_{j(n)})+\log(m_{i(n)}/p_{i(n)})}{\log(m_{i(n)}/p_{j(n)})}$$

and

$$|\widehat{H}_n - H| = \left|\widehat{H}_n - H - \frac{1}{2\log(p_{j(n)}/m_{i(n)})}\log\frac{V_{\pi_{j(n)},T}^{(2)X}}{V_{\pi_{i(n)},T}^{(2)X}}\right.$$
$$\left.+\frac{1}{2\log(p_{j(n)}/m_{i(n)})}\log\frac{V_{\pi_{j(n)},T}^{(2)X}}{V_{\pi_{i(n)},T}^{(2)X}}\right|$$
$$\leq\frac{3}{2}\frac{\log(m_{j(n)}/p_{j(n)})+\log(m_{i(n)}/p_{i(n)})}{\log(m_{i(n)}/p_{j(n)})}+\frac{1}{2\log(m_{i(n)}/p_{j(n)})}\left|\log\frac{V_{\pi_{j(n)},T}^{(2)X}}{V_{\pi_{i(n)},T}^{(2)X}}\right|.$$

For arbitrary partitions $\{\pi_{i(n)}\}$ and $\{\pi_{j(n)}\}$, $i(n) < j(n)$, the first term in above inequality goes to 0 as $\log(p_{i(n)}/p_{j(n)}) \to \infty$, $n \to \infty$. Thus the statement of the theorem holds. □

Chapter 4
Parameter Estimation in the Mixed Models via Power Variations

The purpose of this chapter is to develop parameter estimation for the models that, while being simple enough, already depart from the canons of self-similarity. They can take into account both the independence of process increments over short time intervals and the availability of memory at longer intervals. In fact, this is the simplest version of the multi-fractional Brownian motion. More precisely, we consider so called mixed fractional Brownian motion

$$M_t^H = aB_t^H + bW_t, \quad t \geq 0.$$

This process was introduced by P. Cheridito and studied from the probabilistic point of view, e.g., in [37]. As we understand, process M^H was introduced with the aim to consider models of financial markets that are simultaneously arbitrage-free and have a memory. The applications of mixed fBm to finance and network traffic as well as the existence-uniqueness problems for the SDE involving mixed fBm, were considered in many papers, see [3, 57, 120, 126]. It turned out that in these models, in a sense, the Wiener process predominates. For example, for $H > 3/4$ the mixed fBm is equivalent in measure to the Wiener process and is a semimartingale with respect to the natural filtration. However, the presence of fBm calls for the necessity of estimating the Hurst parameter and scaling parameters a and b. Taking into account numerous articles in which statistical inference for the Wiener process and the fBm, considered separately, is based on the asymptotic behavior of power variations (see, e.g., [136] and [138]), in this chapter we obtain results on both weak and almost sure asymptotic behavior of power variations of a linear combination of independent Wiener process and fBm. Theorems on weak convergence are based on the method of conditioning. Despite incomparably more complex calculations and estimates for variances and covariances, we succeeded not only in obtaining limit theorems, but also in calculating the exact values of numerical characteristics, using the fact that the processes are independent and Gaussian. These results are of independent interest. Then they are used to construct strongly consistent parameter estimators

© Springer International Publishing AG 2017

K. Kubilius et al., *Parameter Estimation in Fractional Diffusion Models*,
B&SS – Bocconi & Springer Series 8,
https://doi.org/10.1007/978-3-319-71030-3_4

in mixed models. More information concerning the model and the description of statistical results is provided in the Sect. 4.1, together with detailed references to the previous related papers. Section 4.2 contains weak and strong theorems on the asymptotic behavior of mixed power variations, together with the calculation of the numerical characteristics of the limit Gaussian and non-Gaussian distribution. The calculation is based on some combinatorial reasonings. The strong limit theorems are used in Sect. 4.4 to construct strongly consistent estimators of parameters H, a, b in model (4.1) and to study asymptotic normality of the estimators of H. Finally, in Sect. 4.5 we present simulation results to illustrate quality of the estimators provided.

4.1 Description of the Mixed Model and Mixed Power Variations

We consider the following stochastic process, that is called a *mixed fractional Brownian motion*,

$$M_t^H = aB_t^H + bW_t, \quad t \geq 0, \tag{4.1}$$

where a and b are some non-zero coefficients, W is a Wiener process, and B^H is a fBm with Hurst index $H \in (0, 1)$. Processes W and B^H are assumed to be independent.

The advantage of the model is that it combines the properties of both the memoryless process and process with memory. Moreover, due to its simple linear structure, we can investigate its properties for any $H \in (0, 1)$.

We consider statistical identification of model (4.1), i. e. the statistical estimation of the model parameters. The principal attention will be given to the estimation of H, though we will also present estimators for a and b. The estimators are partially based on both weak and almost sure asymptotic behavior of mixed power variations that are related to the components of the model.

Note that asymptotic behavior of power variations and, more generally, of non-linear transformations of stationary Gaussian sequences was studied in [32, 47, 61, 163], and statistical estimation for the fBm and multifractional processes with the help of power variations, in [3, 17, 40, 41, 62, 78, 86]. Weighted power variations serving similar purposes for stochastic differential equations driven by fBm, were studied in [135, 138].

There are only few papers considering the parameter estimation in the mixed model. In particular, [35] address the estimation of drift parameter in a model with mixed fBm. The article [57] proposes an estimation procedure for a, b and H, which is based on the empirical moments of M^H; the consistency properties are not investigated, only empirical studies for $H = 0.25$ and $H = 0.75$ are presented. The article [169] proposes maximum likelihood estimators of parameters in the mixed model based on the observations of the process at integer points, this is so called low-frequency data. In [2], the authors construct several estimators based on discrete variation. They also work in the low-frequency setting, which is essentially different from the high-frequency setting we consider. In both settings, the first order difference of the observed series is a stationary sequence. However, in the low-frequency setting the covariance does not depend on the number of observations, while in the high-frequency one, the covariance structure is very different. Mention that for $H > 1/2$, in a small scale a mixed fBm behaves like the Wiener process. Thus, the increments of Wiener process become more and more dominating as the partition becomes finer, which makes estimation of the Hurst parameter much harder in the case where $H > 1/2$.

Our main aim is the estimation of the parameters of the process (4.1) based on its single observation on a uniform partition of a fixed interval. As it was mentioned above, we use power variations of this process. We remark that, in contrast to the pure fractional case, there is no self-similarity property in the mixed model (4.1), so we cannot directly apply the results of [32, 47, 61, 163] concerning the asymptotic behavior of sums of transformed stationary Gaussian sequences. For this reason we need to study the asymptotic behavior as $n \to \infty$ of "mixed" power variations of the form

$$\sum_{i=0}^{n-1} \left(W_{(i+1)/n} - W_{i/n}\right)^p \left(B^H_{(i+1)/n} - B^H_{i/n}\right)^r,$$

involving increments of independent fBm B^H and Wiener process W, where $p \geq 0$, $r \geq 0$ are fixed integer parameters. For statistical purposes, in order to construct strongly consistent estimators, we mainly need the almost sure behavior of the power variations. However, we also study their weak behavior, which is of independent interest. In particular, the calculation of the numerical characteristics of the limit Gaussian distribution and distribution involving standard Rosenblatt random variable, is provided.

4.2 Exact Calculation and Asymptotic Behavior of the Moments of Higher Order of Mixed Power Variations

Let $W = \{W_t, t \geq 0\}$ be a standard Wiener process and $B^H = \{B^H_t, t \geq 0\}$ be an independent of W fBm with Hurst parameter $H \in (0, 1)$ defined on a complete probability space $(\Omega, \mathfrak{F}, \mathsf{P})$.

For a function $X\colon [0,1] \to \mathbb{R}$ and integers $n \geq 1$, we denote $\Delta_i^n X = X_{(i+1)/n} - X_{i/n}$, $i = 0, 1, \dots, n-1$. In this section we will study the asymptotic behavior as $n \to \infty$ of the following mixed power variations

$$\sum_{i=0}^{n-1} \left(\Delta_i^n W\right)^p \left(\Delta_i^n B^H\right)^r,$$

where $p \geq 0$, $r \geq 0$ are fixed integer numbers. Since $\Delta_i^n W$ and $\Delta_i^n B^H$ are centered Gaussian random variables with variances n^{-1} and n^{-2H} respectively, we get that

$$\mathsf{E}\left[\left(\Delta_i^n W\right)^p \left(\Delta_i^n B^H\right)^r\right] = n^{-rH-p/2} \mu_p \mu_r,$$

where for an integer $m \geq 1$

$$\mu_m = \mathsf{E}\left[\mathcal{N}(0,1)^m\right] = (m-1)!!\, \mathbf{1}_{m \text{ is even}}$$

is the mth moment of the standard Gaussian law; $(m-1)!! = (m-1)(m-3)\dots$ is the double factorial.

In view of this, we will study centered sums of the form

$$S_n^{H,p,r} = \sum_{i=0}^{n-1} \left(n^{rH+p/2} \left(\Delta_i^n W\right)^p \left(\Delta_i^n B^H\right)^r - \mu_p \mu_r\right).$$

Denote

$$\rho_H(m) = \mathsf{E}\left[B_1^H \left(B_{m+1}^H - B_m^H\right)\right] = \frac{1}{2}\left(|m+1|^{2H} + |m-1|^{2H} - 2\,|m|^{2H}\right) \tag{4.2}$$

the covariance of the so-called fractional Gaussian noise $\{B_{k+1}^H - B_k^H\}$. The following lemma contains an auxiliary result for calculating the characteristics of limit distributions.

Lemma 4.1 *If ξ, η are centered and jointly Gaussian with unit variance and covariance ρ, then for any $r \geq 1$,*

$$\mathsf{E}\left[\xi^r \eta^r\right] = \sum_{l=0}^{r/2} \frac{(r!)^2}{(2l)!((r-2l)!!)^2} \rho^{2l},$$

if r is even, and

$$\mathsf{E}\left[\xi^r \eta^r\right] = \sum_{l=0}^{(r-1)/2} \frac{(r!)^2}{(2l+1)!((r-2l-1)!!)^2} \rho^{2l+1},$$

if r is odd.

Proof We will use Isserlis' theorem, which says that for jointly Gaussian centered random variables $\xi_1, \dots, \xi_{2r}$,

$$\mathsf{E}[\xi_1 \dots \xi_{2r}] = \sum_{\pi} \prod_{\{i,j\} \in \pi} \mathsf{E}[\xi_i \xi_j],$$

where the sum is taken over partitions of $\{1, 2, \dots, 2r\}$ into doubletons, i.e. $\pi = \{\{i_1, j_1\}, \dots, \{i_r, j_r\}\}$ with $\{i_1, \dots, i_r, j_1, \dots, j_r\} = \{1, 2, \dots, 2r\}$.

Setting $\xi_k = \xi$, $k = 1, \dots, r$, $\xi_k = \eta$, $k = r + 1, \dots, 2r$, we get

$$\mathsf{E}[\xi^r \eta^r] = \sum_{m \geq 0} c(r, m) \rho^m,$$

where $c(r, m)$ is the number of partitions π such that there are m pairs $\{i, j\} \in \pi$ with $i \leq r, j \geq r + 1$.

Let r be even; for odd r the proof is similar. There are $\binom{r}{m}^2 \cdot m!$ ways to choose the m pairs $\{i, j\} \in \pi$ with $i \leq r, j \geq r + 1$. The remaining $r - m$ numbers in each group should be divided into pairs, which means that m should be even, say $m = 2l$, and there are $(r - m - 1)!!$ ways to divide $r - m$ numbers in each group into pairs. Therefore,

$$c(r, m) = \frac{(r!)^2}{(m!)^2((r - m)!)^2} \cdot m! \cdot ((r - m - 1)!!)^2 = \frac{(r!)^2}{m!((r - m)!!)^2}$$

for even m and $c(r, m) = 0$ for odd m, whence we get the required formula. □

Now we can control the limit behavior of the variance of $S_n^{H,p,r}$ for different values of H.

Proposition 4.2

1. *If p and r are even, $r \geq 2$, then*

   (i) *for $H \in (0, 3/4)$*

$$\mathsf{E}\left[\left(S_n^{H,p,r}\right)^2\right] \sim \left(\sigma_{H,r}^2 \mu_p^2 + \sigma_{p,r}^2\right) n, \ n \to \infty,$$

where

$$\sigma_{H,r}^2 = \sum_{l=1}^{r/2} \frac{(r!)^2}{(2l)!((r - 2l)!!)^2} \sum_{m=-\infty}^{\infty} \rho_H(m)^{2l}, \qquad \sigma_{p,r}^2 = \mu_{2r}\left(\mu_{2p} - \mu_p^2\right);$$

(ii) for $H = 3/4$

$$\mathsf{E}\left[\left(S_n^{3/4,p,r}\right)^2\right] \sim \sigma_{3/4,r}^2 \mu_p^2\, n \log n, \; n \to \infty,$$

where $\sigma_{3/4,r} = 3r(r-1)!!/8$;

(iii) for $H \in (3/4, 1)$

$$\mathsf{E}\left[\left(S_n^{H,p,r}\right)^2\right] \sim \sigma_{H,r}^2 \mu_p^2\, n^{4H-2}, \quad n \to \infty,$$

where $\sigma_{H,r}^2 = \frac{H^2(2H-1)r^2((r-1)!!)^2}{2(4H-3)}$.

2. *If* p *is odd and* $r \geq 1$ *is arbitrary, then for any* $H \in (0, 1)$

$$\mathsf{E}\left[\left(S_n^{H,p,r}\right)^2\right] = n\mu_{2p}\mu_{2r}.$$

3. *If* p *is even and* r *is odd, then*

(i) for $H \in (0, 1/2]$

$$\mathsf{E}\left[\left(S_n^{H,p,r}\right)^2\right] \sim n\left(\sigma_{H,r}^2 \mu_p^2 + \sigma_{p,r}^2\right), \quad n \to \infty,$$

where $\sigma_{H,1} = \mathbf{1}_{H=1/2}$,

$$\sigma_{H,r}^2 = \sum_{l=1}^{(r-1)/2} \frac{(r!)^2}{(2l+1)!((r-2l-1)!!)^2} \sum_{m=-\infty}^{\infty} \rho_H(m)^{2l+1} + (r!!)^2 \mathbf{1}_{H=\frac{1}{2}}, \; r \geq 3;$$

(ii) for $H \in (1/2, 1)$,

$$\mathsf{E}\left[\left(S_n^{H,p,r}\right)^2\right] \sim n^{2H} \mu_p^2 \mu_{r+1}^2, \quad n \to \infty.$$

Proof Denote $\eta_i = n^{rH+p/2}\left(\Delta_i^n W\right)^p \left(\Delta_i^n B^H\right)^r - \mu_p\mu_r$. In any case we have, thanks to independence of increments of W, the independence of B^H and W, and the self-similarity of B^H,

$$\mathsf{E}\left[\left(S_n^{H,p,r}\right)^2\right] = \sum_{i,j=0}^{n-1} \mathsf{E}\left[\eta_i\eta_j\right]$$

$$= \sum_{i,j=0}^{n-1} \left(n^{2rH+p}\mathsf{E}\left[(\Delta_i^n W)^p (\Delta_j^n W)^p\right]\mathsf{E}\left[(\Delta_i^n B^H)^r (\Delta_j^n B^H)^r\right] - \mu_p^2\mu_r^2\right)$$

$$= \sum_{i=0}^{n-1} \left(n^{2rH+p}\mathsf{E}\left[(\Delta_i^n W)^{2p}\right]\mathsf{E}\left[(\Delta_i^n B^H)^{2r}\right] - \mu_p^2\mu_r^2\right)$$

$$+ \sum_{\substack{i,j=0\\ i\neq j}}^{n-1} \left(n^{2rH+p}\mathsf{E}\left[(\Delta_i^n W)^p\right]\mathsf{E}\left[(\Delta_j^n W)^p\right]\mathsf{E}\left[(\Delta_i^n B^H)^r (\Delta_j^n B^H)^r\right] - \mu_p^2\mu_r^2\right)$$

$$= \sum_{i=0}^{n-1} \left(\mu_{2p}\mu_{2r} - \mu_p^2\mu_r^2\right) + \sum_{\substack{i,j=0\\ i\neq j}}^{n-1} \left(\mu_p^2\mathsf{E}\left[(\Delta_i B^H)^r (\Delta_j B^H)^r\right] - \mu_p^2\mu_r^2\right),$$

$$= n\left(\mu_{2p}\mu_{2r} - \mu_p^2\mu_r^2\right) + n\mu_p^2\left(\mu_r^2 - \mu_{2r}\right) + \mu_p^2 V_r(n) = n\sigma_{p,r}^2 + \mu_p^2 V_r(n), \tag{4.3}$$

where $\Delta_i B^H = B^H(i+1) - B^H(i)$, $i = 0, \ldots, n-1$,

$$V_r(n) = \sum_{i,j=0}^{n-1} \operatorname{cov}\left((\Delta_i B^H)^r, (\Delta_j B^H)^r\right).$$

1. Let r and p be even. Using Lemma 4.1, we have

$$V_r(n) = \sum_{i,j=0}^{n-1}\sum_{l=1}^{r/2} \frac{(r!)^2}{(2l)!((r-2l)!!)^2}\rho_H(i-j)^{2l}$$

$$= \sum_{l=1}^{r/2} \frac{(r!)^2}{(2l)!((r-2l)!!)^2} \sum_{j=0}^{n-1}\sum_{m=-j}^{n-1-j} \rho_H(m)^{2l}$$

$$= \sum_{l=1}^{r/2} \frac{(r!)^2}{(2l)!((r-2l)!!)^2} \sum_{m=-n+1}^{n-1} (n-|m|)\rho_H(m)^{2l}$$

$$=: \sum_{l=1}^{r/2} \frac{(r!)^2}{(2l)!((r-2l)!!)^2} D_{2l}(n).$$

Since $\rho_H(m) \sim H(2H-1)m^{2H-2}$, $m \to \infty$, the series $\sum_{m=-\infty}^{\infty} \rho_H(m)^{2l}$ converges for $H < 1 - 1/(4l)$, in this case, by the dominated convergence theorem

$$D_{2l}(n) = n \sum_{m=-n+1}^{n-1} \left(1 - \frac{|m|}{n}\right) \rho_H(m)^{2l} \sim n \sum_{m=-\infty}^{\infty} \rho_H(m)^{2l}, \ n \to \infty.$$

If $H > 1 - 1/(4l)$,

$$D_{2l}(n+1) - D_{2l}(n) = \sum_{m=-n}^{n} \rho_H(m)^{2l} \sim \frac{2H^{2l}(2H-1)^{2l}}{4l(H-1)+1} n^{4l(H-1)+1}, \quad n \to \infty,$$

so by the Stolz–Cesàro theorem,

$$D_{2l}(n) \sim \frac{H^{2l}(2H-1)^{2l}}{(2l(H-1)+1)(4l(H-1)+1)} n^{4l(H-1)+2}, \quad n \to \infty;$$

similarly, if $H = 1 - 1/(4l)$, then

$$D_{2l}(n) \sim 2H^{2l}(2H-1)^{2l} n \log n, \ n \to \infty.$$

Therefore, for $H < 3/4$,

$$V_r(n) \sim n \sum_{l=1}^{r/2} \sum_{m=-\infty}^{\infty} \rho_H(m)^{2l}, \ n \to \infty.$$

For $H \geq 3/4$ the first term of the sum dominates the other terms. Therefore,

$$V_r(n) \sim \frac{(r!)^2}{2((r-2)!!)^2} \cdot 2 \cdot \frac{9}{16} \cdot \frac{1}{4} \cdot n \log n = \sigma_{3/4,r}^2 n \log n, \ n \to \infty,$$

if $H = 3/4$, and

$$V_r(n) \sim \frac{(r!)^2}{2((r-2)!!)^2} \cdot \frac{H^2(2H-1)^2}{(2H-1)(4H-3)} n^{4H-2} = \sigma_{H,r}^2 n^{4H-2}, \ n \to \infty,$$

if $H > 3/4$. Combining these findings with (4.3) yields the first part of the statement.

2. If r is even, p is odd, then from (4.3)

$$\mathsf{E}\left[\left(S_n^{H,p,r}\right)^2\right] = n\sigma_{p,r}^2 = n\mu_{2p}\mu_{2r}.$$

3. Similarly to the first case,

$$V_r(n) = \sum_{l=0}^{(r-1)/2} \frac{(r!)^2}{(2l+1)!((r-2l-1)!!)^2} D_{2l+1}(n),$$

where

$$D_{2l+1}(n) = \sum_{m=-n+1}^{n-1} (n-|m|)\rho_H(m)^{2l+1}.$$

Observe that $D_1(n) = \mathsf{E}\left[\left(\sum_{i=0}^{n-1} \Delta_i B^H\right)^2\right] = \mathsf{E}\left[B^H(n)^2\right] = n^{2H}$.

For $l \geq 1$, the series $\sum_{m=-\infty}^{\infty} \rho_H(m)^{2l+1}$ converges for $H < 1 - 1/(2l+1)$ and, similarly to the first case,

$$D_{2l+1}(n) \sim n \sum_{m=-\infty}^{\infty} \rho_H(m)^{2l+1}$$

for $H < 1 - 1/(2l+1)$,

$$D_{2l+1}(n) \sim \frac{H^{2l+1}(2H-1)^{2l+1}}{((2l+1)(H-1)+1)(2(2l+1)(H-1)+1)} n^{2(2l+1)(H-1)+2}, \ n \to \infty,$$

for $H > 1 - 1/(2l+1)$, and

$$D_{2l+1}(n) \sim 2H^{2l+1}(2H-1)^{2l+1} n \log n, \ n \to \infty,$$

for $H = 1 - 1/(2l+1)$. As a result, we have

$$V_r(n) \sim \frac{(r!)^2}{((r-1)!!)^2} n^{2H} + n \sum_{l=1}^{(r-1)/2} \frac{(r!)^2}{(2l+1)!((r-2l-1)!!)^2} \sum_{m=-\infty}^{\infty} \rho_H(m)^{2l+1}$$

$$\sim \sigma_{H,r}^2 n, \ n \to \infty,$$

for $H \leq 1/2$, and

$$V_r(n) \sim \frac{(r!)^2}{((r-1)!!)^2} n^{2H} = (r!!)^2 n^{2H} = \mu_{r+1}^2 n^{2H}, n \to \infty,$$

for $H > 1/2$. Combining this with (4.3), we finish the proof. □

4.3 Weak and Strong Limit Theorems for the Centered and Normalized Mixed Power Variations

The following theorem summarizes the weak limit behavior of $S_n^{H,p,r}$.

Theorem 4.3 *We have the following weak convergence, with the variances defined in Proposition 4.2.*

1. *If p and r are even, $r \geq 2$, then*
   (i) *for $H \in (0, 3/4)$*

$$n^{-1/2} S_n^{H,p,r} \xrightarrow{d} \mathcal{N}(0, \sigma_{H,r}^2 \mu_p^2 + \sigma_{p,r}^2), \ n \to \infty; \tag{4.4}$$

   (ii) *for $H = 3/4$*

$$\frac{S_n^{3/4,p,r}}{\sqrt{n \log n}} \xrightarrow{d} \mathcal{N}(0, \sigma_{3/4,r}^2 \mu_p^2), \ n \to \infty; \tag{4.5}$$

   (iii) *for $H \in (3/4, 1)$*

$$n^{1-2H} S_n^{H,p,r} \xrightarrow{d} \sigma_{H,r} \mu_p \zeta_{2H-1}, \quad n \to \infty, \tag{4.6}$$

   where ζ_{2H-1} is the standard Rosenblatt random variable with Hurst parameter $2H - 1$, defined and described in Sect. B.2.1.

2. *If p is odd and $r \geq 1$ is arbitrary, then for any $H \in (0, 1)$*

$$n^{-1/2} S_n^{H,p,r} \xrightarrow{d} \mathcal{N}(0, \mu_{2p} \mu_{2r}). \tag{4.7}$$

3. *If p is even and r is odd, then*
   (i) *for $H \in (0, 1/2]$*

$$n^{-1/2} S_n^{H,p,r} \xrightarrow{d} \mathcal{N}\left(0, \sigma_{H,r}^2 \mu_p^2 + \sigma_{p,r}^2\right), \quad n \to \infty; \tag{4.8}$$

   (ii) *for $H \in (1/2, 1)$*

$$n^{-H} S_n^{H,p,r} \xrightarrow{d} \mathcal{N}\left(0, \mu_p^2 \mu_{r+1}^2\right), \quad n \to \infty. \tag{4.9}$$

Remark 4.4 For $r = 0$ we have the pure Wiener case, so for any $H \in (0, 1)$

$$n^{-1/2} S_n^{H,p,r} \xrightarrow{d} \mathcal{N}\left(0, \mu_{2p} - \mu_p^2\right), \quad n \to \infty.$$

Also note that in the case $p = 0$, $r = 1$ the limit variance in (4.8) vanishes. Obviously, in this case

$$n^{-H} S_n^{H,0,1} = B_1^H,$$

so it has the standard normal distribution.

Proof We study different cases in the same order as they appear in the formulation.

1) Assume first that p and $r \geq 2$ are even. The principal idea in this case is to rewrite mixed power variation as

$$S_n^{H,p,r} = S_n' + S_n'',$$

where

$$S_n' = n^{rH} \sum_{i=0}^{n-1} \left(\Delta_i^n B^H\right)^r \left(n^{p/2} \left(\Delta_i^n W\right)^p - \mu_p\right),$$

$$S_n'' = \mu_p \sum_{i=0}^{n-1} \left(n^{rH} \left(\Delta_i^n B^H\right)^r - \mu_r\right).$$

Then we apply known results concerning the asymptotic behavior of S_n'', since it contains only an fBm, and consider S_n' conditionally on the fBm. Further we realize this idea.

(i) For $H \in (0, 3/4)$, write

$$n^{-1/2} S_n^{H,p,r} = A_n' + A_n''$$

with $A_n' = n^{-1/2} S_n'$, $A_n'' = n^{-1/2} S_n''$. Thanks to Proposition 4.2, for r even, $H \in (0, 3/4)$, the assumptions of Theorem B.8 are satisfied, so

$$n^{-1/2} \sum_{i=0}^{n-1} \left(n^{rH} \left(\Delta_i^n B^H\right)^r - \mu_r\right) \xrightarrow{d} \mathcal{N}\left(0, \sigma_{H,r}^2\right), \tag{4.10}$$

as $n \to \infty$. Consequently, $A_n'' \xrightarrow{d} \mathcal{N}\left(0, \sigma_{H,r}^2 \mu_p^2\right)$, $n \to \infty$. Further,

$$A_n' = n^{rH-1/2} R_n k_n,$$

where

$$R_n = \frac{1}{k_n} \sum_{i=0}^{n-1} \lambda_{i,n} \left(n^{p/2} \left(\Delta_i^n W\right)^p - \mu_p\right),$$

$$k_n = \left(\left(\mu_{2p} - \mu_p^2\right) \sum_{i=0}^{n-1} \lambda_{i,n}^2\right)^{1/2}, \quad \lambda_{i,n} = \left(\Delta_i^n B^H\right)^r.$$

Since B^H is uniformly continuous, $\max_{1\le k\le n}\lambda_{k,n}\to 0$, $n\to\infty$. Thus, taking into account the independence of B^H and W and an evident fact that $\operatorname{var} R_n = 1$, we get by the CLT that the conditional distribution of R_n given B^H converges to the standard normal distribution as $n\to\infty$. Further, from the ergodic theorem (Theorem B.1)

$$n^{2rH-1}\sum_{i=0}^{n-1}\left(\Delta_i^n B^H\right)^{2r}\to\mu_{2r},\quad n\to\infty,$$

hence

$$n^{rH-1/2}k_n\to\left(\mu_{2r}\left(\mu_{2p}-\mu_p^2\right)\right)^{1/2},\quad n\to\infty.$$

By Slutsky's theorem in the form of Theorem B.5, the conditional distribution of A_n' given B^H converges to $\mathcal{N}\left(0,\sigma_{p,r}^2\right)$, that is, for any $\lambda\in\mathbb{R}$ we have

$$\int_{\mathbb{R}} e^{i\lambda x}\mathsf{P}\left(A_n'\in dx\mid B^H\right)\to e^{-\lambda^2\sigma_{p,r}^2/2} \tag{4.11}$$

as $n\to\infty$. Now write

$$\mathsf{E}\left[e^{i\lambda(A_n'+A_n'')}\right]=\mathsf{E}\left[\mathsf{E}\left[e^{i\lambda(A_n'+A_n'')}\middle|B^H\right]\right]=\mathsf{E}\left[\int_{\mathbb{R}} e^{i\lambda x}\mathsf{P}\left(A_n'\in dx\middle|B^H\right)e^{i\lambda A_n''}\right],$$

whence

$$\left|\mathsf{E}\left[e^{i\lambda(A_n'+A_n'')}\right]-e^{-\lambda^2(\sigma_{H,r}^2\mu_p^2+\sigma_{p,r}^2)/2}\right|\le E_1+E_2,$$

where

$$E_1=\left|\mathsf{E}\left[\left(\int_{\mathbb{R}} e^{i\lambda x}\mathsf{P}\left(A_n'\in dx\middle|B^H\right)-e^{-\lambda^2\sigma_{p,r}^2/2}\right)e^{i\lambda A_n''}\right]\right|\to 0,\quad n\to\infty$$

by (4.11) and dominated convergence;

$$E_2=e^{-\lambda^2\sigma_{p,r}^2/2}\left|\mathsf{E}\left[e^{i\lambda A_n''}\right]-e^{-\lambda^2\sigma_{H,r}^2\mu_p^2/2}\right|\to 0,\quad n\to\infty$$

by (4.10). It follows that

$$n^{-1/2}S_n^{H,p,r}\xrightarrow{d}\mathcal{N}\left(0,\sigma_{H,r}^2\mu_p^2+\sigma_{p,r}^2\right),\quad n\to\infty,$$

as required in this case.

(ii) In the case where $H=3/4$, thanks to the proof of Proposition 4.2 we have

$$\sum_{m=-n}^{n}\rho_H(n)^2\sim\frac{9}{64}n\log n,\ n\to\infty,$$

and

$$\sum_{m=-n}^{n} \rho_H(n)^{2l} = O(n), \ n \to \infty$$

for $l \geq 2$. Then, by Theorem B.9,

$$\frac{1}{\sqrt{n \log n}} \sum_{i=0}^{n-1} \left(n^{rH} \left(\Delta_i^n B^H\right)^r - \mu_r\right) \xrightarrow{d} \mathcal{N}\left(0, \sigma_{3/4,r}^2\right),$$

as $n \to \infty$, whence (4.5) can be deduced using the same reasoning as above.

(iii) For $H \in (3/4, 1)$, write

$$n^{1-2H} S_n^{H,p,r} = n^{1-2H} S_n' + n^{1-2H} S_n'' = n^{1-2H+rH} R_n k_n + n^{1-2H} S_n'',$$

where R_n, k_n, S_n', S_n'' are defined above. As before, $R_n \xrightarrow{d} \mathcal{N}(0,1)$ conditionally given B^H as $n \to \infty$. However, this time $n^{1-2H+rH} k_n \to 0$, $n \to \infty$, since $n^{rH-1/2} k_n$ has a finite limit and $n^{3/2-2H} \to 0$, $n \to \infty$. Therefore, $n^{1-2H} S_n' \to 0$, $n \to \infty$. Further, the function $G(x) = x^r - \mu_r$ has the Hermite rank 2 (see Sect. B.1.4). Indeed, for a standard Gaussian ζ, $\mathsf{E}\left[G(\zeta)\right] = 0$, $\mathsf{E}\left[\zeta G(\zeta)\right] = \mu_{r+1} = 0$, and $\mathsf{E}\left[(\zeta^2 - 1) G(\zeta)\right] = \mathsf{E}\left[\zeta^2 G(\zeta)\right] = \mu_{r+2} - \mu_r = r(r-1)!!$. Therefore, by Theorem B.10,

$$n^{1-2H} S_n'' \xrightarrow{d} \sigma_{H,r} \mu_p \zeta_{2H-1}, \ n \to \infty,$$

where ζ_{2H-1} is the standard Rosenblatt random variable considered in Sect. B.2.1. Thus, we get (4.6) using Theorem B.5. This finishes the case where p and r are even.

2), 3) Now assume that p or r is odd. In this case $S_n^{H,p,r}$ has the form

$$S_n^{H,p,r} = \sum_{k=0}^{n-1} n^{rH+p/2} \left(\Delta_k^n W\right)^p \left(\Delta_k^n B^H\right)^r.$$

Write

$$n^{-1/2} S_n^{H,p,r} = n^{rH-1/2} R_n k_n + Z_n,$$

where R_n and k_n are defined above, $Z_n = n^{rH-1/2} \mu_p \sum_{k=0}^{n-1} \left(\Delta_k^n B^H\right)^r$. As before, given B^H, $n^{rH-1/2} R_n k_n \xrightarrow{d} \mathcal{N}\left(0, \sigma_{p,r}^2\right)$, $n \to \infty$.

Now if p is odd, we have $Z_n = 0$ irrespective of the value of H, whence (4.7) immediately follows.

Further, assume that p is even and $r \geq 3$ is odd. For $H \in (0, 1/2]$, similarly to above, it is established in Theorem B.8 that

$$n^{rH-1/2} \sum_{k=0}^{n-1} \left(\Delta_k^n B^H\right)^r \xrightarrow{d} \mathcal{N}\left(0, \sigma_{H,r}^2\right), \quad n \to \infty.$$

Therefore, $Z_n \xrightarrow{d} \mathcal{N}\left(0, \mu_p^2 \sigma_{H,r}^2\right)$, $n \to \infty$. For $r = 1$

$$n^{rH-1/2} \sum_{k=0}^{n-1} \left(\Delta_k^n B^H\right)^r = n^{H-1/2} B_1^H \to 0, \quad n \to \infty.$$

Arguing as in deriving of (4.4), we get (4.8). For $H \in (1/2, 1)$, in order to apply Theorem B.10, note that the Hermite rank of $G(x) = x^r$ equals 1. Indeed, if ζ is a standard Gaussian random variable, then $\mathsf{E}\left[G(\zeta)\right] = 0$ and $\mathsf{E}\left[\zeta G(\zeta)\right] = \mathsf{E}\left[\zeta^{r+1}\right] = \mu_{r+1}$. Furthermore, as it was shown in Proposition 4.2, the covariance function ρ_H satisfies the assumptions of Theorem B.10. Therefore,

$$n^{(r-1)H} \sum_{k=0}^{n-1} \left(\Delta_k^n B^H\right)^r \xrightarrow{d} \mathcal{N}\left(0, \mu_{r+1}^2\right), \quad n \to \infty,$$

so $n^{1/2-H} Z_n \xrightarrow{d} \mathcal{N}\left(0, \mu_p^2 \mu_{r+1}^2\right)$, $n \to \infty$. But

$$n^{(r-1)H} R_n k_n = n^{1/2-H} n^{rH-1/2} R_n k_n \to 0, \quad n \to \infty,$$

whence (4.9) follows.

The proof is now complete. □

The next result explains the almost sure behavior of $S_n^{H,p,r}$.

Theorem 4.5 *Let $\varepsilon > 0$ be arbitrary. Then a. s.:*

1. *If $r = 0$, then $S_n^{H,p,r} = o(n^{1/2+\varepsilon})$, $n \to \infty$.*
2. *If p and $r \geq 2$ are even, then*
   (i) *for $H \in (0, 3/4]$ $S_n^{H,p,r} = o(n^{1/2+\varepsilon})$, $n \to \infty$.*
   (ii) *for $H \in (3/4, 1)$ $S_n^{H,p,r} = o(n^{2H-1+\varepsilon})$, $n \to \infty$.*
3. *If p is odd and $r \geq 1$ is arbitrary, then for any $H \in (0, 1)$ $S_n^{H,p,r} = o(n^{1/2+\varepsilon})$, $n \to \infty$.*
4. *If p is even and r is odd, then*
   (i) *for $H \in (0, 1/2]$ $S_n^{H,p,r} = o(n^{1/2+\varepsilon})$, $n \to \infty$.*
   (ii) *for $H \in (1/2, 1)$ $S_n^{H,p,r} = o(n^{H+\varepsilon})$, $n \to \infty$.*

In particular, for any $H \in (0, 1)$ the following version of the ergodic theorem takes place: $n^{-1} S_n^{H,p,r} \to 0$ a. s., $n \to \infty$.

Proof We shall provide the proof only for the case when p and r are even, and $H \in (0, 3/4)$, in other cases the argument is similar. Abbreviate $Q_n = n^{-1/2} S_n^{H,p,r}$. We need to show that $Q_n = o(n^\varepsilon)$, $n \to \infty$. By Proposition 4.2, $\mathsf{E}\left[Q_n^2 \right]$ has a finite limit. It follows that $\sup_{n\geq 1} \mathsf{E}\left[Q_n^2 \right] < \infty$.

Now consider $\mathcal{H} = L^2([0,1]) \times L_H^2([0,1])$ with

$$\langle (f_1, g_1), (f_2, g_2) \rangle_{\mathcal{H}} = \langle f_1, f_2 \rangle_{L^2([0,1])} + \langle g_1, g_2 \rangle_H$$

and $\widetilde{W}(f,g) = \int_0^1 f(s)dW(s) + \int_0^1 g(s)dB^H(s)$. This is an isonormal Gaussian process in $\mathcal{H}$, and it is evident that Q_n can be represented as a combination of multiple stochastic integrals of orders between 1 and $p + r$ with respect to $\widetilde{W}$ (see Appendix B.2). Then, by Corollary B.18, for any integer $l \geq 1$, there exists a constant C_l such that for all $n \geq 1$ $\mathsf{E}\left[Q_n^{2l} \right] \leq C_l \left(\mathsf{E}\left[Q_n^2 \right] \right)^l$.

Now take any integer $l \geq \varepsilon^{-1}$ and write

$$\begin{aligned}\mathsf{E}\left[\sum_{n=1}^{\infty} \frac{Q_n^{2l}}{n^2} \right] &= \sum_{n=1}^{\infty} \mathsf{E}\left[\frac{Q_n^{2l}}{n^2} \right] \leq C_l \sum_{n=1}^{\infty} \frac{\left(\mathsf{E}\left[Q_n^2 \right]\right)^l}{n^2} \\ &\leq C_l \left(\sup_{n\geq 1} \mathsf{E}\left[Q_n^2 \right] \right)^l \sum_{n=1}^{\infty} \frac{1}{n^2} < \infty.\end{aligned}$$

Therefore, the series $\sum_{n=1}^{\infty} Q_n^{2l}/n^2$ converges a. s.; in particular, $Q_n = o(n^{1/l})$, as $n \to \infty$, whence the statement follows. □

4.4 Statistical Estimation in Mixed Model

Now we turn to the question of parametric estimation in the mixed model

$$M_t^H = aB_t^H + bW_t, \quad t \in [0, T], \tag{4.12}$$

where a, b are non-zero numbers, which we assume to be positive, without loss of generality. Our primary goal is to construct a strongly consistent estimator for the Hurst parameter H, given a single observation of M^H.

Basing on the method of relative entropies, it was established in [37] that for $H \in (3/4, 1)$ the measure induced by M^H in $C[0, T]$ is equivalent to that of bW. Therefore, the property of almost sure convergence in this case is independent of H. Consequently, no strongly consistent estimator for $H \in (3/4, 1)$ based on a single observation of M^H exists.

In this section we denote $\Delta_i^n X = X_{T(i+1)/n} - X_{Ti/n}$ and

$$V_n^{H,p,r} = \sum_{i=0}^{n-1} \left(\Delta_i^n W\right)^p \left(\Delta_i^n B^H\right)^r.$$

4.4.1 Statistical Estimation Based on Quadratic Variation

Consider the quadratic variation of M^H, i. e.

$$V_n^{H,2} := \sum_{i=0}^{n-1} \left(\Delta_i^n M^H\right)^2 = a^2 V_n^{H,0,2} + 2ab V_n^{H,1,1} + b^2 V_n^{H,2,0}.$$

Note that $V_n^{H,2}$ depends only on the observed process but not on H. We use this notation to specify the distribution. Namely, we will use it to refer to the limit behavior of the quadratic variation for a specified value of the Hurst parameter H.

By Theorem 4.5, we have that

$$V_n^{H,0,2} \sim T^{2H} n^{1-2H}, \quad V_n^{H,2,0} \to T, \quad V_n^{H,1,1} = o(n^{1/2-H}), \quad n \to \infty.$$

Therefore, the asymptotic behavior of $V_n^{H,2}$ depends on whether $H < 1/2$ or not. Precisely, for $H \in (0, 1/2)$,

$$V_n^{H,2} \sim a^2 T^{2H} n^{1-2H}, \quad n \to \infty, \tag{4.13}$$

so the quadratic variation behaves similarly to that of a scaled fBm.

For $H \in (1/2, 1)$,

$$V_n^{H,2} \to b^2 T, \quad n \to \infty, \tag{4.14}$$

so the quadratic variation behaves similarly to that of a scaled Wiener process.

Let us consider the cases $H < 1/2$ and $H > 1/2$ individually in more detail.

4.4.1.1 $H \in (0, 1/2)$

We have seen above that this case is similar to the pure fBm case. Unsurprisingly, the same estimators work, which is precisely stated below.

Theorem 4.6 *For $H \in (0, 1/2)$, the following statistics*

$$\widehat{H}_k = \frac{1}{2}\left(1 - \frac{1}{k}\log_2 V_{2^k}^{H,2}\right)$$

and

$$\widetilde{H}_k = \frac{1}{2}\left(\log_2 \frac{V_{2^{k-1}}^{H,2}}{V_{2^k}^{H,2}} + 1\right)$$

are strongly consistent estimators of the Hurst parameter H.

Proof Write

$$\log_2 V_{2^k}^{H,2} = \log_2\left(a^2 T^{2H} 2^{(1-2H)k}\right) + \log_2\left(1 + \frac{b^2}{a^2} T^{1-2H} 2^{-(1-2H)k} + \zeta_k\right),$$

where

$$\zeta_k = \frac{a^2\left(V_{2^k}^{H,0,2} - T^{2H}2^{(1-2H)k}\right) + b^2\left(V_{2^k}^{H,2,0} - T\right) + 2abV_{2^k}^{H,1,1}}{a^2 T^{2H} 2^{(1-2H)k}}.$$

It follows from Theorem 4.5 that for any $\varepsilon > 0$

$$\zeta_k = o\left(2^{(-1/2+\varepsilon)k}\right) + o\left(2^{(2H-3/2+\varepsilon)k}\right) + o\left(2^{(H-1+\varepsilon)k}\right) = o\left(2^{(-1/2+\varepsilon)k}\right),\ k \to \infty.$$

Hence we have

$$\begin{aligned}\log_2 V_{2^k}^{H,2} &= 2\log_2 a + 2H\log_2 T + (1-2H)k \\ &\quad + O\left(2^{(2H-1)k}\right) + o\left(2^{(-1/2+\varepsilon)k}\right), \quad k \to \infty.\end{aligned} \tag{4.15}$$

In particular,

$$\log_2 V_{2^k}^{H,2} \sim (1-2H)k, \quad k \to \infty,$$

whence the result immediately follows. □

Remark 4.7 At first sight, there is no clear advantage of $\widehat{H}_k$ or $\widetilde{H}_k$. But a careful analysis shows that $\widetilde{H}_k$ is better. Indeed, from (4.15) it is easy to see that

$$\widehat{H}_k = H - \frac{\log_2 a + H\log_2 T}{k} + o\left(k^{-1}\right), \quad k \to \infty, \tag{4.16}$$

while

$$\widetilde{H}_k = H + O\left(2^{k(2H-1)}\right) + o\left(2^{k(-1/2+\varepsilon)}\right), \quad k \to \infty. \tag{4.17}$$

Now it is absolutely clear that $\widetilde{H}_k$ performs much better (unless one hits the jackpot by having $aT^H = 1$).

Now we turn to the question of asymptotic normality of the estimators. Note that in the purely fractional case, the estimator $\widetilde{H}_k$ is asymptotically normal for all $H \in (0, 3/4)$. In the mixed case, the asymptotic normality ends at $H = 1/4$.

Proposition 4.8 *For $H \in (0, 1/4)$,*

$$2^{k/2}\left(\widetilde{H}_k - H\right) \xrightarrow{d} \mathcal{N}\left(0, (\sigma'_H)^2\right), \quad k \to \infty,$$

where

$$\sigma'_H = \frac{1}{\sqrt{2}\log 2}\left(\rho'_{H,0} + 2\sum_{m=1}^{\infty}\rho'_{H,m}\right)^{1/2},$$

$$\rho'_{H,m} = \mathsf{E}\Bigg[\left(\left(B_1^H\right)^2 - 2^{2H-1}\left(B_{1/2}^H\right)^2 - 2^{2H-1}\left(B_1^H - B_{1/2}^H\right)^2\right)$$
$$\times\left(\left(B_{m+1}^H - B_m^H\right)^2 - 2^{2H-1}\left(B_{m+1/2}^H - B_m^H\right)^2 - 2^{2H-1}\left(B_{m+1}^H - B_{m+1/2}^H\right)^2\right)\Bigg].$$

Proof Write

$$\widetilde{H}_k - H = \frac{1}{2}\left(\log_2\frac{V_{2^{k-1}}^{H,2}}{V_{2^k}^{H,2}} - (2H-1)\right) = \frac{1}{2}\log_2\frac{V_{2^{k-1}}^{H,2}}{2^{2H-1}V_{2^k}^{H,2}}$$
$$= \frac{1}{2}\log_2\left(\frac{V_{2^{k-1}}^{H,2} - 2^{2H-1}V_{2^k}^{H,2}}{2^{2H-1}V_{2^k}^{H,2}} + 1\right).$$

Since by (4.13)

$$\zeta_k := \frac{V_{2^{k-1}}^{H,2} - 2^{2H-1}V_{2^k}^{H,2}}{2^{2H-1}V_{2^k}^{H,2}} \to 0, \quad k \to \infty,$$

we obtain

$$\widetilde{H}_k - H = \zeta_k\left(\frac{1}{2\log 2} + o(1)\right), \quad k \to \infty.$$

Now write

$$V_{2^{k-1}}^{H,2} - 2^{2H-1} V_{2^k}^{H,2} = a^2 R_k^{H,0,2} + 2ab R_k^{H,1,1} + b^2 R_k^{H,2,0},$$

where

$$R_k^{H,i,j} = V_{2^{k-1}}^{H,i,j} - 2^{2H-1} V_{2^k}^{H,i,j}, \quad i,j \in \{0,1,2\}.$$

By Theorem 4.5 we have for any $\varepsilon \in (0, H)$ $V_n^{H,1,1} = o(n^{-H+\varepsilon})$, $n \to \infty$, whence $R_k^{H,1,1} = o(2^{(-H+\varepsilon)k}) = o(1)$, $k \to \infty$. Therefore,

$$\frac{2abR_k^{H,1,1}}{2^{2H-1} V_{2^k}^{H,2}} \sim \frac{2bR_k^{H,1,1}}{aT^{2H} 2^{(1-2H)(k-1)}} = o\left(2^{(2H-1)k}\right), \quad k \to \infty.$$

Further, by Theorem 4.5, $V_n^{H,2,0} \to T$, $n \to \infty$, so

$$\frac{b^2 R_k^{H,2,0}}{2^{2H-1} V_{2^k}^{H,2}} \sim \frac{b^2 R_k^{H,2,0}}{a^2 T^{2H} 2^{(k-1)(1-2H)}} = O\left(2^{(2H-1)k}\right), \quad k \to \infty.$$

Thus, we get

$$\begin{aligned} 2^{k/2}\left(\widetilde{H}_k - H\right) &= \left(\frac{a^2 2^{k/2} R_k^{H,0,2}}{2^{2H-1} V_{2^k}^{H,2}} + O(2^{(2H-1/2)k})\right)\left(\frac{1}{2\log 2} + o(1)\right) \\ &= \frac{2^{(2H-1/2)(k-1)} R_k^{H,0,2}}{\sqrt{2} T^{2H} \log 2} + o(1), \quad k \to \infty. \end{aligned} \tag{4.18}$$

Now write

$$R_k^{H,0,2} = \sum_{m=0}^{2^{k-1}-1} \left(\left(\Delta_m^{2^{k-1}} B^H\right)^2 - 2^{2H-1}\left(\Delta_{2m}^{2^k} B^H\right)^2 - 2^{2H-1}\left(\Delta_{2m+1}^{2^k} B^H\right)^2 \right).$$

In view of the self-similarity of B^H,

$$R_k^{H,0,2} \stackrel{d}{=} 2^{2H(1-k)} T^{2H} \sum_{m=0}^{2^{k-1}-1} \eta_m,$$

where

$$\eta_m = \left(B_{m+1}^H - B_m^H\right)^2 - 2^{2H-1}\left(B_{m+1/2}^H - B_m^H\right)^2 - 2^{2H-1}\left(B_{m+1}^H - B_{m+1/2}^H\right)^2.$$

So we can apply Theorem B.11 and deduce that

$$\frac{2^{(2H-1/2)(k-1)}R_k^{H,0,2}}{T^{2H}} \stackrel{d}{=} 2^{(1-k)/2}\sum_{m=0}^{2^{k-1}-1}\eta_m \xrightarrow{d} \mathcal{N}(0,\sigma^2), \quad k\to\infty,$$

where

$$\sigma^2 = \mathsf{E}\left[\eta_0^2\right] + 2\sum_{m=0}^{\infty}\mathsf{E}\left[\eta_0\eta_m\right] = \rho'_{H,0} + 2\sum_{m=0}^{\infty}\rho'_{H,m}.$$

Using this convergence and (4.18), we get the required statement with the help of Theorem B.5. □

Now let $H \in (1/4, 1/2)$. (We omit $H = 1/4$ for two reasons: first, it is hard to distinguish this case statistically from $H \neq 1/4$; second, in this case it is shown exactly as in Proposition 4.8 that $2^{k/2}(\widetilde{H}_k - H)$ converges to a non-central limit law.) In this case neither $\widehat{H}_k$ nor $\widetilde{H}_k$ is asymptotically normal. In fact, a careful analysis of the proof of Proposition 4.8 shows that $2^{(1-2H)k}(\widetilde{H}_k - H)$ converges to some constant. Nevertheless, it is possible to construct an asymptotically normal estimator by canceling this constant out. To this end, one has to consider

$$U_k^{H,2} = V_{2^{k-1}}^{H,2} - V_{2^k}^{H,2}$$

instead of $V_{2^k}^{H,2}$. For well-definiteness we introduce the notation

$$\log_{2+} x = \begin{cases} \log_2 x, & x > 0, \\ 0, & x \le 0. \end{cases}$$

Theorem 4.9 *For $H \in (0, 1/2)$, the statistics*

$$\widetilde{H}_k^{(2)} = \frac{1}{2}\left(\log_{2+}\frac{U_{k-1}^{H,2}}{U_k^{H,2}} + 1\right)$$

is a strongly consistent estimator of H, moreover, for any $\varepsilon > 0$,

$$\widetilde{H}_k^{(2)} = H + o\left(2^{k(-1/2+\varepsilon)}\right), \quad k\to\infty. \tag{4.19}$$

The estimator $\widetilde{H}_k^{(2)}$ is asymptotically normal:

$$2^{k/2}\left(\widetilde{H}_k^{(2)} - H\right) \xrightarrow{d} \mathcal{N}\left(0, (\sigma''_H)^2\right), \quad k\to\infty,$$

with

$$\sigma''_H = \frac{1}{(1-2^{2H-1})\log 2}\left(\rho''_{H,0} + 2\sum_{m=1}^{\infty}\rho''_{H,m}\right)^{1/2},$$

$$\rho''_{H,m} = \mathsf{E}\Big[\Big(s_0^1 - (c_H+1)\big(s_0^{1/2} + c_H(s_{1/2}^{1/2}\big) + c_H\big(s_0^{1/4} + s_{1/4}^{1/4} + s_{1/2}^{1/4} + s_{3/4}^{1/4}\big)\Big)$$

$$\times\Big(s_m^1 - (c_H+1)\big(s_m^{1/2} + s_{m+1/2}^{1/2}\big) + c_H\big(s_m^{1/4} + s_{m+1/4}^{1/4} + s_{m+1/2}^{1/4} + s_{m+3/4}^{1/4}\big)\Big)\Big];$$

here $s_t^h = \left(B_{t+h}^H - B_t^H\right)^2$, $c_H = 2^{2H-1}$.

Proof The proof is similar to that of Proposition 4.8, so we will omit some details. Using the same transformations as there, we get

$$\widetilde{H}_k^{(2)} - H = \frac{U_{k-1}^{H,2} - 2^{2H-1}U_k^{H,2}}{2^{2H-1}U_k^{H,2}}\left(\frac{1}{2\log 2} + o(1)\right), \quad k\to\infty.$$

Expand

$$U_{k-1}^{H,2} - 2^{2H-1}U_k^{H,2} = a^2P_k^{H,0,2} + 2abP_k^{H,1,1} + b^2P_k^{H,2,0},$$

where

$$P_k^{H,i,j} = V_{2^{k-2}}^{H,i,j} - (c_H+1)V_{2^{k-1}}^{H,i,j} + c_H V_{2^k}^{H,i,j}, \quad i,j\in\{0,1,2\}.$$

Similarly to $R_k^{H,1,1}$ in Proposition 4.8, for any $\varepsilon>0$ $P_k^{H,1,1} = o(2^{(-H+\varepsilon)k})$, $k\to\infty$. Further, $P_k^{H,2,0}$ has a generalized chi-square distribution with $\mathsf{E}\left[P_k^{H,2,0}\right] = 0$ and $\mathsf{E}\left[\left(P_k^{H,2,0}\right)^2\right] = O(2^{-k})$, $k\to\infty$. As in Theorem 4.5, we deduce that for any $\varepsilon>0$ $P_k^{H,0,2} = o(2^{(-1/2+\varepsilon)k})$, $k\to\infty$.

Further, from (4.13) $U_k^{H,2} \sim a^2T^{2H}(2^{2H-1}-1)2^{(1-2H)k}$, $k\to\infty$. Combining the obtained asymptotics, we can write

$$2^{k/2}\left(\widetilde{H}_k^{(2)} - H\right) = \frac{2^{(2H-1/2)(k-2)}P_k^{H,0,2}}{T^{2H}(1-2^{1-2H})\log 2} + o(1), \quad k\to\infty,$$

whence we deduce the asymptotic normality exactly as in Proposition 4.8.

The estimate (4.19) is obtained as in Theorem 4.5. □

Remark 4.10 Despite $\widetilde{H}_k^{(2)}$ has asymptotically a better rate of approximation than $\widetilde{H}_k$ for $H\in(1/4,1/2)$, we still do not recommend using it, as the asymptotic variance is high; it is practically useless for $k\le 10$.

Now we turn to estimation of the scale coefficients a and b. As it is known from [167], for $H \in (0, 1/4)$ the measure induced by M^H in $C[0, T]$ is equivalent to that of aB^H. This not only gives another explanation why the results for $H \in (0, 1/4)$ are essentially the same as for fBm alone, but also has another important consequence: for $H \in (0, 1/4)$ it is not possible to estimate b consistently.

Proposition 4.11 *For $H \in (0, 1/2)$, the statistics*

$$\widetilde{a_k^2} = 2^{k(2\widetilde{H}_k - 1)} T^{-2\widetilde{H}_k} V_{2^k}^{H,2}$$

is a strongly consistent estimator of a^2.

For $H \in (1/4, 1/2)$ the statistics

$$\widetilde{b_k^2} = \frac{2^{1-2\widetilde{H}_k^{(2)}} V_{2^{k-1}}^{H,2} - V_{2^k}^{H,2}}{\left(2^{1-2\widetilde{H}_k^{(2)}} - 1\right) T}$$

is a strongly consistent estimator of b^2.

Proof First, observe that

$$\frac{\widetilde{a_k^2}}{a^2} \sim 2^{2(\widetilde{H}_k - H)k} T^{2(H - \widetilde{H}_k)} \to 1, \quad k \to \infty,$$

since $(\widetilde{H}_k - H)k \to 0$, $k \to \infty$, by (4.17). Hence we get the strong consistency of $\widetilde{a_k^2}$.

Concerning $\widetilde{b_k^2}$, define

$$\widehat{b_k^2} = \frac{2^{1-2H} V_{2^{k-1}}^{H,2} - V_{2^k}^{H,2}}{(2^{1-2H} - 1) T}.$$

It easily follows from (4.13) that $\widehat{b_k^2} \to b^2$, $k \to \infty$. So it is enough to show that $\widetilde{b_k^2} - \widehat{b_k^2} \to 0$, $k \to \infty$. To this end, write

$$\widetilde{b_k^2} - b^2 = \frac{\left(2^{1-2\widetilde{H}_k^{(2)}} - 2^{1-2H}\right) V_{2^{k-1}}^{H,2}}{(2^{1-2\widetilde{H}_k^{(2)}} - 1) T} + T^{-1} \left(2^{1-2H} V_{2^{k-1}}^{H,2} - V_{2^k}^{H,2}\right) \left(\left(2^{1-2\widetilde{H}_k^{(2)}} - 1\right)^{-1} - \left(2^{1-2H} - 1\right)^{-1}\right).$$

Obviously, the second term converges to zero. Due to (4.13) and (4.19), for any $\varepsilon > 0$

$$\left(2^{1-2\widetilde{H}_k^{(2)}} - 2^{1-2H}\right)V_{2^k}^{H,2} \sim -2^{2-2H}\left(\widetilde{H}_k^{(2)} - H\right)a^2T^{2H}2^{(1-2H)(k-1)}\log 2$$
$$= 2^{(1-2H)k}o\left(2^{(-1/2+\varepsilon)k}\right), \quad k \to \infty,$$

whence we deduce the strong consistency of $\widetilde{b}_k^2$ for $H \in (1/4, 1/2)$, since $1 - 2H < \frac{1}{2}$. □

4.4.1.2 $H \in (1/2, 3/4)$

Now we move to the case $H \in (1/2, 1)$. In view of (4.14), both $\widehat{H}_k$ and $\widetilde{H}_k$ converge to $1/2$ for $H \in (1/2, 1)$, so they are not suitable for estimating H. The solution is to use $U_k^{H,2} = V_{2^{k-1}}^{H,2} - V_{2^k}^{H,2}$, rather than $V_{2^k}^{H,2}$, for the construction of estimators. The resulting estimators work also for $H \in (0, 1/2)$.

Theorem 4.12 *For $H \in (0, 1/2) \cup (1/2, 3/4)$, the statistics*

$$\widehat{H}_k^{(2)} = \frac{1}{2}\left(1 - \frac{1}{k}\log_{2+} U_k^{H,2}\right)$$

and

$$\widetilde{H}_k^{(2)} = \frac{1}{2}\left(\log_{2+}\frac{U_{k-1}^{H,2}}{U_k^{H,2}} + 1\right)$$

are strongly consistent estimators of the Hurst parameter H.

Proof Write

$$U_k^{H,2} = a^2Q_k^{H,0,2} + 2abQ_k^{H,1,1} + b^2Q_k^{H,2,0},$$

where $Q_k^{H,i,j} = V_{2^{k-1}}^{H,i,j} - V_{2^k}^{H,i,j}$, $i, j \in 0, 1, 2$. By Theorem 4.5,

$$Q_k^{H,0,2} \sim T^{2H}(2^{2H-1} - 1)2^{(1-2H)k}, \quad k \to \infty$$

and for any $\varepsilon > 0$ $Q_k^{H,1,1} = o(2^{(-H+\varepsilon)k})$, $k \to \infty$, and

$$Q_k^{H,2,0} = \left(V_{2^{k-1}}^{H,2,0} - T\right) - \left(V_{2^k}^{H,2,0} - T\right) = o\left(2^{(-1/2+\varepsilon)k}\right), \quad k \to \infty.$$

Thus, we have

$$U_k^{H,2} \sim a^2T^{2H}(2^{2H-1} - 1)2^{(1-2H)k}, \quad k \to \infty, \tag{4.20}$$

which yields the proof. □

Remark 4.13 We will see in Sect. 4.5 that $\widetilde{H}_k^{(2)}$ performs very poorly, and $\widehat{H}_k^{(2)}$ performs somewhat better, despite having worse asymptotic rate of convergence.

As in the case $H \in (0, 1/2)$, the estimator $\widetilde{H}_k^{(2)}$ is asymptotically normal for $H \in (1/2, 3/4)$; however, the limit Gaussian law comes out of the quadratic variation of the Wiener process, so the convergence rate is different, and the expression for the asymptotic variance is much simpler.

Theorem 4.14 *For $H \in (1/2, 3/4)$ and any $\varepsilon > 0$, the statistics*

$$\widetilde{H}_k^{(2)} = \frac{1}{2}\left(\log_{2+} \frac{U_{k-1}^{H,2}}{U_k^{H,2}} + 1\right)$$

is a strongly consistent estimator of H, moreover, it satisfies

$$\widetilde{H}_k^{(2)} = H + o\left(2^{k(2H-3/2+\varepsilon)}\right), \quad k \to \infty. \tag{4.21}$$

It is asymptotically normal:

$$2^{k(3/2-2H)}\left(\widetilde{H}_k^{(2)} - H\right) \xrightarrow{d} \mathcal{N}\left(0, (\sigma_H'')^2\right), \quad k \to \infty,$$

with

$$\sigma_H''^2 = \frac{b^2 T^{1-2H}\sqrt{2^{4H-3}+1}}{a^2 2^{2H-1}(2^{2H-1}-1)\log 2}.$$

Proof As in the proof of Theorem 4.9, write

$$\widetilde{H}_k^{(2)} - H = \frac{U_{k-1}^{H,2} - 2^{2H-1}U_k^{H,2}}{2^{2H-1}U_k^{H,2}}\left(\frac{1}{2\log 2} + o(1)\right), \quad k \to \infty,$$

and expand

$$U_{k-1}^{H,2} - 2^{2H-1}U_k^{H,2} = a^2 P_k^{H,0,2} + 2ab P_k^{H,1,1} + b^2 P_k^{H,2,0},$$

where

$$P_k^{H,i,j} = V_{2^{k-2}}^{H,i,j} - (c_H + 1)V_{2^{k-1}}^{H,i,j} + c_H V_{2^k}^{H,i,j}, \quad i,j \in \{0,1,2\},$$

and $c_H = 2^{2H-1}$.

As in the proof of Theorem 4.9, we have that for any $\varepsilon > 0$ $P_k^{H,1,1} = o(2^{(-H+\varepsilon)k})$, $P_k^{H,0,2} = o(2^{(1/2-2H+\varepsilon)k})$, $k \to \infty$. Therefore, using (4.20), we get

$$2^{(3/2-2H)k}\left(\widetilde{H}_k^{(2)} - H\right) = \frac{b^2 2^{(k-2)/2} P_k^{H,2,0}}{a^2 2^{2H-1} T^{2H}(2^{2H-1}-1)\log 2} + o(1), \quad k \to \infty.$$

We can write $P_k^{H,2,0} = \sum_{m=0}^{2^{k-2}-1} \kappa_{k,m}$, where

$$\kappa_{k,m} = \left(\Delta_m^{2^{k-2}} W\right)^2 - (c_H + 1)\left(\left(\Delta_{2m}^{2^{k-1}} W\right)^2 + \left(\Delta_{2m+1}^{2^{k-1}} W\right)^2\right)$$
$$+ c_H \left(\left(\Delta_{4m}^{2^k} W\right)^2 + \left(\Delta_{4m+1}^{2^k} W\right)^2 + \left(\Delta_{4m+2}^{2^k} W\right)^2 + \left(\Delta_{4m+3}^{2^k} W\right)^2\right).$$

The random variables $\{\kappa_{k,m}, m = 0, \ldots, 2^k - 1\}$ are i. i. d. with

$$\mathsf{E}[\kappa_{k,m}] = 0 \quad \text{and} \quad \mathsf{E}\left[\kappa_{k,m}^2\right] = T^2 2^{-2(k-2)}\left(2^{4H-3} + 1\right).$$

Therefore, by the classical CLT,

$$2^{(k-2)/2} P_k^{H,2,0} \xrightarrow{d} \mathcal{N}\left(0, T^2\left(2^{4H-3} + 1\right)\right), \quad k \to \infty,$$

whence we get by Theorem B.5,

$$2^{(3/2-2H)k}\left(\widetilde{H}_k^{(2)} - H\right) \xrightarrow{d} \mathcal{N}\left(0, (\sigma_H'')^2\right), \quad k \to \infty.$$

Again, the estimate (4.21) is obtained as in Theorem 4.5. □

The estimation of the scale coefficient a is similar to the case $H \in (0, 1/2)$, but we have to use $U_k^{H,2}$ and $\widetilde{H}_k^{(2)}$ instead of $V_{2^k}^{H,2}$ and $\widetilde{H}_k$; the resulting estimator works also for $H \in (0, 1/2)$. Estimating b^2 is a lot easier, thanks to (4.14).

Proposition 4.15 *For $H \in (0, 1/2) \cup (1/2, 3/4)$, the statistic*

$$\hat{a}_k^2 = 2^{k\left(2\widetilde{H}_k^{(2)} - 1\right)} T^{-2\widetilde{H}_k^{(2)}} \left(2^{2\widetilde{H}_k^{(2)} - 1} - 1\right)^{-1} U_k^{H,2}$$

is a strongly consistent estimator of a^2.

For $H \in (1/2, 1)$, the statistic

$$\hat{b}_k^2 = \frac{V_{2^k}^{H,2}}{T}$$

is a strongly consistent estimator of b^2.

Proof In view of (4.20),

$$\frac{\hat{a}_k^2}{a^2} \sim 2^{2\left(\widetilde{H}_k^{(2)} - H\right)k} T^{2\left(H - \widetilde{H}_k^{(2)}\right)} \frac{2^{2H-1} - 1}{2^{2\widetilde{H}_k^{(2)} - 1} - 1} \to 1, \quad k \to \infty,$$

since $\left(\widetilde{H}_k^{(2)} - H\right) k \to 0$, $k \to \infty$, by (4.21). Hence we get the strong consistency of $\hat{a}_k^2$. The strong consistency of $\hat{b}_k^2$ is obvious from (4.14). □

4.4.1.3 $H \in (3/4, 1)$

As we have already mentioned in the beginning of this section, it is impossible to make conclusions about the value of H in this case. In fact, we have

$$n^{1/2}(V_n^{H,2} - b^2T) \xrightarrow{d} b^2T\mathcal{N}(0,2), \quad n \to \infty.$$

Indeed, $n^{1/2}(V_n^{H,2,0} - T) \xrightarrow{d} \mathcal{N}(0, 2T^2)$, $n \to \infty$, by the classical CLT; $V_n^{H,0,2} \sim T^{2H}n^{1-2H} = o\left(n^{-1/2}\right)$, $n \to \infty$, and for any $\varepsilon > 0$ $V_n^{H,1,1} = o\left(n^{-H+\varepsilon}\right)$, $n \to \infty$, due to Theorem 4.5. This means that the behavior of $V_n^{H,2}$ is essentially the same as that of the quadratic power variation of Wiener process, in particular, it says nothing about H.

Nevertheless, we will study the behavior of quadratic variation in more detail in order to be able to distinguish between the cases $H < 3/4$, considered above, and $H > 3/4$ statistically.

Define

$$Z_k = \frac{2^{(k-1)/2}}{b^2T} U_k^{H,2}.$$

Proposition 4.16 *For $H \in (3/4, 1)$, the sequence $(Z_k, Z_{k+1}, \dots)$ converges in distribution as $k \to \infty$ to a sequence $(\zeta_1, \zeta_2, \dots)$ of independent standard Gaussian variables.*

Proof Define

$$\xi_k = \frac{2^{k/2}}{\sqrt{2}T}\left(V_{2^k}^{H,2,0} - T\right) = \frac{2^{-k/2}}{\sqrt{2}T}\sum_{i=0}^{2^k-1}\left(2^k\left(\Delta_i^{2^k} W\right)^2 - T\right).$$

By the classical CLT, $\xi_k \xrightarrow{d} \mathcal{N}(0,1)$, $k \to \infty$, so we need to study the collective behaviour. To this end, observe that the vector $(\xi_k, \xi_{k+1}, \dots, \xi_{k+m})$ can be represented as a sum of independent vectors

$$(\xi_k, \xi_{k+1}, \dots, \xi_{k+m}) = \sum_{i=0}^{2^k-1} \zeta_{k,i},$$

where the jth coordinate of $\zeta_{k,i}$, $j = 0, 1, 2, \ldots, m$, is

$$\zeta_{k,i,j} = \frac{2^{-k/2}}{\sqrt{2}T} \sum_{l=0}^{2^j-1} \left(2^{k+j} \left(\Delta_{l+i2^j}^{2^{k+j}} W \right)^2 - T \right).$$

(We simply group terms on the intervals of the partition $\{ iT2^{-k}, i = 0, \ldots, 2^k \}$.) Therefore, we can apply a vector CLT and deduce that for every $m \geq 0$ the vector $(\xi_k, \xi_{k+1}, \ldots, \xi_{k+m})$ converges in distribution to an $(m+1)$-dimensional centered Gaussian vector as $k \to \infty$. Consequently, the sequence $(\xi_k, \xi_{k+1}, \xi_{k+2}, \ldots)$ converges to a centered stationary Gaussian sequence as $k \to \infty$. Observe that

$$\begin{aligned}
\sqrt{2}\xi_{k-1} - \xi_k &= \frac{2^{-(k-1)/2}}{T} \sum_{i=0}^{2^{k-1}-1} \left(2^{k-1} \left(\Delta_i^{2^{k-1}} W \right)^2 - T \right) \\
&\quad - \frac{2^{-(k+1)/2}}{T} \sum_{j=0}^{2^k-1} \left(2^k \left(\Delta_j^{2^k} W \right)^2 - T \right) \\
&= \frac{2^{(k-1)/2}}{T} \sum_{i=0}^{2^{k-1}-1} \left(\left(\Delta_{2i}^{2^k} W + \Delta_{2i+1}^{2^k} W \right)^2 - \left(\Delta_{2i}^{2^k} W \right)^2 - \left(\Delta_{2i+1}^{2^k} W \right)^2 \right) \\
&= \frac{2^{(k+1)/2}}{T} \sum_{i=0}^{2^{k-1}-1} \Delta_{2i}^{2^k} W \Delta_{2i+1}^{2^k} W.
\end{aligned}$$

Therefore,

$$\begin{aligned}
\mathsf{E}\left[\left(\sqrt{2}\xi_{k-1} - \xi_k \right)^2 \right] &= \frac{2^{k+1}}{T^2} \sum_{i=0}^{2^{k-1}-1} \mathsf{E}\left[\left(\Delta_{2i}^{2^k} W \right)^2 \left(\Delta_{2i+1}^{2^k} W \right)^2 \right] \\
&= \frac{2^{k+1}}{T^2} \cdot 2^{k-1} \cdot \frac{T^2}{2^{2k}} = 1,
\end{aligned}$$

and for any $m \geq 1$

$$\begin{aligned}
&\operatorname{cov}(\sqrt{2}\xi_{k-1} - \xi_k, \sqrt{2}\xi_{k+m-1} - \xi_{k+m}) \\
&= \frac{2^{k+1+m/2}}{T} \sum_{i=0}^{2^{k-1}-1} \sum_{j=0}^{2^{k+m-1}-1} \mathsf{E}\left[\Delta_{2i}^{2^k} W \Delta_{2i+1}^{2^k} W \Delta_{2j}^{2^{k+m}} W \Delta_{2j+1}^{2^{k+m}} W \right] = 0,
\end{aligned}$$

since in the product one of the increments $\Delta_{2i}^{2^k} W$ or $\Delta_{2i+1}^{2^k} W$ is independent of all others. Consequently, the sequence $(\sqrt{2}\xi_{k-1} - \xi_k, \sqrt{2}\xi_k - \xi_{k+1}, \sqrt{2}\xi_{k+1} - \xi_{k+2}, \ldots)$ converges weakly as $k \to \infty$ to an i. i. d. standard Gaussian sequence.

We have seen above that $V_n^{H,2} = b^2 V_n^{H,2,0} + o(n^{-1/2})$, $n \to \infty$. Therefore,

$$Z_k = \left(\sqrt{2}\xi_{k-1} - \xi_k\right) + o(1), \quad k \to \infty,$$

so by Theorem B.5 the sequence $(Z_k, Z_{k+1}, Z_{k+2}, \dots)$ also converges to an i. i. d. standard Gaussian sequence. □

Remark 4.17 We emphasize a sharp contrast with the case $H \in (1/2, 3/4)$, where the sequence $\{Z_k, k \geq 1\}$ diverges to $+\infty$ due to (4.20), hence, it eventually becomes positive. This clearly gives a possibility to distinguish statistically between cases $H \in (1/2, 3/4)$ and $H \in (3/4, 1)$. (See 4.5.1.3 for comparative simulations.)

Remark 4.18 For $H = 3/4$, an analogue of Proposition 4.16 can be proved, that is, $(Z_k, Z_{k+1}, \dots)$ converges in distribution as $k \to \infty$ to a sequence $(\zeta_1, \zeta_2, \dots)$ of independent Gaussian variables with unit variance. However, it can be checked that the limiting distribution now has a positive mean, namely, $\mathsf{E}[\zeta_1] = a^2 b^{-2} T^{\frac{1}{2}} \left(1 - 2^{-\frac{1}{2}}\right)$. As long as this value depends on how big is a compared to b, we might be unable to distinguish this case from $H > 3/4$. On the other hand, if b is small relative to a, it might be hard to distinguish this case from $H < 3/4$.

4.4.2 Statistical Estimation Using 4th Power Variation

It was mentioned in the previous section that the performance of quadratic variation estimators in the case $H \in (1/2, 3/4)$ is not very satisfactory. One could try to improve it by considering quartic variation of M^H

$$V_n^{H,4} := \sum_{k=0}^{n-1} \left(\Delta_k^n M^H\right)^4 = \sum_{i=0}^{4} \binom{4}{i} a^i b^{4-i} V_n^{H,4-i,i}.$$

As for the quadratic variation, we have to cancel out the leading term, considering

$$U_k^{H,4} = V_{2^{k-1}}^{H,4} - 2V_{2^k}^{H,4}.$$

Theorem 4.19 *The statistics*

$$\widehat{H}_k^{(4)} = -\frac{1}{2k} \log_{2+} U_k^{H,4}$$

and

$$\widetilde{H}_k^{(4)} = \frac{1}{2} \log_{2+} \frac{U_{k-1}^{H,4}}{U_k^{H,4}}$$

are strongly consistent estimators of the Hurst parameter $H \in (1/2, 3/4)$ *in the mixed model* (4.12).

Proof By Theorem 4.5, $V_n^{H,2,2} \sim T^{2H+1}n^{-2H}$, $V_n^{H,0,4} \sim 3T^{4H}n^{1-4H}$, $n \to \infty$ and for any $\varepsilon > 0$ $V_n^{H,4,0} - 3T^2n^{-1} = o(n^{-3/2+\varepsilon})$, $V_n^{H,3,1} = o(n^{-1-H+\varepsilon})$, $V_n^{H,1,3} = o(n^{-3H+\varepsilon})$, $n \to \infty$.

Now write

$$U_k^{H,4} = \sum_{i=0}^{4} \binom{4}{i} a^i b^{4-i} U_k^{H,4-i,i},$$

where $U_k^{H,4-i,i} = V_{2^{k-1}}^{H,4-i,i} - 2V_{2^k}^{H,4-i,i}$, $i = 0, \ldots, 4$. We have

$$U_k^{H,2,2} \sim T^{2H+1}\left(2^{2H} - 2\right)2^{-2Hk}, \quad U_k^{H,0,4} = O\left(2^{(1-4H)k}\right) = o\left(2^{-2Hk}\right), \quad k \to \infty,$$

and for any $\varepsilon > 0$

$$U_k^{H,4,0} = \left(V_{2^{k-1}}^{H,4,0} - T^2 2^{-(k-1)}\right) - 2\left(V_{2^k}^{H,4,0} - T^2 2^{-k}\right) = o\left(2^{(-3/2+\varepsilon)k}\right),$$

$$U_k^{H,3,1} = o\left(2^{(-1-H+\varepsilon)k}\right), \quad U_k^{H,1,3} = o\left(2^{(-3H+\varepsilon)k}\right), \quad k \to \infty.$$

Collecting all the terms, we get

$$U_k^{H,4} \sim 4T^{2H+1}\left(2^{2H} - 2\right)2^{-2Hk}, \quad k \to \infty.$$

Hence, the assertion follows. □

Remark 4.20 Both these estimators are quite poor. A regression of several values of $\log_2 U_k^{H,4}$ on k leads to a much better estimator. However, as numerical experiments in Sect. 4.5 suggest, it is better to use the quadratic variation based estimators (which are not very efficient as well).

4.4.3 Estimation of Hurst Parameter for Known Scale Coefficients

When the scale coefficients a and b are known, the estimation procedure significantly simplifies, and the quality of estimators is improved. It may seem unnatural at first glance that the scale coefficients are known while H is not. However, the case where b is known is quite natural, as we can have known white noise amplitude with unknown long-range perturbation of this white noise. The cases of known a or known both coefficients are less natural, but there is no reason to omit these cases considering only the case of known b.

Theorem 4.21 *If a is known, then the statistic*

$$\widehat{H}_k(a) = \frac{k + 2\log_2 a - \log_2 V_{2^k}^{H,2}}{2(k - \log_2 T)}$$

is a strongly consistent estimator of $H \in (0, 1/2)$, moreover, for any $\varepsilon > 0$,

$$\widehat{H}_k(a) = H + O\left(2^{k(2H-1)}\right) + o\left(2^{k(-1/2+\varepsilon)}\right), \quad k \to \infty.$$

If b is known, then the statistic

$$\widetilde{H}_k(b) = \frac{1}{2}\left(\log_{2+} \frac{V_{2^{k-1}}^{H,2} - b^2T}{V_{2^k}^{H,2} - b^2T} + 1\right)$$

is a strongly consistent estimator of $H \in (0, 3/4)$, moreover, for any $\varepsilon > 0$,

$$\widetilde{H}_k(b) = H + o\left(2^{k(-1/2+\varepsilon)}\right) + o\left(2^{k(2H-3/2+\varepsilon)}\right), \quad k \to \infty.$$

If a and b are known, then the statistic

$$\widehat{H}_k(a,b) = \frac{k + 2\log_2 a - \log_{2+}\left(V_{2^k}^{H,2} - b^2T\right)}{2(k - \log_2 T)}$$

is a strongly consistent estimator of $H \in (0, 3/4)$, moreover, for any $\varepsilon > 0$,

$$\widehat{H}_k(a,b) = H + o\left(2^{k(-1/2+\varepsilon)}\right) + o\left(2^{k(2H-3/2+\varepsilon)}\right), \quad k \to \infty.$$

Proof The statement for $\widehat{H}_k(a)$ follows immediately from (4.15). To prove the statement for $\widetilde{H}_k(b)$ and $\widehat{H}_k(a,b)$, note that, in view of (4.13), $V_{2^k}^{H,2} > b^2T$ for sufficiently large k. Therefore, as in the proof of Theorem 4.6, we can write

$$\log_2\left(V_{2^k}^{H,2} - b^2T\right) = \log_2\left(a^2T^{2H}2^{k(1-2H)}\right) + \log_2\left(1 + \zeta_k\right),$$

with the same ζ_k; in particular, for $H \in (0, 1/2]$ and any $\varepsilon > 0$, $\zeta_k = o(2^{k(-1/2+\varepsilon)})$, $k \to \infty$. For $H \in (1/2, 3/4)$,

$$\zeta_k = o\left(2^{k(-H+\varepsilon)}\right) + o\left(2^{k(2H-3/2+\varepsilon)}\right) + o\left(2^{k(H-1+\varepsilon)}\right) = o\left(2^{k(2H-3/2+\varepsilon)}\right), \quad k \to \infty.$$

This implies the statement for both $\widetilde{H}_k(b)$ and $\widehat{H}_k(a,b)$. □

Remark 4.22 It can be shown that $\widehat{H}_k(a)$ is asymptotically normal for $H \in (0, 1/4)$, $\widetilde{H}_k(b)$, for $H \in (1/2, 3/4)$, $\widehat{H}_k(a,b)$, for $H \in (0, 3/4)$. This is not our main concern here, so we skip the asymptotic normality results.

4.5 Simulations

In each procedure we take $T = 3, a = b = 1, n = 2^{20}$ and use the circulant method to simulate values of B^H on the uniform partition $\{ iT/n, i = 0, 1, \dots, n\}$ of $[0, T]$. For each value of the Hurst parameter, we simulate 1,000 trajectories of the fBm. Then for each estimator $\check{H}$ we compute the average $\mu\check{H}$ of 1,000 obtained values and the mean square error $\sigma\check{H}$, i. e. the square root of the average of values $(\check{H} - H)^2$, and compare it to the theoretical standard deviation $\sigma_t\check{H}$, if the latter is available. Where possible, we make similar procedure for a and/or b.

Each simulation takes about 300 ms on Intel Core i5-3210M processor, computing all estimators takes about 20 ms.

4.5.1 Estimators Based on Quadratic Variation

4.5.1.1 $H \in (0, 1/2)$

In Table 4.1, we compare the estimators $\widehat{H}_{20}$, $\widetilde{H}_{20}$, $\widetilde{H}^{(2)}_{20}$ (observe that all these estimators are based on the values of fBm on the chosen partition). We also give values of the estimator $\widetilde{a}_{20}$; the estimator $\widetilde{b}_{20}$ is quite bad: 2–5 values of $\widetilde{b}^2_{20}$ out of 10 are negative, others are quite away from the true value, so we do not give its values.

The results show that the estimator $\widetilde{H}_{20}$ has consistently the best performance. For $H > 1/4$, a positive bias is visible, which is not surprising as it can be checked using the same transformations as in the proof of Proposition 4.8 that in this case

$$\widetilde{H}_k - H \sim (1 - 2^{2H-1})a^{-2}b^2T^{1-2H}2^{(2H-1)k}, \quad k \to \infty.$$

The estimator $\widehat{H}_{20}$ underestimates all values of H by around 8 %. The underestimation follows from (4.16), since $aT^H > 1$. Finally, the relative error of $\widetilde{H}^{(2)}_{20}$ is larger than that of $\widetilde{H}_{20}$. The mean square error reflects the theoretical standard deviation quite good for all values of H except 0.4 and 0.45. The latter divergence from the theoretical values is not surprising. Indeed, a careful check of the proof of Proposition 4.8 reveals that the error of the normal approximation is of order 2^{-Hk}, which for values of H close to 1/2 is comparable with the order $2^{-k/2}$ of the Gaussian term.

The estimator $\widetilde{a}_{20}$ is quite reliable, especially for smaller values of H; for $H > 1/4$ it has a positive bias (inherited from $\widetilde{H}_k$).

Table 4.1 Values of the quadratic variation based estimators for $H \in (0, 1/2)$

H	0.05	0.1	0.15	0.2	0.25	0.3	0.35	0.4	0.45
$\mu\widehat{H}_{20}$	0.0460	0.0921	0.1381	0.1841	0.2301	0.2760	0.3215	0.3656	0.4055
$\sigma\widehat{H}_{20}$	0.0040	0.0079	0.0119	0.0159	0.0199	0.0240	0.0285	0.0344	0.0445
$\mu\widetilde{H}_{20}$	0.0500	0.1000	0.1500	0.2002	0.2504	0.3013	0.3535	0.4077	0.4612
$\sigma\widetilde{H}_{20}$	0.0016	0.0015	0.0014	0.0014	0.0014	0.0019	0.0037	0.0078	0.0112
$\sigma_t\widetilde{H}_{20}$	0.0015	0.0015	0.0014	0.0014					
$\mu\widetilde{H}^{(2)}_{20}$	0.0498	0.1000	0.1498	0.2001	0.2498	0.2996	0.3497	0.4006	0.4496
$\sigma\widetilde{H}^{(2)}_{20}$	0.0047	0.0051	0.0053	0.0058	0.0069	0.0082	0.0103	0.0155	0.0345
$\sigma_t\widetilde{H}^{(2)}_{20}$	0.0046	0.0050	0.0054	0.0060	0.0069	0.0082	0.0102	0.0144	0.0269
$\mu\widetilde{a}_{20}$	1.000	1.000	1.000	1.003	1.007	1.020	1.057	1.146	1.305
$\sigma\widetilde{a}_{20}$	0.0200	0.0196	0.0180	0.0175	0.0181	0.0263	0.0591	0.1468	0.3050

4.5.1.2 $H \in (1/2, 3/4)$

Table 4.2 compares estimators $\widehat{H}_{20}^{(2)}$ and $\widetilde{H}_{20}^{(2)}$ of Hurst parameter H. It also contains a "regression" estimator $\bar{H}^{(2)}$ obtained in the following way: we consider the linear regression of $\left\{\log_{2+} U_j^{H,2}, j = m, m+1, \dots, 19\right\}$ on $\{m, m+1, \dots, 19\}$, where $m = 11, 12, \dots, 15$, and take the best regression (in terms of the coefficient of determination). If $\bar{r}^{(2)}$ is the coefficient of the best linear regression, we set $\bar{H}^{(2)} = (1 - \bar{r}^{(2)})/2$. We also give the estimator $\hat{b}_{20}$. Due to uselessness of the estimator $\hat{a}_{20}$, we do not present its values.

It is clear that none of the estimators is reliable: average errors are in most cases comparable to the length of the range $(1/2, 3/4)$, so they are quite useless. Only the performance of $\widehat{H}_{20}^{(2)}$ in the range 0.575–0.7 is acceptable, but one should be aware of a positive bias.

It is interesting to note that the errors of both $\widetilde{H}_{20}^{(2)}$ and $\bar{H}^{(2)}$ explode for $H > 5/8$. We admit that we found no explanation for this phenomenon.

4.5.1.3 $H \in (3/4, 1)$ Versus $H \in (1/2, 3/4)$

Table 4.3 contains values of $\left\{\left[10^4 U_k^{H,2}\right], k = 10, 11, 12, \dots, 19\right\}$ for $H = 0.7$ and $H = 0.8$. The difference is clearly visible: for $H = 0.7$ the sequence is positive, while for $H = 0.8$ there is a plenty of negative values.

4.5.2 *Estimators Based on Quartic Variation*

4.5.2.1 $H \in (1/2, 3/4)$

Table 4.4 contains estimators $\widehat{H}_{20}^{(4)}$ and $\widetilde{H}_{20}^{(4)}$ of Hurst parameter H, the values of H range from 0.525 to 0.725 with step 0.025. We also give a "regression" estimator $\bar{H}^{(4)}$. It is obtained in the following way: we consider the linear regression of $\left\{\log_{2+} U_j^{H,4}, j = m, m+1, \dots, 19\right\}$ on $\{m, m+1, \dots, 20\}$, where $m = 11, 12, \dots, 16$, and take the best regression (in terms of the coefficient of determination). If $\bar{r}^{(4)}$ is the coefficient of the best linear regression, we set $\bar{H}^{(4)} = -\bar{r}^{(4)}/2$.

We see that the estimators based on the quartic variation are quite useless and definitely worse than those based on the quadratic variation. Again, the errors of $\widetilde{H}_{20}^{(4)}$ and $\mu\bar{H}^{(4)}$ explode for $H \geq 5/8$. In contrast to the quadratic variation case, now this phenomenon can be easily explained. The fact is that the nature of the error changes at $H = 5/8$: for $H < 5/8$, the error comes from the term $U_k^{H,0,4}$

Table 4.2 Values of the quadratic variation based estimators for $H \in (1/2, 3/4)$

H	0.525	0.55	0.575	0.6	0.625	0.65	0.675	0.7	0.725
$\mu\widehat{H}^{(2)}_{20}$	0.6099	0.6068	0.6152	0.6279	0.6433	0.6615	0.6785	0.6789	0.6713
$\sigma\widehat{H}^{(2)}_{20}$	0.0850	0.0569	0.0403	0.0284	0.0197	0.0172	0.0328	0.0683	0.1067
$\mu\widetilde{H}^{(2)}_{20}$	0.5234	0.5440	0.5784	0.5926	0.6150	0.6377	0.6535	0.6250	0.6517
$\sigma\widetilde{H}^{(2)}_{20}$	0.1401	0.1041	0.1167	0.1533	0.2218	0.3854	0.6446	0.7568	0.8520
$\sigma_1\widetilde{H}^{(2)}_{20}$	0.0905	0.0824	0.0999	0.1365	0.1989	0.3021	0.4723	0.7540	1.2234
$\mu\bar{H}^{(2)}$	0.5148	0.5477	0.5705	0.5920	0.6150	0.6469	0.7178	0.7065	0.7071
$\sigma\bar{H}^{(2)}$	0.0830	0.0543	0.0549	0.0717	0.1014	0.1607	0.3272	0.5247	0.7174
$\mu\hat{b}_{20}$	1.236	1.131	1.071	1.038	1.020	1.011	1.006	1.003	1.002
$\sigma\hat{b}_{20}$	0.2362	0.1310	0.0713	0.0382	0.0204	0.0109	0.0058	0.0031	0.0018

Table 4.3 Scaled values of $U_k^{H,2}$ for $H = 0.7$ and $H = 0.8$

$H = 0.7$	869	649	523	3	260	18	78	98	53	50
$H = 0.8$	665	−620	482	−475	8	−29	−104	−71	−78	−28

(in the notation of the proof of Theorem 4.19), which behaves quite smoothly, but for $H \geq 5/8$, the main contribution comes from the fluctuations of $U_k^{H,4,0}$, which are much wilder.

4.5.3 Estimation When a and b Are Known

Table 4.5 gives the estimators $\widehat{H}_{20}(a)$ and $\widehat{H}_{20}(a, b)$ for H from 0.05 to 0.45 with the step 0.05. Since the errors are very small, we multiply them by 100. We can see that the estimator $\widehat{H}_{20}(a)$ is comparable to $\widetilde{H}_{20}(a, b)$ for $H \leq 1/4$; then it becomes worse, but it uses only the knowledge of a.

Table 4.6 contains the estimators $\widetilde{H}_{20}(b)$ and $\widetilde{H}_{20}^{(2)}(a, b)$ of Hurst parameter $H \in [1/2, 1)$. We multiply average errors by 10 to make them visible.

We see that $\widehat{H}_{20}(a, b)$ outperforms $\widetilde{H}_{20}(b)$ by a good margin, but the advantage of the latter is that it uses only the knowledge of b.

4.5.4 Comparison of Estimators and Some Recommendations

To facilitate the usage of the estimators, we summarize our findings about them.

For $H \in (0, 1/2)$, it is better to use the estimator $\widetilde{H}$ for the Hurst parameter. The estimator for the scale coefficient a is quite reliable, but always overestimates the coefficient for $H \in (1/4, 1/2)$. The estimator for b is virtually useless.

For $H \in (1/2, 3/4)$, there is no good estimator for the Hurst parameter. Only the regression estimator $\bar{H}^{(2)}$ is useful for values of H between 0.55 and 0.6, but still the error is comparable with the length of this integral. The coefficient b can be estimated efficiently, while the estimator for a is useless. Nevertheless, it is possible to construct efficient estimators for H using the knowledge of b or of the both scale coefficients.

Finally, for $H > 3/4$, the estimation of H is not possible (even the knowledge of the scale coefficients is not helpful). However, it is possible to distinguish statistically between the cases $H > 3/4$ and $H < 3/4$ by looking at the statistic $U_k^{H,2}$.

Table 4.4 Values of the quartic variation based estimators for $H \in (1/2, 3/4)$

H	0.525	0.55	0.575	0.6	0.625	0.65	0.675	0.7	0.725
$\mu\widehat{H}_{20}^{(4)}$	0.4840	0.4876	0.4997	0.5157	0.5321	0.5509	0.5460	0.5001	0.4450
$\sigma\widehat{H}_{20}^{(4)}$	0.0414	0.0626	0.0754	0.0846	0.0934	0.1021	0.1793	0.2899	0.3838
$\mu\widetilde{H}_{20}^{(4)}$	0.5313	0.5527	0.5799	0.6122	0.6103	0.6230	0.6085	0.5064	0.3915
$\sigma\widetilde{H}_{20}^{(4)}$	0.1994	0.1489	0.1595	0.2177	0.3067	0.5534	0.8112	0.8873	0.9655
$\mu\bar{H}^{(4)}$	0.5164	0.5484	0.5751	0.6013	0.6249	0.7566	0.8640	0.7561	0.5888
$\sigma\bar{H}^{(4)}$	0.1531	0.0749	0.0905	0.1309	0.2300	0.4827	0.844	1.155	1.421

Table 4.5 Values of the estimators for $H \in (0, 1/2)$ and known scale coefficients

H	0.05	0.1	0.15	0.2	0.25	0.3	0.35	0.4	0.45
$\mu\widehat{H}_{20}(a)$	0.05	0.1	0.15	0.2	0.25	0.3	0.349	0.397	0.44
$100\sigma\widehat{H}_{20}(a)$	0.007	0.007	0.006	0.006	0.009	0.024	0.085	0.294	0.964
$\mu\widehat{H}_{20}(a,b)$	0.05	0.1	0.15	0.2	0.25	0.3	0.35	0.4	0.45
$100\sigma\widehat{H}_{20}(a,b)$	0.007	0.007	0.006	0.006	0.006	0.006	0.006	0.006	0.007

Table 4.6 Values of the estimators for $H \in [1/2, 3/4)$ and known scale coefficients

H	0.5	0.525	0.55	0.575	0.6	0.625	0.65	0.675	0.7	0.725
$\mu\widetilde{H}_{20}(b)$	0.4999	0.5249	0.5501	0.5749	0.5999	0.6248	0.648	0.6744	0.6934	0.7307
$10\sigma\widetilde{H}_{20}(b)$	0.0196	0.0288	0.0432	0.0713	0.1251	0.2136	0.4055	0.7316	1.48	2.402
$\mu\widehat{H}_{20}(a,b)$	0.5	0.525	0.55	0.575	0.6	0.625	0.65	0.6752	0.7008	0.7306
$10\sigma\widehat{H}_{20}(a,b)$	0.0011	0.0015	0.0025	0.0043	0.0073	0.0140	0.0252	0.0485	0.0957	0.2398

Chapter 5
Drift Parameter Estimation in Diffusion and Fractional Diffusion Models

The present chapter is devoted to the drift parameter estimation in the diffusion, fractional diffusion and mixed Brownian-fractional Brownian diffusion models. More precisely, we consider the solutions of SDEs involving Wiener process, fractional or mixed fractional Brownian motion with the linear unknown drift parameter to be estimated. In the case when fBm is involved, we always assume Hurst index to be known.

The statistical inference for diffusion models driven by the standard Brownian motion has been thoroughly studied by now, presented in many papers and summarized in several books, see e.g., [26, 69, 75, 87, 106, 112, 145] and references cited therein. At the same time, the drift parameter estimation in models involving fBm was developed only for several particular cases. In particular, a simple linear fractional model was considered in the papers [23, 73, 107, 131]. A fractional Ornstein–Uhlenbeck process (fO-U) is another popular and well-studied model with fBm. The maximum likelihood estimator (MLE) of the drift parameter for this process was constructed in [90] and investigated in [21, 161, 166]. Several non-standard estimates for the drift parameter of an ergodic fO-U were proposed in [71] and further studied in [74]. The corresponding non-ergodic case was treated in [15, 53, 162]. In the papers [36, 55, 56, 72, 169, 170] drift parameter estimators were constructed via discrete observations. In the present chapter we focus on rather general fractional diffusion models, and study drift parameter estimators in continuous and discrete settings. However, we also pay special attention to the fractional Ornstein–Uhlenbeck model mentioned above. Also in this chapter we consider the drift parameter estimation for several extensions of fractional models such as a mixed fractional diffusion model, a linear model involving two independent fBms and multifractional models. In all the cases we concentrate on the strong consistency results.

We start in Sect. 5.1 with the standard diffusion model involving a Wiener process and give rather general condition for the strong consistency of the MLE of the drift parameter. In Sect. 5.2 we consider the parameter estimation in the

© Springer International Publishing AG 2017

K. Kubilius et al., *Parameter Estimation in Fractional Diffusion Models*,
B&SS – Bocconi & Springer Series 8,
https://doi.org/10.1007/978-3-319-71030-3_5

fractional diffusion model. At first, we construct the MLE and present sufficient conditions for its strong consistency. Then we introduce a transferred MLE. We call it so, because it has the same form as the MLE for the diffusion model. However, reader is advised that the stochastic representation of this estimator is different. The advantages of this estimator compared to the MLE are described in Sect. 5.2.6, see Examples 5.27 and 5.29. In Sect. 5.2.7 we consider a mixed model, which includes both standard and fractional Brownian motion. The MLE for this model does not exist. Therefore we construct an analog of the transferred MLE and prove its strong consistency. In Sect. 5.3 we investigate the estimation of the drift parameter in the fractional diffusion model by discrete observations. We consider the dicretization of the transferred MLE. The strong consistency of this discretized estimator holds under quite a complicated additional condition. Therefore, we introduce two other discrete estimators, for which this condition is not required. All three estimators are compared by simulations. Section 5.4 studies in detail the fractional Ornstein–Uhlenbeck process (fO-U). This is one of the most investigated examples of the fractional diffusion model, which is quite simple, because it contains additive fBm. Moreover, this process is well-defined for all $H \in (0, 1)$, while in the general case we usually assume that $H > \frac{1}{2}$. We start with the asymptotic properties of the MLE for the fO-U. Since it has rather complicated form, we consider two other strong consistent estimators depending on the drift parameter sign. For the negative parameter we study the estimator that is based on ergodic properties of the stationary process related to the model. For the positive parameter we prove the strong consistency of the least squares estimator, which in this case coincides with the transferred MLE. In Sect. 5.4.4 we study the hypotheses testing of the drift parameter sign for the fO-U. In Sect. 5.5 we construct the MLE for a simple linear model involving two independent fBms and prove its strong consistency. Section 5.6 deals with multifractional generalizations of linear and (non-ergodic) Ornstein–Uhlenbeck models. We apply the rate of asymptotic growth with probability 1 of mfBm and its increments studied in Appendix B.4 in order to get the strong consistency of the transferred MLE of the drift parameter.

5.1 Drift Parameter Estimation in the Homogeneous Diffusion Model: Standard MLE Is Always Strongly Consistent

There is an extended literature devoted to standard and nonstandard approaches to the drift parameter estimation in the diffusion models that involve a Wiener process, both for discrete and continuous observations. We mention only the books of Heyde [69], Kessler, Lindner, Sørensen (eds.) [87], Liptser and Shiryaev [112], Prakasa Rao [145], and references therein. Many complicated models have been studied. However, there was a curious gap even in the case of simplest homogeneous diffusion model: there were no conditions for the strong consistency of the standard MLE that are close to be necessary and are sufficiently mild. This section fills the gap. We prove that, in some sense, the standard MLE is always strongly consistent unless the drift coefficient is identically zero. To achieve this goal, we apply the results of Sect. 1.2.3 about the integral functionals of diffusion process.

In Sect. 5.1.1 we prove that the denominator in the stochastic representation of the standard MLE tends to infinity with probability 1 under very mild conditions and deduce from here the strong consistency of the standard MLE, unless the drift coefficient is identically zero. In view of this result, we say that the standard MLE is always strongly consistent. In Sect. 5.1.2 we establish the sufficient conditions for the strong consistency of the discretized version of the MLE. Some simulation results are included into Sect. 5.1.3.

5.1.1 Strong Consistency of the Drift Parameter Maximum-Likelihood Estimator Constructed for Continuous Observations

Consider SDE (1.22), but with unknown drift parameter. So, consider the equation of the form

$$dX_t^x = \theta a(X_t^x)dt + b(X_t^x)dW_t, \quad t \geq 0, \tag{5.1}$$

where θ is unknown parameter to be estimated. We assume that (5.1) admits the unique strong global solution and consider this solution X^x. From now on we fix the initial value x of the homogeneous diffusion process X^x that solves (5.1), and omit superscript x. So, we assume that the process X is observable at any point $t \geq 0$. Introduce the functions $c(x) = \frac{a(x)}{b^2(x)}$ and $d(x) = \frac{a(x)}{b(x)}$, so that $c^2(x) = \frac{d^2(x)}{b^2(x)}$. In what follows we suppose that the following conditions hold:

(D$_1$) $b(x) \neq 0$ for $x \in \mathbb{R}$;
(D$_2$) functions $\frac{1}{b^2}$, c, and $\frac{d^2}{b^2}$ are locally integrable.

Consider the local time $L_t(y)$ of the process X at the point y on the interval $[0, t]$, $t \geq 0$, defined by formula (1.24), where, as we decided, initial value is omitted. Applying formula (1.29), we get that under the condition of local integrability, the integral $\int_0^t d(X_s)ds$ is finite a. s. for any $t > 0$. Now produce the formula for likelihood function to construct the estimator of parameter θ. In some sense, it will be pseudo-likelihood function since we shall not check if we can apply the Girsanov theorem. So, let P_θ be the probability measure that corresponds to the process X, and P_0 be the probability measure that corresponds to the process satisfying equation (5.1) with $\theta = 0$. Then according to the Girsanov theorem, if we consider the process of the form $\widetilde{W}_t = W_t + \int_0^t d(X_s)ds$, then it will be a Wiener process w. r. t. a probability measure P_0, where

$$\frac{d\mathsf{P}_0(t)}{d\mathsf{P}_\theta(t)} = \exp\left\{ -\theta \int_0^t d(X_s)dW_s - \frac{\theta^2}{2}\int_0^t d^2(X_s)ds \right\}, \tag{5.2}$$

if this relation indeed defines a new probability measure, which can be verified, for instance, with the help of Novikov's condition. However, as we said before, we are not interested if we indeed get a new probability measure, i. e., if $\mathsf{E}\left(\frac{d\mathsf{P}_0(t)}{d\mathsf{P}_\theta(t)}\right) = 1$ because our goal is the construction of the likelihood function for the estimator. Recall the notation $c(x) = \frac{a(x)}{b^2(x)}$. Now we transform the formula (5.2) to the form

$$\frac{d\mathsf{P}_\theta(t)}{d\mathsf{P}_0(t)} = \exp\left\{ \theta \int_0^t d(X_s)dW_s + \frac{\theta^2}{2}\int_0^t d(X_s)ds \right\}$$
$$= \exp\left\{ \theta \int_0^t c(X_s)dX_s - \frac{\theta^2}{2}\int_0^t d(X_s)ds \right\},$$

and after the differentiation in θ we get that the MLE of parameter θ constructed by the observations of X on the interval $[0, t]$, has the form

$$\hat{\theta}_t = \frac{\int_0^t c(X_s)dX_s}{\int_0^t d^2(X_s)ds} = \theta + \frac{\int_0^t d(X_s)dW_s}{\int_0^t d^2(X_s)ds}. \tag{5.3}$$

In order to establish the criteria of the strong consistency of $\hat{\theta}_t$ in terms of the coefficients a and b, we apply Theorem 1.33 and consider the asymptotic behavior of the integral $\int_0^t d^2(X_s)ds$ under the fixed value of parameter $\theta \neq 0$. Two cases can be considered, depending on the behavior as $y \to \pm\infty$ of the function

$$\Phi(x, y) = \int_x^y \exp\left\{-2\theta \int_x^z c(u)du\right\} dz,$$

which is the function Φ from (1.23) for the parametrized equation (5.1).

Assume that for some value of parameter $\theta \in \mathbb{R}$ we have that for some (consequently any) $x \in \mathbb{R}$ $\Phi(x, +\infty) = -\Phi(x, -\infty) = +\infty$. In this case the diffusion process X is recurrent and its trajectories have the following property: $\overline{\lim}_{t\to\infty} X_t = +\infty$ a. s. and $\underline{\lim}_{t\to\infty} X_t = -\infty$ a. s.

Furthermore, equality (1.30) and the fact that $L_\infty(y) = \infty$ P-a. s. for any $y \in \mathbb{R}$ and recurrent process X, imply that $\int\limits_0^\infty f(X_s)ds = \infty$ a. s. for any f that is not identically 0, and in this case

$$\int\limits_0^\infty d^2(X_s)ds = \infty \quad \mathsf{P}\text{-a. s.}, \tag{5.4}$$

unless $a(y) \equiv 0$. Now, let at least one of the integrals $\Phi(x, +\infty)$ or $\Phi(x, -\infty)$ be finite. In this case the process X is transient.

Theorem 5.1 *Assume that conditions* (D_1)–(D_2) *hold.*

(i) Let for some $\theta \neq 0$ $\Phi(x, +\infty) < +\infty$. *Then*

$$I_1(d^2) = \int\limits_0^{+\infty} \frac{d^2(x)}{b^2(x)} \Phi(x, +\infty)dx = +\infty.$$

(ii) Let for some $\theta \neq 0$ $\Phi(x, -\infty) > -\infty$. *Then*

$$I_2(d^2) := \int\limits_{-\infty}^{0} \frac{d^2(x)}{b^2(x)} \Phi(x, -\infty)dx = -\infty.$$

Proof We prove only the first statement since the second one can be proved similarly. Note that $\Phi(x, +\infty) = \int_x^{+\infty} \exp\Big\{ - 2\theta \int_x^y c(u)du\Big\}dy$. Without loss of generality, we can put $\theta = 1$. Therefore, applying the Fubini theorem for nonnegative integrands and the Schwartz inequality, we get

$$\begin{aligned} I_1(d^2) &= \int_0^\infty c^2(x) \int_x^\infty \exp\Big\{-2\int_x^y c(u)du\Big\}dydx \\ &= \int_0^\infty \int_0^y c^2(x) \exp\Big\{-2\int_x^y c(u)du\Big\}dxdy \\ &\geq \int_0^\infty \Big(\int_0^y c(x) \exp\Big\{-\int_x^y c(u)du\Big\}dx\Big)^2 \frac{dy}{y} \\ &\geq \int_1^\infty \Big(\int_1^y c(x) \exp\Big\{-\int_x^y c(u)du\Big\}dx\Big)^2 \frac{dy}{y} \\ &= \int_1^\infty \Big(1 - \exp\Big\{-\int_1^y c(u)du\Big\}\Big)^2 \frac{dy}{y}. \end{aligned} \tag{5.5}$$

It is sufficient to prove that the last integral in (5.5) diverges. However, it consists of three terms, one of which, $\int_1^\infty \frac{dy}{y}$ diverges, and two other converge:

$$\int_1^\infty \exp\Big\{-\int_1^y c(u)du\Big\}\frac{dy}{y} \le (\Phi(1,\infty))^{\frac{1}{2}}\Big(\int_1^\infty \frac{dy}{y^2}\Big)^{\frac{1}{2}} < \infty$$

and

$$\int_1^\infty \exp\Big\{-2\int_1^y c(u)du\Big\}\frac{dy}{y} \le \Phi(1,\infty) < \infty.$$

Corollary 5.2 *As an immediate consequence of Theorems 1.33, 5.1 for transient processes and formula* (5.4) *for recurrent ones, we get the following statement: let Eq.* (5.1) *have a unique strong solution, the coefficients a and b satisfy the conditions* (D_1) *and* (D_2), *and a is not identically zero. Then* $\int_0^\infty d^2(X_s)ds = +\infty$ P*-a. s.*

The next theorem is the main statistical result of this section.

Theorem 5.3 *Let Eq.* (5.1) *have a unique strong solution, coefficients a and b satisfy conditions* (D_1) *and* (D_2), *and a is not identically zero. Then MLE* $\hat{\theta}_t$ *is strongly consistent as* $t \to \infty$.

Proof According to representation (5.3), $\hat{\theta}_t$ is strongly consistent if for locally square-integrable martingale $M_t = \int_0^t \frac{a(X_s)}{b(X_s)}dW_s$ we have that $\frac{M_t}{\langle M\rangle_t} \to 0$ P-a. s. However, according to the strong law of large numbers for martingales (Theorem B.3), under condition $\langle M\rangle_t \to \infty, t \to \infty$ P-a. s., we have that $\frac{M_t}{\langle M\rangle_t} \to 0$ P-a. s. The proof immediately follows now from Corollary 5.2. □

Remark 5.4 As to asymptotic normality, note that from (5.3)

$$\left(\int_0^t d^2(X_s)ds\right)^{\frac{1}{2}}\left(\hat{\theta}_t - \theta\right) = \frac{\int_0^t d(X_s)dW_s}{\left(\int_0^t d^2(X_s)ds\right)^{\frac{1}{2}}}.$$

Consider the stopping times

$$\tau_u = \inf\left\{t \ge 0 : \int_0^t d^2(X_s)ds = t\right\}.$$

Under conditions of Theorem 5.3 $\tau_u < \infty$ a. s. Furthermore, stochastic process $\int_0^{\tau_t} d(X_s)dW_s = \widetilde{W}_t$, where $\widetilde{W}$ is a Wiener process. It means that

$$\frac{\int_0^{\tau_t} d(X_s)dW_s}{\left(\int_0^{\tau_t} d^2(X_s)ds\right)^{\frac{1}{2}}} \stackrel{d}{=} \mathcal{N}(0,1),$$

i. e., has standard Gaussian distribution. In the particular case when the function $d(y) = d$ is constant, we have that

$$\left(\int_0^t d^2(X_s)ds\right)^{\frac{1}{2}} (\hat{\theta}_t - \theta) = t^{\frac{1}{2}} \cdot d \cdot (\hat{\theta}_t - \theta) = t^{-\frac{1}{2}} W_t \stackrel{d}{=} \mathcal{N}(0,1).$$

5.1.2 Discretization and Strong Consistency

Since real observations are discrete, consider the properties of dicretized estimator. Assume additionally that the coefficients a, b and c are bounded and Lipschitz. More precisely, they satisfy the condition: for some $a_0 > 0$ and $K > 0$ and for any $x, y \in \mathbb{R}$

(D$_3$) $|a(x)|+|b(x)|+|c(x)|+|d(x)| \le a_0, \quad |a(x)-a(y)|+|b(x)-b(y)| \le K|x-y|.$

Let $0 < \alpha < \frac{1}{2}$. Suppose that we observe the process X that is the solution of Eq. (5.1) only at discrete moments of time $t_k^n = \frac{k}{n}$, $0 \le k \le \lfloor n^{1+\alpha} \rfloor$, where $\lfloor a \rfloor$ means the nearest integer that is less than or equal to a. Consider a discretized version of the estimate $\hat{\theta}_t$:

$$\hat{\theta}_n^{discr} = \frac{\sum_{k=0}^{\lfloor n^{1+\alpha}\rfloor - 1} c(X_{\frac{k}{n}}) \bigtriangleup X_k^n}{\frac{1}{n} \sum_{k=0}^{\lfloor n^{1+\alpha}\rfloor - 1} d^2(X_{\frac{k}{n}})}, \tag{5.6}$$

where $\triangle X_k^n = X_{\frac{k+1}{n}} - X_{\frac{k}{n}}$. Then

$$\hat{\theta}_n^{discr} = \left(\frac{1}{n} \sum_{k=0}^{\lfloor n^{1+\alpha} \rfloor - 1} d^2(X_{\frac{k}{n}}) \right)^{-1}$$

$$\times \left(\sum_{k=0}^{\lfloor n^{1+\alpha} \rfloor - 1} c(X_{\frac{k}{n}}) \left(\theta \int_{\frac{k}{n}}^{\frac{k+1}{n}} a(X_s) ds + \int_{\frac{k}{n}}^{\frac{k+1}{n}} b(X_s) dW_s \right) \right) =$$

$$= \theta + \left(\frac{1}{n} \sum_{k=0}^{\lfloor n^{1+\alpha} \rfloor - 1} d^2(X_{\frac{k}{n}}) \right)^{-1} \left(\sum_{k=0}^{\lfloor n^{1+\alpha} \rfloor - 1} c(X_{\frac{k}{n}}) \theta \int_{\frac{k}{n}}^{\frac{k+1}{n}} (a(X_s) - a(X_{\frac{k}{n}})) ds + \right.$$

$$\left. + \sum_{k=0}^{\lfloor n^{1+\alpha} \rfloor - 1} d(X_{\frac{k}{n}}) \triangle W_k^n + \sum_{k=0}^{\lfloor n^{1+\alpha} \rfloor - 1} c(X_{\frac{k}{n}}) \int_{\frac{k}{n}}^{\frac{k+1}{n}} (b(X_s) - b(X_{\frac{k}{n}})) dW_s \right)$$

$$=: \theta + I_1^n + I_2^n + I_3^n.$$

We can consider also the "binary option" $\hat{\theta}_{2^n}^{discr}$ of $\hat{\theta}_n^{discr}$, replacing n for 2^n everywhere in (5.6). Let us start with the denominator $D_n := \frac{1}{n} \sum_{k=0}^{\lfloor n^{1+\alpha} \rfloor - 1} d^2(X_{\frac{k}{n}})$ and its "binary option" D_{2^n}.

Lemma 5.5

(i) Let the following condition hold:

(D_4) *Function $d = d(x)$ is bounded away from zero: $d(x) \geq d_0 > 0$.*

Then both D_n and D_{2^n} tend to ∞ a. s. as $n \to \infty$.

(ii) Let the following condition hold:

(D_5) *Function d is Lipschitz: there exists $K > 0$ such that for any $x, y \in \mathbb{R}$*

$$|d(x) - d(y)| \leq K|x - y|.$$

Then D_n tends to ∞ in probability as $n \to \infty$, D_{2^n} tends to ∞ a. s. as $n \to \infty$, and there exists $m > 0$ such that D_{n^m} tends to ∞ a. s. as $n \to \infty$.

Proof Under condition (D_4) it is obvious since in this case $D_n \geq d_0^2 \cdot n^{-1}\lfloor n^{1+\alpha}\rfloor$. Under condition ($D_5$) the following relations hold:

$$\left|\int_0^{n^{-1}\lfloor n^{1+\alpha}\rfloor} d^2(X_s)\,ds - D_n\right| \leq \sum_{k=0}^{\lfloor n^{1+\alpha}\rfloor-1} \int_{\frac{k}{n}}^{\frac{k+1}{n}} \left|d^2(X_s) - d^2\left(X_{\frac{k}{n}}\right)\right| ds$$

$$\leq 2a_0K \sum_{k=0}^{\lfloor n^{1+\alpha}\rfloor-1} \int_{\frac{k}{n}}^{\frac{k+1}{n}} \left|X_s - X_{\frac{k}{n}}\right| ds$$

$$\leq 2a_0K \sum_{k=0}^{\lfloor n^{1+\alpha}\rfloor-1} \int_{\frac{k}{n}}^{\frac{k+1}{n}} \left|\int_{\frac{k}{n}}^{s} a(X_u)\,du\right| ds + 2a_0K \sum_{k=0}^{\lfloor n^{1+\alpha}\rfloor-1} \int_{\frac{k}{n}}^{\frac{k+1}{n}} \left|\int_{\frac{k}{n}}^{s} b(X_u)\,dW_u\right| ds$$

$$= 2a_0K\left(D_n^{(1)} + D_n^{(2)}\right). \tag{5.7}$$

Evidently,

$$D_n^{(1)} \leq a_0\left\lfloor n^{1+\alpha}\right\rfloor \frac{1}{n^2} \leq a_0 n^{\alpha-1}, \tag{5.8}$$

so, $D_n^{(1)} \to 0$ a. s. as $n \to \infty$.

Further, denote

$$\varphi_n(s) = \sum_{k=0}^{\lfloor n^{1+\alpha}\rfloor-1} \left|\int_{\frac{k}{n}}^{s} b(X_u)\,dW_u\right| \mathbf{1}_{\left\{s\in\left[\frac{k}{n},\frac{k+1}{n}\right)\right\}}.$$

Then $D_n^{(2)} = \int_0^{n^{-1}\lfloor n^{1+\alpha}\rfloor} \varphi_n(s)\,ds$, and for any $\varepsilon > 0$

$$\mathsf{P}\left\{D_n^{(2)} \geq \varepsilon\right\} \leq \varepsilon^{-2}\left(\int_0^{n^{-1}\lfloor n^{1+\alpha}\rfloor} \mathsf{E}\varphi_n^2(s)\,ds\right)^{\frac{1}{2}} \left(n^{-1}\left\lfloor n^{1+\alpha}\right\rfloor\right)^{\frac{1}{2}}$$

$$= \varepsilon^{-2}\left(\sum_{k=0}^{\lfloor n^{1+\alpha}\rfloor-1} \int_{\frac{k}{n}}^{\frac{k+1}{n}} \mathsf{E}\left|\int_{\frac{k}{n}}^{s} b(X_u)\,dW_u\right|^2 ds\right)^{\frac{1}{2}} n^{\frac{\alpha}{2}}$$

$$
\begin{aligned}
&\le \varepsilon^{-2}\left(\sum_{k=0}^{\lfloor n^{1+\alpha}\rfloor-1}\int_{\frac{k}{n}}^{\frac{k+1}{n}}\left(\int_{\frac{k}{n}}^{s}\mathsf{E}b^2(X_u)\,du\right)ds\right)^{\frac{1}{2}} n^{\frac{\alpha}{2}}\\
&\le \varepsilon^{-2}a_0\left(\sum_{k=0}^{\lfloor n^{1+\alpha}\rfloor-1} n^{-2}\right)^{\frac{1}{2}} n^{\frac{\alpha}{2}} \le \varepsilon^{-2}a_0\left(n^{\alpha-1}\right)^{\frac{1}{2}} n^{\frac{\alpha}{2}} \qquad (5.9)\\
&= \varepsilon^{-2}a_0 n^{\alpha-\frac{1}{2}} \to 0, \quad n\to\infty.
\end{aligned}
$$

Taking into account that $\int_0^{n^{-1}\lfloor n^{1+\alpha}\rfloor} d^2(X_s)\,ds \to \infty$, a. s., $n\to\infty$, we get from (5.7)–(5.9) that $D_n \to \infty$ in probability as $n\to\infty$.

Now, let us replace n with 2^n in (5.7)–(5.9). Then for any $\varepsilon > 0$

$$
\mathsf{P}\left\{D_{2^n}^{(2)} \ge \varepsilon\right\} \le \varepsilon^{-2}a_0 2^{n(\alpha-\frac{1}{2})}.
$$

Since $\sum_{n=1}^{\infty} 2^{n(\alpha-\frac{1}{2})} < \infty$, we get from the Borel–Cantelli lemma that

$$
\int_0^{2^{-n}\lfloor 2^{n(1+\alpha)}\rfloor} d^2(X_s)\,ds - D_{2^n} \to 0 \quad \text{a. s. as } n\to\infty,
$$

whence $D_{2^n} \to \infty$ a. s. as $n\to\infty$.

In the case when we replace n with n^m, we get that

$$
\mathsf{P}\left\{D_{n^m}^{(2)} \ge \varepsilon\right\} \le \varepsilon^{-2}a_0 n^{m(\alpha-\frac{1}{2})}.
$$

If $m(\alpha-\frac{1}{2}) < -1$, i. e. $m > (\frac{1}{2}-\alpha)^{-1}$, we conclude as above. □

Theorem 5.6 *Let Eq.* (5.1) *have a unique strong solution, the coefficients a and b satisfy conditions* (D₁)–(D₃)*, a is not identically zero. Then:*

(i) Under condition (D₄) *both* $\hat{\theta}_n^{discr}$ *and* $\hat{\theta}_{2^n}^{discr}$ *are strongly consistent as* $n\to\infty$.
(ii) Under condition (D₅) $\hat{\theta}_n^{discr}$ *is consistent,* $\hat{\theta}_{2^n}^{discr}$ *is strongly consistent and there exists* $m>0$ *such that* $\hat{\theta}_{n^m}^{discr}$ *is strongly consistent as* $n\to\infty$.

Proof Consider asymptotic behavior of I_r^n, $r = 1, 2, 3$ as $n \to \infty$ separately. Denote J_1^n the numerator of I_1^n. Since

$$|J_1^n| = \left| \sum_{k=0}^{\lfloor n^{1+\alpha} \rfloor - 1} c(X_{\frac{k}{n}})\theta \int_{\frac{k}{n}}^{\frac{k+1}{n}} \left(a(X_s) - a(X_{\frac{k}{n}})\right) ds \right|$$

$$\leq |\theta|\, a_0 K \sum_{k=0}^{\lfloor n^{1+\alpha} \rfloor - 1} \int_{\frac{k}{n}}^{\frac{k+1}{n}} \left| X_s - X_{\frac{k}{n}} \right| ds,$$

it can be bounded similarly to (5.7)–(5.9), and we conclude that $J_1^n \to 0$ in probability as $n \to \infty$, $J_1^{2^n} \to 0$ a. s. as $n \to \infty$, and $J_1^{n^m} \to 0$ a. s. as $n \to \infty$ for $m > (\frac{1}{2} - \alpha)^{-1}$. Therefore $I_1^n \to 0$ a. s. as $n \to \infty$ under condition (D_4) and $I_1^n \to 0$ in probability under condition (D_5). Further, $I_1^{2^n}$ and $I_1^{n^m}$ tend to zero a. s. under any of conditions (D_4), (D_5).

Consider the term I_2^n. Denote martingale

$$N_n := \sum_{k=0}^{\lfloor n^{1+\alpha} \rfloor - 1} d(X_{\frac{k}{n}}) \,\Delta\, W_k^n,$$

so that $I_2^n = \frac{N_n}{D_n}$. Then under condition (D_4) for any $\varepsilon > 0$ and any $p > 1$ it follows from the Burkholder–Gandy inequality that

$$\mathsf{P}\left\{\frac{N_n}{D_n} \geq \varepsilon\right\} \leq \mathsf{P}\left\{N_n \geq \varepsilon d_0^2 n^{-1} \lfloor n^{1+\alpha} \rfloor\right\} \leq \frac{\mathsf{E}\,|N_n|^p}{\varepsilon^p d_0^{2p} n^{-p} \lfloor n^{1+\alpha} \rfloor^p}$$

$$= \frac{\mathsf{E}\left| \int_0^{n^{-1}\lfloor n^{1+\alpha} \rfloor} \psi_n(s)\, dW_s \right|^p}{\varepsilon^p d_0^{2p} n^{-p} \lfloor n^{1+\alpha} \rfloor^p} \leq \frac{C_p \mathsf{E}\left(\int_0^{n^{-1}\lfloor n^{1+\alpha} \rfloor} \psi_n^2(s)\, ds \right)^{\frac{p}{2}}}{\varepsilon^p d_0^{2p} n^{-p} \lfloor n^{1+\alpha} \rfloor^p}$$

$$\leq \frac{C_p d_0^p n^{\frac{\alpha p}{2}}}{\varepsilon^p d_0^{2p} n^{\alpha p}} + o\left(n^{-\frac{\alpha p}{2}}\right) = C_p^{(1)} n^{-\frac{\alpha p}{2}} + o\left(n^{-\frac{\alpha p}{2}}\right),$$

where

$$\psi_n(s) = \sum_{k=0}^{\lfloor n^{1+\alpha} \rfloor - 1} d(X_{\frac{k}{n}}) \mathbf{1}_{\left\{s \in \left[\frac{k}{n}, \frac{k+1}{n}\right)\right\}}.$$

Choosing $p > \frac{2}{\alpha}$ and applying the Borel–Cantelli lemma we get that $\frac{N_n}{D_n} \to 0$ a. s. as $n \to \infty$.

Under condition (D_5) let us do the following. Denote

$$\widetilde{N}_n = \int_0^{n^{-1}\lfloor n^{1+\alpha}\rfloor} d(X_s)\,dW_s, \quad \widetilde{D}_n = \int_0^{n^{-1}\lfloor n^{1+\alpha}\rfloor} d^2(X_s)\,ds,$$

and consider the difference

$$\left|\frac{\widetilde{N}_n}{\widetilde{D}_n} - \frac{N_n}{D_n}\right| \le \frac{\left|\widetilde{N}_n - N_n\right|}{\widetilde{D}_n} + \frac{\left|N_n - \widetilde{N}_n\right|\left|D_n - \widetilde{D}_n\right|}{D_n\widetilde{D}_n} + \frac{\left|\widetilde{N}_n\right|\left|D_n - \widetilde{D}_n\right|}{D_n\widetilde{D}_n}.$$

Since

$$\left|\widetilde{N}_n - N_n\right| \le \left|\sum_{k=0}^{\lfloor n^{1+\alpha}\rfloor-1} \int_{\frac{k}{n}}^{\frac{k+1}{n}} \left(d(X_s) - d\left(X_{\frac{k}{n}}\right)\right) dW_s\right|,$$

we have that

$$\mathsf{E}\left|\widetilde{N}_n - N_n\right|^2 \le \mathsf{E}\sum_{k=0}^{\lfloor n^{1+\alpha}\rfloor-1} \int_{\frac{k}{n}}^{\frac{k+1}{n}} \left(d(X_s) - d\left(X_{\frac{k}{n}}\right)\right)^2 ds$$

$$\le K^2 \sum_{k=0}^{\lfloor n^{1+\alpha}\rfloor-1} \int_{\frac{k}{n}}^{\frac{k+1}{n}} \mathsf{E}\left|X_s - X_{\frac{k}{n}}\right|^2 ds \le 2a_0^2K^2n^{1+\alpha-2} \to 0 \quad \text{as } n \to \infty.$$

Therefore, $\frac{\left|\widetilde{N}_n - N_n\right|}{\widetilde{D}_n} \to 0$ in probability as $n \to \infty$. Furthermore, according to the strong law of large numbers for martingales (Theorem B.3), $\frac{\widetilde{N}_n}{\widetilde{D}_n} \to 0$ a. s. as $n \to \infty$. It follows from Lemma 5.5 that $\left|D_n - \widetilde{D}_n\right| \to 0$ and $D_n \to \infty$ in probability as $n \to \infty$. Therefore, $I_2^n = \frac{N_n}{D_n} \to 0$ in probability as $n \to \infty$. Similarly to above estimates, we can apply the Borel–Cantelli lemma in order to prove that $I_2^{2^n}$ and $I_2^{n^m} \to 0$ a. s. as $n \to \infty$ for $m > \frac{1}{\frac{1}{2}-\alpha}$.

Finally, consider I_3^n. Its numerator

$$J_3^n := \sum_{k=0}^{\lfloor n^{1+\alpha}\rfloor-1} c\left(X_{\frac{k}{n}}\right) \int_{\frac{k}{n}}^{\frac{k+1}{n}} \left(b(X_s) - b\left(X_{\frac{k}{n}}\right)\right) dW_s$$

is bounded in probability; moreover if we denote

$$\zeta_n(s) = \sum_{k=0}^{\lfloor n^{1+\alpha} \rfloor - 1} c\left(X_{\frac{k}{n}}\right)\left(b(X_s) - b\left(X_{\frac{k}{n}}\right)\right) \mathbf{1}_{\left\{s \in \left[\frac{k}{n}, \frac{k+1}{n}\right)\right\}},$$

then

$$\mathsf{E}\,|J_3^n|^2 = \mathsf{E}\left|\int_0^{n^{-1}\lfloor n^{1+\alpha} \rfloor} \zeta_n(s)\,dW_s\right|^2 = \int_0^{n^{-1}\lfloor n^{1+\alpha} \rfloor} \mathsf{E}\zeta_n^2(s)\,ds$$

$$\leq a_0^2 K^2 \sum_{k=0}^{\lfloor n^{1+\alpha} \rfloor - 1} \int_{\frac{k}{n}}^{\frac{k+1}{n}} \mathsf{E}\left|X_s - X_{\frac{k}{n}}\right|^2 ds \leq a_0^2 K^2 n^{1+\alpha} n^{-2} \to 0, \quad \text{as } n \to \infty,$$

so that $J_3^n \to 0$ in probability as $n \to \infty$. Hence $I_3^n \to 0$ in probability as $n \to \infty$, and $I_3^{2^n}, I_3^{n^m} \to 0$ a. s. as $n \to \infty$. Theorem is proved. □

5.1.3 Some Simulation Results

We have simulated the model with the following four examples of the coefficients:

1. $a(x) = 2 + \sin x$, $b(x) = 2 + \cos x$,
2. $a(x) = 1 + x$, $b(x) = 2 + \sin x$,
3. $a(x) = 1 + x$, $b(x) = \sqrt{1 + x^2}$,
4. $a(x) = \log\left(1 + x^2\right)$, $b(x) = 1$.

We set $X_0 = 1$ and consider two values of the parameter to be estimated: $\theta = 2$ and $\theta = -2$. The discretization interval is $\Delta t = 5 \cdot 10^{-5}$; number of the simulated trajectories for each set of parameters is 100.

Recall that we observe the process X at the moments $t_k^n = \frac{k}{n}$, $0 \leq k \leq \lfloor n^{1+\alpha} \rfloor$, i.e., the observation interval is $[0, T]$, where $T = \lfloor n^{1+\alpha} \rfloor n^{-1}$. In particular, for $n = 10{,}000$ we have $T = 6.3095$ in the case $\alpha = 0.2$, and $T = 39.8107$ in the case $\alpha = 0.4$.

We calculate the values of the discretized estimator $\hat{\theta}_n^{discr}$ defined by (5.6) for $\alpha = 0.2$ and $\alpha = 0.4$ and various values of n. The means and standard deviations are reported in Tables 5.1, 5.2, 5.3, and 5.4. We see that the convergence to the true value of parameter holds in all cases. However the rate of convergence differs substantially depending on the coefficients as well as on the sign of parameter. For example, for $\theta = 2$ the estimator demonstrates the best performance in the case $a(x) = 1 + x$, $b(x) = 2 + \sin x$, but for $\theta = -2$ the rate of convergence is the slowest for these coefficients.

Table 5.1 $\theta = 2, \alpha = 0.2$

			n					
$a(x)$	$b(x)$		50	100	500	1000	5000	10,000
$2 + \sin x$	$2 + \cos x$	Mean	1.9171	1.9252	1.9401	1.9521	1.9535	1.9609
		Std.dev.	0.5350	0.5154	0.4373	0.4003	0.3422	0.3168
$1 + x$	$2 + \sin x$	Mean	1.8924	1.8835	1.9427	1.9845	1.9991	2.0001
		Std.dev.	0.9655	0.8896	0.4440	0.1871	0.0124	0.0002
$1 + x$	$\sqrt{1 + x^2}$	Mean	2.1034	2.0656	2.0580	2.0377	2.0221	2.0065
		Std.dev.	0.6387	0.5637	0.4567	0.4256	0.3932	0.3886
$\log(1 + x^2)$	1	Mean	1.8219	1.8806	1.9636	1.9964	1.9980	1.9982
		Std.dev.	0.5724	0.4185	0.2395	0.1574	0.1181	0.0994

Table 5.2 $\theta = 2, \alpha = 0.4$

			n					
$a(x)$	$b(x)$		50	100	500	1000	5000	10,000
$2 + \sin x$	$2 + \cos x$	Mean	1.9235	1.9483	1.9769	1.9850	2.0005	2.0071
		Std.dev.	0.3580	0.3113	0.2109	0.1995	0.1602	0.1322
$1 + x$	$2 + \sin x$	Mean	2.0376	2.0200	2.0039	2.0019	2.0003	2.0001
		Std.dev.	0.0399	0.0006	$1 \cdot 10^{-8}$	$4 \cdot 10^{-12}$	$2 \cdot 10^{-13}$	$6 \cdot 10^{-13}$
$1 + x$	$\sqrt{1 + x^2}$	Mean	2.0679	2.0249	2.0009	1.9967	2.0029	2.0111
		Std.dev.	0.4375	0.3982	0.2760	0.2538	0.1945	0.1729
$\log(1 + x^2)$	1	Mean	2.0003	2.0007	1.9953	1.9977	1.9995	1.9998
		Std.dev.	0.1421	0.0919	0.0398	0.0332	0.0198	0.0153

Table 5.3 $\theta = -2, \alpha = 0.2$

			n					
$a(x)$	$b(x)$		50	100	500	1000	5000	10,000
$2 + \sin x$	$2 + \cos x$	Mean	−1.9653	−1.9928	−1.9779	−1.9754	−2.0134	−2.0180
		Std.dev.	0.5517	0.4875	0.3729	0.3361	0.2917	0.2750
$1 + x$	$2 + \sin x$	Mean	−2.7707	−2.6468	−2.4752	−2.4336	−2.3662	−2.3113
		Std.dev.	1.7097	1.5527	1.2282	1.1972	1.0513	0.9522
$1 + x$	$\sqrt{1 + x^2}$	Mean	−2.2843	−2.2955	−2.2907	−2.2435	−2.2701	−2.2790
		Std.dev.	0.9114	0.8945	0.8056	0.8266	0.7912	0.7681
$\log(1 + x^2)$	1	Mean	−2.1506	−2.0081	−1.9403	−1.9314	−1.9538	−1.9273
		Std.dev.	1.1211	0.9518	0.7274	0.6381	0.5392	0.4728

Table 5.4 $\theta = -2, \alpha = 0.4$

$a(x)$	$b(x)$		n = 50	100	500	1000	5000	10,000
$2 + \sin x$	$2 + \cos x$	Mean	−2.0170	−2.0298	−2.0055	−2.0060	−2.0302	−2.0107
		Std.dev.	0.3341	0.2742	0.2377	0.2083	0.1624	0.1410
$1 + x$	$2 + \sin x$	Mean	−2.4224	−2.2797	−2.1821	−2.1454	−2.0573	−2.0532
		Std.dev.	1.1524	0.9692	0.6149	0.5230	0.3861	0.3473
$1 + x$	$\sqrt{1+x^2}$	Mean	−2.2071	−2.2454	−2.1052	−2.0826	−2.0347	−2.0418
		Std.dev.	0.8203	0.7614	0.5767	0.5147	0.3876	0.3404
$\log(1 + x^2)$	1	Mean	−1.9355	−1.9329	−1.9792	−1.9986	−2.0013	−2.0010
		Std.dev.	0.5271	0.4483	0.1260	0.0362	0.0195	0.0152

5.2 Estimation in Fractional Diffusion Model by Continuous Observations

Modern mathematical statistics tends to shift away from the standard statistical schemes based on independent random variables; besides, these days many statistical models are based on continuous time. Therefore, the corresponding statistical problems (e.g., parameter estimation) can be handled by methods of the theory of stochastic processes in addition to the standard statistical methods. Statistics for stochastic processes is well-developed for diffusion processes and even for semimartingales (see, for instance, [110]) but is still developing for the processes with long-range dependence. The latter is an integral part of stochastic processes, featuring a wide spectrum of applications in economics, physics, finance and other fields. The present section is devoted to the parameter estimation in such models involving fractional fBm with Hurst parameter $H > \frac{1}{2}$ which is a well-known long-memory process.

In Sect. 5.2.1 we describe the fractional diffusion model and derive the standard MLE for the unknown drift parameter. In Sect. 5.2.2 we propose a non-standard "transferred maximum likelihood estimator" and give sufficient conditions for its strong consistency. The proofs are based on auxiliary results for the asymptotic growth of Gaussian processes and, in particular, asymptotic growth of the fractional derivative of fBm. These results are established in Appendix B.3. Some examples of models satisfying the strong consistency conditions are investigated in Sect. 5.2.3. Moreover, for these examples we study the convergence rate of the remainder term to zero. The numerical illustrations of the results mentioned above are given in Sect. 5.2.4. In Sect. 5.2.5 we introduce sequential versions of the MLE and the transferred MLE and establish their properties. The linear model is studied in detail in Sect. 5.2.6. Using this simple model, we compare properties of the estimators and show advantages of the transferred MLE, see Examples 5.27–5.29.

In Sect. 5.2.7 we study a mixed model based on both standard and fractional Brownian motion which turns out to be more flexible. One of the reasons to consider such model comes from the modern mathematical finance where it has become very popular to assume that the underlying random noise consists of two parts: the fundamental part, describing the economical background for the stock price, and the trading part, related to the randomness inherent to the stock market. In our case the fundamental part of the noise has a long memory while the trading part is a white noise. In the mixed model the standard MLE does not exist but the transferred MLE works. A sequential version of the latter estimator is also given. As an example we consider a linear mixed model.

5.2.1 *Model Description: The Standard MLE*

Assume that $H > \frac{1}{2}$ and consider a one-dimensional SDE driven by an fBm $B^H = \{B_t^H, t \in \mathbb{R}^+\}$

$$X_t = x_0 + \theta \int_0^t a(s, X_s)\, ds + \int_0^t b(s, X_s)\, dB_s^H,\ t \in \mathbb{R}^+, \tag{5.10}$$

where $x_0 \in \mathbb{R}$ is the initial value, θ is the unknown parameter to be estimated, the first integral in the right-hand side of (5.10) is the Lebesgue–Stieltjes integral, the second integral is the generalized Lebesgue–Stieltjes integral introduced in Definition 1.13. From now on, we shall assume that the coefficients of Eq. (5.10) satisfy the assumptions of Theorem 1.42 on any interval $[0, T]$. Then, according to that theorem, solution to Eq. (5.10) exists on any interval $[0, T]$ and is unique in the class of processes satisfying

$$\|X\|_{\infty,\alpha,T} < \infty \quad \text{a.s.} \tag{5.11}$$

for some $\alpha > 1 - H$.

In addition, suppose that the following assumption holds:

(E$_1$) $b(t, X_t) \neq 0, t \in [0, T]$ and $\frac{a(t,X_t)}{b(t,X_t)}$ is a. s. Lebesgue integrable on $[0, T]$ for any $T > 0$.

Denote $\psi(t, x) = \frac{a(t,x)}{b(t,x)}$, $\varphi(t) := \psi(t, X_t)$, and introduce the integral

$$J_t = \int_0^t l_H(t, s)\varphi(s)\, ds = c_H \int_0^t (t - s)^{\frac{1}{2}-H} s^{\frac{1}{2}-H} \varphi(s)\, ds. \tag{5.12}$$

where $l_H(t, s)$ and c_H are defined in (1.11)–(1.12). Also, let $M_t^H = \int_0^t l_H(t, s)\, dB_s^H$ be the corresponding Molchan martingale, see (1.13).

Consider two processes:

$$Y_t = \int_0^t b^{-1}(s, X_s)\, dX_s = \theta \int_0^t \varphi(s)\, ds + B_t^H \tag{5.13}$$

and

$$Z_t = \int_0^t l_H(t,s)\, dY_s = \theta J_t + M_t^H.$$

Remark 5.7 Note that the transformation from X to Z does not lead to loss of information since we can present Y (consequently, X) via Z and Volterra kernel (1.6). So, these processes generate the same filtration.
Also, note that we can rewrite process Z as

$$Z_t = \int_0^t l_H(t,s) b^{-1}(s, X_s)\, dX_s,$$

so Z is a functional of the observable process X. The following smoothness condition for the function ψ ensures the semimartingale property of Z.

Lemma 5.8 ([120, Lemma 6.3.2]) *Let $\psi = \psi(t,x) \in C^1(\mathbb{R}^+) \times C^2(\mathbb{R})$. Then for any $t > 0$*

$$\begin{aligned} J'(t) &= (2-2H) C_H \psi(0, x_0) t^{1-2H} + \int_0^t l_H(t,s) \left(\psi_t'(s, X_s) + \theta \psi_x'(s, X_s) a(s, X_s)\right) ds \\ &\quad - \left(H - \frac{1}{2}\right) c_H \int_0^t s^{-\frac{1}{2}-H} (t-s)^{\frac{1}{2}-H} \int_0^s \left(\psi_t'(u, X_u) + \theta \psi_x'(u, X_u) a(u, X_u)\right) du\, ds \\ &\quad + (2-2H) c_H t^{1-2H} \int_0^t s^{2H-3} \int_0^s u^{\frac{3}{2}-H} (s-u)^{\frac{1}{2}-H} \psi_x'(u, X_u) b(u, X_u)\, dB_u^H\, ds \\ &\quad + c_H t^{-1} \int_0^t u^{\frac{3}{2}-H} (t-u)^{\frac{1}{2}-H} \psi_x'(u, X_u) b(u, X_u)\, dB_u^H, \end{aligned} \tag{5.14}$$

where $C_H = \mathrm{B}(\frac{3}{2} - H, \frac{3}{2} - H) c_H = \left(\frac{\Gamma(\frac{3}{2}-H)}{2H\Gamma(H+\frac{1}{2})\Gamma(3-2H)}\right)^{\frac{1}{2}}$, and all of the involved integrals exist a. s.

Remark 5.9 Suppose that $\psi(t,x) \in C^1(\mathbb{R}^+) \times C^2(\mathbb{R})$ and limit $\varsigma(0) = \lim_{s \to 0} \varsigma(s)$ exists a. s., where $\varsigma(s) = s^{\frac{1}{2}-H} \varphi(s)$. In this case $J(t)$ can be presented as

$$J(t) = c_H \int_0^t (t-s)^{\frac{1}{2}-H} \varsigma(s)\, ds = \frac{c_H t^{\frac{3}{2}-H}}{\frac{3}{2} - H} \varsigma(0) + c_H \int_0^t \frac{(t-s)^{\frac{3}{2}-H}}{\frac{3}{2} - H} \varsigma'(s)\, ds,$$

and $J'(t)$ from (5.14) can be simplified to

$$J'(t) = c_H t^{\frac{1}{2}-H} \varsigma(0) + \int_0^t l_H(t,s)\Big(\left(\tfrac{1}{2} - H\right) s^{-1} \varphi(s) + \psi_t'(s, X_s) + \theta \psi_x'(s, X_s) a(s, X_s)\Big)\, ds + \int_0^t l_H(t,s) \psi_x'(s, X_s) b(s, X_s)\, dB_s^H.$$

Same way as Z, the processes J and J' are functionals of X. It is more convenient to consider process $\chi(t) = (2-2H)^{-1} J'(t) t^{2H-1}$, so that

$$Z_t = (2-2H)\theta \int_0^t \chi(s) s^{1-2H} ds + M_t^H = \theta \int_0^t \chi(s) d\langle M^H \rangle_s + M_t^H.$$

Suppose that the following conditions hold:

(E_2) $\mathsf{E} I_T := \mathsf{E} \int_0^T \chi_s^2\, d\langle M^H \rangle_s < \infty$ for any $T > 0$,
(E_3) $I_\infty := \int_0^\infty \chi_s^2\, d\langle M^H \rangle_s = \infty$ a. s.

Since we have reduced the problem of parameter estimation to the semimartingale model, we can construct the MLE, applying the standard Girsanov theorem. According to it, the likelihood function has the form

$$\exp\left\{ \theta \int_0^T \chi_s\, dM_s + \frac{\theta^2}{2} \int_0^T \chi_s^2\, d\langle M^H \rangle_s \right\} = \exp\left\{ \theta \int_0^T \chi_s\, dZ_s - \frac{\theta^2}{2} \int_0^T \chi_s^2\, d\langle M^H \rangle_s \right\}.$$

Therefore, the MLE has the form

$$\hat{\theta}_T^{(1)} = \frac{\int_0^T \chi_s\, dZ_s}{\int_0^T \chi_s^2\, d\langle M^H \rangle_s} = \theta + \frac{\int_0^T \chi_s\, dM_s^H}{\int_0^T \chi_s^2\, d\langle M^H \rangle_s}. \tag{5.15}$$

Condition (E_2) ensures that process $\int_0^t \chi_s\, dM_s^H$, $t > 0$ is a square integrable martingale, and condition (E_3) alongside with the strong law of large numbers for martingales (Theorem B.3) ensures that $\frac{\int_0^T \chi_s\, dM_s^H}{\int_0^T \chi_s^2\, d\langle M^H \rangle_s} \to 0$ a. s. as $T \to \infty$. Summarizing, we arrive at the following result.

Proposition 5.10 *Let the coefficients a and b satisfy the assumptions of Theorem 1.42 on any interval $[0, T]$. Let $\psi(t, x) \in C^1(\mathbb{R}^+) \times C^2(\mathbb{R})$ and assumptions (E_1)–(E_3) hold. Then estimator $\hat{\theta}_T^{(1)}$ is strongly consistent as $T \to \infty$.*

5.2.2 A Non-standard Transferred MLE

Taking into account the complicated form of the estimator $\hat{\theta}_T^{(1)}$, it is possible to construct another estimator for parameter θ, preserving the structure of the standard MLE for the standard diffusion process. In this connection, we call this estimator "transferred maximum likelihood estimator". Similar approach was applied in [71] to the fO-U with constant coefficients. We shall use process Y introduced in (5.13) to define the transferred MLE as

$$\hat{\theta}_T^{(2)} = \frac{\int_0^T \varphi_s \, dY_s}{\int_0^T \varphi_s^2 \, ds} = \theta + \frac{\int_0^T \varphi_s \, dB_s^H}{\int_0^T \varphi_s^2 \, ds}, \tag{5.16}$$

where $\varphi_s = \psi(s, X_s) = \frac{a(s,X_s)}{b(s,X_s)}$.

Now we shall establish conditions for the strong consistency of $\hat{\theta}_T^{(2)}$.

Theorem 5.11 *Let the coefficients a and b satisfy the assumptions of Theorem 1.42 on any interval* $[0, T]$. *Let the function* φ *satisfy the following assumption:*

(E$_4$) *There exist* $\alpha > 1 - H$ *and* $p > 1$ *such that*

$$\varrho_{\alpha,p,T} := \frac{T^{H+\alpha-1} (\log T)^p \int_0^T \left|(\mathcal{D}_{0+}^{\alpha} \varphi)(s)\right| \, ds}{\int_0^T \varphi_s^2 \, ds} \to 0 \quad a.\,s., \, as \, T \to \infty. \tag{5.17}$$

Then the estimator $\hat{\theta}_T^{(2)}$ *is correctly defined and strongly consistent as* $T \to \infty$.

Proof We must prove that $\frac{\int_0^T \varphi_s \, dB_s^H}{\int_0^T \varphi_s^2 \, ds} \to 0$ a. s., as $T \to \infty$. According to (1.21),

$$\left| \int_0^T \varphi_s \, dB_s^H \right| \le \sup_{0 \le t \le T} \left| (\mathcal{D}_{T-}^{1-\alpha} B_{T-}^H)(t) \right| \int_0^T \left| (\mathcal{D}_{0+}^{\alpha} \varphi)(s) \right| ds.$$

Furthermore, according to Theorem B.41, for any $p > 1$ there exists a random variable $\xi = \xi(p)$ independent of T such that for any $T > 0$

$$\sup_{0 \le t \le T} \left| (\mathcal{D}_{T-}^{1-\alpha} B_{T-}^H)(t) \right| \le \xi(p) T^{H+\alpha-1} (\log T)^p,$$

which concludes the proof. □

Relation (5.17) ensures convergence $\frac{\int_0^T \varphi_s dB_s^H}{\int_0^T \varphi_s^2 ds} \to 0$ a. s. in the general case. In a particular case when the function φ is non-random and consequently integral $\int_0^T \varphi_s dB_s^H$ is a Wiener integral w. r. t. the fBm, the conditions for existence of this integral are simpler since assumption (5.17) can be simplified.

Theorem 5.12 *Let the coefficients a and b satisfy the assumptions of Theorem 1.42 on any interval* $[0, T]$. *Let the assumptions* (E_1) *and* (E_2) *hold and let the function* φ *be non-random and satisfy the following assumption:*

$(\mathbf{E}_5)$ *There exists* $p > 0$ *such that*

$$\limsup_{T\to\infty} \frac{T^{2H-1+p}}{\int_0^T \varphi^2(t)\,dt} < \infty.$$

Then estimator $\hat{\theta}_T^{(2)}$ *is strongly consistent as* $T \to \infty$.

Proof It follows from Proposition 1.6 and the Hölder inequality that for any $r > 0$

$$\mathsf{E}\left|\int_0^T \varphi(s)\,dB_s^H\right|^r \le C(H,r)\,\|\varphi\|^r_{L_{\frac{1}{H}}[0,T]} \le C(H,r)\,\|\varphi\|^r_{L_2[0,T]}\,T^{(H-\frac{1}{2})r}.$$

Denote $F_T = \frac{\left|\int_0^T \varphi(t)\,dB_t^H\right|}{\int_0^T \varphi^2(t)\,dt}$. Also, for any $N > 1$ and any $\varepsilon > 0$ define event $A_N = \{F_N > \varepsilon\}$. Then

$$\mathsf{P}(A_N) \le \varepsilon^{-r} \frac{\mathsf{E}\left|\int_0^N \varphi(s)\,dB_s^H\right|^r}{\left(\int_0^N \varphi^2(t)\,dt\right)^r} \le \varepsilon^{-r} C(H,r) \frac{\|\varphi\|^r_{L_{\frac{1}{H}}[0,N]}}{\|\varphi\|^{2r}_{L_2[0,N]}} \le \varepsilon^{-r} C(H,r) \frac{N^{(H-\frac{1}{2})r}}{\|\varphi\|^r_{L_2[0,N]}}.$$

Under condition (E_5) we have $\mathsf{P}(A_N) \le C(H,r,p)N^{-\frac{rp}{2}}$. If $r > \frac{2}{p}$, then it follows immediately from the Borel–Cantelli lemma that series $\sum \mathsf{P}(A_N)$ converges, whence $F_N \to 0$ a. s., as $N \to \infty$. Now estimate the residual

$$R_N = \sup_{T\in[N,N+1]} |F_T - F_N|.$$

Evidently,

$$R_N \le \sup_{T\in[N,N+1]} \left|\frac{\int_N^T \varphi(t)\,dB_t^H}{\int_0^T \varphi^2(t)\,dt}\right| + F_N,$$

and it is sufficient to estimate

$$R_N^{(1)} = \sup_{T\in[N,N+1]} \left|\frac{\int_N^T \varphi(t)\,dB_t^H}{\int_0^T \varphi^2(t)\,dt}\right| \le \frac{\sup_{T\in[N,N+1]}\left|\int_N^T \varphi(t)\,dB_t^H\right|}{\int_0^N \varphi^2(t)\,dt} := R_N^{(2)}.$$

According to Proposition 1.7 and the Hölder inequality,

$$\mathsf{E}\left(\sup_{T\in[N,N+1]}\left|\int_N^T \varphi(t)\,dB_t^H\right|\right)^r \le C(H,r)\,\|\varphi\|^r_{L_{\frac{1}{H}}[N,N+1]} \le C(H,r)\,\|\varphi\|^r_{L_2[N,N+1]}\,.$$

Now we can use condition (E_5) to conclude that for any $\varepsilon > 0$

$$\begin{aligned} P\left(R_N^{(2)} > \varepsilon\right) &\le C(H,r)\varepsilon^{-r}\frac{\|\varphi\|^r_{L_2[N,N+1]}}{\|\varphi\|^{2r}_{L_2[0,N]}} \\ &\le C(H,r)\varepsilon^{-r}\|\varphi\|^{-r}_{L_2[0,N]} \le C(H,r)\varepsilon^{-r}N^{-r(2H-1+p)}. \end{aligned}$$

We can set $r > \frac{1}{2H-1+p}$ and apply the Borel–Cantelli lemma again. Then we obtain that $R_N^{(2)} \to 0$ a. s., as $N \to 0$, which means that $\hat{\theta}_T^{(2)}$ is strongly consistent. □

In the next subsection, we consider some examples of φ and establish not only the convergence to zero, but the rate of convergence as well. The examples from Sect. 5.2.3 are based on the results of the paper [14].

5.2.3 Examples of the Remainder Terms with the Estimation of the Rate of Convergence to Zero

Assume that the conditions of Theorem 5.11 are satisfied. Then the estimator $\hat{\theta}_T^{(2)}$ is correctly defined and we can study the properties of the remainder term $\varrho_{\alpha,p,T}$ defined in (5.17).

We start with the simplest case when φ is a power function, $\varphi(t) = t^a$, $a \ge 0$, $t \ge 0$. It means that $a(t,x) = b(t,x)t^a$.

Lemma 5.13 *Let $\varphi(t) = t^a$, $a \ge 0$, $t \ge 0$. Then $\varrho_{\alpha,p,T} = C_a T^{H-a-1}(\log T)^p \to 0$ as $T \to \infty$, where*

$$C_a = \frac{(2a+1)\Gamma(a+1)}{\Gamma(a-\alpha+2)}.$$

Proof For the power function $\varphi(t) = t^a$ we can use the definition of fractional derivative and get that for any $0 < \alpha < 1$

$$\begin{aligned} (\mathcal{D}_{0+}^{\alpha}\varphi)(x) &= \frac{1}{\Gamma(1-\alpha)}\frac{d}{dx}\int_0^x t^a(x-t)^{-\alpha}\,dt \\ &= \mathrm{B}(a+1,1-\alpha)\frac{1}{\Gamma(1-\alpha)}\left(x^{a+1-\alpha}\right)' = \frac{\Gamma(a+1)}{\Gamma(a+1-\alpha)}x^{a-\alpha}. \end{aligned}$$

Therefore, for $p > 1$ we have that

$$\begin{aligned}\varrho_{\alpha,p,T} &= \frac{T^{H+\alpha-1}(\log T)^p \int_0^T \left|(\mathcal{D}_{0+}^{\alpha}\varphi)(x)\right| dx}{\int_0^T \varphi^2(x)\,dx} \\ &= \frac{T^{H+\alpha-1}(\log T)^p \int_0^T \frac{\Gamma(a+1)}{\Gamma(a+1-\alpha)} x^{a-\alpha}\,dx}{\frac{T^{2a+1}}{2a+1}} \\ &= \frac{(2a+1)\Gamma(a+1)}{\Gamma(a+2-\alpha)} T^{H-a-1}(\log T)^p.\end{aligned}$$

Thus

$$\varrho_{\alpha,p,T} = C_a \frac{(\log T)^p}{T^{a+1-H}} \to 0 \text{ as } T \to +\infty.$$

□

Remark 5.14 As to the rate of convergence to zero, we can say that

$$\varrho_{\alpha,p,T} = O\left(T^{H-1-a+\varepsilon}\right)$$

as $T \to \infty$ for any $\varepsilon > 0$.

Now, we can consider φ that is a polynomial function. As an immediate generalization of Lemma 5.13, we get the following statement.

Lemma 5.15 *Let $N \in \mathbb{N} \setminus \{0\}$ and $\varphi_N(t) = \sum_{k=0}^{N} \alpha_k t^{a_k}$, $t \geq 0$, (a_k) be a sequence of non-negative power coefficients, $0 \leq a_0 < a_1 < \ldots < a_N$, and (α_k) be a sequence of non-negative coefficients, $\alpha_N > 0$. Then $\varrho_{\alpha,p,T} \to 0$ as $T \to \infty$, and the rate of convergence to zero is $\varrho_{\alpha,p,T} = O\left(T^{H-1-a_N+\varepsilon}\right)$ for any $\varepsilon > 0$.*

Proof The linearity of the operator $\mathcal{D}_{0+}^{\alpha}$ implies that

$$\left|\left(\mathcal{D}_{0+}^{\alpha}\varphi_N\right)(x)\right| \leq \sum_{k=0}^{N} \alpha_k \left|\mathcal{D}_{0+}^{\alpha}(x^{a_k})\right| = \sum_{k=0}^{N} \alpha_k \frac{\Gamma(a_k+1)}{\Gamma(a_k+1-\alpha)} |x|^{a_k-\alpha}.$$

So, on the one hand, similarly to the calculations performed in the proof of Lemma 5.13, we get

$$\varrho_{\alpha,p,T} = \frac{T^{H+\alpha-1}(\log T)^p \int_0^T \left| \left(\mathcal{D}_{0+}^{\alpha}\varphi_N\right)(x)\right| dx}{\int_0^T \varphi_N^2(x)\,dx}$$

$$\leq \frac{T^{H+\alpha-1}(\log T)^p \int_0^T \sum_{k=0}^N \frac{\alpha_k \Gamma(a_k+1)}{\Gamma(a_k+1-\alpha)} |x|^{a_k-\alpha}\, dx}{\int_0^T \varphi_N^2(x)\,dx}$$

$$= \frac{T^{H+\alpha-1}(\log T)^p \sum_{k=0}^N \frac{\alpha_k \Gamma(a_k+1)}{\Gamma(a_k+1-\alpha)(a_k-\alpha+1)} T^{a_k-\alpha+1}}{\int_0^T \varphi_N^2(x)\,dx}.$$

On the other hand,

$$\int_0^T \varphi_N^2(x)dx = \int_0^T \left(\sum_{k=0}^N \alpha_k^2 t^{2a_k} + \sum_{k\neq j} \alpha_k\alpha_j t^{a_k} t^{a_j}\right) dt$$

$$= \sum_{k=0}^N \frac{\alpha_k^2}{2a_k+1} T^{2a_k+1} + \sum_{k\neq j} \frac{\alpha_k\alpha_j}{a_k+a_j+1} T^{a_k+a_j+1}$$

$$\sim \frac{\alpha_N^2}{2a_N+1} T^{2a_N+1}, \quad \text{as } T\to\infty,$$

and

$$\sum_{k=0}^N \alpha_k \frac{\Gamma(a_k+1)}{\Gamma(a_k+1-\alpha)(a_k-\alpha+1)} T^{a_k-\alpha+1} \sim \frac{\alpha_N\Gamma(a_N+1)}{\Gamma(a_N+2-\alpha)} T^{a_N-\alpha+1}, \quad \text{as } T\to\infty,$$

whence

$$\varrho_{\alpha,p,T} \sim T^{H-a_N-1}(\log T)^p \frac{(2a_N+1)\Gamma(a_N+1)}{\Gamma(a_N+2-\alpha)\alpha_N} \to 0, \quad \text{as } T\to\infty.$$

We clearly see that $\varrho_{\alpha,p,T} = O\left(T^{H-a_N-1+\varepsilon}\right)$ as $T\to\infty$, for any $\varepsilon>0$. □

Now consider the case of trigonometric function.

Lemma 5.16 *Let* $\varphi(t) = \sin(\lambda t)$, $\lambda \geq 0$. *Then the estimator* $\hat{\theta}_T^{(2)}$ *is strongly consistent as* $T\to\infty$.

Proof In this case we apply Theorem 5.12. According to this result, if there exists $p > 0$ such that

$$\limsup_{T\to\infty} \frac{T^{2H-1+p}}{\int_0^T \varphi^2(t)\,dt} < \infty,$$

then the estimator $\hat{\theta}_T^{(2)}$ is strongly consistent as $T \to \infty$. In our case

$$\int_0^T \varphi^2(t)\,dt = \int_0^T \sin^2(\lambda t)\,dt = \frac{1}{2}\left(T - \frac{\sin(2\lambda T)}{2\lambda}\right),$$

which means that we can choose any $0 < p < 2 - 2H$, and the proof follows. □

Remark 5.17 We see that in the case of power and polynomial functions (Remark 5.14 and Lemma 5.15) we can get not only convergence to zero but the rate of convergence, but in the case of trigonometric function we only get the former convergence. The difference can be seen from the following result.

Lemma 5.18 *Let* $\varphi(t) = \sin(\lambda t)$, $\lambda \geq 0$. *Then*

$$\lim_{T\to+\infty} \varrho_{\alpha,p,T} = \lim_{T\to+\infty} \frac{T^{H+\alpha-1}(\log T)^p \int_0^T \left|\left(\mathcal{D}_{0+}^{\alpha}\varphi\right)(x)\right|\,dx}{\int_0^T \varphi^2(x)\,dx} = +\infty.$$

Proof First, consider the fractional derivative. Since φ is absolutely continuous on the interval $[0, T]$, for any $T > 0$, by Proposition 1.10, we have that

$$\begin{aligned}\left(\mathcal{D}_{0+}^{\alpha}\varphi\right)(x) &= \frac{1}{\Gamma(1-\alpha)}\left[\frac{\varphi(0)}{(x-0)^{\alpha}} + \int_0^x \frac{\varphi'(s)\,ds}{(x-s)^{\alpha}}\right] \\ &= \frac{1}{\Gamma(1-\alpha)}\int_0^x \frac{\lambda\cos(\lambda s)}{(x-s)^{\alpha}}\,ds = \frac{\lambda}{\Gamma(1-\alpha)}\int_0^x \frac{\cos(\lambda(x-v))}{v^{\alpha}}\,dv \\ &= \frac{\lambda}{\Gamma(1-\alpha)}\left[\cos(\lambda x)\int_0^x \frac{\cos(\lambda v)}{v^{\alpha}}\,dv + \sin(\lambda x)\int_0^x \frac{\sin(\lambda v)}{v^{\alpha}}\,dv\right] \\ &= \frac{\lambda}{\Gamma(1-\alpha)}\left[\cos(\lambda x)J_1(x) + \sin(\lambda x)J_2(x)\right],\end{aligned} \tag{5.18}$$

where

$$J_1(x) = \int_0^x \frac{\cos(\lambda v)}{v^{\alpha}}\,dv, \quad \text{and} \quad J_2(x) = \int_0^x \frac{\sin(\lambda v)}{v^{\alpha}}\,dv.$$

According to [64, p. 893], for any $0 < \alpha < 1$ and $\lambda > 0$ we have that

$$\int_0^{\infty} \cos(\lambda t)t^{-\alpha}\,dt = \Gamma(1-\alpha)\sin\left(\frac{\pi\alpha}{2}\right)\lambda^{\alpha-1} > 0,$$

and

$$\int_0^\infty \sin(\lambda t)t^{-\alpha}\,dt = \Gamma(1-\alpha)\cos\left(\frac{\pi\alpha}{2}\right)\lambda^{\alpha-1} > 0.$$

Denote

$$J(\alpha,\lambda) = \left(\Gamma(1-\alpha)\sin\left(\frac{\pi\alpha}{2}\right)\lambda^{\alpha-1}\right) \wedge \left(\Gamma(1-\alpha)\cos\left(\frac{\pi\alpha}{2}\right)\lambda^{\alpha-1}\right) > 0.$$

Then there exists $x_0 > 0$ such that for any $x > x_0$, $J_i(x) > \frac{J(\alpha,\lambda)}{2}$, $i = 1, 2$. Additionally, for $x \in \left[\frac{2\pi k}{\lambda}, \frac{2\pi k + \frac{\pi}{2}}{\lambda}\right]$, with $k > 1 \vee \left(\frac{\lambda x_0}{2\pi}\right)$ we have that

$$\begin{aligned} \left|\left(\mathcal{D}^\alpha_{0+}\varphi\right)(x)\right| &= \left|\frac{\lambda}{\Gamma(1-\alpha)}[\cos(\lambda x)J_1(x) + \sin(\lambda x)J_2(x)]\right| \\ &\geq \frac{\lambda}{\Gamma(1-\alpha)}\frac{J(\alpha,\lambda)}{2}(\cos(\lambda x) + \sin(\lambda x)) \\ &\geq \frac{\lambda}{\Gamma(1-\alpha)}\frac{J(\alpha,\lambda)}{2}(\cos^2(\lambda x) + \sin^2(\lambda x)) \\ &= \frac{\lambda}{\Gamma(1-\alpha)}\frac{J(\alpha,\lambda)}{2} = \frac{\lambda^\alpha}{2}\sin\left(\frac{\pi\alpha}{2}\right) \wedge \cos\left(\frac{\pi\alpha}{2}\right) =: J_1(\alpha,\lambda). \end{aligned} \tag{5.19}$$

Now, consider $I(T) = \int_0^T \left|\left(\mathcal{D}^\alpha_{0+}\varphi\right)(x)\right| dx$. Evidently, for

$$T > \left(\frac{5\pi}{2\lambda}\right) \vee \left(x_0 + \frac{\pi}{2\lambda}\right) \quad \text{and} \quad \left(1 \vee \left(\frac{\lambda x_0}{2\pi}\right)\right) < k < \frac{\lambda T}{2\pi} - \frac{1}{4}$$

we have from (5.19) that

$$\int_{\frac{2\pi k}{\lambda}}^{\frac{2\pi k + \frac{\pi}{2}}{\lambda}} \left|\left(\mathcal{D}^\alpha_{0+}\varphi\right)(x)\right| dx > \frac{\pi}{2\lambda}J_1(\alpha,\lambda).$$

Finally, denoting $\lfloor a \rfloor$ the integer part of the number a, we get for $T > \left(\frac{5\pi}{2\lambda}\right) \vee \left(x_0 + \frac{\pi}{2\lambda}\right)$ the following bound from below:

$$\begin{aligned} I(T) &\geq \sum_{k=\left\lfloor 1 \vee \left(\frac{\lambda x_0}{2\pi}\right)\right\rfloor + 1}^{\left\lfloor \frac{\lambda T}{2\pi} - \frac{1}{4} \right\rfloor} \int_{\frac{2\pi k}{\lambda}}^{\frac{2\pi k + \frac{\pi}{2}}{\lambda}} \left|\left(\mathcal{D}^\alpha_{0+}\varphi\right)(x)\right| dx \\ &> \left(\left\lfloor \frac{\lambda T}{2\pi} - \frac{1}{4} \right\rfloor - \left\lfloor 1 \vee \left(\frac{\lambda x_0}{2\pi}\right)\right\rfloor\right)\frac{\pi}{2\lambda}J_1(\alpha,\lambda) \sim CT \end{aligned}$$

as $T \to \infty$ for some $C > 0$. Recall that $\alpha + H > 1$ and we immediately get that

$$\lim_{T\to+\infty} \varrho_{\alpha,p,T} \geq \lim_{T\to+\infty} \frac{T^{H+\alpha-1}(\log T)^p CT}{T - \frac{\sin(2\lambda T)}{2\lambda}} = +\infty,$$

and the lemma is proved. □

Lemma 5.19 *Let $\varphi(t) = \exp(-\lambda t)$, $\lambda > 0$. Then*

$$\lim_{T\to+\infty} \varrho_{\alpha,p,T} = \lim_{T\to+\infty} \frac{T^{H+\alpha-1}(\log T)^p \int_0^T \left|\left(\mathcal{D}^\alpha_{0+}\varphi\right)(x)\right| dx}{\int_0^T \varphi^2(x)\, dx} = 0.$$

Proof Note that the function

$$G(x) = \int_0^x e^{-\lambda t}(x-t)^{-\alpha}\, dt = e^{-\lambda x}\int_0^x e^{\lambda t} t^{-\alpha}\, dt$$

has the derivative

$$\begin{aligned} g(x) = G'(x) &= x^{-\alpha} - \lambda e^{-\lambda x}\int_0^x e^{\lambda z} z^{-\alpha}\, dz \\ &= e^{-\lambda x}\left(e^{\lambda x}x^{-\alpha} - \lambda\int_0^x e^{\lambda z}z^{-\alpha} dz\right) =: e^{-\lambda x} g_1(x). \end{aligned}$$

Obviously, $\lim_{x\to 0} g(x) = +\infty$, $\lim_{x\to+\infty} g(x) = 0$ and the derivative

$$g_1'(x) = -\alpha e^{\lambda x} x^{-1-\alpha} < 0.$$

It means that g is the product of two decreasing non-negative functions, so, it is decreasing and therefore is non-negative. Then it follows that

$$\begin{aligned} \int_0^T \left|\left(\mathcal{D}^\alpha_{0+}\varphi\right)(x)\right| dx = \int_0^T \left(\mathcal{D}^\alpha_{0+}\varphi\right)(x)\, dx &= \frac{1}{\Gamma(1-\alpha)}\int_0^T e^{-\lambda t}(T-t)^{-\alpha}\, dt \\ &= \frac{1}{\Gamma(1-\alpha)} e^{-\lambda T}\int_0^T e^{\lambda t} t^{-\alpha}\, dt. \end{aligned} \tag{5.20}$$

Calculate

$$\lim_{T\to+\infty} \frac{\int_0^T e^{\lambda t} t^{-\alpha}\, dt}{e^{\lambda T} T^{-\alpha}} = \lim_{T\to+\infty} \frac{e^{\lambda T} T^{-\alpha}}{\lambda e^{\lambda T} T^{-\alpha} - \alpha e^{\lambda T} T^{-\alpha-1}} = \frac{1}{\lambda}.$$

Moreover, $\int_0^T \varphi^2(x)\,dx = \frac{1-e^{-2\lambda T}}{2\lambda} \sim \frac{1}{2\lambda}$ as $T \to \infty$. Therefore

$$\lim_{T\to+\infty} \varrho_{\alpha,p,T} = \lim_{T\to+\infty} \frac{2}{\Gamma(1-\alpha)} T^{H-1} (\log T)^p = 0.$$

Lemma is proved. □

Remark 5.20 It is easy to deduce from the previous calculations that in the latter case

$$\varrho_{\alpha,p,T} = O\left(T^{H-1+\varepsilon}\right)$$

as $T \to \infty$ for any $\varepsilon > 0$.

Lemma 5.21 *Let $\varphi(t) = \exp(\lambda t)$, $\lambda > 0$. Then*

$$\lim_{T\to+\infty} \varrho_{\alpha,p,T} = \lim_{T\to+\infty} \frac{T^{H+\alpha-1}(\log T)^p \int_0^T \left|\left(\mathcal{D}_{0+}^{\alpha}\varphi\right)(x)\right| dx}{\int_0^T \varphi^2(x)\,dx} = 0.$$

Proof It is easy to check that for every $x > 0$ we have the relations

$$\left(\mathcal{D}_{0+}^{\alpha}\varphi\right)(x) = \frac{1}{\Gamma(1-\alpha)}\left(x^{-\alpha} + \lambda e^{\lambda x}\int_0^x e^{-\lambda z} z^{-\alpha}\,dz\right) \geq 0.$$

Since for any $T > 0$, $\int_0^T e^{-\lambda t} t^{-\alpha}\,dt \leq \lambda^{\alpha-1}\Gamma(1-\alpha)$, then it follows that

$$\begin{aligned}\int_0^T \left|\left(\mathcal{D}_{0+}^{\alpha}\varphi\right)(x)\right| dx &= \int_0^T \left(\mathcal{D}_{0+}^{\alpha}\varphi\right)(x)\,dx = \frac{1}{\Gamma(1-\alpha)}\int_0^T e^{\lambda t}(T-t)^{-\alpha}\,dt \\ &= \frac{1}{\Gamma(1-\alpha)} e^{\lambda T}\int_0^T e^{-\lambda t} t^{-\alpha}\,dt \leq \lambda^{\alpha-1} e^{\lambda T}.\end{aligned} \tag{5.21}$$

Thus,

$$\varrho_{\alpha,p,T} = \frac{T^{H+\alpha-1}(\log T)^p \int_0^T \left|\left(\mathcal{D}_{0+}^{\alpha}\varphi\right)(x)\right| dx}{\int_0^T \varphi^2(x)\,dx} \leq \frac{\lambda^{\alpha-1} T^{H+\alpha-1}(\log T)^p e^{\lambda T}}{\int_0^T \varphi^2(x)\,dx}.$$

Moreover, $\int_0^T \varphi^2(x)\,dx = \frac{e^{2\lambda T}-1}{2\lambda} \sim \frac{1}{2\lambda} e^{2\lambda T}$ as $T \to \infty$. Therefore

$$\lim_{T\to+\infty} \varrho_{\alpha,p,T} = \lim_{T\to+\infty} \frac{2\lambda^{\alpha} T^{H+\alpha-1}(\log T)^p}{e^{\lambda T}} = 0.$$

Lemma is proved. □

Remark 5.22 In this case it is easy to deduce from the previous calculations that

$$\varrho_{\alpha,p,T} = O\left(e^{-(\lambda-\varepsilon)T}\right) = o\left(T^{-\varepsilon}\right)$$

as $T \to \infty$ for any $\varepsilon > 0$.

Lemma 5.23 *Let* $\varphi(t) = \log(1+t)$. *Then*

$$\lim_{T\to+\infty} \varrho_{\alpha,p,T} = \lim_{T\to+\infty} \frac{T^{H+\alpha-1}(\log T)^p \int_0^T \left|\left(\mathcal{D}^{\alpha}_{0+}\varphi\right)(x)\right| dx}{\int_0^T \varphi^2(x)\, dx} = 0.$$

Proof By integration by parts, it is easy to get that for every $x > 0$

$$\left(\mathcal{D}^{\alpha}_{0+}\varphi\right)(x) = \frac{1}{\Gamma(1-\alpha)} \int_0^x \frac{(x-z)^{-\alpha}}{1+z}\, dz \geq 0.$$

Thus,

$$\int_0^T \left|\left(\mathcal{D}^{\alpha}_{0+}\varphi\right)(x)\right| dx = \int_0^T \left(\mathcal{D}^{\alpha}_{0+}\varphi\right)(x)\, dx = \frac{1}{\Gamma(1-\alpha)} \int_0^T \int_0^x \frac{(x-z)^{-\alpha}}{1+z}\, dz\, dx$$
$$= \frac{1}{\Gamma(1-\alpha)} \int_0^T \int_z^T (x-z)^{-\alpha}\, dx \frac{1}{1+z}\, dz = \frac{1}{\Gamma(2-\alpha)} \int_0^T (T-z)^{1-\alpha} \frac{1}{1+z}\, dz$$
$$\leq \frac{T^{1-\alpha}}{\Gamma(2-\alpha)} \int_0^T \frac{1}{1+z}\, dz = \frac{T^{1-\alpha}\log(1+T)}{\Gamma(2-\alpha)}.$$

On the other hand we have that

$$\int_0^T \varphi^2(t)\, dt = \int_0^T \log^2(1+t)\, dt$$
$$= (T+1)\log^2(1+T) - 2(T+1)\log(1+T) + 2T \sim T(\log T)^2$$

as $T \to \infty$.

Therefore

$$\varrho_{\alpha,p,T} \leq \frac{T^H (\log T)^p \log(T+1)}{\Gamma(2-\alpha)\left((T+1)\log^2(1+T) - 2(T+1)\log(1+T) + 2T\right)}$$
$$\sim \frac{T^{H-1}(\log T)^{p-1}}{\Gamma(2-\alpha)}, \quad \text{as} \quad T \to \infty,$$

which allows to deduce that $\lim_{T\to\infty} \varrho_{\alpha,p,T} = 0$. □

Remark 5.24 In this case

$$\varrho_{\alpha,p,T} = O\left(T^{H-1+\varepsilon}\right)$$

as $T \to \infty$ for any $\varepsilon > 0$.

5.2.4 Simulations

Now, we illustrate our results by some simulations. For some fixed step $h = 0.005$, we simulate 10 paths of the process Y on the interval $[0, T]$, for different values of T, with $\theta = 1$ then $\theta = -1$, $H = 0.6$ then $H = 0.75$, and with some polynomial, logarithmic, trigonometric and exponential particular expressions of φ. Simulated results for unknown parameter θ are given in the Tables 5.5 and 5.6.

From these tables we see that with increase of T the estimator $\hat{\theta}_T^{(2)}$ tends to the real value of θ. This clearly illustrates the strong consistency of our estimator. In the particular case of trigonometric expressions of φ, it is obvious that the rate of convergence to the true value of θ is not very high.

Table 5.5 Mean values of $\hat{\theta}_T^{(2)}$ for $\theta = 1$

			T				
H	$a(t,x)$	$b(x)$	50	100	300	500	1000
0.6	$t^2\sqrt{x^2+1}$	$\sqrt{x^2+1}$	1.0043	1.00169	1.00047	1.00027	1.00013
	$e^t\sqrt{x^2+1}$	$\sqrt{x^2+1}$	1.00501	1.00501	1.00501	1.00501	1.00501
	$\sin t\sqrt{x^2+1}$	$\sqrt{x^2+1}$	0.97704	1.00815	1.04516	1.04630	1.00408
	$\log(1+t)(2+\sin x)$	$2+\sin x$	1.19095	1.13984	1.09693	1.08801	1.07777
	$\cos t(2+\sin x)$	$2+\sin x$	0.88322	0.96351	0.98973	1.00807	1.00474
0.75	$t^2\sqrt{x^2+1}$	$\sqrt{x^2+1}$	1.00363	1.00153	1.00045	1.00026	1.00013
	$e^t\sqrt{x^2+1}$	$\sqrt{x^2+1}$	1.00501	1.00501	1.00501	1.00501	1.00501
	$\sin t\sqrt{x^2+1}$	$\sqrt{x^2+1}$	1.03370	1.01917	1.02307	1.02432	1.00383
	$\log(1+t)(2+\sin x)$	$2+\sin x$	1.18920	1.14153	1.09654	1.08931	1.07922
	$\cos t(2+\sin x)$	$2+\sin x$	0.84565	0.92377	0.98510	0.99827	1.00263

Table 5.6 Mean values of $\hat{\theta}_T^{(2)}$ for $\theta = -1$

			T				
H	$a(t,x)$	$b(x)$	50	100	300	500	1000
0.6	$t^2\sqrt{x^2+1}$	$\sqrt{x^2+1}$	−1.00076	−1.00081	−1.00037	−1.00023	−1.00012
	$e^t\sqrt{x^2+1}$	$\sqrt{x^2+1}$	−1.00501	−1.00501	−1.00501	−1.00501	−1.00501
	$\sin t\sqrt{x^2+1}$	$\sqrt{x^2+1}$	−1.00951	−1.00288	−0.95999	−0.94766	−0.98908
	$\log(1+t)(2+\sin x)$	$2+\sin x$	−0.88573	−0.90086	−0.90960	−0.91602	−0.92648
	$\cos t(2+\sin x)$	$2+\sin x$	−1.08132	−1.01240	−0.99469	−1.00395	−1.00607
0.75	$t^2\sqrt{x^2+1}$	$\sqrt{x^2+1}$	−1.00137	−1.00097	−1.00039	−1.00024	−1.00012
	$e^t\sqrt{x^2+1}$	$\sqrt{x^2+1}$	−1.00501	−1.00501	−1.00501	−1.00501	−1.00501
	$\sin t\sqrt{x^2+1}$	$\sqrt{x^2+1}$	−1.00180	−0.99471	−0.97423	−0.97601	−1.00121
	$\log(1+t)(2+\sin x)$	$2+\sin x$	−0.89430	−0.90122	−0.91074	−0.91552	−0.92564
	$\cos t(2+\sin x)$	$2+\sin x$	−1.08129	−1.07844	−1.05366	−1.02158	−1.01580

5.2.5 Sequential Drift Parameter Estimators

Return to model (5.10) and suppose that conditions (E_1)–(E_3) hold. For any $h > 0$ consider the stopping time

$$\tau(h) = \inf\left\{t > 0 : \int_0^t \chi_s^2 \, d\langle M^H\rangle_s = h\right\}.$$

Under conditions (E_1)–(E_2) we have $\tau(h) < \infty$ a. s. and $\int_0^{\tau(h)} \chi_s^2 \, d\langle M^H\rangle_s = h$. The sequential MLE has the form

$$\hat{\theta}^{(1)}_{\tau(h)} = \frac{\int_0^{\tau(h)} \chi_s \, dZ_s}{h} = \theta + \frac{\int_0^{\tau(h)} \chi_s \, dM_s^H}{h}. \tag{5.22}$$

A sequential version of the estimator $\hat{\theta}^{(2)}_T$ has the form

$$\hat{\theta}^{(2)}_{\upsilon(h)} = \theta + \frac{\int_0^{\upsilon(h)} \varphi_s \, dB_s^H}{h},$$

where

$$\upsilon(h) = \inf\left\{t > 0 : \int_0^t \varphi^2(s) \, ds = h\right\}.$$

Theorem 5.25

(a) Let assumptions (E_1)–(E_3) *hold. Then the estimator* $\hat{\theta}^{(1)}_{\tau(h)}$ *is unbiased, strongly consistent, and* $\mathsf{E}\left(\hat{\theta}^{(1)}_{\tau(h)} - \theta\right)^2 = \frac{1}{h}$.

(b) Let the function φ *be bounded away from zero,* $|\varphi(s)| \geq c > 0$ *a. s. and satisfy the assumption: for some* $1 - H < \alpha < 1$ *and* $p > 0$

$$\frac{\int_0^{\upsilon(h)} \left|(\mathcal{D}^{\alpha}_{0+}\varphi)(s)\right| ds}{(\upsilon(h))^{2-\alpha-H-p}} \to 0 \quad a.\,s. \tag{5.23}$$

as $h \to \infty$*. Then the estimator* $\hat{\theta}^{(2)}_{\upsilon(h)}$ *is strongly consistent.*

Proof

(a) Process $\int_0^{\tau(h)} \chi_s \, dM_s^H$ is a square-integrable martingale which implies that the estimator $\hat{\theta}^{(1)}_{\tau(h)}$ is unbiased. The strong consistency follows from Theorem B.3.

(b) We have that $\left|\int_0^{\upsilon(h)} \varphi(s) \, dB_s^H\right| \leq (\upsilon(h))^{H+\alpha-1+p} \int_0^{\upsilon(h)} \left|(\mathcal{D}^{\alpha}_{0+}\varphi)(s)\right| ds$. It is sufficient to note that $h = \int_0^{\upsilon(h)} \varphi_s^2 \, ds \geq c^2 \upsilon(h)$. The proof of statement (b) is now evident. □

5.2.6 Linear Fractional Models and Strong Consistency

Consider the linear version of model (5.10):

$$dX_t = \theta a(t)X_t\,dt + b(t)X_t\,dB^H_t,$$

where a and b are locally bounded non-random measurable functions. In this case solution X exists, it is unique and can be presented in the integral form

$$X_t = x_0+\theta\int_0^t a(s)X_s\,ds+\int_0^t b(s)X_s\,dB^H_s = x_0\exp\left\{\theta\int_0^t a(s)\,ds + \int_0^t b(s)\,dB^H_s\right\}.$$

Suppose that the function b is non-zero and note that in this model $\varphi(t) = \frac{a(t)}{b(t)}$. Suppose that $\varphi(t)$ is also locally bounded and consider MLE $\hat{\theta}^{(1)}_T$. According to (5.12), to guarantee the existence of process J', we have to assume that the fractional derivative of order $\frac{3}{2} - H$ for function $\varsigma(s) := \varphi(s)s^{\frac{1}{2}-H}$ exists and is integrable. By Proposition 1.10, the following condition is sufficient for the existence of this fractional derivative:

(E_6) The functions φ and ς are differentiable and their derivatives are locally integrable.

So, it is hard to conclude what is the behavior of the MLE for an arbitrary locally bounded function φ. Suppose that condition (E_6) holds and limit $\varsigma_0 = \lim_{s\to 0}\varsigma(s)$ exists. In this case, according to Lemma 5.8 and Remark 5.9, process J' admits both of the following representations:

$$\begin{aligned} J'(t) &= (2-2H)C_H\varphi(0)t^{1-2H} + \int_0^t l_H(t,s)\varphi'(s)ds \\ &\quad - \left(H-\frac{1}{2}\right)c_H\int_0^t s^{-\frac{1}{2}-H}(t-s)^{\frac{1}{2}-H}\int_0^s \varphi'(u)duds \\ &= c_H\varsigma_0 t^{\frac{1}{2}-H} + c_H\int_0^t (t-s)^{\frac{1}{2}-H}\varsigma'(s)ds, \end{aligned}$$

and assuming (E_3) also holds true, the estimator $\hat{\theta}^{(1)}_T$ is strongly consistent. Let us formulate some simple conditions sufficient for the strong consistency. The proof is obvious and therefore is omitted.

Lemma 5.26 *If the function φ is non-random, locally bounded, satisfies* (E_6), *the limit $\varsigma(0)$ exists and one of the following assumptions hold:*

(a) function φ is not identically zero and φ' is non-negative and non-decreasing;
(b) derivative ς' preserves the sign and is bounded away from zero;
(c) derivative ς' is non-decreasing and has a non-zero limit,

then the estimator $\hat{\theta}^{(1)}_T$ is strongly consistent as $T\to\infty$.

Example 5.27 Assume that the coefficients are constant, $a(s) = a \neq 0$ and $b(s) = b \neq 0$. Then the estimator $\hat{\theta}_T^{(1)}$ has the form $\hat{\theta}_T^{(1)} = \theta + \frac{bM_T^H}{aC_H T^{2-2H}}$ and is strongly consistent. In this case assumption (a) holds. In addition, power functions $\varphi(s) = s^\rho$ are appropriate for $\rho > H - 1$: this can be verified directly from (5.12).

Let us now apply the estimator $\hat{\theta}_T^{(2)}$ to the same model. It has the form (5.16). We can use Theorem 5.12 directly and under assumption (E_5) the estimator $\hat{\theta}_T^{(2)}$ is strongly consistent. Note that we do not need any assumptions on the smoothness of φ, which is a clear advantage of $\hat{\theta}_T^{(2)}$. We shall consider two more examples.

Example 5.28 If the coefficients are constant, $a(s) = a \neq 0$ and $b(s) = b \neq 0$, then the estimator $\hat{\theta}_T^{(2)}$ has the form $\hat{\theta}_T^{(2)} = \theta + \frac{bB_T^H}{aT}$. We can refer to Theorem 5.12 and conclude that $\hat{\theta}_T^{(2)}$ is strongly consistent. Alternatively, we can use Remark B.45 which states that $\left|B_T^H\right| \leq \xi T^H (\log T)^p$ for any $p > 1$ and some random variable ξ, therefore $\frac{B_T^H}{T} \to 0$ a. s., as $T \to \infty$. In this case both estimators $\hat{\theta}_T^{(1)}$ and $\hat{\theta}_T^{(2)}$ are strongly consistent and $\mathsf{E}\left(\theta - \hat{\theta}_T^{(1)}\right)^2 = \frac{\gamma^2 T^{2H-2}}{a^2 C_H^2}$ has the same asymptotic behavior as $\mathsf{E}\left(\theta - \hat{\theta}_T^{(2)}\right)^2 = \frac{\gamma^2 T^{2H-2}}{a^2}$.

Example 5.29 If non-random functions φ and ς are bounded on some fixed interval $[0, t_0]$ but ς is sufficiently irregular on this interval and has no fractional derivative of order $\frac{3}{2} - H$ or higher then we can not even calculate $J'(t)$ on this interval and it is hard to analyze the behavior of the MLE. However, if we assume that $\varphi(t) \sim t^{H-1+\rho}$ at infinity with some $\rho > 0$, then assumption (E_5) holds and the estimator $\hat{\theta}_T^{(2)}$ is strongly consistent as $T \to \infty$. In this sense the estimator $\hat{\theta}_T^{(2)}$ is more flexible.

5.2.7 *Estimation in the Mixed Model by Continuous Observations: Transferred MLE*

Let us take a Wiener process $W = \{W_t, t \in \mathbb{R}^+\}$ possibly correlated with an fBm $B^H = \{B^H, t \in \mathbb{R}^+\}$. Assume that $H > \frac{1}{2}$ and consider a one-dimensional mixed SDE involving both the Wiener process and the fBm

$$X_t = x_0 + \theta \int_0^t a(s, X_s)\, ds + \int_0^t b(s, X_s)\, dB_s^H + \int_0^t c(s, X_s)\, dW_s, \quad t \in \mathbb{R}^+, \tag{5.24}$$

where $x_0 \in \mathbb{R}$ is the initial value, θ is the unknown parameter to be estimated, the first integral in the right-hand side of (5.24) is the Lebesgue–Stieltjes integral, the second integral is the generalized Lebesgue–Stieltjes integral introduced in Definition 1.13, and the third one is the Itô integral. From now on, we shall assume that the coefficients of Eq. (5.24) satisfy the assumptions (B_1)–(B_4) of Theorem 1.58

on any interval $[0, T]$. Then there exists a solution $X = \{X_t, \mathfrak{F}_t, t \in [0, T]\}$ for the Eq. (5.24) on any interval $[0, T]$ which satisfies (5.11) for some $\alpha > 1 - H$.

For general equation (5.24) with non-zero c, it is impossible to construct reasonable MLE of the parameter θ. Therefore we construct the estimator of the same type as $\hat{\theta}_T^{(2)}$ and call it transferred MLE as well. More exactly, suppose that the following assumption holds:

(F_1) $c(t, X_t) \neq 0, t \in [0, T]$, $\frac{a(t,X_t)}{c(t,X_t)}$ is a. s. Lebesgue integrable on $[0, T]$ for any $T > 0$ and there exists generalized Lebesgue–Stieltjes integral $\int_0^T \frac{b(t,X_t)}{c(t,X_t)} dB_t^H$.

Define functions $\psi_1(t, x) = \frac{a(t,x)}{c(t,x)}$ and $\psi_2(t, x) = \frac{b(t,x)}{c(t,x)}$, processes $\varphi_i(t) = \psi_i(t, X_t)$, $i = 1, 2$ and process

$$Y_t = \int_0^t c^{-1}(s, X_s)\, dX_s = \theta \int_0^t \varphi_1(s)\, ds + \int_0^t \varphi_2(s)\, dB_s^H + W_t.$$

Evidently, Y is a functional of X and is observable. Assume additionally that the generalized Lebesgue–Stieltjes integral $\int_0^T \varphi_1(t)\varphi_2(t)\, dB_t^H$ exists and

(F_2) for any $T > 0$, $\mathsf{E} \int_0^T \varphi_1^2(s)\, ds < \infty$.

Denote $\vartheta(s) = \varphi_1(s)\varphi_2(s)$. We can consider the following estimator of parameter θ:

$$\hat{\theta}_T = \frac{\int_0^T \varphi_1(s)\, dY_s}{\int_0^T \varphi_1^2(s)\, ds} = \theta + \frac{\int_0^T \vartheta(s)\, dB_s^H}{\int_0^T \varphi_1^2(s)\, ds} + \frac{\int_0^T \varphi_1(s)\, dW_s}{\int_0^T \varphi_1^2(s)\, ds}. \tag{5.25}$$

The estimator $\hat{\theta}_T$ preserves the traditional form of MLEs for diffusion models. The right-hand side of (5.25) provides a stochastic representation of $\hat{\theta}_T$. We shall use it to investigate the strong consistency of this estimator.

Theorem 5.30 *Let assumptions* (F_1) *and* (F_2) *hold, and, in addition,*

(F_3) $\int_0^T \varphi_1^2(s)\, ds = \infty$ *a. s.*
(F_4) *There exist* $\alpha > 1 - H$ *and* $p > 1$ *such that*

$$\frac{T^{H+\alpha-1} (\log T)^p \int_0^T \left| \left(\mathcal{D}_{0+}^{\alpha} \vartheta\right)(s) \right| ds}{\int_0^T \varphi_1^2(s)\, ds} \to 0 \quad a.\,s.,\ as \quad T \to \infty. \tag{5.26}$$

Then the estimator $\hat{\theta}_T$ *is strongly consistent as* $T \to \infty$.

Proof The last term in the right-hand side of (5.25) tends to zero under condition (F_3). The proof of convergence of the second term repeats the proof of Theorem 5.11.

Similarly to Theorem 5.12, the conditions stated in Theorem 5.30 can be simplified in the case when function ϑ is non-random.

Theorem 5.31 *Let assumptions* (F_1) *and* (F_2) *hold. Then, if the functions* φ_1 *and* φ_2 *are non-random, function* φ_1 *satisfies condition* (E_5)*, function* φ_2 *is bounded, then the estimator* $\hat{\theta}_T$ *is strongly consistent as* $T \to \infty$*.*

A sequential version of estimator $\hat{\theta}_T$ has the form

$$\hat{\theta}_{\upsilon_1(h)} = \theta + \frac{\int_0^{\upsilon_1(h)} \vartheta(s)\, dB_s^H}{h} + \frac{\int_0^{\upsilon_1(h)} \varphi_1(s)\, dW_s}{h},$$

where

$$\upsilon_1(h) = \inf\left\{t > 0 : \int_0^t \varphi_1^2(s)\, ds = h\right\}.$$

Theorem 5.32

(a) Let the function φ_1 *be bounded away from zero,* $|\varphi_1(s)| \geq c > 0$ *a. s. and let the function* ϑ *satisfy the assumption: for some* $1 - H < \alpha < 1$ *and* $p > 0$

$$\frac{\int_0^{\upsilon_1(h)} \left|\left(\mathcal{D}_{0+}^{\alpha} \vartheta\right)(s)\right|\, ds}{(\upsilon_1(h))^{2-\alpha-H-p}} \to 0 \quad a.\,s., \tag{5.27}$$

as $h \to \infty$*. Then the estimator* $\hat{\theta}_{\upsilon_1(h)}$ *is strongly consistent.*

(b) Let the function ϑ *be non-random, bounded and positive,* φ_1 *be bounded away from zero. Then the estimator* $\hat{\theta}_{\upsilon(h)}$ *is consistent in the following sense: for any* $p > 0$, $\mathsf{E}\left|\theta - \hat{\theta}_{\upsilon_1(h)}\right|^p \to 0$ *as* $h \to \infty$*.*

Proof Statement (a) is proved similarly to statement (b) of Theorem 5.25.

(b) It was proved in [120] that in the case of non-random bounded positive function $0 \leq \vartheta(s) \leq \vartheta^*$ and for any stopping time υ

$$\left(\mathsf{E}\left(\sup_{0 \leq t \leq \upsilon}\left|\int_0^t \vartheta(s)\, dB_s^H\right|\right)^p\right)^{\frac{1}{p}} \leq C(H,p)\vartheta^* \left(\mathsf{E}\upsilon^{pH}\right)^{\frac{1}{p}}.$$

Furthermore, the same as in the proof of Theorem 5.25, we have that

$$h = \int_0^{\upsilon_1(h)} \varphi_1^2(s)\, ds \geq c^2 \upsilon_1(h),$$

because $|\varphi_1(s)| \geq c$. This inequality together with the Burkholder–Gundy inequality yields

$$\mathsf{E}\left|\theta - \hat{\theta}_{\upsilon_1(h)}\right|^p \leq C(H,p)\left(\frac{\vartheta^*}{c^2} h^{H-1} + h^{-\frac{p}{2}}\right) \to 0, \quad \text{as} \quad h \to \infty.$$

Remark 5.33 Assumption (5.27) holds, for example, for bounded and Lipschitz function ϑ.

Consider a mixed linear model of the form

$$dX_t = X_t \left(\theta a(t)\, dt + b(t)\, dB_t^H + c(t)\, dW_t\right), \tag{5.28}$$

where a, b and c are non-random measurable functions. Assume that they are locally bounded. In this case solution X for Eq. (5.28) exists, it is unique and can be presented in the integral form

$$X_t = x_0 \exp\left\{\theta \int_0^t a(s)\, ds + \int_0^t b(s)\, dB_s^H + \int_0^t c(s)\, dW_s - \frac{1}{2}\int_0^t c^2(s)\, ds\right\}.$$

In what follows assume that $c(s) \neq 0$. We have that $\varphi_1(t) = \frac{a(t)}{c(t)}$ and $\varphi_2(t) = \frac{b(t)}{c(t)}$. The estimator $\hat{\theta}_T$ has the form

$$\hat{\theta}_T = \frac{\int_0^T \varphi_1(s)\, dY_s}{\int_0^T \varphi_1^2(s)\, ds} = \theta + \frac{\int_0^T \varphi_1(s)\varphi_2(s)\, dB_s^H}{\int_0^T \varphi_1^2(s)\, ds} + \frac{\int_0^T \varphi_1(s)\, dW_s}{\int_0^T \varphi_1^2(s)\, ds}. \tag{5.29}$$

In accordance with Theorem 5.31, assume that the function φ_1 satisfies (E_5) and φ_2 is bounded. Then estimator $\hat{\theta}_T$ is strongly consistent. Evidently, these assumptions hold for the constant coefficients.

5.3 Estimation in Homogeneous Fractional Diffusion Model by Discrete Observations

This section deals with statistical estimation of drift parameter for an SDE with fBm by discrete observation of its solution. We propose three estimators and prove their strong consistency under the so-called "high-frequency data" assumption that the horizon of observations tends to infinity, while the interval between them goes to zero. Moreover, we obtain almost sure upper bounds for the rate of convergence of the estimators. The estimators proposed go far away from being MLEs, and this is their crucial advantage, because they keep strong consistency but they are not complicated technically and are convenient for the simulations. This section is organized as follows. In Sect. 5.3.1, we give some auxiliary results, which are also of independent interest: global estimates on increments of solution to SDE with fBm. In Sect. 5.3.2 we construct estimators for drift parameter, prove their strong consistency and establish their rate of convergence.

In this section we assume that the Hurst parameter $H \in (1/2, 1)$ is fixed. In this case, the integral with respect to the fBm B^H will be understood in the generalized Lebesgue–Stieltjes sense (see Definition 1.13).

Throughout the section, the symbol C will denote a generic constant, whose value is not important and may change from one line to another. If a constant depends on some variable parameters, we will put them in subscripts.

5.3.1 Bounds for Solution of SDE Driven by fBm

Consider an SDE

$$X_t = x_0 + \int_0^t a(X_s)\,ds + \int_0^t b(X_s)\,dB_s^H, \tag{5.30}$$

where $x_0 \in \mathbb{R}$ is nonrandom.

Assume that there exist constants $\delta \in (1/H - 1, 1]$, $K > 0, L > 0$ and for every $N \ge 1$ there exists $R_N > 0$ such that

(G$_1$**)** $|a(x)| + |b(x)| \le K$ for all $x, y \in \mathbb{R}$,
(G$_2$**)** $|a(x) - a(y)| + |b(x) - b(y)| \le L\,|x - y|$ for all $x, y \in \mathbb{R}$,
(G$_3$**)** $|b'(x) - b'(y)| \le R_N\,|x - y|^\delta$ for all $x \in [-N, N], y \in [-N, N]$.

Then according to Theorem 1.42, the Eq. (5.30) has a unique solution.

Remark 5.34 Note that the conditions (G$_1$)–(G$_3$) are more restrictive then the assumptions of Theorem 1.42. Here we assume additionally the boundedness of the coefficients, and global Lipschitz continuity of a instead of local.

Fix some $\alpha \in (1 - H, 1/2)$ and $\beta \in (1/2, H)$. Denote for $t_1 < t_2$

$$Z(t_1, t_2) = \frac{1}{\Gamma(\alpha)}\left(\frac{B_{t_1}^H - B_{t_2}^H}{(t_2 - t_1)^{1-\alpha}} + (1 - \alpha)\int_{t_1}^{t_2} \frac{B_{t_1}^H - B_u^H}{(u - t_1)^{2-\alpha}}du\right),$$

$$\Lambda_\beta(t_1, t_2) = 1 \vee \sup_{t_1 \le u < v \le t_2} \frac{|Z(u, v)|}{(v - u)^{\beta+\alpha-1}}.$$

Define for $a < b$

$$\|f\|_{a,b,\beta} = \sup_{a \le s < t \le b} \frac{|f(t) - f(s)|}{|t - s|^\beta}.$$

Theorem 5.35 *There exists a constant $M_{\alpha,\beta}$ depending on α, β, K, and L such that for any $t_1 \ge 0$, $t_2 \in (t_1, t_1 + 1]$*

$$|X_{t_2} - X_{t_1}| \le M_{\alpha,\beta}\left(\Lambda_\beta(t_1, t_2)(t_2 - t_1)^\beta + \Lambda_\beta(t_1, t_2)^{1/\beta}(t_2 - t_1)\right).$$

Proof Fix $t_1 \ge 0$ and $t_2 \in (t_1, t_1 + 1]$. Take any s, t such that $t_1 \le s < t \le t_2$. Write

$$|X_t - X_s| \le \int_s^t |a(X_u)|\,du + \left|\int_s^t b(X_u)\,dB_u^H\right| \le K(t - s) + \left|\int_s^t b(X_u)\,dB_u^H\right|.$$

Abbreviate $\Lambda = \Lambda_\beta(t_1, t_2)$ and estimate

$$\left| \int_s^t b(X_u) dB_u^H \right| \le \int_s^t \left| \left(\mathcal{D}_{s+}^{\alpha} b(X)\right)(u) \right| \left| \left(\mathcal{D}_{t-}^{1-\alpha} B_{t-}^H\right)(u) \right| du$$
$$\le \Lambda \int_s^t \left| \left(\mathcal{D}_{s+}^{\alpha} b(X)\right)(u) \right| (u-s)^{\beta+\alpha-1} du. \tag{5.31}$$

Now for $s \le u \le t$

$$\left| \left(\mathcal{D}_{s+}^{\alpha} b(X)\right)(u) \right| \le \left(\frac{|b(X_u)|}{(u-s)^{\alpha}} + \int_s^u \frac{|b(X_u) - b(X_v)|}{(u-v)^{1+\alpha}} dv \right)$$
$$\le K(u-s)^{-\alpha} + L \|X\|_{s,t,\beta} \int_s^u (u-v)^{\beta-\alpha-1} dv$$
$$\le C_{\alpha,\beta} \left((u-s)^{-\alpha} + \|X\|_{s,t,\beta} (u-s)^{\beta-\alpha} \right).$$

Hence,

$$\left| \int_s^t b(X_s) dB_u^H \right| \le C_{\alpha,\beta} \Lambda \left((t-s)^{\beta} + \|X\|_{s,t,\beta} (t-s)^{2\beta} \right)$$

and

$$\|X\|_{s,t,\beta} \le K_{\alpha,\beta} \Lambda \left(1 + \|X\|_{s,t,\beta} (t-s)^{\beta}\right)$$

with a constant $K_{\alpha,\beta}$ depending only on α, β, K, and L. Setting $\Delta = (2K_{\alpha,\beta}\Lambda)^{-1/\beta}$, we obtain $\|X\|_{s,t,\beta} \le 2K_{\alpha,\beta}\Lambda$ whenever $t - s < \Delta$.

Now, if $t_2 - t_1 \le \Delta$, then

$$|X_{t_2} - X_{t_1}| \le \|X\|_{t_1,t_2,\beta} (t_2 - t_1)^{\beta} \le 2K_{\alpha,\beta} \Lambda (t_2 - t_1)^{\beta}.$$

On the other hand, if $t_2 - t_1 > \Delta$, then, partitioning the interval $[t_1, t_2]$ into $k = [(t_2 - t_1)/\Delta]$ parts of length Δ and, possibly, an extra smaller part, we obtain

$$|X_{t_2} - X_{t_1}| \le |X_{t_1+\Delta} - X_{t_1}| + \cdots + \left|X_{t_1+k\Delta} - X_{t_1+(k-1)\Delta}\right| + |X_{t_2} - X_{t_1+k\Delta}|$$
$$\le (k+1)\, 2K_{\alpha,\beta} \Lambda \Delta^{\beta} \le 4kK_{\alpha,\beta} \Lambda \Delta^{\beta} \le 4K_{\alpha,\beta} \Lambda (t_2 - t_1) \Delta^{\beta-1}$$
$$= 2(2K_{\alpha,\beta}\Lambda)^{1/\beta} (t_2 - t_1).$$

The proof is now complete. □

Corollary 5.36 *For any $\gamma > 1/2$, there exist random variables ξ and ζ such that for all $t_1 \ge 0$, $t_2 \in (t_1, t_1 + 1]$*

$$|X_{t_2} - X_{t_1}| \le \zeta (t_2 - t_1)^{\beta} \left(\log(t_2 + 2)\right)^{\kappa}, \quad \Lambda_\beta(t_1, t_2) \le \xi \left(\log(t_2 + 2)\right)^{\kappa\beta},$$

where $\kappa = \gamma/\beta$. *Moreover, there exists some* $c > 0$ *such that* $\mathsf{E}\left[\exp\left\{x\xi^2\right\}\right] < \infty$ *and* $\mathsf{E}\left[\exp\left\{x\zeta^{2\beta}\right\}\right] < \infty$ *for* $x < c$. *In particular, all moments of* ξ *and* ζ *are finite.*

Proof From Theorem B.46 we have for all $u < v$

$$\begin{aligned}|Z(u,v)| &\le \eta_{H,\alpha,\gamma}(v-u)^{H+\alpha-1}\left(|\log(v-u)|^{1/2}+1\right)\left(\log(v+2)\right)^{\gamma}\\ &\le C_{H,\beta}\eta_{H,\alpha,\gamma}(v-u)^{\beta+\alpha-1}\left(\log(v+2)\right)^{\gamma},\end{aligned}$$

Dividing by $(v-u)^{\beta+\alpha-1}$ and taking supremum over u, v such that $t_1 \le u < v \le t_2$, we get

$$\Lambda_\beta(t_1,t_2) \le 1 \vee \left(C_{H,\beta}\eta_{H,\alpha,\gamma}\left(\log(t_2+2)\right)^{\gamma}\right) \le \left(1 \vee C_{H,\beta}\eta_{H,\alpha,\gamma}\right)\left(\log(t_2+2)\right)^{\gamma}.$$

Further, since $\Lambda_\beta(t_1,t_2) \ge 1$ and $t_2 - t_1 \le 1$, it follows from Theorem 5.35 that

$$|X_{t_2} - X_{t_1}| \le 2M_{\alpha,\beta}\Lambda_\beta(t_1,t_2)^{1/\beta}(t_2-t_1)^{\beta}.$$

Hence, the desired statement holds with $\xi = 1 \vee C_{H,\beta}\eta_{H,\alpha,\gamma}$ and $\zeta = 2M_{\alpha,\beta}\xi^{1/\beta}$. □

The following lemma gives a particular case of Corollary 5.36, suitable for proving of the strong consistency results. Let $\gamma > 1/2$ and $\kappa = \gamma/\beta$ be fixed, ξ and ζ be the corresponding random variables from Corollary 5.36.

Lemma 5.37 *For any* $n \ge 3$ *and any* $t_1, t_2 \in [0, 2^n]$ *such that* $t_1 < t_2 \le t_1 + 1$

$$|X_{t_2} - X_{t_1}| \le \zeta n^{\kappa}(t_2-t_1)^{\beta}, \quad \Lambda_\beta(t_1,t_2) \le \xi n^{\gamma}.$$

Proof In this case

$$\log(t_2+2) \le \log(2^n+2) \le \log 2^{n+1} = (n+1)\log 2 \le n,$$

whence the statement follows. □

5.3.2 Drift Parameter Estimation by Discretized Observations

Consider a parametrized version of (5.30)

$$X_t = x_0 + \theta\int_0^t a(X_s)\,ds + \int_0^t b(X_s)\,dB_s^H. \tag{5.32}$$

Assume that assumptions (G_1)–(G_3) of Sect. 5.3.1 hold.

Our main problem is the following: to construct an estimator for θ based on discrete observations of X. Specifically, we will assume that for some $n \ge 1$ we observe the values $X_{t_k^n}$ at the following uniform partition of $[0, 2^n]$: $t_k^n = k2^{-n}$, $k = 0, 1, \dots, 2^{2n}$.

In the following we will fix the parameters $\alpha \in (1 - H, 1/2)$, $\beta \in (1 - \alpha, H)$, $\gamma > 1/2$ and $\kappa = \gamma/\beta$. Let ξ and ζ be the random variables from Corollary 5.36 applied to Eq. (5.32).

In order to construct a consistent estimator, we need a lemma concerning the discrete approximation of integrals in (5.32).

Lemma 5.38 *For all $n \geq 1$ and $k = 1, 2, \dots, 2^{2n}$*

$$\left| \int_{t_{k-1}^n}^{t_k^n} \left(a(X_u) - a(X_{t_{k-1}^n})\right) du \right| \leq C\zeta n^\kappa 2^{-n(\beta+1)}$$

and

$$\left| \int_{t_{k-1}^n}^{t_k^n} \left(b(X_u) - b(X_{t_{k-1}^n})\right) dB_u^H \right| \leq C\xi\zeta n^{\gamma+\kappa} 2^{-2n\beta}.$$

Proof Write

$$\left| \int_{t_{k-1}^n}^{t_k^n} \left(a(X_u) - a(X_{t_{k-1}^n})\right) du \right| \leq \int_{t_{k-1}^n}^{t_k^n} \left|a(X_u) - a(X_{t_{k-1}^n})\right| du$$

$$\leq K\zeta n^\kappa \int_{t_{k-1}^n}^{t_k^n} \left(u - t_{k-1}^n\right)^\beta du \leq C\zeta n^\kappa \left(t_k^n - t_{k-1}^n\right)^{\beta+1} = C\zeta n^\kappa 2^{-n(\beta+1)}.$$

Similarly to (5.31),

$$\left| \int_{t_{k-1}^n}^{t_k^n} \left(b(X_u) - b(X_{t_{k-1}^n})\right) dB_u^H \right|$$

$$\leq \Lambda_\beta\left(t_{k-1}^n, t_k^n\right) \int_{t_{k-1}^n}^{t_k^n} \left| \mathcal{D}_{t_{k-1}^n+}^\alpha \left(b(X) - b(X_{t_{k-1}^n})\right)(u) \right| \left(u - t_{k-1}^n\right)^{\beta+\alpha-1} du$$

$$\leq \xi n^\gamma \int_{t_{k-1}^n}^{t_k^n} \left| \mathcal{D}_{t_{k-1}^n+}^\alpha \left(b(X) - b(X_{t_{k-1}^n})\right)(u) \right| \left(u - t_{k-1}^n\right)^{\beta+\alpha-1} du,$$

and

$$\left| \mathcal{D}_{t_{k-1}^n+}^\alpha \left(b(X) - b(X_{t_{k-1}^n})\right)(u) \right| \leq \frac{\left|b(X_u) - b(X_{t_{k-1}^n})\right|}{\left(u - t_{k-1}^n\right)^\alpha} + \int_{t_{k-1}^n}^u \frac{|b(X_u) - b(X_v)|}{(u - v)^{1+\alpha}} dv$$

$$\leq K\zeta n^\kappa \left(u - t_{k-1}^n\right)^{\beta-\alpha} + K\zeta n^\kappa \int_{t_{k-1}^n}^u (u - v)^{\beta-\alpha-1} dv \leq C\zeta \left(u - t_{k-1}^n\right)^{\beta-\alpha}.$$

Then we can write the upper bound

$$\left|\int_{t_{k-1}^n}^{t_k^n} \left(b(X_u) - b\left(X_{t_{k-1}^n}\right)\right) dB_u^H\right| \le C\xi\zeta n^{\gamma+\kappa} \int_{t_{k-1}^n}^{t_k^n} \left(u - t_{k-1}^n\right)^{2\beta-1} du$$
$$\le C\xi\zeta n^{\gamma+\kappa} \left(t_k^n - t_{k-1}^n\right)^{2\beta} = C\xi\zeta n^{\gamma+\kappa} 2^{-2n\beta},$$

which finishes the proof. □

Now we are ready to construct consistent estimators for θ. In order to proceed, we need a technical assumption, in addition to conditions (G_1)–(G_3) of Sect. 5.3.1:

(G_4) There exists a constant $M > 0$ such that for all $x \in \mathbb{R}$

$$|a(x)| \ge M, \qquad |b(x)| \ge M.$$

First, we define a discretized version of the estimator (5.16) for the equation (5.32). Put

$$\hat{\theta}_n^{(1)} := \frac{2^n \sum_{k=1}^{2^{2n}} a\left(X_{t_{k-1}^n}\right) b^{-2}\left(X_{t_{k-1}^n}\right)\left(X_{t_k^n} - X_{t_{k-1}^n}\right)}{\sum_{k=1}^{2^{2n}} a^2\left(X_{t_{k-1}^n}\right) b^{-2}\left(X_{t_{k-1}^n}\right)},$$

$$\widehat{\varphi}_n(t) := \sum_{k=0}^{2^{2n}-1} \varphi(t_k^n) \mathbf{1}_{[t_k^n, t_{k+1}^n)}(t), \quad \varphi(t) = \frac{a(X_t)}{b(X_t)}.$$

Theorem 5.39 *Let assumptions* (G_1)–(G_4) *hold, and suppose additionally that there exist constants $\nu > 1 - H$ and $p > 1$ such that*

$$\frac{2^{n(H+\nu)} n^p \int_0^{2^n} \left|\left(\mathcal{D}_{0+}^{\nu} \widehat{\varphi}_n\right)(s)\right| ds}{\sum_{k=1}^{2^{2n}} \varphi^2\left(t_{k-1}^n\right)} \to 0 \quad \text{a. s., as } n \to \infty.$$

Then with probability one, $\hat{\theta}_n^{(1)} \to \theta$, *as* $n \to \infty$.

Proof It follows from (5.32) that

$$X_{t_k^n} - X_{t_{k-1}^n} = \theta \int_{t_{k-1}^n}^{t_k^n} a(X_v)\, dv + \int_{t_{k-1}^n}^{t_k^n} b(X_v)\, dB_v^H$$
$$= \theta \int_{t_{k-1}^n}^{t_k^n} a\left(X_{t_{k-1}^n}\right) dv + \theta \int_{t_{k-1}^n}^{t_k^n} \left(a(X_v) - a\left(X_{t_{k-1}^n}\right)\right) dv$$
$$+ \int_{t_{k-1}^n}^{t_k^n} \left(b(X_v) - b\left(X_{t_{k-1}^n}\right)\right) dB_v^H + \int_{t_{k-1}^n}^{t_k^n} b\left(X_{t_{k-1}^n}\right) dB_v^H.$$

Then

$$\hat{\theta}_n^{(1)} = \theta + \frac{\theta B_n + E_n + D_n}{A_n},$$

where

$$A_n = 2^{-n} \sum_{k=1}^{2^{2n}} \varphi^2\left(t_{k-1}^n\right),$$

$$B_n = \sum_{k=1}^{2^{2n}} a(X_{t_{k-1}^n}) b^{-2}(X_{t_{k-1}^n}) \int_{t_{k-1}^n}^{t_k^n} \left(a(X_v) - a(X_{t_{k-1}^n})\right) dv,$$

$$E_n = \sum_{k=1}^{2^{2n}} a(X_{t_{k-1}^n}) b^{-2}(X_{t_{k-1}^n}) \int_{t_{k-1}^n}^{t_k^n} \left(b(X_v) - b(X_{t_{k-1}^n})\right) dB_v^H,$$

$$D_n = \sum_{k=1}^{2^{2n}} \varphi\left(t_{k-1}^n\right) \left(B_{t_k^n}^H - B_{t_{k-1}^n}^H\right).$$

The value D_n can be represented in the form

$$D_n = \int_0^{2^n} \widehat{\varphi}_n(s)\, dB_s^H.$$

Applying Theorem B.41 we can get the upper bound

$$\sup_{0 \le t \le 2^n} \left| \left(\mathcal{D}_{2^n-}^{1-\nu} B_{2^n-}^H\right)(t) \right| \le \xi(p) 2^{n(H+\nu-1)} n^p (\log 2)^p.$$

Therefore

$$\begin{aligned} |D_n| &\le \sup_{0 \le t \le 2^n} \left| \left(\mathcal{D}_{2^n-}^{1-\nu} B_{2^n-}^H\right)(t) \right| \int_0^{2^n} \left| \left(\mathcal{D}_{0+}^{\nu} \widehat{\varphi}_n\right)(s) \right| ds \\ &\le \xi(p) (\log 2)^p 2^{n(H+\nu-1)} n^p \int_0^{2^n} \left| \left(\mathcal{D}_{0+}^{\nu} \widehat{\varphi}_n\right)(s) \right| ds. \end{aligned}$$

Then

$$\frac{|D_n|}{A_n} \le \xi(p) (\log 2)^p \frac{2^{n(H+\nu)} n^p \int_0^{2^n} \left| \left(\mathcal{D}_{0+}^{\nu} \widehat{\varphi}_n\right)(s) \right| ds}{\sum_{k=1}^{2^{2n}} \varphi^2\left(t_{k-1}^n\right)} \to 0 \quad \text{a. s., as } n \to \infty.$$

Using the condition (G_1) we can write

$$\frac{|B_n|}{A_n} \le K^{-1}2^{-n}\sum_{k=1}^{2^{2n}}\int_{t_{k-1}^n}^{t_k^n} \left|a(X_v)-a\left(X_{t_{k-1}^n}\right)\right| dv,$$

$$\frac{|E_n|}{A_n} \le K^{-1}2^{-n}\sum_{k=1}^{2^{2n}}\int_{t_{k-1}^n}^{t_k^n} \left|b(X_v)-b\left(X_{t_{k-1}^n}\right)\right| dB_v^H.$$

It now follows from Lemma 5.38 that $\frac{|B_n|}{A_n}\to 0$, $\frac{|E_n|}{A_n}\to 0$, as $n\to\infty$. □

Example 5.40 Consider the following model:

$$dX_t = \theta b(X_t)\,dt + b(X_t)\,dB_t^H,$$

where b is a function satisfying the assumptions (G_1)–(G_4).

In this case the estimator $\hat{\theta}_n^{(1)}$ has the form

$$\hat{\theta}_n^{(1)} = 2^{-n}\sum_{k=1}^{2^{2n}} b^{-1}\left(X_{t_{k-1}^n}\right)\left(X_{t_k^n}-X_{t_{k-1}^n}\right), \tag{5.33}$$

$\widehat{\varphi}_n(t) = 1$. Then $\left(\mathcal{D}_{0+}^{\nu}\widehat{\varphi}_n\right)(s) = \frac{1}{\Gamma(1-\nu)}\cdot s^{-\nu}$ and

$$\frac{2^{n(H+\nu)}n^p\int_0^{2^n}\left|\left(\mathcal{D}_{0+}^{\nu}\widehat{\varphi}_n\right)(s)\right|ds}{\sum_{k=1}^{2^{2n}}\varphi^2(t_{k-1}^n)} = \frac{n^p}{\Gamma(2-\nu)\cdot 2^{n(1-H)}} \to 0, \quad n\to\infty.$$

Consequently the conditions of Theorem 5.39 are satisfied and the estimator (5.33) is strongly consistent.

Let us return to the general form of Eq. (5.32) and consider the estimator that is a version of the discretization of MLE:

$$\hat{\theta}_n^{(2)} = \frac{2^n\sum_{k=1}^{2^{2n}-1}\left(t_k^n\right)^{\lambda}\left(2^n-t_k^n\right)^{\lambda}b^{-1}\left(X_{t_{k-1}^n}\right)\left(X_{t_k^n}-X_{t_{k-1}^n}\right)}{\sum_{k=1}^{2^{2n}-1}\left(t_k^n\right)^{\lambda}\left(2^n-t_k^n\right)^{\lambda}b^{-1}\left(X_{t_{k-1}^n}\right)a\left(X_{t_{k-1}^n}\right)},$$

where $\lambda = 1/2 - H$.

Theorem 5.41 *Under assumptions* (G_1)–(G_4), $\hat{\theta}_n^{(2)} \to \theta$ *a. s., as* $n\to\infty$. *Moreover, there exists a random variable* η *with all finite moments such that* $\left|\hat{\theta}_n^{(2)}-\theta\right| \le \eta n^{\kappa+\gamma}2^{-\rho n}$, *where* $\rho = (1-H)\wedge(2\beta-1)$.

Proof It follows from (5.32) that

$$\begin{aligned}X_{t_k^n} - X_{t_{k-1}^n} &= \theta \int_{t_{k-1}^n}^{t_k^n} a(X_v)dv + \int_{t_{k-1}^n}^{t_k^n} b(X_v)\, dB_v^H \\ &= \theta \int_{t_{k-1}^n}^{t_k^n} a\big(X_{t_{k-1}^n}\big)\, dv + \theta \int_{t_{k-1}^n}^{t_k^n} \big(a(X_v) - a\big(X_{t_{k-1}^n}\big)\big)\, dv \\ &+ \int_{t_{k-1}^n}^{t_k^n} b\big(X_{t_{k-1}^n}\big)\, dB_v^H + \int_{t_{k-1}^n}^{t_k^n} \big(b(X_v) - b\big(X_{t_{k-1}^n}\big)\big)\, dB_v^H.\end{aligned}$$

Then

$$\hat{\theta}_n^{(2)} = \theta + \frac{B_n + C_n + D_n}{A_n},$$

where

$$A_n = 2^{n(2H-3)} \sum_{k=1}^{2^{2n}-1} \big(t_k^n\big)^\lambda \big(2^n - t_k^n\big)^\lambda a\big(X_{t_{k-1}^n}\big) b^{-1}\big(X_{t_{k-1}^n}\big),$$

$$B_n = 2^{2n(H-1)}\theta \sum_{k=1}^{2^{2n}-1} \big(t_k^n\big)^\lambda \big(2^n - t_k^n\big)^\lambda b^{-1}\big(X_{t_{k-1}^n}\big) \int_{t_{k-1}^n}^{t_k^n} \big(a(X_v) - a\big(X_{t_{k-1}^n}\big)\big)\, dv,$$

$$C_n = 2^{2n(H-1)} \sum_{k=1}^{2^{2n}-1} \big(t_k^n\big)^\lambda \big(2^n - t_k^n\big)^\lambda \Big(B_{t_k^n}^H - B_{t_{k-1}^n}^H\Big),$$

$$D_n = 2^{2n(H-1)} \sum_{k=1}^{2^{2n}-1} \big(t_k^n\big)^\lambda \big(2^n - t_k^n\big)^\lambda b^{-1}\big(X_{t_{k-1}^n}\big) \int_{t_{k-1}^n}^{t_k^n} \big(b(X_v) - b\big(X_{t_{k-1}^n}\big)\big)\, dB_v^H.$$

It is not hard to show that the sequence

$$\gamma_n := 2^{n(2H-3)} \sum_{k=1}^{2^{2n}-1} \big(t_k^n\big)^\lambda \big(2^n - t_k^n\big)^\lambda = 2^{-2n} \sum_{k=1}^{2^{2n}-1} \left(\frac{k}{2^{2n}}\right)^\lambda \left(1 - \frac{k}{2^{2n}}\right)^\lambda \frac{1}{2^{2n}}$$

converges to $\int_0^1 x^\lambda (1-x)^\lambda dx = \mathrm{B}(1+\lambda, 1+\lambda)$, and, hence, is bounded and uniformly positive.

Indeed, $h(x) = x^\lambda (1-x)^\lambda$ increases for $x \in \big(0, \frac{1}{2}\big]$; then

$$\int_0^{\frac{1}{2}} h(x)dx = \sum_{k=0}^{2^{2n-1}-1} \int_{\frac{k}{2^{2n}}}^{\frac{k+1}{2^{2n}}} h(x)\, dx < \int_0^{\frac{1}{2^{2n}}} h(x)\, dx + \sum_{k=1}^{2^{2n-1}} h\left(\frac{k}{2^{2n}}\right) \frac{1}{2^{2n}}.$$

On the other hand,

$$\int_0^{\frac{1}{2}} h(x)dx = \sum_{k=1}^{2^{2n-1}} \int_{\frac{k-1}{2^{2n}}}^{\frac{k}{2^{2n}}} h(x)\,dx > \sum_{k=1}^{2^{2n-1}} h\left(\frac{k}{2^{2n}}\right)\frac{1}{2^{2n}}.$$

So

$$0 < \int_0^{\frac{1}{2}} h(x)\,dx - \sum_{k=1}^{2^{2n-1}} h\left(\frac{k}{2^{2n}}\right)\frac{1}{2^{2n}} < \int_0^{\frac{1}{2^{2n}}} h(x)\,dx \to 0, \quad \text{as } n \to \infty.$$

Hence,

$$\sum_{k=1}^{2^{2n-1}} h\left(\frac{k}{2^{2n}}\right)\frac{1}{2^{2n}} \to \int_0^{\frac{1}{2}} h(x)\,dx, \quad \text{as } n \to \infty.$$

Similarly one can prove that

$$\sum_{k=2^{2n-1}+1}^{2^{2n}-1} h\left(\frac{k}{2^{2n}}\right)\frac{1}{2^{2n}} \to \int_{\frac{1}{2}}^{1} h(x)\,dx, \quad \text{as } n \to \infty.$$

By assumption (G_4), $a(x)b^{-1}(x)$ is bounded away from zero and keeps its sign. Therefore,

$$\liminf_{n\to\infty} |A_n| \ge MK^{-1} \lim_{n\to\infty} \gamma_n = MK^{-1}\mathrm{B}(1+\lambda, 1+\lambda) > 0.$$

So it is sufficient to estimate B_n, E_n, and D_n.

By assumption (G_4) and Lemma 5.38,

$$|B_n| \le C\,|\theta|\,\zeta n^{\kappa} M^{-1} 2^{n(2H-\beta-3)} \sum_{k=1}^{2^{2n}-1} \left(t_k^n\right)^{\lambda}\left(2^n - t_k^n\right)^{\lambda} \le C\zeta n^{\kappa} 2^{-n\beta};$$

$$|D_n| \le C\xi\zeta n^{\gamma+\kappa} M^{-1} 2^{n(2H-2-2\beta)} \sum_{k=1}^{2^{2n}-1} \left(t_k^n\right)^{\lambda}\left(2^n - t_k^n\right)^{\lambda} \le C\xi\zeta n^{\gamma+\kappa} 2^{n(1-2\beta)}.$$

Finally we estimate E_n. Start by writing

$$\mathsf{E}\left[E_n^2\right] = 2^{4n(H-1)}\mathsf{E}\left[\left(\sum_{k=1}^{2^{2n}-1} \int_{t_{k-1}^n}^{t_k^n} \left(t_k^n\right)^{\lambda}\left(2^n - t_k^n\right)^{\lambda} dB_s^H\right)^2\right].$$

According to Proposition 1.6, for $f \in L_{1/H}[0, t]$ there exists a constant $C_H > 0$ such that

$$\mathsf{E}\left[\left(\int_0^t f(s)\, dB_s^H\right)^2\right] \le C_H \left(\int_0^t |f(s)|^{1/H} ds\right)^{2H}.$$

Hence,

$$\begin{aligned}\mathsf{E}\left[E_n^2\right] &\le C2^{4n(H-1)}\left(\sum_{k=1}^{2^{2n}-1}\int_{t_{k-1}^n}^{t_k^n} (t_k^n)^{\lambda/H}\left(2^n - t_k^n\right)^{\lambda/H} ds\right)^{2H}\\ &= C2^{2n(H-1)}\left(\sum_{k=1}^{2^{2n}-1}\left(\frac{k}{2^{2n}}\right)^{\lambda/H}\left(1-\frac{k}{2^{2n}}\right)^{\lambda/H}\frac{1}{2^{2n}}\right)^{2H}.\end{aligned}$$

As above,

$$\sum_{k=1}^{2^{2n}-1}\left(\frac{k}{2^{2n}}\right)^{\lambda/H}\left(1-\frac{k}{2^{2n}}\right)^{\lambda/H}\frac{1}{2^{2n}} \to \mathrm{B}\left(1+\lambda/H, 1+\lambda/H\right), \quad \text{as } n \to \infty,$$

which implies that $\mathsf{E}\left[E_n^2\right] \le C2^{2n(H-1)}$. Since E_n is Gaussian, we have $\mathsf{E}\left[|E_n|^p\right] \le C_p 2^{pn(H-1)}$ for any $p \ge 1$. Therefore, for any $\nu > 1$

$$\mathsf{E}\left[\sum_{n=1}^{\infty}\frac{|E_n|^p}{n^\nu 2^{pn(H-1)}}\right] = \sum_{n=1}^{\infty}\frac{\mathsf{E}\left[|E_n|^p\right]}{n^\nu 2^{pn(H-1)}} \le C_p\sum_{n=1}^{\infty} n^{-\nu} < \infty.$$

Consequently,

$$\xi' := \sup_{n\ge 1}\frac{|E_n|}{n^{\nu/p}2^{n(H-1)}} < \infty$$

a. s.; moreover, by Fernique's theorem, all moments of ξ' are finite.

Let us summarize the upper bounds:

$$|B_n| \le C\zeta n^\kappa 2^{-n\beta}, \quad |D_n| \le C\xi\zeta n^{\gamma+\kappa}2^{n(1-2\beta)}, \quad |E_n| \le \xi' n^\delta 2^{n(H-1)},$$

where $\delta > 0$ can be taken arbitrarily small. We have $-\beta < -1/2 < H-1$ and $-\beta < 1-2\beta$, so $|B_n|$ is of the smallest order. Which of the remaining two estimates wins, depends on values of β and H: for H close to $1/2$, $1-2\beta$ is close to 0, while $H-1$ is close to $-1/2$; for β close to 1, $1-2\beta$ is close to -1, while $H-1$

is close to 0. Thus, we arrive to

$$|B_n| + |E_n| + |D_n| \leq \eta n^{\gamma+\kappa} 2^{-\rho n},$$

where $\eta \leq C(\zeta + \xi\zeta + \xi')$, so all its moments are finite. The proof is now complete. □

Consider a simpler estimator:

$$\hat{\theta}_n^{(3)} = \frac{2^n \sum_{k=1}^{2^{2n}-1} b^{-1}(X_{t_{k-1}^n})(X_{t_k^n} - X_{t_{k-1}^n})}{\sum_{k=1}^{2^{2n}-1} b^{-1}(X_{t_{k-1}^n}) a(X_{t_{k-1}^n})}.$$

This is a discretized MLE for θ in Eq. (5.30), where B^H is replaced by Wiener process. Nevertheless, this estimator is consistent as well. Namely, we have the following result, whose proof is similar to that of Theorem 5.41, but is much simpler, so we omit it.

Theorem 5.42 *Under assumptions* (G$_1$)–(G$_4$), $\hat{\theta}_n^{(3)} \to \theta$ *a. s., as* $n \to \infty$. *Moreover, there exists a random variable* η' *with all finite moments such that* $\left|\theta_n^{(3)} - \theta\right| \leq \eta' n^{\kappa+\gamma} 2^{-\rho n}$.

Remark 5.43 Using Theorem B.46, it can be shown with some extra technical work that

$$\left|\hat{\theta}_n^{(i)} - \theta\right| \leq \eta_1 n^{\mu} 2^{-\tau n}, \quad i = 2, 3,$$

where $\mu = 1/2 + \gamma(1 + 1/H)$, $\tau = (2H - 1) \wedge (1 - H)$; η_1 is a random variable, for which there exists some $c_\theta > 0$ such that $\mathsf{E}\left[\exp\left\{x\eta_1^{1+1/H}\right\}\right] < \infty$ for $x < c_\theta$.

5.3.3 Simulations

In this subsection we illustrate quality of the estimators with the help of simulation experiments. For each set of parameters, we simulate 20 trajectories of the solution to (5.32). Then for each of estimators $\hat{\theta}_n^{(i)}$, $i = 1, 2, 3$, we compute the average relative error $\delta_n^{(i)}$, i. e. the average of values $\left|\hat{\theta}_n^{(i)} - \theta\right| / \theta$. We remind that for a particular value of n we take 2^{2n} equidistant observations of the process on the interval $[0, 2^n]$.

We start with a case of relatively "tame" coefficients $a(x) = 2\sin x + 3$, $b(x) = 2\cos x + 3$. We choose $\theta = 2$. The relative errors are presented in Table 5.7.

The first observation is that the estimators $\hat{\theta}_n^{(2)}$ and $\hat{\theta}_n^{(3)}$ have similar performance. They clearly outperform the estimator $\hat{\theta}_n^{(1)}$. This means that $\hat{\theta}_n^{(3)}$ is preferable to $\hat{\theta}_n^{(2)}$, since it does not involve H (which might be unknown) and is computable faster (for $n = 6$, computation of $\hat{\theta}_n^{(2)}$ takes 473 μs on Intel Core i5-3210M processor, while that of $\hat{\theta}_n^{(3)}$ takes 32 μs).

Table 5.7 Relative errors of estimators $\hat{\theta}_n^{(i)}$, $i = 1, 2, 3$, for $a(x) = 2\sin x + 3$, $b(x) = 2\cos x + 3$, $\theta = 2$

	$H = 0.6$			$H = 0.7$			$H = 0.8$			$H = 0.9$		
n	$\delta_n^{(1)}$	$\delta_n^{(2)}$	$\delta_n^{(3)}$	$\delta_n^{(1)}$	$\delta_n^{(2)}$	$\delta_n^{(3)}$	$\delta_n^{(1)}$	$\delta_n^{(2)}$	$\delta_n^{(3)}$	$\delta_n^{(1)}$	$\delta_n^{(2)}$	$\delta_n^{(3)}$
3	0.150	0.093	0.093	0.159	0.097	0.094	0.152	0.098	0.096	0.161	0.091	0.092
4	0.080	0.043	0.044	0.079	0.047	0.047	0.085	0.046	0.046	0.080	0.048	0.047
5	0.042	0.025	0.024	0.040	0.027	0.027	0.042	0.029	0.029	0.042	0.028	0.028
6	0.020	0.011	0.011	0.021	0.012	0.012	0.021	0.016	0.016	0.020	0.016	0.016

Table 5.8 Relative errors of $\hat{\theta}_n^{(i)}$, $i = 1, 2, 3$, for $a(x) = 2\sin x + 2.1$, $b(x) = 2\cos x + 2.1$, $\theta = 2$

	$H = 0.6$			$H = 0.7$			$H = 0.8$			$H = 0.9$		
n	$\delta_n^{(1)}$	$\delta_n^{(2)}$	$\delta_n^{(3)}$	$\delta_n^{(1)}$	$\delta_n^{(2)}$	$\delta_n^{(3)}$	$\delta_n^{(1)}$	$\delta_n^{(2)}$	$\delta_n^{(3)}$	$\delta_n^{(1)}$	$\delta_n^{(2)}$	$\delta_n^{(3)}$
3	0.20	0.17	0.18	0.20	0.18	0.19	0.20	0.18	0.18	0.20	0.17	0.17
4	0.111	0.096	0.097	0.109	0.099	0.102	0.110	0.099	0.106	0.109	0.095	0.099
5	0.058	0.045	0.045	0.057	0.052	0.052	0.057	0.051	0.053	0.057	0.046	0.046
6	0.029	0.024	0.024	0.029	0.021	0.021	0.029	0.027	0.028	0.029	0.033	0.033

Table 5.9 Ten values of $\hat{\theta}_n^{(i)}$, $i = 1, 2, 3$, for $a(x) = 2\cos x + 1$, $b(x) = 2\sin x + 2.1$, $\theta = 2$, $H = 0.7$, $n = 6$

$\theta_n^{(1)}$	0.9484	0.9207	0.9370	0.8193	0.7345	0.9088	1.0404	0.8379	0.8706	0.8280
$\theta_n^{(2)}$	0.8297	1.512	2.7865	2.4480	0.4196	−217.3	−78.14	0.5107	−3.466	1.308
$\theta_n^{(3)}$	0.4875	1.206	12.759	7.795	0.218	−1.402	−3.574	0.299	−0.86	0.8188

The second observation is that the estimate (5.43) of the convergence rate is probably not optimal; it seems that the convergence rate is around 2^{-n}, in particular, it is independent of H.

Now take worse coefficients $a(x) = 2\sin x + 2.1$, $b(x) = 2\cos x + 2.1$; again $\theta = 2$. The results are reported in Table 5.8.

The relative errors have increased two to three times due to the coefficients approaching zero closer. Also observe that in this case the convergence rate seems better than the estimate (5.43).

Further we show that, despite condition (G_4) might seem too restrictive, certain condition that the coefficients are non-zero is required.

To illustrate this, take first $a(x) = 2\cos x + 1$, $b(x) = 2\sin x + 2.1$, $\theta = 2$. From the first sight, it seems that the estimators should work fine here. Such intuition is based on the observation that the proof of Theorem 5.41 relies on sufficiently fast convergence of the denominator to $+\infty$, which somehow should follow from the fact that positive values of the ratio $a(x)/b(x)$ are overwhelming. Unfortunately, this intuition is wrong. Ten values of $\hat{\theta}_n^{(i)}$, $i = 1, 2, 3$, for $H = 0.7$, $n = 6$ are given in Table 5.9. We see that values of $\hat{\theta}_n^{(1)}$ are substantially less than the true value $\theta = 2$. Values of the estimators $\hat{\theta}_n^{(2)}$ and $\hat{\theta}_n^{(3)}$ are more extreme and useless.

Now take $a(x) = 2\cos x + 2.1$, $b(x) = 2\sin x$, $\theta = 2$, $H = 0.7$, $n = 6$. Ten values of each $\hat{\theta}_n^{(i)}$ are given in Table 5.10. In this case the estimator $\hat{\theta}_n^{(1)}$

Table 5.10 Ten values of $\theta_n^{(i)}$, $i = 1, 2, 3$, for $a(x) = 2\cos x + 2.1$, $b(x) = 2\sin x$, $\theta = 2$, $H = 0.7, n = 6$

$\hat{\theta}_n^{(1)}$	1.9987	2.0017	2.0041	1.9933	1.9946	2.0069	1.9960	2.0071	2.0031	2.0026
$\theta_n^{(2)}$	1.556	1.126	1.583	1.396	1.191	1.237	−0.4028	1.179	1.245	8.417
$\theta_n^{(3)}$	−0.4774	0.0342	0.5043	0.5077	−0.1168	−5.081	−1.253	0.2040	0.1162	53.3152

performs well. Although other estimators perform poorly, it is clear that $\hat{\theta}_n^{(2)}$ has improved significantly. Further, if we take $b(x) = 2\sin x + 1$, keeping other parameters, estimators' performance becomes much better and comparable to the case $b(x) = 2\sin x + 2.1$, namely, $\delta_6^{(1)} = 0.015$, $\delta_6^{(2)} = 0.039$, $\delta_6^{(3)} = 0.041$. We can conclude that sign changes of the coefficient a affect the performance much stronger than those of the coefficient b.

5.4 Statistical Inference for the Fractional Ornstein–Uhlenbeck Model

The present section deals with the inference problem associated with the Langevin equation (1.34). In Sect. 5.4.1 we consider the standard MLE. This estimator is hardly discretized because of singular kernels and one should choose the nonstandard estimators for discretization. For this reason, in Sects. 5.4.2 and 5.4.3, we study more simple estimators and their discretized versions and prove the corresponding strong consistency results. These estimators are different for the ergodic ($\theta < 0$) and non-ergodic ($\theta > 0$) cases. Therefore we study the hypothesis testing of the drift parameter sign, see Sect. 5.4.4. The maximum likelihood estimation for more general model than (1.34), where the coefficients may evolve in time, is considered in Sect. 5.4.5.

5.4.1 *Maximum Likelihood Estimation*

According to (5.15), the standard MLE $\hat{\theta}_T^{(1)}$ for the model (1.34) has the following form

$$\hat{\theta}_T^{(1)} = \frac{\int_0^T J'(t)\, t^{2H-1}\, dZ_t}{\int_0^T \left(J'(t)\right)^2 t^{2H-1}\, dt}, \tag{5.34}$$

where

$$J(t) = \int_0^t l_H(t,s) X_s\, ds, \qquad Z_t = \frac{1}{\sigma} \int_0^t l_H(t,s)\, dX_s,$$

the kernel $l_H(t,s)$ is defined in (1.11).

The maximum likelihood estimation for the fO-U is now well developed. It was first studied in [90] (under an assumption $\sigma = 1$), where the Laplace transform

$$\Theta_T(\lambda) = \mathsf{E}\exp\left\{-\lambda \int_0^T (J_t')^2 t^{2H-1}\, dt\right\}$$

was calculated explicitly in the following form

$$\Theta_T(\lambda) = \left(\frac{4(\sin \pi H)\sqrt{\theta^2 + 2\lambda(2-2H)}e^{-\theta T}}{\pi T D_T^H\left(\theta, \sqrt{\theta^2 + 2\lambda(2-2H)}\right)} \right)^{1/2}, \quad \lambda > 0,$$

where

$$D_T^H(\theta, \alpha) = \left[\alpha \cosh \frac{\alpha}{2}T - \theta \sinh \frac{\alpha}{2}T\right]^2 I_{-H}\left(\frac{\alpha}{2}T\right) I_{H-1}\left(\frac{\alpha}{2}T\right)$$
$$- \left[\alpha \sinh \frac{\alpha}{2}T - \theta \cosh \frac{\alpha}{2}T\right]^2 I_{1-H}\left(\frac{\alpha}{2}T\right) I_H\left(\frac{\alpha}{2}T\right)$$

(here the standard notation I_ν is used for the modified Bessel function of the first kind and order ν). This enabled not only to establish strong consistency, but also to get the asymptotic properties of the bias and the mean square error. The asymptotic normality in the case $\theta < 0$ was established in [21]. The asymptotic distributions for $\theta = 0$ and $\theta > 0$ were obtained in [161] and [162]. The following theorem summarizes these results.

Theorem 5.44 ([21, 90, 161, 162]) *Let $H \in [\frac{1}{2}, 1)$, $\sigma = 1$.*

1. *For any $\theta \in \mathbb{R}$ the estimator $\hat{\theta}_T^{(1)}$ defined by (5.34) is strongly consistent.*
2. *Denote $B(\theta, T) = \mathsf{E}\left(\hat{\theta}_T^{(1)} - \theta\right)$, $V(\theta, T) = \mathsf{E}\left(\hat{\theta}_T^{(1)} - \theta\right)^2$. The following properties hold:*

   (i) If $\theta < 0$, then, as $T \to \infty$,

$$B(\theta, T) \sim -2T^{-1}; \quad V(\theta, T) \sim 2\,|\theta|\, T^{-1}, \tag{5.35}$$

   (ii) If $\theta = 0$, then, for all T,

$$B(0, T) = B(0, 1)T^{-1}; \quad V(0, T) = V(0, 1)T^{-2},$$

   (iii) If $\theta > 0$, then, as $T \to \infty$,

$$B(\theta, T) \sim -2\sqrt{\pi}\sin \pi H \theta^{3/2} e^{-\theta T}\sqrt{T}; \tag{5.36}$$
$$V(\theta, T) \sim 2\sqrt{\pi}\sin \pi H \theta^{5/2} e^{-\theta T}\sqrt{T}. \tag{5.37}$$

3. *(i) If $\theta < 0$, then, as $T \to \infty$,*

$$\sqrt{T}\left(\hat{\theta}_T^{(1)} - \theta\right) \xrightarrow{d} \mathcal{N}(0, -2\theta),$$

(ii) If $\theta = 0$, then, for all T,

$$T\hat{\theta}_T^{(1)} \stackrel{d}{=} \hat{\theta}_1^{(1)},$$

(iii) If $\theta > 0$, then, as $T \to \infty$,

$$\frac{e^{\theta T}}{2\theta}\left(\hat{\theta}_T^{(1)} - \theta\right) \xrightarrow{d} \sqrt{\sin \pi H}\, \mathcal{C}(1),$$

where $\mathcal{C}(1)$ is the standard Cauchy distribution.

Remark 5.45 It holds that $\hat{\theta}_{H,T}^{(1)} \stackrel{d}{=} \hat{\theta}_{1-H,T}^{(1)}$, where $\hat{\theta}_{H,T}^{(1)}$ is the MLE under the Hurst parameter H and the time span T (see [33] for $\theta < 0$, [161] for $\theta = 0$, and [162] for $\theta > 0$). The MLE for $H < 1/2$ was also considered in [166], where the relations (5.35)–(5.37) were proved for $H < 1/2$.

5.4.2 Least Square Estimator for the Case $\theta > 0$

In this subsection we study the least square estimator for a non-ergodic fO-U proposed in [15]. In this case it coincides with the transferred MLE introduced in Sect. 5.2.2 for a general fractional diffusion model. We also consider two discretized versions of this estimator.

5.4.2.1 Continuous Observations

Assume that a trajectory of $X = X(t)$ is observed over a finite time interval $[0, T]$.

First we assume that $H \in (\frac{1}{2}, 1)$ and consider the non-standard estimator (5.16), which in our case has the following form:

$$\hat{\theta}_T^{(2)} = \frac{\int_0^T X_t\, dX_t}{\int_0^T X_t^2\, dt}. \tag{5.38}$$

By Proposition 1.15, the estimator $\hat{\theta}_T^{(2)}$ admits the following representation:

$$\hat{\theta}_T^{(2)} = \frac{X_T^2 - x_0^2}{2\int_0^T X_t^2 dt}. \tag{5.39}$$

Note that this form of the estimator is well defined for all $H \in (0, 1)$. In order to establish its strong consistency for all $H \in (0, 1)$, we need the following lemma.

Lemma 5.46 *Let $H \in (0, 1)$. For $\theta > 0$*

$$e^{-\theta t} X_t \to \xi_\theta \quad a.\,s.,\ as\ t \to \infty,$$

where $\xi_\theta = x_0 + \theta \int_0^\infty e^{-\theta s} B_s^H\, ds \overset{d}{=} \mathcal{N}\left(x_0, \frac{H\Gamma(2H)}{\theta^{2H}}\right)$.

Proof Note that (B.56) implies the a. s. convergence $e^{-\theta t} B_t^H \to 0$, as $t \to \infty$. Therefore, by (1.35),

$$e^{-\theta t} X_t = x_0 + \theta \int_0^t e^{-\theta s} B_s^H\, ds + e^{-\theta t} B_t^H \to x_0 + \theta \int_0^\infty e^{-\theta s} B_s^H\, ds$$

a. s., as $t \to \infty$. It follows from Corollary 1.44 and Lemma 1.45 that the limit has the distribution $\mathcal{N}\left(x_0, \frac{H\Gamma(2H)}{\theta^{2H}}\right)$. □

Theorem 5.47 *Let $H \in (0, 1)$. For $\theta > 0$, the estimator $\hat{\theta}_T^{(2)}$ is strongly consistent, as $T \to \infty$.*

Proof Lemma 5.46 implies the a. s. convergence

$$\frac{X_T^2}{e^{2\theta T}} \to \xi_\theta^2, \quad \text{as } T \to \infty. \tag{5.40}$$

Therefore, by L'Hôpital's rule,

$$\lim_{T\to\infty} \frac{\int_0^T X_t^2\, dt}{e^{2\theta T}} = \lim_{T\to\infty} \frac{X_T^2}{2\theta e^{2\theta T}} = \frac{\xi_\theta^2}{2\theta}. \tag{5.41}$$

Note that $0 < \xi_\theta^2 < \infty$ with probability 1, since ξ_θ is a normal random variable. Combining (5.40) and (5.41), we get the convergence $\hat{\theta}_T^{(2)} \to \theta$ a. s., as $T \to \infty$.

□

Remark 5.48 It was shown in [15, 53] that under the assumptions of Theorem 5.47,

$$e^{\theta T}\left(\hat{\theta}_T^{(2)} - \theta\right) \xrightarrow{d} 2\theta \mathcal{C}(1),$$

as $T \to \infty$, where $\mathcal{C}(1)$ is the standard Cauchy distribution.

5.4.2.2 Discrete Observations

Now we consider discrete counterparts of the estimators (5.38) and (5.39), and prove their strong consistency. Assume that a trajectory of $X = X(t)$ is observed at the points $t_{k,n} = \frac{k}{n}$, $0 \le k \le n^m$, $n \ge 1$, where $m > 1$ is some fixed integer. Consider the estimator

$$\tilde{\theta}_n^{(1)}(m) = \frac{\sum_{k=0}^{n^m-1} X_{k,n} \Delta X_{k,n}}{\frac{1}{n}\sum_{k=0}^{n^m-1} X_{k,n}^2}, \tag{5.42}$$

where $X_{k,n} = X_{t_{k,n}}$, $\Delta X_{k,n} = X_{k+1,n} - X_{k,n}$.

According to (1.34), the estimator $\tilde{\theta}_n^{(1)}(m)$ from (5.42) can be represented in the following form, which is more convenient for evaluation:

$$\tilde{\theta}_n^{(1)}(m) = \theta + \frac{\theta \sum_{k=0}^{n^m-1} X_{k,n} \int_{\frac{k}{n}}^{\frac{k+1}{n}} (X_s - X_{k,n})\, ds + \sigma \sum_{k=0}^{n^m-1} X_{k,n} \Delta B_{k,n}^H}{\frac{1}{n}\sum_{k=0}^{n^m-1} X_{k,n}^2}. \tag{5.43}$$

Let us also consider the discretized version of the estimator (5.39):

$$\tilde{\theta}_n^{(2)}(m) = \frac{n\left(X_{n^{m-1}}^2 - x_0^2\right)}{2\sum_{k=0}^{n^m-1} X_{k,n}^2}.$$

Theorem 5.49 *Let $H \in (0,1)$, $\theta > 0$. Then for any $m > 1$ the estimators $\tilde{\theta}_n^{(1)}(m)$ and $\tilde{\theta}_n^{(2)}(m)$ are strongly consistent.*

The proof of Theorem 5.49 consists of several steps. First, we establish an auxiliary result concerning the bounds for several sums of integral type that will participate in the bounds for the numerator of (5.43).

Lemma 5.50 *Let $H \in (0,1)$. For any $m > 1$ and $n \geq 2$ there exists $C > 0$ not depending on n such that*

(i) $$\sum_{k=0}^{n^m-1} \left(\tfrac{k+1}{n}\right)^H \log^2 \tfrac{k+1}{n} \leq C n^{(m-1)H+m} \log^2 n,$$

(ii) $$\sum_{k=0}^{n^m-1} \left(\tfrac{k+1}{n}\right)^{2H} \log^4 \tfrac{k+1}{n} \leq C n^{2H(m-1)+m} \log^4 n.$$

Proof We base the proof of both statements on the following evident relation: for any function $f : [0,1] \to \mathbb{R}$ that is Riemann integrable on $[0,1]$, and for any $m \geq 1$ the integral sums $S(f(x), n^m) := \frac{1}{n^m}\sum_{k=0}^{n^m-1} f(\frac{k+1}{n^m})$ tend to the integral $\int_0^1 f(x)dx$ as $n \to \infty$. In particular, these integral sums are bounded. Consider the statement (i). Evidently,

$$\sum_{k=0}^{n^m-1} \left(\tfrac{k+1}{n}\right)^H \log^2 \tfrac{k+1}{n} = n^{(m-1)H} \sum_{k=0}^{n^m-1} \left(\tfrac{k+1}{n^m}\right)^H \log^2\left(\tfrac{k+1}{n^m} n^{m-1}\right)$$

$$\leq 2n^{(m-1)H} \sum_{k=0}^{n^m-1} \left(\tfrac{k+1}{n^m}\right)^H \left(\log^2\left(\tfrac{k+1}{n^m}\right) + (m-1)^2 \log^2 n\right)$$

$$= 2n^{(m-1)H+m}S(x^H \log^2 x, n^m) + 2(m-1)^2 n^{(m-1)H+m} \log^2 n \cdot S\left(x^H, n^m\right)$$
$$\le Cn^{(m-1)H+m} + Cn^{(m-1)H+m} \log^2 n \le Cn^{(m-1)H+m} \log^2 n$$

for $n \ge 2$. Statement (ii) is established similarly. □

Now we will establish almost sure bounds for the values of the fO-U X and its increments. Considering functions of the form $t^p \log t$, $p > 0$ we suppose that $0 \cdot \infty = 0$.

Lemma 5.51 *We have the following bounds for the fO-U X in terms of the underlying fBm:*

(i) Let $\theta > 0$. Then for any $t > 0$

$$\sup_{0\le s\le t} |X_s| \le |x_0| e^{\theta t} + \theta e^{\theta t} \int_0^t e^{-\theta s} \sup_{0\le u\le s} \left|B_u^H\right| ds + \sup_{0\le s\le t} \left|B_s^H\right| \tag{5.44}$$

and for any $s \in [\frac{k}{n}, \frac{k+1}{n})$

$$\sup_{\frac{k}{n}\le u\le s} |X_u - X_{k,n}| \le \int_{\frac{k}{n}}^{s} \left(e^{\theta u} \left(|x_0| + \theta \int_0^u e^{-\theta v} \sup_{0\le z\le v} \left|B_z^H\right| dv \right) + \sup_{0\le z\le u} \left|B_z^H\right| \right) du + \sup_{\frac{k}{n}\le u\le s} \left|B_u^H - B_{k,n}^H\right| . \tag{5.45}$$

(ii) Let $\theta \le 0$. Then for any $t > 0$

$$\sup_{0\le s\le t} |X_s| \le |x_0| + 2 \sup_{0\le s\le t} \left|B_s^H\right| \tag{5.46}$$

and for any $s \in [\frac{k}{n}, \frac{k+1}{n})$

$$\sup_{\frac{k}{n}\le u\le s} |X_u - X_{k,n}| \le \frac{|\theta|\,|x_0|}{n} + \frac{2\,|\theta|}{n} \sup_{0\le u\le s} \left|B_u^H\right| + \sup_{\frac{k}{n}\le u\le s} \left|B_u^H - B_{k,n}^H\right| . \tag{5.47}$$

Proof

(i) The bound (5.44) follows immediately from (1.35), and the bound (5.45) follows immediately from (5.44) and (1.34).
(ii) The bound (5.46) follows from (1.35):

$$|X_t| \le |x_0| e^{\theta t} + |\theta| e^{\theta t} \sup_{0\le s\le t} \left|B_s^H\right| \cdot \int_0^t e^{-\theta s} ds + \left|B_t^H\right| \le |x_0| + 2 \sup_{0\le s\le t} \left|B_s^H\right| .$$

To establish the bound (5.47), we substitute (5.46) into the following inequality that can be easily obtained from (1.34): for $s \geq \frac{k}{n}$

$$|X_s - X_{k,n}| \leq |\theta| \int_{\frac{k}{n}}^{s} |X_u|\,du + \left|B_s^H - B_{k,n}^H\right|. \qquad \square$$

Plugging $p = 2$ and $\gamma = 1$ into the formulae (B.56) and (B.58), we get the following bounds:

$$\sup_{0 \leq s \leq t} \left|B_s^H\right| \leq \left(t^H \log^2 t + 1\right) \xi(2), \tag{5.48}$$

and for $s \in [\frac{k}{n}, \frac{k+1}{n}]$

$$\begin{aligned} \left|B_s^H - B_{\frac{k}{n}}^H\right| &\leq \left(s - \tfrac{k}{n}\right)^H \left(\left|\log\left(s - \tfrac{k}{n}\right)\right|^{1/2} + 1\right) \log(s+2)\eta_{H,1} \\ &\leq \left(\left(s - \tfrac{k}{n}\right)^H \left|\log\left(s - \tfrac{k}{n}\right)\right|^{1/2} + \left(s - \tfrac{k}{n}\right)^H\right) \log(s+2)\eta_{H,1}. \end{aligned} \tag{5.49}$$

The function $f(x) = x^r |\log x|^{\frac{1}{2}}$ is bounded on the interval $(0, 1]$ for any $r > 0$. Therefore

$$\left(s - \tfrac{k}{n}\right)^H \left|\log\left(s - \tfrac{k}{n}\right)\right|^{1/2} \leq C\left(s - \tfrac{k}{n}\right)^{H-r}$$

for any $0 < r < H$. Furthermore, for $s \in [\frac{k}{n}, \frac{k+1}{n}]$ we have that $(s - \frac{k}{n})^H \leq (s - \frac{k}{n})^{H-r}$. Therefore, we get from (5.49) that for any $0 < r < H$ and for $s \in [\frac{k}{n}, \frac{k+1}{n}]$

$$\left|B_s^H - B_{\frac{k}{n}}^H\right| \leq C\left(s - \tfrac{k}{n}\right)^{H-r} \log\left(n^{m-1} + 2\right) \eta_{H,1}. \tag{5.50}$$

It follows immediately from (5.48) that for $\theta > 0$

$$\int_0^t e^{-\theta s} \sup_{0 \leq u \leq s} \left|B_u^H\right| ds \leq \xi(2) \int_0^t e^{-\theta s}\left(s^H \log^2 s + 1\right) ds \leq C\xi(2),$$

and therefore both integrals $\int_0^\infty e^{-\theta s} B_s^H ds$ and $\int_0^\infty e^{-\theta s} \sup_{0 \leq u \leq s} \left|B_u^H\right| ds$ exist with probability 1 and admit the same upper bound $C\xi(2)$. Combining (5.44)–(5.47), (5.48) and (5.50), we get that for $\theta > 0$

$$\sup_{0 \leq u \leq s} |X_u| \leq |x_0| e^{\theta s} + C\theta e^{\theta s}\xi(2) + \left(s^H \log^2 s + 1\right) \xi(2),$$

and for $s \in [\frac{k}{n}, \frac{k+1}{n}]$

$$\sup_{\frac{k}{n} \le u \le s} |X_u - X_{k,n}| \le \theta \int_{\frac{k}{n}}^{s} (e^{\theta u} (|x_0| + C\theta\xi(2)) + (u^H \log^2 u + 1)\xi(2))\, du + \left(n^{-H+r} \log n\right) \eta_{H,1},$$

while for $\theta \le 0$

$$\sup_{0 \le u \le s} |X_u| \le |x_0| + 2\left(s^H \log^2 s + 1\right) \xi(2),$$

and for $s \in [\frac{k}{n}, \frac{k+1}{n}]$

$$\sup_{\frac{k}{n} \le u \le s} |X_u - X_{k,n}| \le \frac{|\theta|\,|x_0|}{n} + \frac{2\,|\theta|}{n} \sup_{0 \le u \le s} \left|B_u^H\right| + \sup_{\frac{k}{n} \le u \le s} \left|B_u^H - B_{k,n}^H\right| \le \frac{|\theta|\,|x_0|}{n} + \frac{2\,|\theta|}{n} \left(s^H \log^2 s + 1\right) \xi(2) + \left(n^{-H+r} \log n\right) \eta_{H,1}.$$

To simplify the notations, we denote by C any constant whose value is not important for our bounds. Furthermore, we denote by $\mathfrak{Z}$ the class of nonnegative random variables with the following property: there exists $C > 0$ not depending on n such that $\mathsf{E}\exp\{x\zeta^2\} < \infty$ for any $0 < x < C$. Then, for any constant C, r. vs. $\xi(2) + C$, $\eta_{H,1} + C$, $C\xi(2)$, and $C\eta_{H,1}$ belong to $\mathfrak{Z}$. Also, note that for fixed $m > 1$ and $n > 3$ we have the upper bound $\log(n^{m-1} + 3) \le C \log n$. Moreover, for any $\alpha > 0$ there exists $n(\alpha)$ such that for $n \ge n(\alpha)$ we have $\log n < n^{\alpha}$. Taking this into account, we get the bounds with the same $\zeta \in \mathfrak{Z}$.

Lemma 5.52 *There exists $\zeta \in \mathfrak{Z}$ such that for $\theta > 0$ we have for any fixed $\alpha > 0$, starting with $n \ge n(\alpha)$:*

$$\sup_{0 \le u \le s} |X_u| \le \left(e^{\theta s} + s^H \log^2 s\right) \zeta, \tag{5.51}$$

and for $s \in [\frac{k}{n}, \frac{k+1}{n}]$

$$\sup_{\frac{k}{n} \le u \le s} |X_u - X_{k,n}| \le \left(\tfrac{1}{n} e^{\theta s} + \tfrac{1}{n} s^H \log^2 s + n^{-H+\alpha}\right) \zeta, \tag{5.52}$$

while for $\theta \le 0$

$$\sup_{0 \le u \le s} |X_u| \le \left(1 + s^H \log^2 s\right) \zeta, \tag{5.53}$$

and for $s \in [\frac{k}{n}, \frac{k+1}{n}]$

$$\sup_{\frac{k}{n} \le u \le s} |X_u - X_{k,n}| \le \left(\tfrac{1}{n} + \tfrac{1}{n} s^H \log^2 s + n^{-H+\alpha}\right) \zeta. \tag{5.54}$$

Now we are in the position to bound both terms in the numerator of the right-hand side of (5.43). First, we present the bound with probability 1 for the 1st term in the numerator of (5.43). The inequality claimed in Lemma 5.53 holds for any $\alpha > 0$ starting with some non-random number $n(\alpha)$.

Lemma 5.53 *Let* $H \in (0, 1)$, $\theta > 0$. *Then for any* $m > 1$ *there exists* $\zeta \in \mathfrak{Z}$ *such that*

$$\left| \sum_{k=0}^{n^m - 1} X_{k,n} \int_{\frac{k}{n}}^{\frac{k+1}{n}} (X_s - X_{k,n})\, ds \right| \le \zeta^2 n^{-1} e^{2\theta n^{m-1}}.$$

Proof It follows immediately from (5.51) that

$$|X_{k,n}| \le \sup_{0 \le u \le \frac{k+1}{n}} |X_u| \le \left(e^{\theta \frac{k+1}{n}} + \left(\tfrac{k+1}{n}\right)^H \log^2 \left(\tfrac{k+1}{n}\right)\right) \zeta. \tag{5.55}$$

Now we take into account (5.55), substitute $\frac{k+1}{n}$ instead of s into (5.52) and apply Lemma 5.50 to get the following relations for any $\alpha > 0$:

$$\begin{aligned}
\left| \sum_{k=0}^{n^m - 1} X_{k,n} \int_{\frac{k}{n}}^{\frac{k+1}{n}} (X_s - X_{k,n})\, ds \right| &\le \zeta^2 \frac{1}{n} \sum_{k=0}^{n^m-1} \left(e^{\theta \frac{k+1}{n}} + \left(\tfrac{k+1}{n}\right)^H \log^2 \left(\tfrac{k+1}{n}\right)\right) \\
&\quad \times \left(\frac{1}{n} e^{\theta \frac{k+1}{n}} + \frac{1}{n} \left(\tfrac{k+1}{n}\right)^H \log^2 \left(\tfrac{k+1}{n}\right) + n^{-H+\alpha} \log n\right) \\
&= \zeta^2 \left(\frac{1}{n^2} \sum_{k=0}^{n^m-1} e^{2\theta \frac{k+1}{n}} + \frac{2}{n^2} \sum_{k=0}^{n^m-1} e^{\theta \frac{k+1}{n}} \left(\tfrac{k+1}{n}\right)^H \log^2 \left(\tfrac{k+1}{n}\right) \right. \\
&\quad + \frac{1}{n^2} \sum_{k=0}^{n^m-1} \left(\tfrac{k+1}{n}\right)^{2H} \log^4 \left(\tfrac{k+1}{n}\right) + n^{-1-H+\alpha} \log n \\
&\quad \left. \times \left(\sum_{k=0}^{n^m-1} e^{\theta \frac{k+1}{n}} + \sum_{k=0}^{n^m-1} \left(\tfrac{k+1}{n}\right)^H \log^2 \tfrac{k+1}{n} \right) \right) \\
&\le \zeta^2 \left(\frac{1}{n} e^{2\theta n^{m-1}} + e^{\theta n^{m-1}} \left(n^{H(m-1)+m-2+\alpha} + n^{-H+\alpha}\right) \right. \\
&\quad \left. + n^{m-1-H+(m-1)H+\alpha} \right).
\end{aligned} \tag{5.56}$$

Evidently, the term $\frac{1}{n}e^{2\theta n^{m-1}}$ dominates and the other terms are negligible, whence the proof follows. □

Lemma 5.54 *Let $H \in (0,1)$, $\theta > 0$. Then for any $m > 1$ there exists $\zeta \in \mathfrak{Z}$ such that*

$$\left|\sum_{k=0}^{n^m-1} X_{k,n}\Delta B_{k,n}^H\right| \le \zeta^2 e^{\theta n^{m-1}} n^{m+mH-H}\log^2 n.$$

Proof By (5.51) and (5.48), we have

$$\begin{aligned}\left|\sum_{k=0}^{n^m-1} X_{k,n}\Delta B_{k,n}^H\right| &\le n^m \sup_{0\le t\le n^{m-1}} |X_t| \cdot 2 \sup_{0\le t\le n^{m-1}} \left|B_t^H\right| \\ &\le \zeta^2 n^m \left(e^{\theta n^{m-1}} + n^{(m-1)H}\log^2 n^{m-1}\right)\left(n^{(m-1)H}\log^2 n^{m-1} + 1\right).\end{aligned}$$

Evidently, the term $n^m e^{\theta n^{m-1}} n^{(m-1)H}\log^2 n^{m-1}$ dominates. □

Now our goal is to present the denominator of (5.43) in a more convenient form. First, we compare the sum $\frac{1}{n}\sum_{k=0}^{n^m-1} X_{k,n}^2$ to the corresponding integral $\int_0^{n^{m-1}} X_s^2\,ds$. The reason to replace the sum with the corresponding integral is that for some values of H and m we can prove the consistency with the help of some kind of L'Hôpital's rule, however, the application of the L'Hôpital's rule or the Stolz–Cesàro theorem to the sum $\sum_{k=0}^{n^m-1} X_{k,n}^2$ is problematic because not only the upper bound but also the terms in the sum depend on n.

Lemma 5.55 *Let $H \in (0,1)$, $\theta > 0$. Then there exists $\zeta_1 \in \mathfrak{Z}$ such that*

$$\left|\int_0^{n^{m-1}} X_s^2\,ds - \frac{1}{n}\sum_{k=0}^{n^m-1} X_{k,n}^2\right| \le \frac{\zeta_1^2}{n} e^{2\theta n^{m-1}}.$$

Proof Evidently, the difference between the integral and the corresponding integral sum can be bounded as

$$\left|\int_0^{n^{m-1}} X_s^2\,ds - \frac{1}{n}\sum_{k=0}^{n^m-1} X_{k,n}^2\right| \le \int_0^{n^{m-1}} |\varphi_n(s)|\,ds,$$

where the integrand has the form

$$\varphi_n(s) = \left(X_s^2 - X_{k,n}^2\right)\mathbf{1}_{s\in\left[\frac{k}{n},\frac{k+1}{n}\right)},\quad 0 \le k \le n^m - 1.$$

Furthermore, the integrand can be bounded as

$$\begin{aligned}|\varphi_n(s)| &\le |X_s - X_{k,n}|\left(|X_s| + |X_{k,n}|\right)\mathbf{1}_{s\in\left[\frac{k}{n},\frac{k+1}{n}\right)} \\ &\le 2\,|X_s - X_{k,n}| \sup_{0\le u\le s}|X_u|\,\mathbf{1}_{s\in\left[\frac{k}{n},\frac{k+1}{n}\right)}.\end{aligned}$$

Then from (5.51), (5.52) and similarly to (5.56),

$$|X_s - X_{k,n}| \sup_{0\le u\le s} |X_u| \mathbf{1}_{s\in\left[\frac{k}{n},\frac{k+1}{n}\right)} \le \Big(n^{-1}e^{2\theta s} + 2n^{-1}e^{\theta s}s^H \log^2 s + e^{\theta s}n^{-H+\alpha} + n^{-1}s^{2H}\log^4 s + n^{-H+r}s^H \log^2 s\Big)\zeta_1^2. \tag{5.57}$$

Integrating over $[0, n^{m-1}]$, we see that the integral of the first term in the right-hand side of (5.57) dominates, whence the proof follows. □

Corollary 5.56 *Let $H \in (0,1)$, $\theta > 0$. Then there exists $\zeta_1 \in \mathfrak{Z}$ such that*

$$\frac{1}{n}\sum_{k=0}^{n^m-1} X_{k,n}^2 = \int_0^{n^{m-1}} X_s^2\, ds + \vartheta_n,$$

where

$$|\vartheta_n| \le \frac{\zeta_1^2}{n} e^{2\theta n^{m-1}}.$$

Proof of Theorem 5.49

(1) First, we prove the strong consistency of $\tilde{\theta}_n^{(1)}(m)$. According to Lemmas 5.53–5.54, and Corollary 5.56, it is sufficient to prove that

$$\psi_n := \frac{\zeta^2 n^{-1} e^{2\theta n^{m-1}} + \sigma\zeta^2 e^{\theta n^{m-1}} n^{m+mH-H}\log^2 n}{\int_0^{n^{m-1}} X_s^2\, ds + \vartheta_n} \to 0$$

a.s. as $n \to \infty$, where $|\vartheta_n| \le \frac{\zeta_1^2}{n} e^{2\theta n^{m-1}}$. Rewrite ψ_n as

$$\psi_n := \frac{\zeta^2\left(n^{-1} + \sigma e^{-\theta n^{m-1}} n^{m+mH-H}\log^2 n\right)}{e^{-2\theta n^{m-1}}\int_0^{n^{m-1}} X_s^2\, ds + e^{-2\theta n^{m-1}}\vartheta_n}.$$

Evidently, $\zeta^2\left(n^{-1} + \sigma e^{-\theta n^{m-1}} n^{m+mH-H}\log^2 n\right) \to 0$ a. s. as $n \to \infty$. Furthermore, $e^{-2\theta n^{m-1}}|\vartheta_n| \le \frac{\zeta_1^2}{n} \to 0$ a. s. as $n \to \infty$. At last, according to (5.41),

$$e^{-2\theta n^{m-1}}\int_0^{n^{m-1}} X_s^2\, ds \to \frac{\xi_\theta^2}{2\theta}, \quad \text{a. s., as } n \to \infty,$$

where $0 < \xi_\theta^2 < \infty$ with probability 1. Therefore, $\psi_n \to 0$, as $n \to \infty$, and the estimator $\tilde{\theta}_n^{(1)}(m)$ is strongly consistent.

(2) Now let us consider $\tilde{\theta}_n^{(2)}(m)$. It follows from Corollary 5.56 that for $\theta > 0$,

$$\frac{1}{n}\sum_{k=0}^{n^m-1} X_{k,n}^2 = \int_0^{n^{m-1}} X_t^2\,dt + \vartheta_n,$$

where $\frac{\vartheta_n}{e^{2\theta n^{m-1}}} \to 0$ a. s., as $n \to \infty$. Combining this with Theorem 5.47 and (5.40), we get

$$\tilde{\theta}_n^{(2)}(m) = \frac{X_{n^{m-1}}^2 - x_0^2}{2\int_0^{n^{m-1}} X_t^2\,dt + 2\vartheta_n} = \left(\frac{1}{\hat{\theta}_{n^{m-1}}^{(2)}} + 2\cdot\frac{e^{2\theta n^{m-1}}}{X_{n^{m-1}}^2 - x_0^2}\cdot\frac{\vartheta_n}{e^{2\theta n^{m-1}}}\right)^{-1} \to \theta$$

a. s., as $n \to \infty$. □

5.4.3 Ergodic-Type Estimator for the Case $\theta < 0$

In this section we consider an ergodic-type estimator proposed in [71] and its discrete counterpart.

5.4.3.1 Continuous Observations

Assume that a trajectory of $X = X(t)$ is observed over a finite time interval $[0, T]$.

Theorem 5.57 *Let $H \in (0, 1)$. For $\theta < 0$, the estimator*

$$\hat{\theta}_T^{(3)} = -\left(\frac{1}{\sigma^2 H\Gamma(2H)T}\int_0^T X_t^2 dt\right)^{-\frac{1}{2H}} \tag{5.58}$$

is strongly consistent, as $T \to \infty$.

Proof Let Y be the stationary ergodic process, introduced in (1.38). It follows from the ergodic theorem that

$$\frac{1}{T}\int_0^T Y_t^2\,dt \to \mathsf{E}Y_0^2,$$

as $T \to \infty$ a. s. and in L^2. The process X_t can be expressed as

$$X_t = Y_t - e^{\theta t}\eta_\theta, \tag{5.59}$$

where

$$\eta_\theta = \sigma\theta \int_{-\infty}^{0} e^{-\theta s} B_s^H \, ds - x_0$$

is a normal random variable. Using this representation, it is easy to show that

$$\lim_{T\to\infty} \frac{1}{T} \int_0^T X_t^2 \, dt = \lim_{T\to\infty} \frac{1}{T} \int_0^T Y_t^2 \, dt = \mathsf{E} Y_0^2 = \frac{\sigma^2 H \Gamma(2H)}{(-\theta)^{2H}},$$

where the last equality was established in (1.41). □

Remark 5.58 According to [74], under the assumptions of Theorem 5.58, we have the following central limit theorem (for $H \le 3/4$) and noncentral limit theorem (for $H > 3/4$):

(i) If $H \in (0, \frac{3}{4})$, then $\sqrt{T}\left(\hat{\theta}_T^{(3)} - \theta\right) \xrightarrow{d} \mathcal{N}\left(0, -\frac{\theta\sigma_H^2}{(2H)^2}\right)$, as $T \to \infty$, where

$$\sigma_H^2 = \begin{cases} 4H - 1 + \frac{2\Gamma(2-4H)\Gamma(4H)}{\Gamma(1-2H)\Gamma(2H)}, & H \in \left(0, \frac{1}{2}\right), \\ (4H-1)\left(1 + \frac{\Gamma(3-4H)\Gamma(4H-1)}{\Gamma(2-2H)\Gamma(2H)}\right), & H \in \left[\frac{1}{2}, \frac{3}{4}\right). \end{cases}$$

(ii) If $H = \frac{3}{4}$, then $\frac{\sqrt{T}}{\log T}\left(\hat{\theta}_T^{(3)} - \theta\right) \xrightarrow{d} \mathcal{N}\left(0, -\frac{16\theta}{9\pi}\right)$, as $T \to \infty$.

(iii) If $H \in (\frac{3}{4}, 1)$, then $T^{2-2H}\left(\hat{\theta}_T^{(3)} - \theta\right) \xrightarrow{d} -\frac{(-\theta)^{2H-1}\sqrt{4H-2}}{\sqrt{4H-3}\,\Gamma(2H+1)} \zeta_{2H-1}$, as $T \to \infty$, where ζ_{2H-1} is a Rosenblatt random variable, see Sect. B.2.1.

5.4.3.2 Discrete Observations

The estimator $\hat{\theta}_T^{(3)}$ can be discretized as follows. Let $h > 0$. Assume that a trajectory of $X = X(t)$ is observed at times $t_k = kh$, $k = 0, 1, \dots, n$.

Theorem 5.59 *Let $H \in (0, 1)$. For $\theta < 0$, the estimator*

$$\tilde{\theta}_n^{(3)} = -\left(\frac{1}{\sigma^2 H \Gamma(2H) n} \sum_{k=0}^{n-1} X_{kh}^2\right)^{-\frac{1}{2H}}$$

is strongly consistent, as $n \to \infty$.

Proof Similarly to the proof of Theorem 5.57, the ergodic theorem implies that

$$\frac{1}{n} \sum_{k=0}^{n-1} Y_{kh}^2 \to \mathsf{E} Y_0^2 = \frac{\sigma^2 H \Gamma(2H)}{(-\theta)^{2H}}. \tag{5.60}$$

a. s., as $n \to \infty$, where $Y = \{Y_t, t \geq 0\}$ is the ergodic process defined by (1.38). Then, using the representation (5.59) and the convergence (5.60), it is not hard to obtain that

$$\frac{1}{n}\sum_{k=0}^{n-1} X_{kh}^2 \to \frac{\sigma^2 H\Gamma(2H)}{(-\theta)^{2H}},$$

a. s., as $n \to \infty$, whence the proof follows. □

Remark 5.60 The strong consistency of the discretized estimator $\tilde{\theta}_n^{(3)}$ for $H \in [\frac{1}{2}, 1)$ was first obtained by Hu and Song [72]. Moreover, they proved that if $H \in [\frac{1}{2}, \frac{3}{4})$, then

$$\sqrt{n}\left(\tilde{\theta}_n^{(3)} - \theta\right) \xrightarrow{d} \mathcal{N}\left(0, \frac{\theta^2}{2H^2}\right),$$

as $n \to \infty$.

5.4.4 *Hypothesis Testing of the Drift Parameter Sign for Fractional Ornstein–Uhlenbeck Process*

We see from the Sects. 5.4.1, 5.4.2, and 5.4.3 that the methods of construction of drift parameter estimators for the fO-U and their asymptotic properties essentially depend on the sign of unknown drift parameter θ. In particular, the estimator (5.58) is based on the ergodicity and does not work in the non-ergodic case. Similarly, the estimator (5.38) converges to zero if $\theta < 0$, see remark at the end of Sect. 5.3 in [71]. This motivates the hypothesis testing of the sign of drift parameter in the model (1.34). Interest to this problem is also connected with the stability properties of the solution to the Eq. (1.34), which also depend on the sign of θ.

In the present subsection we propose a comparatively simple test for testing the null hypothesis $H_0: \theta \leq 0$ against the alternative $H_1: \theta > 0$. The main advantage of our approach is that it can be used for any $H \in (0, 1)$. The distribution of the test statistic is computed explicitly, and the power of test can be found numerically for any given simple alternative. Also we consider the hypothesis testing $H_0: \theta \geq \theta_0$ against $H_1: \theta \leq 0$, where $\theta_0 \in (0, 1)$ is some fixed number. Unfortunately, our approach does not enable testing the hypothesis $H_0: \theta = 0$ against the two-sided alternative $H_1: \theta \neq 0$. The test is based on the observations of the process X at two points: 0 and T. Therefore, it is applicable for both continuous and discrete cases.

For the sake of simplicity, we assume that $\sigma = 1$ throughout this subsection.

Remark 5.61 For $H \geq 1/2$, this problem was studied by Moers [130]. He constructed a test using the estimator

$$\hat{\theta}_T^{(4)} = \frac{X_T^2 - x_0^2}{2\int_0^T X_t^2 dt} - \left(\frac{1}{H\Gamma(2H)T}\int_0^T X_t^2 dt\right)^{-\frac{1}{2H}}, \tag{5.61}$$

a combination of the estimators (5.39) and (5.58). The exact distribution of $\hat{\theta}_T^{(4)}$ is not known, and the test is based on the asymptotic distribution of $T\hat{\theta}_T^{(4)}$. The values of the corresponding test statistic should be compared with quantiles of the random variable

$$\frac{\left(B_1^H\right)^2}{2\int_0^1 \left(B_t^H\right)^2 dt} - \left(\frac{1}{H\Gamma(2H)}\int_0^1 \left(B_t^H\right)^2 dt\right)^{-\frac{1}{2H}}, \tag{5.62}$$

and the quantiles can be obtained by Monte Carlo simulation. The test can be used for testing three types of hypothesis: $H_0{:}\,\theta \geq 0$ against $H_1{:}\,\theta < 0$, $H_0{:}\,\theta \leq 0$ against $H_1{:}\,\theta > 0$, and $H_0{:}\,\theta = 0$ against $H_1{:}\,\theta \neq 0$. The consistency of the test is proved only for $H \in [1/2, 3/4)$ for a simple alternative $\theta_1 < 0$, and for $H \in [1/2, 1)$ for $\theta_1 > 0$.

Tanaka [161, 162] considered the testing of the hypothesis $H_0{:}\,\theta = 0$ against the alternatives $H_1{:}\,\theta < 0$ and $H_1{:}\,\theta > 0$. He proposed tests based on the MLE (for both alternatives) and on the minimum contrast estimator (only for the ergodic case). Those tests were considered also for $H \geq 1/2$.

5.4.4.1 Test Statistic

For hypothesis testing of the sign of the parameter θ we construct a test based on the asymptotic behavior of the random variable

$$Z(t) = \frac{\log^+ \log |X_t|}{\log t}, \quad t > 1. \tag{5.63}$$

The following result explains the main idea. It is based on the different asymptotic behavior of the fO-U with positive drift parameter and with negative one.

Lemma 5.62 *The value of $Z(t)$ converges a. s. to 1 for $\theta > 0$, and to 0 for $\theta \leq 0$, as $t \to \infty$.*

Proof For $\theta > 0$, Lemma 5.46 implies the convergence

$$\log |X_t| - \theta t \to \log |\xi_\theta| \quad \text{a. s., as } t \to \infty,$$

where ξ_θ is a normal random variable, hence, $0 < |\xi_\theta| < \infty$ a. s. Therefore,

$$\frac{\log |X_t|}{t} \to \theta \quad \text{a. s., as } t \to \infty.$$

It means that there exists $\Omega' \subset \Omega$ such that $\mathsf{P}(\Omega') = 1$ and for any $\omega \in \Omega'$ there exists $t(\omega)$ such that for $t \geq t(\omega)$ $\log |X_t| > 0$. Hence, for $t \geq t(\omega)$ we have that

$$\left|\frac{\log^+ \log |X_t|}{\log t} - 1\right| = \left|\frac{\log\log |X_t|}{\log t} - 1\right| = \left|\frac{\log\log |X_t| - \log t}{\log t}\right|$$
$$= \left|\frac{\log \frac{\log |X_t|}{t}}{\log t}\right| \to 0$$

a. s., as $t \to \infty$. For $\theta \leq 0$, it follows from (5.53) that

$$|Z(t)| \leq \left|\frac{\log^+ \left(\log\left(1 + t^H \log^2 t\right) + \log \zeta\right)}{\log t}\right|$$
$$= \left|\frac{\log\left(\log\left(1 + t^H \log^2 t\right) + \log \zeta\right)}{\log t}\right| \sim \left|\frac{\log\left(\log\left(t^H \log^2 t\right)\right)}{\log t}\right| \to 0,$$

as $t \to \infty$. □

The next result gives the cdf of $Z(t)$. Let Φ and φ denote the cdf and pdf, respectively, of the standard normal variable. Recall that $v(\theta, t)$ is the variance of the fO-U process, see (1.37).

Lemma 5.63 *For $t > 1$ the probability $g(\theta, x_0, t, c) = \mathsf{P}(Z(t) \leq c)$ is given by*

$$g(\theta, x_0, t, c) = \Phi\left(\frac{e^{t^c} - x_0 e^{\theta t}}{\sqrt{v(\theta, t)}}\right) + \Phi\left(\frac{e^{t^c} + x_0 e^{\theta t}}{\sqrt{v(\theta, t)}}\right) - 1, \tag{5.64}$$

and g is a decreasing function of $\theta \in \mathbb{R}$.

Proof Using Corollary 1.44 and taking into account that $\log^+ x$ is a non-decreasing function, we obtain

$$\mathsf{P}(Z(t) \leq c) = \mathsf{P}\left(|X_t| \leq e^{t^c}\right) = \Phi\left(\frac{e^{t^c} - x_0 e^{\theta t}}{\sqrt{v(\theta, t)}}\right) - \Phi\left(\frac{-e^{t^c} - x_0 e^{\theta t}}{\sqrt{v(\theta, t)}}\right)$$
$$= \Phi\left(\frac{e^{t^c} - x_0 e^{\theta t}}{\sqrt{v(\theta, t)}}\right) + \Phi\left(\frac{e^{t^c} + x_0 e^{\theta t}}{\sqrt{v(\theta, t)}}\right) - 1.$$

Let us prove the monotonicity of the function g with respect to θ. Note that g is an even function with respect to x_0. Therefore, it suffices to consider only the case $x_0 \geq 0$. The partial derivative equals

$$\begin{aligned}\frac{\partial g}{\partial \theta} &= \varphi\left(\frac{e^{t^c} - x_0 e^{\theta t}}{\sqrt{v(\theta,t)}}\right)\left(-x_0 t e^{\theta t} v^{-\frac{1}{2}}(\theta,t) - \frac{1}{2} v^{-\frac{3}{2}}(\theta,t) v'_\theta(\theta,t)\left(e^{t^c} - x_0 e^{\theta t}\right)\right)\\
&\quad + \varphi\left(\frac{e^{t^c} + x_0 e^{\theta t}}{\sqrt{v(\theta,t)}}\right)\left(x_0 t e^{\theta t} v^{-\frac{1}{2}}(\theta,t) - \frac{1}{2} v^{-\frac{3}{2}}(\theta,t) v'_\theta(\theta,t)\left(e^{t^c} + x_0 e^{\theta t}\right)\right)\\
&= -\frac{1}{2} v^{-\frac{3}{2}}(\theta,t) v'_\theta(\theta,t) e^{t^c}\left(\varphi\left(\frac{e^{t^c} - x_0 e^{\theta t}}{\sqrt{v(\theta,t)}}\right) + \varphi\left(\frac{e^{t^c} + x_0 e^{\theta t}}{\sqrt{v(\theta,t)}}\right)\right)\\
&\quad - x_0 e^{\theta t} v^{-\frac{3}{2}}(\theta,t)\left(t v(\theta,t) - \frac{1}{2} v'_\theta(\theta,t)\right)\\
&\quad \times\left(\varphi\left(\frac{e^{t^c} - x_0 e^{\theta t}}{\sqrt{v(\theta,t)}}\right) - \varphi\left(\frac{e^{t^c} + x_0 e^{\theta t}}{\sqrt{v(\theta,t)}}\right)\right).\end{aligned} \tag{5.65}$$

Since

$$v'_\theta(\theta,t) = H\int_0^t s^{2H-1}\left(s e^{\theta s} + (2t-s)e^{\theta(2t-s)}\right) ds > 0, \tag{5.66}$$

we see that the first term in the right-hand side of (5.65) is negative. Let us consider the second term. From (1.37) and (5.66) it follows that

$$t v(\theta,t) - \tfrac{1}{2} v'_\theta(\theta,t) = H\int_0^t s^{2H-1}\left(\left(t - \tfrac{1}{2}s\right) e^{\theta s} + \tfrac{1}{2} s e^{\theta(2t-s)}\right) ds > 0.$$

Since $\left|e^{t^c} - x_0 e^{\theta t}\right| \leq e^{t^c} + x_0 e^{\theta t}$ for $x_0 \geq 0$, we have

$$\varphi\left(\frac{e^{t^c} - x_0 e^{\theta t}}{\sqrt{v(\theta,t)}}\right) - \varphi\left(\frac{e^{t^c} + x_0 e^{\theta_1 t}}{\sqrt{v(\theta,t)}}\right) \geq 0.$$

Thus, the second term in the right-hand side of (5.65) is non-positive. Hence, $\frac{\partial g}{\partial \theta} < 0$. □

5.4.4.2 Testing the Hypothesis $H_0\colon \theta \leq 0$ Against $H_1\colon \theta > 0$

We consider the test with the following procedure of testing the hypothesis $H_0\colon \theta \leq 0$ against the alternative $H_1\colon \theta > 0$. For a given significance level α, and for sufficiently large value of t we choose a threshold $c = c_t \in (0,1)$, see Lemma 5.64.

Further, when $Z(t) \le c$ the hypothesis H_0 is accepted, and when $Z(t) > c$ it is rejected. Below we will propose a technically simpler version of this test, without the computation of c, see Remark 5.65. The threshold c can be chosen as follows.

Fix a number $\alpha \in (0, 1)$, the significance level of the test. This level gives the maximal probability of a type I error, that is in our case the probability to reject the hypothesis $H_0\colon \theta \le 0$ when it is true. By Lemma 5.63, for a threshold $c \in (0, 1)$ and $t > 1$ this probability equals

$$\sup_{\theta\le 0} \mathsf{P}(Z(t) > c) = 1 - g(0, x_0, t, c).$$

Therefore, we determine c_t as a solution to the equation

$$g(0, x_0, t, c_t) = 1 - \alpha. \tag{5.67}$$

The following result shows that for any $\alpha \in (0, 1)$, it is possible to choose a sufficiently large t such that $c_t \in (0, 1)$.

Lemma 5.64 *Let $\alpha \in (0, 1)$. Then there exists $t_0 \ge 1$ such that for all $t > t_0$ there exists a unique $c_t \in (0, 1)$ such that $g(0, x_0, t, c_t) = 1 - \alpha$. Moreover $c_t \to 0$, as $t \to \infty$.*

The constant t_0 can be chosen as the largest $t \ge 1$ that satisfies at least one of the following two equalities

$$g(0, x_0, t, 0) = 1 - \alpha \quad \textit{or} \quad g(0, x_0, t, 1) = 1 - \alpha. \tag{5.68}$$

Proof By Lemma 1.45 (*iii*), $v(0, t) = t^{2H}$. Then for $\theta = 0$ the formula (5.64) becomes

$$g(0, x_0, t, c) = \Phi\left(\frac{e^{t^c} - x_0}{t^H}\right) + \Phi\left(\frac{e^{t^c} + x_0}{t^H}\right) - 1. \tag{5.69}$$

For any $t > 1$, the function $g(0, x_0, t, c)$ is strictly increasing with respect to c. For $c = 0$ we have

$$g(0, x_0, t, 0) = \Phi\left(\frac{e - x_0}{t^H}\right) + \Phi\left(\frac{e + x_0}{t^H}\right) - 1 \to 2\Phi(0) - 1 = 0, \quad \text{as } t \to \infty.$$

Therefore, there exists $t_1 > 1$ such that $g(0, x_0, t, 0) < 1 - \alpha$ for all $t \ge t_1$.

Similarly, for $c = 1$

$$g(0, x_0, t, 1) = \Phi\left(\frac{e^t - x_0}{t^H}\right) + \Phi\left(\frac{e^t + x_0}{t^H}\right) - 1 \to 2\Phi(\infty) - 1 = 1, \quad \text{as } t \to \infty.$$

Therefore, there exists $t_2 > 1$ such that $g(0, x_0, t, 1) > 1 - \alpha$ for all $t \ge t_2$.

Thus, for any $t \geq t_0 = \max\{t_1, t_2\}$ there exists a unique $c_t \in (0, 1)$ such that $g(0, x_0, t, c_t) = 1 - \alpha$.

To prove the convergence $c_t \to 0$, $t \to \infty$, consider an arbitrary $\varepsilon \in (0, 1)$. Then

$$g(0, x_0, t, \varepsilon) = \Phi\left(\frac{e^{t^\varepsilon} - x_0}{t^H}\right) + \Phi\left(\frac{e^{t^\varepsilon} + x_0}{t^H}\right) - 1 \to 2\Phi(\infty) - 1 = 1, \text{ as } t \to \infty.$$

Arguing as above, we see that there exists $t_3 > 1$ such that for any $t > t_3$ the unique $c_t \in (0, 1)$, for which $g(0, x_0, t, c_t) = 1 - \alpha$, belongs to the interval $(0, \varepsilon)$. This implies the convergence $c_t \to 0$, as $t \to \infty$.

It follows from (5.69) that $g(0, x_0, t, 0) = g(0, x_0, t, 1)$ for $t = 1$. As $t \to \infty$, we have $g(0, x_0, t, 0) \to 0$, $g(0, x_0, t, 1) \to 1$. Hence, at least one of the equalities (5.68) is satisfied for some $t \geq 1$ and the set of such t's is bounded. □

Remark 5.65 Since the function $g(0, x_0, t, c)$ is strictly increasing with respect to c for $t > 1$, we see that the inequality $Z(t) \leq c_t$ is equivalent to the inequality $g(0, x_0, t, Z(t)) \leq g(0, x_0, t, c_t) = 1 - \alpha$. Therefore, we do not need to compute the value of c_t. It is sufficient to compare $g(0, x_0, t, Z(t))$ with the level $1 - \alpha$.

Algorithm 5.66 The hypothesis $H_0 \colon \theta \leq 0$ against the alternative $H_1 \colon \theta > 0$ can be tested as follows.

1. Find t_0 defined in Lemma 5.64. The algorithm can be applied only in the case $t > t_0$.
2. Evaluate the statistic $Z(t)$ defined by (5.63).
3. Compute the value of $g(0, x_0, t, Z(t))$, see (5.69).
4. Do not reject the hypothesis H_0 if $g(0, x_0, t, Z(t)) \leq 1-\alpha$, and reject it otherwise.

Remark 5.67 In fact, the condition $t > t_0$ is not too restrictive, since for reasonable values of α, the values of t_0 are quite small, see Table 5.11.

Let us summarize the properties of the test in the following theorem.

Theorem 5.68 *The test described in Algorithm 5.66 is unbiased and consistent, as $t \to \infty$. For a simple alternative $\theta_1 > 0$ and moment $t > t_0$, the power of the test equals $1 - g(\theta_1, x_0, t, c_t)$, where c_t can be found from* (5.67).

Proof It follows from the monotonicity of g with respect to θ (see Lemma 5.63) that for any $\theta_1 > 0$

$$\mathsf{P}(Z(t) > c_t) = 1 - g(\theta_1, x_0, t, c_t) > 1 - g(0, x_0, t, c_t) = \alpha.$$

This means that the test is unbiased. Evidently, for a simple alternative $\theta_1 > 0$ the power of the test equals $1 - g(\theta_1, x_0, t, c_t)$.

It follows from the convergence $c_t \to 0$, as $t \to \infty$ (see Lemma 5.64), that $c_t < c$ for sufficiently large t and some constant $c \in (0, 1)$. Taking into account the formula (5.64) and Lemma 1.45 (*i*), we get, as $t \to \infty$:

$$1 \geq 1 - g(\theta_1, x_0, t, c_t) \geq 1 - g(\theta_1, x_0, t, c)$$

$$= 2 - \Phi\left(\frac{e^{t^c} - x_0 e^{\theta_1 t}}{\sqrt{v(\theta_1, t)}}\right) - \Phi\left(\frac{e^{t^c} + x_0 e^{\theta_1 t}}{\sqrt{v(\theta_1, t)}}\right)$$

$$\to 2 - \Phi\left(-\frac{x_0 \theta_1^H}{\sqrt{H\Gamma(2H)}}\right) - \Phi\left(\frac{x_0 \theta_1^H}{\sqrt{H\Gamma(2H)}}\right) = 1.$$

Hence, the test is consistent. □

Remark 5.69 For the composite alternative $H_1 : \theta > 0$, the power of the test of Algorithm 5.66 is small and equals

$$\inf_{\theta_1 > 0} (1 - g(\theta_1, x_0, t, c_t)) = 1 - g(0, x_0, t, c_t) = \alpha.$$

We can consider testing of $H_0 : \theta \leq 0$ against the alternative $H_1 : \theta \geq \theta_0$, where $\theta_0 > 0$ is a fixed number. For this composite alternative the power of the test of Algorithm 5.66 equals

$$\inf_{\theta_1 \geq \theta_0} (1 - g(\theta_1, x_0, t, c_t)) = 1 - g(\theta_0, x_0, t, c_t)$$

and tends to 1, as $t \to \infty$.

5.4.4.3 Testing the Hypothesis $H_0 : \theta \geq \theta_0$ Against $H_1 : \theta \leq 0$

Fix $\theta_0 \in (0, 1)$. Let us consider the problem of testing the hypothesis $H_0 : \theta \geq \theta_0$ against alternative $H_1 : \theta \leq 0$.

Algorithm 5.70 The hypothesis $H_0 : \theta \geq \theta_0$ against the alternative $H_1 : \theta \leq 0$ can be tested as follows.

1. Find $\tilde{t}_0$ defined in Lemma 5.71. The algorithm can be applied only in the case $t > \tilde{t}_0$.
2. Evaluate the statistic $Z(t)$ defined by (5.63).
3. Compute the value of $g(\theta_0, x_0, t, Z(t))$, see (5.64).
4. Do not reject the hypothesis H_0 if $g(\theta_0, x_0, t, Z(t)) \geq \alpha$, and reject it otherwise.

This algorithm is based on the following results. They can be proved similarly to the previous subsection.

Lemma 5.71 *Let $\alpha \in (0,1)$. There exists $\tilde{t}_0 > 1$ such that for all $t > \tilde{t}_0$ there exists a unique $\tilde{c}_t \in (0,1)$ such that*

$$g(\theta_0, x_0, t, \tilde{c}_t) = \alpha. \tag{5.70}$$

In this case $\tilde{c}_t \to 1$, as $t \to \infty$.

The constant $\tilde{t}_0$ can be chosen as the largest $t > 1$ that satisfies at least one of the following two equalities

$$g(\theta_0, x_0, t, 0) = \alpha \quad or \quad g(\theta_0, x_0, t, 1) = \alpha.$$

Theorem 5.72 *The test described in Algorithm 5.70 is unbiased and consistent, as $t \to \infty$. For a simple alternative $\theta_1 \le 0$ and moment $t > \tilde{t}_0$, the power of the test equals $g(\theta_1, x_0, t, \tilde{c}_t)$, where $\tilde{c}_t$ can be found from (5.70). For the composite alternative $H_1\colon \theta \le 0$, the power of the test equals $g(0, x_0, t, \tilde{c}_t)$ and tends to 1, as $t \to \infty$.*

Remark 5.73 The values of $\tilde{t}_0$ for various values of θ_0 and H are represented in Table 5.12. We see that if θ_0 is too close to zero, then for small H, the condition $t > \tilde{t}_0$ does not hold for reasonable values of t.

Remark 5.74 If we have a confidence interval for θ, then the value of $\theta_0 \in (0,1)$ can be chosen less than or equal to a lower confidence bound (in the case when it is positive).

5.4.4.4 Simulations

In this section we illustrate the performance of our algorithms by simulation experiments. We choose the initial value $x_0 = 1$ for all simulations.

In Tables 5.11 and 5.12 the values of t_0 and $\tilde{t}_0$ for various H and θ_0 are given.

Table 5.11 Value of t_0 for various H and α ($x_0 = 1$)

H	0.1	0.2	0.3	0.4	0.5	0.6	0.7	0.8	0.9
$\alpha = 0.01$	1.2157	1.2313	1.2492	1.2699	1.2940	1.3224	1.3561	1.3968	1.4462
$\alpha = 0.05$	1.5310	1.2373	1.1526	1.1124	1.0889	1.0736	1.0627	1.0547	1.0485

Table 5.12 Values of $\tilde{t}_0$ for various H and θ_0 ($x_0 = 1, \alpha = 0.05$)

H	0.1	0.2	0.3	0.4	0.5	0.6	0.7	0.8	0.9
$\theta_0 = 0.1$	32.433	32.667	31.994	30.592	28.659	26.375	23.903	21.386	18.946
$\theta_0 = 0.05$	65.242	64.721	61.728	57.078	51.413	45.233	38.967	32.995	27.621
$\theta_0 = 0.01$	326.47	307.43	271.64	227.99	181.64	137.06	98.759	69.618	49.408
$\theta_0 = 0.001$	3193.6	2719.1	2073.5	1387.8	778.94	382.06	189.71	104.11	63.576
$\theta_0 = 0$	$2.336 \cdot 10^{16}$	$1.528 \cdot 10^{8}$	285,900	12,364	1878.1	534.70	217.96	111.19	65.878

Table 5.13 Empirical rejection probabilities of the test of Algorithm 5.66 for the hypothesis testing $H_0\colon \theta \le 0$ against the alternative $H_1\colon \theta > 0$ for $H = 0.3$ and $H = 0.7$

θ	−0.1	−0.05	0	0.05	0.1	0.15	0.2	0.25	0.3
$H = 0.3$									
$T = 20$	0.000	0.003	0.043	0.341	0.701	0.880	0.973	0.982	0.996
$T = 40$	0.000	0.000	0.043	0.675	0.952	0.995	0.999	1.000	1.000
$T = 60$	0.000	0.000	0.039	0.860	0.994	1.000	1.000	1.000	1.000
$T = 80$	0.000	0.000	0.048	0.940	1.000	1.000	1.000	1.000	1.000
$T = 100$	0.000	0.000	0.049	0.986	1.000	1.000	1.000	1.000	1.000
$H = 0.7$									
$T = 20$	0.000	0.001	0.058	0.284	0.540	0.800	0.910	0.967	0.979
$T = 40$	0.000	0.000	0.050	0.581	0.889	0.984	0.998	1.000	1.000
$T = 60$	0.000	0.000	0.042	0.782	0.980	1.000	0.999	1.000	1.000
$T = 80$	0.000	0.000	0.047	0.908	0.995	1.000	1.000	1.000	1.000
$T = 100$	0.000	0.000	0.048	0.959	1.000	1.000	1.000	1.000	1.000

Table 5.14 Empirical rejection probabilities of the test of Moers [130] for the hypothesis testing $H_0\colon \theta \le 0$ against the alternative $H_1\colon \theta > 0$ for $H = 0.7$

θ	−0.1	−0.05	0	0.05	0.1	0.15	0.2	0.25	0.3
$T = 20$	0.001	0.013	0.085	0.370	0.706	0.873	0.947	0.976	0.992
$T = 40$	0.000	0.004	0.095	0.682	0.948	0.993	0.999	1.000	1.000
$T = 60$	0.000	0.002	0.092	0.881	0.995	1.000	1.000	1.000	1.000
$T = 80$	0.000	0.000	0.105	0.948	0.999	1.000	1.000	1.000	1.000
$T = 100$	0.000	0.000	0.089	0.977	1.000	1.000	1.000	1.000	1.000

We simulate fBm at the points $t = 0, h, 2h, 3h, \dots$ and compute the approximate values of the Ornstein–Uhlenbeck process as the solution to the Eq. (1.34), using Euler's approximations. For various values of θ we simulate $n = 1000$ sample paths with the step $h = 1/10,000$. Then we apply our algorithms, choosing the significance level $\alpha = 0.05$. In Table 5.13 the empirical rejection probabilities of the test of Algorithm 5.66 for the hypothesis testing $H_0\colon \theta \le 0$ against the alternative $H_1\colon \theta > 0$ for $H = 0.3$ and $H = 0.7$ are reported.

Then we test the same hypothesis using the test of Moers [130] for $H = 0.7$. By Monte Carlo simulations for 20,000 sample paths of the process $\{B_t^H, t \in [0,1]\}$ we estimate the $(1-\alpha)$-quantile $\psi_{1-\alpha}$ of the distribution (5.62) for $\alpha = 0.05$. Then we compare the statistic $T\hat{\theta}_T^{(4)}$ (see (5.61)) with the value of this quantile and reject the hypothesis $H_0\colon \theta \le 0$ if $T\hat{\theta}_T^{(4)} > \psi_{1-\alpha}$. We obtained that $\psi_{0.95} \approx 0.827946$. The empirical rejection probabilities for this test are given in Table 5.14. We see that comparing to our algorithm, the test of Moers has bigger power, i. e., it works a bit better when the alternative is true. But for $\theta = 0$, the necessary significance level $\alpha = 0.05$ is not achieved.

Tables 5.15 and 5.16 represent empirical rejection probabilities of the test of Algorithm 5.70 for $\theta_0 = 0.1$ and $\theta_0 = 0.05$, respectively. We see that the test

Table 5.15 Empirical rejection probabilities of the test of Algorithm 5.70 for the hypothesis testing $H_0\colon \theta \geq \theta_0$ against the alternative $H_1\colon \theta \leq 0$ for $\theta_0 = 0.1$, $H = 0.3$ and $H = 0.7$

θ	−0.25	−0.2	−0.15	−0.1	−0.05	0	0.1	0.15	0.2
$H = 0.3$									
$T = 40$	1.000	1.000	1.000	1.000	1.000	0.938	0.056	0.008	0.000
$T = 60$	1.000	1.000	1.000	1.000	1.000	1.000	0.052	0.005	0.000
$T = 80$	1.000	1.000	1.000	1.000	1.000	1.000	0.054	0.001	0.000
$T = 100$	1.000	1.000	1.000	1.000	1.000	1.000	0.054	0.001	0.000
$H = 0.7$									
$T = 40$	1.000	1.000	1.000	1.000	0.978	0.689	0.051	0.006	0.001
$T = 60$	1.000	1.000	1.000	1.000	1.000	1.000	0.052	0.004	0.001
$T = 80$	1.000	1.000	1.000	1.000	1.000	1.000	0.051	0.003	0.000
$T = 100$	1.000	1.000	1.000	1.000	1.000	1.000	0.051	0.001	0.000

Table 5.16 Empirical rejection probabilities of the test of Algorithm 5.70 for the hypothesis testing $H_0\colon \theta \geq \theta_0$ against the alternative $H_1\colon \theta \leq 0$ for $\theta_0 = 0.05$, $H = 0.7$

θ	−0.25	−0.2	−0.15	−0.1	−0.05	0	0.05	0.1	0.15
$T = 40$	0.842	0.773	0.661	0.566	0.368	0.149	0.051	0.007	0.000
$T = 604$	0.999	1.000	0.990	0.955	0.799	0.346	0.047	0.002	0.000
$T = 80$	1.000	1.000	1.000	1.000	0.999	0.719	0.044	0.001	0.000
$T = 100$	1.000	1.000	1.000	1.000	1.000	0.980	0.050	0.000	0.000

power increases if θ_0 increases. Also, the test power tends to 1, as the time horizon T increases. Hence these simulation studies confirm the theoretical results on the consistency.

5.4.5 Maximum Likelihood Estimation in the Generalized Fractional Ornstein–Uhlenbeck Model with Time-Dependent Coefficients

Consider the fractional Ornstein–Uhlenbeck model with non-constant coefficients. It has the form

$$dX_t = \theta\big(a(t)X_t + b(t)\big)\,dt + \gamma(t)\,dB_t^H,\ t \geq 0, \tag{5.71}$$

where a, b and γ are non-random measurable functions. Suppose they are locally bounded and $\gamma = \gamma(t) > 0$. The unique solution for this equation is a Gaussian process and it has the form

$$X_t = e^{\theta A(t)}\left(x_0 + \theta \int_0^t b(s)e^{-\theta A(s)}\,ds + \int_0^t \gamma(s)e^{-\theta A(s)}\,dB_s^H\right) := E(t) + G(t),$$

where $A(t) = \int_0^t a(s)\,ds$, $E(t) = e^{\theta A(t)}\left(x_0 + \theta \int_0^t b(s)e^{-\theta A(s)}\,ds\right)$ is a non-random function, $G(t) = e^{\theta A(t)} \int_0^t \gamma(s)e^{-\theta A(s)}\,dB_s^H$ is a Gaussian process with zero mean.

Denote $c(t) = \frac{a(t)}{\gamma(t)}$, $d(t) = \frac{b(t)}{\gamma(t)}$. Now we shall state the conditions for strong consistency of the MLE (5.15). Note that for the model (5.71) it has the following form

$$\hat{\theta}_T^{(1)} = \frac{\int_0^T J'(t)\, t^{2H-1}\, dZ_t}{\int_0^T \left(J'(t)\right)^2 t^{2H-1}\, dt},$$

where

$$J(t) = \int_0^t l_H(t,s)\big(c(s)X_s + d(s)\big)\, ds, \qquad Z_t = \int_0^t l_H(t,s)\gamma^{-1}(s)\, dX_s,$$

$l_H(t,s)$ is defined in (1.11).

Theorem 5.75 *Let $\theta > 0$. Assume that the functions a, c, d and γ satisfy the following assumptions: $0 < a_1 \le a(s) \le a_2$, $0 < c_1 \le c(s) \le c_2$, $0 < \gamma_1 \le \gamma(s) \le \gamma_2$, the functions c and d are continuously differentiable, c' is bounded, $c'(s) \ge 0$ and $c'(s) \to 0$ as $s \to \infty$. Then the MLE $\hat{\theta}_T^{(1)}$ is strongly consistent as $T \to \infty$.*

Proof We shall check the conditions of Proposition 5.10. Obviously, $\psi(t,x) = c(t)x + d(t) \in C^1(\mathbb{R}^+) \times C^2(\mathbb{R})$ and

$$J(t) = \int_0^t l_H(t,s)\big(d(s) + c(s)E(s)\big)\, ds + \int_0^t l_H(t,s)c(s)G(s)\, ds =: F(t) + H(t).$$

Furthermore, (E_1) and the assumptions of Theorem 1.42 hold. Note that the trajectories of process G are a. s. Hölder up to order H, whence

$$\lim_{s\to 0} s^{\frac{1}{2}-H} c(s)G(s) = 0.$$

Therefore

$$J'(t) = F'(t) + H'(t) = F'(t) + \int_0^t l_H(t,s)f(s)G(s)\, ds + \int_0^t l_H(t,s)c(s)\gamma(s)\, dB_s^H,$$

where $f(s) = \left(\frac{1}{2} - H\right)s^{-1}c(s) + c'(s) + \theta a(s)c(s)$. Evidently, J_t' is a Gaussian process with mean and variance that are bounded on any bounded interval. Therefore, condition (E_2) holds. As for condition (E_3), we must verify that

$$I_\infty = \int_0^\infty (J_t')^2 t^{2H-1}\, dt = \infty \quad \text{a. s.}$$

For any $\lambda > 0$ consider the moment generating function

$$\Theta_T(\lambda) = \mathsf{E}\exp\{-\lambda I_T\} = \mathsf{E}\exp\left\{-\lambda \int_0^T (J'_t)^2 t^{2H-1}\,dt\right\} \tag{5.72}$$

and

$$\Theta_\infty(\lambda) = \mathsf{E}\exp\{-\lambda I_\infty\} = \mathsf{E}\exp\left\{-\lambda \int_0^\infty (J'_t)^2 t^{2H-1}\,dt\right\},$$

so that $\Theta_\infty(\lambda) = \lim_{T\to\infty} \Theta_T(\lambda)$. Evidently,

$$\int_0^T (J'_t)^2 t^{2H-1}\,dt \geq T^{-1}\left(\int_0^T J'_t t^{H-\frac{1}{2}}\,dt\right)^2,$$

whence

$$\Theta_T(\lambda) \leq \Theta_T^{(1)}(\lambda) := \mathsf{E}\exp\left\{-\frac{\lambda}{T}\left(\int_0^T J'_t t^{H-\frac{1}{2}}\,dt\right)^2\right\}.$$

Random variable $\int_0^T J'_t t^{H-\frac{1}{2}}\,dt$ is Gaussian with mean $M(T)$ and variance $\sigma^2(T)$, say. Note that for a Gaussian random variable $\xi = m + \sigma N(0,1)$ we can easily calculate

$$\mathsf{E}\exp\left\{-a\xi^2\right\} = \left(2a\sigma^2+1\right)^{-\frac{1}{2}} \exp\left\{-\frac{am^2}{2a\sigma^2+1}\right\}. \tag{5.73}$$

This value attains its maximum at the point $m = 0$. Hence, it is sufficient to prove that

$$\lim_{T\to\infty} \Theta_T^{(2)}(\lambda) := \lim_{T\to\infty} \mathsf{E}\exp\left\{-\frac{\lambda}{T}\left(\int_0^T H'_t t^{H-\frac{1}{2}}\,dt\right)^2\right\} = 0.$$

However, it follows from (5.73) that $\Theta_T^{(2)}(\lambda) = \left(\frac{2\lambda\sigma_T^2}{T}+1\right)^{-\frac{1}{2}}$, therefore in order to prove the strong consistency of the MLE $\theta_T^{(1)}$, we only need to analyze the asymptotic behavior of σ_T^2. More specifically, we need to prove that $\frac{\sigma_T^2}{T} \to \infty$, as $T \to \infty$.

Divide $\int_0^T H'_t t^{H-\frac{1}{2}}dt$ into two parts: $\int_0^T H'_t t^{H-\frac{1}{2}}\,dt = H_T^{(1)} + H_T^{(2)}$, where

$$H_T^{(1)} = \int_0^T t^{H-\frac{1}{2}} \int_0^t l_H(t,s) f(s) G(s)\,ds\,dt$$

and

$$H_T^{(2)} = \int_0^T t^{H-\frac{1}{2}} \int_0^t l_H(t,s) c(s) \gamma(s)\,dB_s^H\,dt.$$

Since the functions c and γ are bounded from below and from above,

$$\mathsf{E}\left(H_T^{(2)}\right)^2 = C(H)\int_0^T\int_0^T (t_1t_2)^{H-\frac{1}{2}}\int_0^{t_1}\int_0^{t_2}\prod_{i=1,2} l_H(t_i,s_i)c(s_i)\gamma(s_i)$$

$$\times |s_1-s_2|^{2H-2}\,ds_1\,ds_2\,dt_1\,dt_2$$

$$\asymp C(H)\int_0^T\int_0^T (t_1t_2)^{H-\frac{1}{2}}\int_0^{t_1}\int_0^{t_2}\prod_{i=1,2} l_H(t_i,s_i)\,|s_1-s_2|^{2H-2}\,ds_1\,ds_2\,dt_1\,dt_2$$

$$\asymp C(H)T^3 \tag{5.74}$$

as $T\to\infty$.

Consider the behavior of f. Under assumptions of the theorem, $s^{-1}c(s)+c'(s)$ vanishes at infinity, $\theta a(s)c(s)$ is positive and bounded away from zero. Therefore, there exist $C_i>0$, $i=1,2$ and $s_0>0$ such that $C_1\le f(s)\le C_2$ for all $s>s_0$. Boundedness of f implies that $\mathsf{E}\left(H_T^{(1)}\right)^2$ has the same asymptotic behavior as

$$\int_{s_0}^T\int_{s_0}^T (t_1t_2)^{H-\frac{1}{2}}\int_{s_0}^{t_1}\int_{s_0}^{t_2}\left(\prod_{i=1,2} l_H(t_i,s_i)f(s_i)\right)$$

$$\times\left(\int_{s_0}^{s_1}\int_{s_0}^{s_2}\gamma(u_1)\gamma(u_2)\exp\left\{\theta\left(\int_{u_1}^{s_1}+\int_{u_2}^{s_2}\right)a(v)\,dv\right\}\right.$$

$$\left.\times |u_1-u_2|^{2H-2}\,du_1\,du_2\right)ds_1\,ds_2\,dt_1\,dt_2 \tag{5.75}$$

$$\ge C(H)\int_{s_0}^T\int_{s_0}^T (t_1t_2)^{H-\frac{1}{2}}\int_{s_0}^{t_1}\int_{s_0}^{t_2}\left(\prod_{i=1,2} l_H(t_i,s_i)\right)$$

$$\times\left(\int_{s_0}^{s_1}\int_{s_0}^{s_2}|u_1-u_2|^{2H-2}du_1du_2\right)ds_1\,ds_2\,dt_1\,dt_2 \asymp C(H)T^5.$$

Relations (5.74) and (5.75) mean that the asymptotic behavior of σ_T^2 is $\sigma_T^2\asymp C(H)T^5$ and $\frac{\sigma_T^2}{T}\to\infty$, as $T\to\infty$. □

5.5 Maximum Likelihood Drift Estimation in the Linear Model Containing Two fBms

Two important properties of an fBm are the stationarity of increments and self-similarity. However, these properties restrict applications of an fBm, and for this reason multifractional processes gained huge attention. Multifractionality can consist both in dependence of memory depth and regularity of a process on the time instance and on the time scale. In this section, we are dealing with the latter kind of multifractionality, where the properties of a process depend on the size of the time interval, on which the process is considered. In other words, we are looking at processes, which are not inherently self-similar.

The simplest approach is to consider a linear combination

$$G_t = \sum_{k=1}^{N} \sigma_k B_t^{H_k}, \quad t \geq 0, \tag{5.76}$$

of independent fBms B^{H_k} with different Hurst parameters $H_1 < H_2 < \cdots < H_N$. It is easy to see that for any $t_0 \geq 0$, $\left\{\delta^{-H_1}\left(G_{t_0+\delta t} - G_{t_0}\right), t \geq 0\right\}$ converges weakly to $\left\{\sigma_1 B_t^{H_1}, t \geq 0\right\}$ as $\delta \to 0+$. Therefore, in a small scale the process G behaves like the fBm with the smallest Hurst parameter. On the other hand, $\left\{c^{-H_N} G_{ct}, t \geq 0\right\}$ converges weakly to $\left\{\sigma_N B_t^{H_N}, t \geq 0\right\}$ as $c \to +\infty$, so in a long scale G behaves like the fBm with the largest Hurst parameter. It is well known that the Hurst parameter measures the pathwise regularity and the strength of the memory, thus, in a large scale the process G is smoother and the long-range dependence is stronger. In the next section we will consider another approach to multifractionality.

To simplify the situation we consider $N = 2$, arriving to the continuous-time linear model

$$Z_t = \theta t + \sigma_1 B_t^{H_1} + \sigma_2 B_t^{H_2}, \quad t \in [0, T], \tag{5.77}$$

where B^{H_1} and B^{H_2} are two independent fBms with different Hurst indices H_1 and H_2 defined on some stochastic basis $(\Omega, \mathfrak{F}, \mathbb{F}, \mathsf{P})$. We assume that the filtration $\mathbb{F} = \{\mathfrak{F}_t\}_{t\geq 0}$ is generated by these processes and completed by P-negligible sets of $\mathfrak{F}_0$.

From now on we suppose that Hurst indices in (5.77) satisfy the inequality:

$$\frac{1}{2} \leq H_1 < H_2 < 1,$$

and we consider the continuous modifications of both processes, which exist due to the Kolmogorov theorem. Assume that the Hurst indices H_1, H_2 and parameters $\sigma_1 \geq 0$, $\sigma_2 \geq 0$ are known, we aim to estimate the unknown drift parameter θ by the continuous observations of the trajectories of Z.

The section is organized as follows. In Sect. 5.5.1, we describe the model and explain how to reduce the solution of the estimation problem to the existence–uniqueness problem for the Fredholm integral equation of the second kind with some non-standard weakly singular kernel. In Sect. 5.5.2, we solve the existence-uniqueness problem. Sect. 5.5.3 is devoted to the basic properties of estimator, that is, we establish its form, consistency, and asymptotic normality.

5.5.1 Preliminaries: The Reduction of the Original Problem to the Integral Equation

Since we suppose that the Hurst parameters H_1, H_2 and scale parameters σ_1, σ_2 are known, for technical simplicity, we consider the case where $\sigma_1 = \sigma_2 = 1$ and, as it was mentioned before, $\frac{1}{2} \le H_1 < H_2 < 1$. If we wish to include the unknown parameter θ into the fBm with the smallest Hurst parameter in order to apply Girsanov's theorem for construction of the estimator, we consider a couple of processes $\left\{\widetilde{B}_t^{H_1}, B_t^{H_2}, t \ge 0\right\}$ defined on the space $(\Omega, \mathfrak{F}, \mathbb{F})$ and let P_θ be a probability measure under which $\widetilde{B}^{H_1}$ and B^{H_2} are independent, B^{H_2} is an fBm with Hurst parameter H_2, and $\widetilde{B}^{H_1}$ is an fBm with Hurst parameter H_1 and drift θ, that is,

$$\widetilde{B}_t^{H_1} = \theta t + B_t^{H_1}.$$

The probability measure P_0 corresponds to the case when $\theta = 0$. Our main problem is the construction of MLE for $\theta \in \mathbb{R}$ by the observations of the process

$$Z_t = \theta t + B_t^{H_1} + B_t^{H_2} = \widetilde{B}_t^{H_1} + B_t^{H_2} \quad t \in [0, T].$$

We apply a linear transformation to Z in order to reduce the construction to the sum with one term being the Wiener process. So, we take the kernel (1.11) and construct the integral

$$\begin{aligned} Y_t &= \int_0^t l_{H_1}(t,s)\, dZ_s \\ &= \theta \int_0^t l_{H_1}(t,s)\, ds + \int_0^t l_{H_1}(t,s)\, dB_s^{H_1} + \int_0^t l_{H_1}(t,s)\, dB_s^{H_2} \\ &= \theta c_{H_1} \mathrm{B}\left(\tfrac{3}{2} - H_1, \tfrac{3}{2} - H_1\right) t^{2-2H_1} + M_t^{H_1} + \int_0^t l_{H_1}(t,s)\, dB_s^{H_2}, \end{aligned} \tag{5.78}$$

where M^{H_1} is a Gaussian martingale (Molchan martingale), admitting the representations (1.13), that is

$$M_t^{H_1} = \int_0^t l_{H_1}(t,s)\, dB_s^{H_1} = (2 - 2H_1)^{1/2} \int_0^t s^{1/2 - H_1}\, dW_s,$$

where W is a standard Wiener process.

Lemma 5.76 *The linear transformation* (5.78) *is correctly defined.*

Proof It is sufficient to establish the existence of the integral $\int_0^t l_{H_1}(t,s)\, dB_s^{H_2}$ for any $t \in [0, T]$. But we have that for any $u, s \in [0, t]$

$$|u - s|^{2H_2 - 2} \le t^{2H_2 - 2H_1} |u - s|^{2H_1 - 2},$$

therefore

$$\begin{aligned}\|l_{H_1}(t,\cdot)\|_{H_2}^2 &:= H_2(2H_2-1)\int_0^t\int_0^t l_{H_1}(t,s)l_{H_1}(t,u)\,|u-s|^{2H_2-2}\,ds\,du \\ &\le H_2(2H_2-1)t^{2H_2-2H_1}\int_0^t\int_0^t l_{H_1}(t,s)l_{H_1}(t,u)\,|u-s|^{2H_1-2}\,ds\,du \\ &= H_2(2H_2-1)t^{2H_2-2H_1}\,\|l_{H_1}(t,\cdot)\|_{H_1}^2 = H_2(2H_2-1)t^{2H_2-2H_1}\mathsf{E}\left|M_t^{H_1}\right|^2 \\ &= H_2(2H_2-1)t^{2H_2-4H_1+2} < \infty,\end{aligned}$$

where $\langle\cdot,\cdot\rangle_H$ is defined by (1.7), whence the proof follows. □

The process Y is more convenient to deal with since it involves martingale with a drift. Furthermore, it follows from the next result that processes Z and Y are observed simultaneously, so, we can reduce the original problem to the equivalent problem of the construction of MLE of $\theta \in \mathbb{R}$ basing on the linear transformation Y.

Lemma 5.77 *The processes Z and Y are observed simultaneously.*

Proof Taking into account (5.78), it is enough to present Z via Y. But it follows from (5.78), from Fubini's theorem for integrals w. r. t. fBm [120, Theorem 2.6.5], and from elementary integral transformations, that for any $t \in [0, T]$

$$\begin{aligned}&\int_0^t (t-s)^{H_1-\frac{3}{2}}\int_0^s l_{H_1}(s,u)\,dZ_u\,ds = c_{H_1}\int_0^t u^{\frac{1}{2}-H_1}\int_u^t (t-s)^{H_1-\frac{3}{2}}(s-u)^{\frac{1}{2}-H_1}ds\,dZ_u \\ &= c_{H_1}\mathrm{B}\left(H_1-\tfrac{1}{2},\tfrac{3}{2}-H_1\right)\int_0^t u^{\frac{1}{2}-H_1}\,dZ_u = \int_0^t (t-s)^{H_1-\frac{3}{2}}\,Y_s\,ds \\ &= \left(H_1-\tfrac{1}{2}\right)^{-1}\int_0^t (t-s)^{H_1-\frac{1}{2}}\,dY_s,\end{aligned}$$

whence

$$Z_t = c_{H_1}^{-1}\mathrm{B}\left(H_1-\tfrac{1}{2},\tfrac{3}{2}-H_1\right)^{-1}\int_0^t\int_s^t (u-s)^{H_1-3/2}u^{H_1-1/2}du\,dY_s,$$

and the proof follows. □

This means that we can reduce the original problem to the equivalent problem of the construction of MLE of $\theta \in \mathbb{R}$ basing on the linear transformation Y.

Denote for simplicity $\delta_{H_1} := c_{H_1}\mathrm{B}\left(\frac{3}{2}-H_1,\frac{3}{2}-H_1\right)$. Now the main problem can be formulated as follows. Let $\frac{1}{2} \le H_1 < H_2 < 1$,

$$\left\{\widetilde{X}_1(t) = \widetilde{M}_t^{H_1},\; X_2(t) := \int_0^t l_{H_1}(t,s)\,dB_s^{H_2},\; t \ge 0\right\}$$

be a couple of processes defined on the space $(\Omega, \mathfrak{F})$ and P_θ be a probability measure under which $\widetilde{X}_1$ and X_2 are independent, B^{H_2} is an fBm with the Hurst parameter H_2, and $\widetilde{X}_1$ is a martingale with square characteristics $\langle \widetilde{X}_1 \rangle(t) = t^{2-2H_1}$ and with drift $\theta \delta_{H_1} t^{2-2H_1}$, i. e.,

$$\widetilde{X}_1(t) = \widetilde{M}_t^{H_1} = \theta \delta_{H_1} t^{2-2H_1} + M_t^{H_1}.$$

Also, denote $X_1(t) = M_t^{H_1}$. Our main problem is the construction of MLE for $\theta \in \mathbb{R}$ by the observations of the process

$$Y_t = \theta \delta_{H_1} t^{2-2H_1} + X_1(t) + X_2(t) = \widetilde{X}_1(t) + X_2(t).$$

Consider the martingale $\widetilde{M}^{H_1}$ with a drift:

$$\begin{aligned}
\widetilde{M}_t^{H_1} &= \theta \delta_{H_1} t^{2-2H_1} + (2-2H_1)^{\frac{1}{2}} \int_0^t s^{1/2-H_1}\, dW_s \\
&= \theta \delta_{H_1} (2-2H_1)^{\frac{1}{2}} (2-2H_1)^{\frac{1}{2}} \int_0^t s^{\frac{1}{2}-H_1} s^{\frac{1}{2}-H_1}\, ds + (2-2H_1)^{\frac{1}{2}} \int_0^t s^{\frac{1}{2}-H_1}\, dW_s \\
&= (2-2H_1)^{\frac{1}{2}} \int_0^t s^{\frac{1}{2}-H_1}\, d\widetilde{W}_s,
\end{aligned}$$

where $\widetilde{W}$ is a Wiener process with a drift defined under the measure P_θ by the equation

$$d\widetilde{W}_t := dW_t + \theta (2-2H_1)^{\frac{1}{2}} \delta_{H_1} t^{\frac{1}{2}-H_1}\, dt.$$

By Girsanov's theorem and independence of X_1 and X_2,

$$\begin{aligned}
\frac{d\mathsf{P}_\theta}{d\mathsf{P}_0} &= \exp\left\{ \theta \delta_{H_1} (2-2H_1)^{\frac{1}{2}} \int_0^T s^{\frac{1}{2}-H_1}\, d\widetilde{W}_s - \frac{1}{2} \theta^2 \delta_{H_1}^2 T^{2-2H_1} \right\} \\
&= \exp\left\{ \theta \delta_{H_1} \widetilde{X}_1(T) - \frac{1}{2} \theta^2 \delta_{H_1}^2 T^{2-2H_1} \right\}.
\end{aligned}$$

The derivative above is not the likelihood function of a parameter θ, because it is not measurable with respect to the observed σ-algebra

$$\mathfrak{F}_T^Y := \sigma\{Y_t, t \in [0,T]\} = \mathfrak{F}_T^X := \sigma\{X(t), t \in [0,T]\},$$

where $X(t) = X_1(t) + X_2(t)$.

We shall proceed as in [35]: let μ_θ be the probability measure induced by Y on the space of continuous functions with the supremum topology under probability P_θ. Then for any measurable set A, $\mu_\theta(A) = \int_A \Phi(x) \mu_0(dx)$, where $\Phi(x)$ is a measurable functional such that $\Phi(X) = \mathsf{E}_0\left(\frac{d\mathsf{P}_\theta}{d\mathsf{P}_0} \mid \mathfrak{F}_T^X\right)$. The latter means that $\mu_\theta \ll \mu_0$ for any $\theta \in \mathbb{R}$. Taking into account that $\widetilde{X}_1 = X_1$ under P_0 and the fact that the vector process (X_1, X) is Gaussian, we get that the corresponding likelihood function is given by

$$L_T(X,\theta) = \mathsf{E}_0\left(\frac{d\mathsf{P}_\theta}{d\mathsf{P}_0} \,\middle|\, \mathfrak{F}_T^X\right) = \mathsf{E}_0\left(\exp\left\{\theta\delta_{H_1}X_1(T) - \frac{1}{2}\theta^2\delta_{H_1}^2 T^{2-2H_1}\right\} \,\middle|\, \mathfrak{F}_T^X\right)$$
$$= \exp\left\{\theta\delta_{H_1}\mathsf{E}_0\left(X_1(T) \mid \mathfrak{F}_T^X\right) + \frac{\theta^2\delta_{H_1}^2}{2}\left(V_T - T^{2-2H_1}\right)\right\}, \tag{5.79}$$

where $V_t = \mathsf{E}_0\left(\left(X_1(t) - \mathsf{E}_0\left(X_1(t) \mid \mathfrak{F}_t^X\right)\right)^2 \,\middle|\, \mathfrak{F}_t^X\right)$, $t \in [0,T]$.

Thus, we arrive at the following problem: to find the projection $P_X X_1(T)$ of $X_1(T)$ onto

$$\{X(t) = X_1(t) + X_2(t),\ t \in [0,T]\}.$$

We recall from Sect. 1.1.2 that

$$W_i(t) = \int_0^t \left(\left(K_{H_i}^*\right)^{-1}\mathbf{1}_{[0,t]}\right)(s)\,dB_s^{H_i}, \quad i = 1,2,$$

where K_H^* is defined by (1.9), are standard Wiener processes, which are obviously independent.

Also, from Sect. 1.1 we have

$$X_1(t) = (2-2H_1)^{\frac{1}{2}}\int_0^t s^{1/2-H_1}\,dW_1(s), \quad B_t^{H_2} = \int_0^t K_{H_2}(t,s)\,dW_2(s), \tag{5.80}$$

where the square integrable kernel $K_H(t,s)$ has the form (1.6). Then, using (1.10), we can write

$$X_2(t) = \int_0^t K_{H_1,H_2}(t,s)\,dW_2(s),$$

where

$$K_{H_1,H_2}(t,s) = c_{H_1}\beta_{H_2}s^{1/2-H_2}\int_s^t (t-u)^{1/2-H_1}u^{H_2-H_1}(u-s)^{H_2-3/2}du. \tag{5.81}$$

Similarly to (1.10), we have for $f \in L_{H_2}^2[0,T]$

$$\int_0^T f(s)\,dX_2(s) = \int_0^T \left(K_{H_1,H_2}^* f\right)(s)\,dW_2(s),$$

where

$$\left(K^*_{H_1,H_2}f\right)(s) = \int_s^T f(t)\partial_t K_{H_1,H_2}(t,s)\,dt.$$

The projection of $X_1(T)$ onto $\{X(t), t \in [0,T]\}$ is a centered X-measurable Gaussian random variable, therefore, it has the form

$$P_X X_1(T) = \int_0^T h_T(t)\,dX(t)$$

with $h_T \in L^2_{H_1}[0,T]$. Note that h_T still can be a distribution. However, as we shall see below, it will be continuous function. The projection for all $u \in [0,T]$ must satisfy

$$\mathsf{E}\left[X(u)P_X X_1(T)\right] = \mathsf{E}\left[X(u)X_1(T)\right]. \tag{5.82}$$

Using (5.82) together with independency of X_1 and X_2, we arrive at

$$\begin{aligned}\mathsf{E}\left[X_1(u)\int_0^T h_T(t)\,dX_1(t) + X_2(u)\int_0^T h_T(t)\,dX_2(t)\right]&\\ = \mathsf{E}\left[X_1(u)X_1(T)\right] = u^{2-2H_1}.&\end{aligned} \tag{5.83}$$

Finally, from (5.80)–(5.83) we get the prototype of a Fredholm integral equation

$$(2-2H_1)\int_0^u h_T(s)s^{1-2H_1}\,ds + \int_0^T h_T(s)r_{H_1,H_2}(s,u)\,ds = u^{2-2H_1}, \quad u \in [0,T], \tag{5.84}$$

where

$$r_{H_1,H_2}(s,u) = \int_0^{s\wedge u} \partial_s K_{H_1,H_2}(s,v)K_{H_1,H_2}(u,v)\,dv.$$

This kernel can be written alternatively as $r_{H_1,H_2}(s,u) = \partial_s R_{H_1,H_2}(s,u)$, where

$$\begin{aligned}R_{H_1,H_2}(s,u) &= \int_0^{s\wedge u} K_{H_1,H_2}(s,v)K_{H_1,H_2}(u,v)\,dv = \mathsf{E}[X_2(s)X_2(u)]\\ &= H_2(2H_2-1)\int_0^s\int_0^u (s-v)^{\frac12-H_1}v^{\frac12-H_1}(u-z)^{\frac12-H_1}z^{\frac12-H_1}\,|v-z|^{2H_2-2}\,dz\,dv.\end{aligned}$$

Differentiating (5.84), we get the Fredholm integral equation of the second kind,

$$(2-2H_1)h_T(u)u^{1-2H_1} + \int_0^T h_T(s)k(s,u)\,ds = (2-2H_1)u^{1-2H_1}, \quad u \in (0,T], \tag{5.85}$$

where

$$k(s,u) = \partial_u r_{H_1,H_2}(s,u) = \int_0^{s\wedge u} \partial_s K_{H_1,H_2}(s,v)\partial_u K_{H_1,H_2}(u,v)\,dv. \tag{5.86}$$

It will be established in Remark 5.82 that for the case $H_1 = \frac{1}{2}$ the Eq. (5.85) can be reduced to the corresponding equation from [35]:

$$h_T(u) + H_2(2H_2-1)\int_0^T h_T(s)\,|s-u|^{2H_2-2}\,ds = 1, \quad u \in [0,T], \tag{5.87}$$

but the difference between (5.87) and (5.85) lies in the fact that (5.87) can be characterized as the equation with standard kernel while (5.85) with two different power exponents is more or less non-standard and therefore, it requires an unconventional approach.

We say that two integral equations are equivalent if they have the same continuous solutions. In this sense equations (5.84) and (5.85) are equivalent and both are equivalent to the equation

$$h_T(u) + \frac{1}{2-2H_1}\int_0^T h_T(s)\kappa(s,u)\,ds = 1, \quad u \in [0,T], \tag{5.88}$$

with continuous right-hand side, where

$$\kappa(s,u) = u^{2H_1-1}k(s,u), \quad s,u \in [0,T]. \tag{5.89}$$

We get that the main problem (i. e. the MLE construction for the drift parameter) is reduced to the existence-uniqueness result for the integral equation (5.84).

5.5.2 *Compactness of the Integral Operator: Existence-Uniqueness Result for the Fredholm Integral Equation*

Consider the integral operator K, generated by the kernel K bearing in mind that the notations of the kernel and of the corresponding operator will always coincide:

$$(Kx)(u) = \int_0^T K(s,u)x(s)\,ds, \quad x \in C[0,T].$$

Now we are in position to establish the properties of the kernel $\kappa(s,u)$ defined by (5.89) and (5.86). Introduce the notation $[0,T]_0^2 = [0,T]^2 \setminus \{(0,0)\}$.

Lemma 5.78 *Up to a set of Lebesgue measure zero, the kernel $\kappa(s,u)$, $s,u \in [0,T]$ admits the following representation on $[0,T]$*

$$\kappa(s,u) = \begin{cases} \kappa_0(s,u)\varphi(s,u), & s \neq u, \\ 0, & s = u, \end{cases} \tag{5.90}$$

where $\varphi(s,u) = (s \wedge u)^{1-2H_1} u^{2H_1-1} |s-u|^{2H_2-2H_1-1}$, and the function κ_0 is bounded and belongs to $C([0,T]_0^2)$.

Proof We take (5.86) and first present the derivative of $K_{H_1,H_2}(t,s)$, defined by (5.81), in the appropriate form. To start, put $u = s + (t-s)z$. This allows us to rewrite $K_{H_1,H_2}(t,s)$ as

$$\begin{aligned} K_{H_1,H_2}(t,s) &= c_{H_1}\beta_{H_2} s^{\frac{1}{2}-H_2}(t-s)^{H_2-H_1} \\ &\quad \times \int_0^1 (1-z)^{\frac{1}{2}-H_1}(s+(t-s)z)^{H_2-H_1} z^{H_2-\frac{3}{2}}\,dz. \end{aligned} \tag{5.91}$$

Differentiating (5.91) w. r. t. t for $0 < s < t \le T$ we get

$$\begin{aligned} \partial_t K_{H_1,H_2}(t,s) &= (H_2-H_1)c_{H_1}\beta_{H_2} s^{\frac{1}{2}-H_2}(t-s)^{H_2-H_1-1} \\ &\quad \times \int_0^1 (1-z)^{\frac{1}{2}-H_1}\left(s+(t-s)z\right)^{H_2-H_1} z^{H_2-\frac{3}{2}}\,dz \\ &\quad + (H_2-H_1)c_{H_1}\beta_{H_2} s^{\frac{1}{2}-H_2} \\ &\quad \times (t-s)^{H_2-H_1} \int_0^1 (1-z)^{\frac{1}{2}-H_1}\left(s+(t-s)z\right)^{H_2-H_1-1} z^{H_2-\frac{1}{2}}\,dz \\ &= (H_2-H_1)c_{H_1}\beta_{H_2} s^{\frac{1}{2}-H_2}(t-s)^{H_2-H_1-1} \\ &\quad \times \left(\int_0^1 (1-z)^{\frac{1}{2}-H_1}\left(s+(t-s)z\right)^{H_2-H_1} z^{H_2-\frac{3}{2}}\,dz\right. \\ &\quad \left. + (t-s)\int_0^1 (1-z)^{\frac{1}{2}-H_1}\left(s+(t-s)z\right)^{H_2-H_1-1} z^{H_2-\frac{1}{2}}\,dz\right) \\ &= (H_2-H_1)c_{H_1}\beta_{H_2} s^{\frac{1}{2}-H_2}(t-s)^{H_2-H_1-1} \\ &\quad \times \left(s^{H_2-H_1}\int_0^1 z^{H_2-\frac{3}{2}}(1-z)^{\frac{1}{2}-H_1}\left(1-\frac{s-t}{s}z\right)^{H_2-H_1} dz\right. \\ &\quad \left. + (t-s)s^{H_2-H_1-1}\int_0^1 (1-z)^{\frac{1}{2}-H_1}\left(1-\frac{s-t}{s}z\right)^{H_2-H_1-1} z^{H_2-\frac{1}{2}}\,dz\right). \end{aligned} \tag{5.92}$$

Denote for technical simplicity $\alpha_i = H_i - \frac{1}{2}$, $i = 1, 2$. Then, according to the definition and properties of the Gauss hypergeometric function, see equalities (A.1) and (A.2), the terms in the right-hand side of (5.92) can be rewritten as follows. For the first term, that is, for

$$I_1(t,s) := s^{H_2-H_1} \int_0^1 z^{\alpha_2-1}(1-z)^{-\alpha_1} \left(1 - \frac{s-t}{s} z\right)^{H_2-H_1} dz, \tag{5.93}$$

the values of parameters for the underlying integral equal $a = H_1 - H_2$, $b = \alpha_2$, $c = H_2 - H_1 + 1$, and $x = \frac{s-t}{s} < 1$, respectively; therefore $\frac{x}{x-1} = \frac{t-s}{t}$, $c - b = 1 - \alpha_1$, and

$$\begin{aligned}
I_1(t,s) &= \mathrm{B}(1-\alpha_1, \alpha_2) s^{H_2-H_1} F\left(H_1 - H_2, \alpha_2, 1 - H_1 + H_2; \frac{s-t}{s}\right) \\
&= \mathrm{B}(1-\alpha_1, \alpha_2) s^{H_2-H_1} \left(\frac{t}{s}\right)^{H_2-H_1} F\left(H_1 - H_2, 1-\alpha_1, 1 - H_1 + H_2; \frac{t-s}{t}\right) \\
&= \mathrm{B}(1-\alpha_1, \alpha_2) t^{H_2-H_1} F\left(H_1 - H_2, 1-\alpha_1, 1 - H_1 + H_2; \frac{t-s}{t}\right).
\end{aligned}$$

Similarly, for the second term, that is, for

$$I_2(t,s) := (t-s) s^{H_2-H_1-1} \int_0^1 z^{\alpha_2}(1-z)^{-\alpha_1} \left(1 - \frac{s-t}{s} z\right)^{H_2-H_1-1} dz, \tag{5.94}$$

the values of parameters for the underlying integral equal $a = H_1 - H_2 + 1$, $b = \alpha_2 + 1$, $c = H_2 - H_1 + 2$, and $x = \frac{s-t}{s}$, respectively; therefore $\frac{x}{x-1} = \frac{t-s}{t}$, $c - b = 1 - \alpha_1$, and

$$\begin{aligned}
I_2(t,s) &= (t-s) s^{H_2-H_1-1} \mathrm{B}(1-\alpha_1, \alpha_2+1) \\
&\quad \times F\left(H_1 - H_2 + 1, \alpha_2 + 1, H_2 - H_1 + 2; \frac{s-t}{s}\right) \\
&= (t-s) s^{H_2-H_1-1} \left(\frac{t}{s}\right)^{H_2-H_1-1} \\
&\quad \times \mathrm{B}(1-\alpha_1, \alpha_2+1) F\left(H_1 - H_2 + 1, 1-\alpha_1, 2 - H_1 + H_2; \frac{t-s}{t}\right) \\
&= (t-s) t^{H_2-H_1-1} \mathrm{B}(1-\alpha_1, \alpha_2+1) \\
&\quad \times F\left(H_1 - H_2 + 1, 1-\alpha_1, 2 - H_1 + H_2; \frac{t-s}{t}\right).
\end{aligned}$$

It is easy to see from the initial representations (5.93) and (5.94) that $I_1(t,s)$ and $I_2(t,s)$ are continuous on the set $0 < s \le t \le T$.

Now, introduce the notations:

$$\Psi_1(t,s) = \mathrm{B}(1-\alpha_1,\alpha_2) F\left(H_1 - H_2, 1-\alpha_1, 1 - H_1 + H_2; \frac{t-s}{t}\right)$$

and

$$\Psi_2(t,s) = \left(\frac{t-s}{t}\right)^{1-H_2+H_1} \mathrm{B}(1-\alpha_1,\alpha_2+1) \times F\left(H_1 - H_2 + 1, 1-\alpha_1, 2 - H_1 + H_2; \frac{t-s}{t}\right),$$

so that $I_1(t,s) = t^{H_2-H_1}\Psi_1(t,s)$ and $I_2(t,s) = (t-s)^{H_2-H_1}\Psi_2(t,s)$. Note that $\frac{t-s}{t} \in [0,1)$, therefore

$$\begin{aligned} F\left(H_1 - H_2, 1-\alpha_1, 1 - H_1 + H_2; \frac{t-s}{t}\right) &= \frac{1}{\mathrm{B}(1-\alpha_1,\alpha_2)} \int_0^1 z^{-\alpha_1}(1-z)^{\alpha_2-1}\left(1 - \frac{t-s}{t}z\right)^{H_2-H_1} dz \\ &\le \frac{1}{\mathrm{B}(1-\alpha_1,\alpha_2)} \int_0^1 z^{-\alpha_1}(1-z)^{\alpha_2-1}\, dz = 1, \end{aligned}$$

whence the function $\Psi_1(t,s)$ is bounded by $\mathrm{B}(1-\alpha_1,\alpha_2)$. In order to establish that $\Psi_2(t,s)$ is bounded, we use Proposition A.1. Its conditions are satisfied: $a = H_1 - H_2 + 1 \in (0,1)$, $b = 1-\alpha_1 > 0$, $c - b = \alpha_2 + 1 > 1$, and $x = \frac{t-s}{t} \in [0,1)$. Therefore,

$$\begin{aligned} & x^{1-H_2+H_1} F\left(H_1 - H_2 + 1, 1-\alpha_1, 2 - H_1 + H_2; x\right) \\ &\le x^{1-H_2+H_1}\left(1 - \frac{1-\alpha_1}{1-H_1+H_2}x\right)^{-1-H_1+H_2} = \left(\frac{1}{x} - \frac{1-\alpha_1}{1-H_1+H_2}\right)^{-1-H_1+H_2} \\ &\le \left(1 - \frac{1-\alpha_1}{1-H_1+H_2}\right)^{-1-H_1+H_2} = \left(\frac{1-H_1+H_2}{\alpha_2}\right)^{H_1-H_2+1}, \end{aligned}$$

whence $\Psi_2(t,s) \le \mathrm{B}(1-\alpha_1,\alpha_2+1)\left(\frac{1-H_1+H_2}{\alpha_2}\right)^{H_1-H_2+1}$. Additionally, both functions are homogeneous:

$$\Psi_i(at,as) = \Psi_i(t,s) \text{ for } a > 0, i = 1,2.$$

Introduce the notation

$$\Phi(t,s) = I_1(t,s) + I_2(t,s) = t^{H_2-H_1}\Psi_1(t,s) + (t-s)^{H_2-H_1}\Psi_2(t,s) \tag{5.95}$$

and note that $\Phi \in C([0,T]_0^2)$ is bounded and homogeneous:

$$\Phi(at,as) = a^{H_2-H_1}\Phi(t,s), \quad a>0. \tag{5.96}$$

In terms of notation (5.95) the representation (5.92) for $\partial_t K_{H_1,H_2}(t,s)$ can be rewritten as

$$\partial_t K_{H_1,H_2}(t,s) = c_{H_1}\beta_{H_2}(H_2-H_1)s^{\frac{1}{2}-H_2}(t-s)^{H_2-H_1-1}\Phi(t,s). \tag{5.97}$$

In turn, the kernel $k(s,u)$ from (5.86) can be rewritten as

$$k(s,u) = \left(c_{H_1}\beta_{H_2}(H_2-H_1)\right)^2 \times \int_0^{s\wedge u} v^{1-2H_2}(s-v)^{H_2-H_1-1}(u-v)^{H_2-H_1-1}\Phi(s,v)\Phi(u,v)\,dv. \tag{5.98}$$

Consider the kernel $k(s,u)$ for $s>u$. Then it evidently equals

$$k(s,u) = \left(c_{H_1}\beta_{H_2}(H_2-H_1)\right)^2 \times \int_0^{u} v^{1-2H_2}(s-v)^{H_2-H_1-1}(u-v)^{H_2-H_1-1}\Phi(s,v)\Phi(u,v)\,dv.$$

Put $z=\frac{u-v}{s-u}$ and transform $k(s,u)$ to

$$\begin{aligned}k(s,u) &= \left(c_{H_1}\beta_{H_2}(H_2-H_1)\right)^2(s-u)^{2H_2-2H_1-1}\int_0^{\frac{u}{s-u}} z^{H_2-H_1-1}(1+z)^{H_2-H_1-1}\\ &\quad\times\left(u-z(s-u)\right)^{1-2H_2}\Phi\left(s,u-z(s-u)\right)\Phi\left(u,u-z(s-u)\right)dz\\ &=: \frac{k_0(s,u)}{(s-u)^{1-2H_2+2H_1}},\end{aligned}$$

where

$$\begin{aligned}k_0(s,u) &= \left(c_{H_1}\beta_{H_2}(H_2-H_1)\right)^2\int_0^{\frac{u}{s-u}} z^{H_2-H_1-1}(1+z)^{H_2-H_1-1}\\ &\quad\times\left(u-z(s-u)\right)^{1-2H_2}\Phi\left(s,u-z(s-u)\right)\Phi\left(u,u-z(s-u)\right)dz.\end{aligned}$$

In turn, transform $k_0(s,u)$ with the change of variables $tu = z$ and apply (5.96):

$$k_0(s,u) = \left(c_{H_1}\beta_{H_2}(H_2 - H_1)\right)^2 \int_0^{\frac{1}{s-u}} (tu)^{H_2-H_1-1}(1+tu)^{H_2-H_1-1}$$
$$\times \left(u - tu(s-u)\right)^{1-2H_2} \Phi\left(s, u - tu(s-u)\right)\Phi\left(u, u - tu(s-u)\right) u\, dt$$
$$= \left(c_{H_1}\beta_{H_2}(H_2 - H_1)\right)^2 u^{1-2H_1} \int_0^{\frac{1}{s-u}} \left(1 - t(s-u)\right)^{1-2H_2}(1+tu)^{H_2-H_1-1}$$
$$\times t^{H_2-H_1-1}\Phi\left(s, u - tu(s-u)\right)\Phi\left(1, 1 - t(s-u)\right) dt.$$

Introducing the kernel $\kappa_0(s,u) = k_0(s,u)u^{2H_1-1}$, we can present $k(s,u)$ as

$$k(s,u) = \frac{\kappa_0(s,u)}{(s-u)^{1-2H_2+2H_1}u^{2H_1-1}}, \tag{5.99}$$

where, for $s > u > 0$,

$$\kappa_0(s,u) = \left(c_{H_1}\beta_{H_2}(H_2 - H_1)\right)^2 \int_0^{\frac{1}{s-u}} \left(1 - (s-u)t\right)^{1-2H_2}(1+ut)^{H_2-H_1-1}$$
$$\times t^{H_2-H_1-1}\Phi\left(s, u - tu(s-u)\right)\Phi\left(1, 1 - t(s-u)\right) dt$$
$$= \left(c_{H_1}\beta_{H_2}(H_2 - H_1)\right)^2 \int_0^{\infty} \mathbf{1}_{t\le\frac{1}{s-u}} \left(1 - (s-u)t\right)^{1-2H_2}(1+ut)^{H_2-H_1-1}$$
$$\times t^{H_2-H_1-1}\Phi\left(s, u - tu(s-u)\right)\Phi\left(1, 1 - t(s-u)\right) dt. \tag{5.100}$$

For the case $u > s > 0$, we can replace s and u in formulas (5.99) and (5.100). Substituting formally $u = s$ into (5.100), for $s > 0$, we get

$$\kappa_0(s,s) = \left(c_{H_1}\beta_{H_2}(H_2 - H_1)\right)^2 \Phi(s,s)\Phi(1,1) \int_0^{\infty} (1+st)^{H_2-H_1-1}t^{H_2-H_1-1}\, dt$$
$$= \left(c_{H_1}\beta_{H_2}(H_2 - H_1)\right)^2 s^{H_2-H_1}\Phi(1,1)^2 \int_0^{\infty} (1+st)^{H_2-H_1-1}t^{H_2-H_1-1}\, dt. \tag{5.101}$$

Note that $\Phi(1,1) = \mathrm{B}(1-\alpha_1, \alpha_2)$ and

$$\int_0^{\infty} (1+st)^{H_2-H_1-1}t^{H_2-H_1-1}\, dt = s^{H_1-H_2}\mathrm{B}(H_2 - H_1, 1 - 2H_2 + H_1).$$

The former equation holds due to (A.4). We get that $\kappa_0(s,s)$ does not depend on s and equals some constant

$$C_H := \left(c_{H_1}\beta_{H_2}(H_2 - H_1)\mathrm{B}(1-\alpha_1, \alpha_2)\right)^2 \mathrm{B}(H_2 - H_1, 1 - 2H_2 + H_1).$$

Therefore we define $\kappa_0(s,s) = C_H$, $s > 0$.

Now the continuity of κ_0 on $(0,T]^2$ follows from the Lebesgue dominated convergence theorem supplied by representation (5.100), equality (5.101), and its consequence $\kappa_0(s,s) = C_H$, $s > 0$, together with the facts that $\Phi \in C([0,T]_0^2)$ and is bounded. Consider $\kappa_0(s,u)$ for $u \downarrow 0$ and let $s > 0$ be fixed. Then

$$\begin{aligned}\lim_{u\downarrow 0} \kappa_0(s,u) = C_H^1 &:= \left(c_{H_1}\beta_{H_2}(H_2 - H_1)\right)^2 \Phi(1,0) \\ &\times \int_0^1 (1-y)^{1-2H_2} y^{H_2-H_1-1} \Phi(1, 1-y)\, dy < \infty,\end{aligned}$$

and we can put $\kappa_0(s,0) = \kappa_0(0,u) = C_H^1$, $s > 0$, $u > 0$, thus extending the continuity of κ_0 to $[0,T]_0^2$.

It is easy to see that values $\kappa_0(s,s)$ and $\kappa_0(s,0)$ do not depend on $s > 0$ and do not coincide: $C_H \neq C_H^1$. Consequently, the limit

$$\lim_{(s,u)\to(0,0)} \kappa_0(s,u)$$

does not exist and depends on the way the variables s and u tend to zero. We can equate $\kappa_0(0,0)$ to any constant; e.g., let $\kappa_0(0,0) = 0$.

In order to prove that κ_0 is bounded, we consider the case $s > u$ (the opposite case is treated similarly) and put $z = (s-u)t$. Then

$$\begin{aligned}&\int_0^{\frac{1}{s-u}} \left(1-(s-u)t\right)^{1-2H_2} (1+ut)^{H_2-H_1-1} t^{H_2-H_1-1} \\ &\quad\times \Phi\left(s, u - tu(s-u)\right)\Phi\left(1, 1-t(s-u)\right) dt \\ &= \frac{1}{(s-u)^{H_2-H_1}} \int_0^1 (1-z)^{1-2H_2} \left(1 + \frac{u}{s-u} z\right)^{H_2-H_1-1} z^{H_2-H_1-1} \\ &\quad\times \Phi\left(s, u(1-z)\right)\Phi(1, 1-z)\, dz =: I_3(s,u).\end{aligned} \tag{5.102}$$

It follows from (5.96) that, for $s \neq 0$,

$$\Phi\left(s, u(1-z)\right) = s^{H_2-H_1} \Phi\left(1, \frac{u}{s}(1-z)\right).$$

Denote $r = \frac{s}{s-u}$ and put $t = \frac{1-z}{1-(1-r)z}$. Then

$$\frac{u}{s-u} = r - 1, \quad t < 1, \quad z = \frac{1-t}{1-t(1-r)} \in (0,1),$$

and the right-hand side of (5.102) can be rewritten as

$$
\begin{aligned}
I_3(s,u) &= r^{H_2-H_1}\int_0^1 (1-z)^{1-2H_2}\big(1-(1-r)z\big)^{H_2-H_1-1} z^{H_2-H_1-1} \\
&\quad \times \Phi\left(1,\frac{u}{s}(1-z)\right)\Phi(1,1-z)\,dz \\
&= r^{1-2H_1}\int_0^1 t^{1-2H_2}(1-t)^{H_2-H_1-1}\big(1-(1-r)t\big)^{2H_1-1} \\
&\quad \times \Phi\left(1,\frac{u}{s}\frac{rt}{1-(1-r)t}\right)\Phi\left(1,\frac{rt}{1-(1-r)t}\right)dt.
\end{aligned}
\tag{5.103}
$$

At last, put $y = 1-t$. Then the right-hand side of (5.103) is transformed to

$$
\begin{aligned}
I_3(s,u) &= r^{1-2H_1} r^{2H_1-1}\int_0^1 (1-y)^{1-2H_2} y^{H_2-H_1-1}\left(1-y\frac{r-1}{r}\right)^{2H_1-1} \\
&\quad \times \Phi\left(1,\frac{u}{s}\frac{r(1-y)}{r-y(r-1)}\right)\Phi\left(1,\frac{r(1-y)}{r-y(r-1)}\right)dy.
\end{aligned}
$$

Recall that $r = \frac{s}{s-u}$. Then it follows from the boundedness of Φ that there exists a constant C_H^1 such that, for $s > u$,

$$
\begin{aligned}
\kappa_0(s,u) &= \big(c_{H_1}\beta_{H_2}(H_2-H_1)\big)^2\int_0^1 (1-y)^{1-2H_2}\left(1-\frac{u}{s}y\right)^{2H_1-1} y^{H_2-H_1-1} \\
&\quad \times \Phi\left(1,\frac{u(1-y)}{s-uy}\right)\Phi\left(1,\frac{s(1-y)}{s-uy}\right)dy \\
&\le C_H^1\int_0^1 (1-y)^{1-2H_2} y^{H_2-H_1-1}\,dy,
\end{aligned}
\tag{5.104}
$$

so κ_0 is bounded, and the lemma is proved. □

Remark 5.79 Figure 5.1 demonstrates the graph of $\kappa_0(s,u)$ for $H_1 = 0.7, H_2 = 0.9$.

Now, consider the properties of the function

$$
\varphi(s,u) = (s\wedge u)^{1-2H_1} u^{2H_1-1}\,|s-u|^{2H_2-2H_1-1}
$$

participating in the kernel representation (5.90).

Lemma 5.80 *The function φ has the following properties:*

(i) for any $u\in[0,T]$, $\varphi(\cdot,u)\in L_1[0,T]$ and $\sup_{u\in[0,T]}\|\varphi(\cdot,u)\|_{L_1} < \infty$,

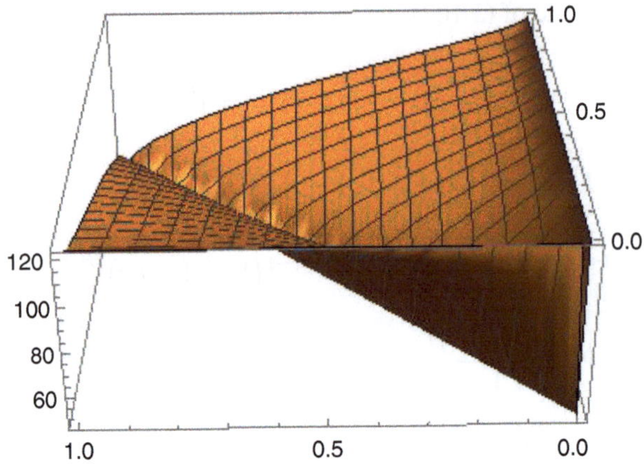

Fig. 5.1 Function $\kappa_0(s,u)$

(ii) for any $u_1 \in [0,T]$, $\int_0^T |\varphi(s,u) - \varphi(s,u_1)|\, ds \to 0$, *as* $u \to u_1$.

Proof

(i) It follows from the evident calculations that

$$\begin{aligned}\int_0^T |\varphi(s,u)|\, ds &= \int_0^T \varphi(s,u)\, ds \\ &= \int_0^u \frac{u^{2H_1-1}\, ds}{s^{2H_1-1}(u-s)^{1+2H_1-2H_2}} + \int_u^T \frac{ds}{(s-u)^{1+2H_1-2H_2}} \\ &= u^{2H_2-2H_1}\mathrm{B}(2-2H_1, 2H_2-2H_1) + \frac{(T-u)^{2H_2-2H_1}}{2H_2-2H_1} \\ &\le C_{H_1,H_2} T^{2H_2-2H_1} < \infty \quad \text{for all } u \in [0,T].\end{aligned}$$

(ii) First, let $u_1 = 0$ and $u \downarrow 0$. Note that $\varphi(s,0) = s^{2H_2-2H_1-1}$. Therefore

$$\begin{aligned}\int_0^T \left|\varphi(s,u) - \frac{1}{s^{1+2H_1-2H_2}}\right| ds &= \int_0^u \left|\frac{u^{2H_1-1}}{s^{2H_1-1}(u-s)^{1+2H_1-2H_2}} - \frac{1}{s^{1+2H_1-2H_2}}\right| ds \\ &+ \int_u^T \frac{ds}{(s-u)^{1+2H_1-2H_2}} - \int_u^T \frac{ds}{s^{1+2H_1-2H_2}} ds \\ &\le \int_0^u \frac{u^{2H_1-1} ds}{s^{2H_1-1}(u-s)^{1+2H_1-2H_2}} + \int_0^u \frac{ds}{s^{1+2H_1-2H_2}} \\ &+ \frac{1}{2H_2-2H_1}\left((s-u)^{2H_2-2H_1} - s^{2H_2-2H_1}\right)\Big|_{s=u}^{s=T}\end{aligned}$$

$$= \mathrm{B}(2-2H_1, 2H_2-2H_1)u^{2H_2-2H_1}$$

$$+ \frac{1}{2H_2-2H_1}\left(2u^{2H_2-2H_1} + (T-u)^{2H_2-2H_1} - T^{2H_2-2H_1}\right) \to 0, \text{ as } u \to 0.$$

From now on suppose that $u_1 > 0$ is fixed. Without loss of generality, suppose that $u \uparrow u_1$. Then

$$\int_0^T |\varphi(s,u) - \varphi(s,u_1)|\, ds = \int_0^u |\varphi(s,u) - \varphi(s,u_1)|\, ds + \int_u^{u_1} |\varphi(s,u) - \varphi(s,u_1)|\, ds$$

$$+ \int_{u_1}^T |\varphi(s,u) - \varphi(s,u_1)|\, ds =: I_1(u,u_1) + I_2(u,u_1) + I_3(u,u_1).$$

Consider the terms separately. First, we establish that $\varphi(s,\cdot)$ is decreasing in the second argument. Indeed, for $0 < s < u < u_1$,

$$\varphi(s,u_1) = \frac{u_1^{2H_1-1}}{s^{2H_1-1}(u_1-s)^{1+2H_1-2H_2}} = \frac{1}{s^{2H_1-1}\left(1-\frac{s}{u_1}\right)^{1+2H_1-2H_2} u_1^{2-2H_2}}$$

$$\le \frac{1}{s^{2H_1-1}\left(1-\frac{s}{u}\right)^{1+2H_1-2H_2} u^{2-2H_2}} = \varphi(s,u).$$

Therefore,

$$I_1(u,u_1) = \int_0^u \left(\varphi(s,u) - \varphi(s,u_1)\right) ds$$

$$= \int_0^u \varphi(s,u)\, ds - \int_0^{u_1} \varphi(s,u_1)\, ds + \int_u^{u_1} \varphi(s,u_1)\, ds$$

$$\le \mathrm{B}(2-2H_1, 2H_2-2H_1)\left(u^{2H_2-2H_1} - u_1^{2H_2-2H_1}\right)$$

$$+ \frac{u_1^{2H_1-1}(u_1-u)^{2H_2-2H_1}}{2H_2-2H_1} \to 0, \quad \text{as } u \uparrow u_1.$$

The second integral vanishes as well:

$$I_2(u,u_1) \le \int_u^{u_1} \varphi(s,u)\, ds + \int_u^{u_1} \varphi(s,u_1)\, ds$$

$$\le \left(\frac{1}{2H_2-2H_1} + \left(\frac{u_1}{u}\right)^{2H_1-1}\right)(u_1-u)^{2H_2-2H_1} \to 0$$

as $u \uparrow u_1$. Finally,

$$I_3(u,u_1) = \int_{u_1}^{T} \frac{ds}{(s-u_1)^{1+2H_1-2H_2}} - \int_{u_1}^{T} \frac{ds}{(s-u)^{1+2H_1-2H_2}}$$

$$= \frac{1}{2H_2-2H_1}\left[(T-u_1)^{2H_2-2H_1} - (T-u)^{2H_2-2H_1} + (u_1-u)^{2H_2-2H_1}\right] \to 0,$$

as $u \uparrow u_1$. Lemma is proved. □

Lemma 5.81 *The kernel κ generates a compact integral operator $\kappa\colon C[0,T] \to C[0,T]$.*

Proof According to [22], it suffices to prove that the kernel κ defined by (5.90) satisfies the following two conditions:

(i) for any $u \in [0,T]$, $\kappa(\cdot,u) \in L_1[0,T]$ and $\sup_{u\in[0,T]} \|\kappa(\cdot,u)\|_{L_1} < \infty$;

(ii) for any $u_1 \in [0,T]$, $\int_0^T |\kappa(s,u) - \kappa(s,u_1)|\, ds \to 0$ as $u \to u_1$.

The first condition follows directly from the fact that $\kappa_0(s,u)$ is bounded, see Lemma 5.78, and from Lemma 5.80(i)

In order to check (ii), consider

$$\int_0^T |\kappa(s,u) - \kappa(s,u_1)|\, ds = \int_0^T |\kappa_0(s,u)\varphi(s,u) - \kappa_0(s,u_1)\varphi(s,u_1)|\, ds$$

$$\leq \int_0^T \kappa_0(s,u)\, |\varphi(s,u) - \varphi(s,u_1)|\, ds + \int_0^T \varphi(s,u_1)\, |\kappa_0(s,u) - \kappa_0(s,u_1)|\, ds.$$

Again, Lemma 5.78 in the part that states that $\kappa_0(s,u)$ is bounded, together with Lemma 5.80 (ii), guarantees that the first term converges to zero as $u \to u_1$. Furthermore, Lemma 5.78 in the part that states that $\kappa_0 \in C([0,T]_0^2)$ guarantees that $\kappa_0(s,u)$ converges to $\kappa_0(s,u_1)$ as $u \to u_1$ for almost every $s \in [0,T]$. Since

$$\varphi(s,u_1)\, |\kappa_0(s,u) - \kappa_0(s,u_1)| \leq C\varphi(s,u_1) \in L_1[0,T],$$

the proof follows from the Lebesgue dominated convergence theorem. □

Remark 5.82 In the case when $H_1 = \frac{1}{2}$, the kernel $\kappa(s,u)$ can be simplified to

$$\kappa(s,u) = H_2(2H_2-1)\, |s-u|^{2H_2-2},$$

and Eq. (5.85) coincides with (5.87). Indeed, let $H_1 = \frac{1}{2}$. Then the function $\kappa_0(s,u)$ equals $H_2(2H_2-1)$. Consider the function $\Phi(s,v)$ defined by (5.95):

$$\Phi(t,s) = t^{H_2-\frac{1}{2}}\left(\int_0^1 \left(1-\frac{t-s}{t}z\right)^{H_2-\frac{1}{2}}(1-z)^{H_2-\frac{3}{2}}\,dz\right.$$
$$\left.+\frac{t-s}{t}\int_0^1 (1-z)^{H_2-\frac{1}{2}}\left(1-\frac{t-s}{t}z\right)^{H_2-\frac{3}{2}}\,dz\right)$$
$$= -\frac{t^{H_2-\frac{1}{2}}}{H_2-\frac{1}{2}}\int_0^1 \left(\left(1-\frac{t-s}{t}z\right)^{H_2-\frac{1}{2}}(1-z)^{H_2-\frac{1}{2}}\right)_z' dz = \frac{t^{H_2-\frac{1}{2}}}{H_2-\frac{1}{2}}. \tag{5.105}$$

Combining (5.104) and (5.105), and taking into account the equality $c_{1/2}=1$, we get

$$\kappa_0(s,u) = \left(\beta_{H_2}(H_2-H_1)\right)^2$$
$$\times\int_0^1 (1-t)^{1-2H_2}t^{H_2-\frac{3}{2}}\Phi\left(1,\frac{u(1-t)}{s-ut}\right)\Phi\left(1,\frac{s(1-t)}{s-ut}\right)dt$$
$$= \beta_{H_2}^2\int_0^1 (1-t)^{1-2H_2}t^{H_2-\frac{3}{2}}\,dt = \beta_{H_2}^2\,\mathrm{B}\left(H_2-\tfrac{1}{2},2-2H_2\right) = H_2(2H_2-1).$$

Theorem 5.83 *There exists a sequence $T_n \to \infty$ such that the integral equation (5.88) has a unique solution $h_{T_n}(u) \in C[0,T_n]$.*

Proof We work on the space $C([0,T])$. Recall that (5.88) is of the form

$$h_T(u) + \frac{1}{2-2H_1}\int_0^T h_T(s)\kappa(s,u)\,ds = 1, \quad u \in [0,T].$$

The corresponding homogeneous equation is of the form

$$\int_0^T h_T(s)\kappa(s,u)\,ds = (2H_1-2)h_T(u), \quad u \in [0,T]. \tag{5.106}$$

Since the integral operator κ is compact, classical Fredholm theory states that Eq. (5.88) has a unique solution if and only if the corresponding homogeneous Eq. (5.106) has only the trivial solution. Now, it is easy to see that, for any $a>0$, the following equalities hold:

$$\kappa_0(sa,ua) = \kappa_0(s,u),$$
$$\varphi(sa,ua) = a^{2H_2-2H_1-1}\varphi(s,u).$$

Consequently, $\kappa(sa, ua) = a^{2H_2-2H_1-1}\kappa(s,u)$. We can change the variable of integration $s = s'T$ and put $u = u'T$ in (5.106). Therefore, the equation will be reduced to the equivalent form

$$\int_0^1 h_T(Ts)\kappa(s,u)\,ds = (2H_1-2)T^{2H_1-2H_2}h_T(Tu), \quad u \in [0,1].$$

Denote $\lambda = (2H_1-2)T^{2H_1-2H_2}$. Note that λ depends continuously on T. At the same time, the compact operator κ has no more than countable number of eigenvalues. Therefore we can take the sequence $T_n \to \infty$ in such a way that

$$\lambda_n = (2H_1-2)T_n^{2H_1-2H_2}$$

will be not an eigenvalue. Consequently, the homogeneous equation has only the trivial solution, whence the proof follows. □

5.5.3 Statistical Results: The Form of the MLE, Its Consistency and Asymptotic Normality

The following result establishes the way MLE for the drift parameter θ can be calculated.

Theorem 5.84 *The likelihood function is of the form*

$$L_{T_n}(X,\theta) = \exp\left\{\theta\delta_{H_1}N(T_n) - \frac{1}{2}\theta^2\delta_{H_1}^2\langle N\rangle(T_n)\right\}, \tag{5.107}$$

and the MLE is of the form

$$\hat{\theta}(T_n) = \frac{N(T_n)}{\delta_{H_1}\langle N\rangle(T_n)}, \tag{5.108}$$

where $N(t) = \mathsf{E}_0\left(X_1(t) \mid \mathfrak{F}_t^X\right)$ is a square integrable Gaussian $\mathfrak{F}_t^X$-martingale,

$$N(T_n) = \int_0^{T_n} h_{T_n}(t)\,dZ(t)$$

with $h_{T_n}(t)t^{\frac{1}{2}-H_1} \in L_2[0,T_n]$, $h_{T_n}(t)$ is a unique solution to (5.88) *and*

$$\langle N\rangle(T_n) = (2-2H_1)\int_0^{T_n} h_{T_n}(t)t^{1-2H_1}\,dt. \tag{5.109}$$

Proof We start with (5.79). Consider a Gaussian process $N(t) = \mathsf{E}_0\left(X_1(t) \mid \mathfrak{F}_t^X\right)$. Since $X_1(t)$ is a $\mathfrak{F}_t$-martingale and $\mathfrak{F}_t^X \subset \mathfrak{F}_t$, the process N is a $\mathfrak{F}_t^X$-martingale with respect to probability measure P_0. Furthermore, we can present $V(t)$ as

$$V(t) = \mathsf{E}_0\left(X_1^2(t) \mid \mathfrak{F}_t^X\right) - N^2(t).$$

Note that $X_1^2(t) - t^{2-2H_1}$ is a $\mathfrak{F}_t$-martingale. Therefore,

$$\begin{aligned}
\mathsf{E}_0\left(N^2(t) - \left(t^{2-2H_1} - V(t)\right) \,\middle|\, \mathfrak{F}_s^X\right) &= \mathsf{E}_0\left(\mathsf{E}_0\left(X_1^2(t) \mid \mathfrak{F}_t^X\right) - t^{2-2H_1} \,\middle|\, \mathfrak{F}_s^X\right)\\
&= \mathsf{E}_0\left(X_1^2(t) - t^{2-2H_1} \,\middle|\, \mathfrak{F}_s^X\right) = \mathsf{E}_0\left(X_1^2(s) \mid \mathfrak{F}_s^X\right) - s^{2-2H_1}\\
&= N^2(s) - \left(s^{2-2H_1} - V(s)\right),
\end{aligned}$$

and this means that the quadratic variation of the martingale N equals $\langle N\rangle(t) = t^{2-2H_1} - V(t)$, and the likelihood function (5.79) is reduced to

$$L_{T_n}(X, \theta) = \exp\left\{\theta\delta_{H_1} N(T_n) - \frac{1}{2}\theta^2\delta_{H_1}^2 \langle N\rangle(T_n)\right\},$$

so, we get (5.107) and (5.108). Now, taking (5.85) into account, we get that

$$\begin{aligned}
\langle N\rangle(T_n) &= \mathsf{E}_0\left(N^2(T_n)\right) = \mathsf{E}_0\left(\int_0^{T_n} h_{T_n}(u)\, dX(u)\right)^2\\
&= \mathsf{E}_0\left(\int_0^{T_n} h_{T_n}(u)\, d\big(X_1(u) + X_2(u)\big)\right)^2\\
&= \mathsf{E}_0\left(\int_0^{T_n} h_{T_n}(u)\, dX_1(u)\right)^2 + \mathsf{E}_0\left(\int_0^{T_n} h_{T_n}(u)\, dX_2(u)\right)^2\\
&= (2-2H_1)\int_0^{T_n} h_{T_n}^2(u) t^{1-2H_1}\, du\\
&\quad + \int_0^{T_n}\int_t^{T_n} h_{T_n}(u)\partial_u K_{H_1,H_2}(u,t)\, du \int_t^{T_n} h_{T_n}(s)\partial_s K_{H_1,H_2}(s,t)\, ds\, dt\\
&= (2-2H_1)\int_0^{T_n} h_{T_n}^2(u) u^{1-2H_1}\, du\\
&\quad + \int_0^{T_n} h_{T_n}(u)\int_0^{T_n} h_{T_n}(s)\int_0^{s\wedge u} \partial_s K_{H_1,H_2}(s,t)\partial_u K_{H_1,H_2}(u,t)\, dt\, ds\, du\\
&= \int_0^{T_n} h_{T_n}(u)\left((2-2H_1)h_{T_n}(u)u^{1-2H_1} + \int_0^{T_n} h_{T_n}(s)k(s,u)\, ds\right) du\\
&= (2-2H_1)\int_0^{T_n} h_{T_n}(u) u^{1-2H_1}\, du,
\end{aligned}$$

whence the proof follows. □

Theorem 5.85 *The estimator $\hat{\theta}_{T_n}$ is unbiased and the corresponding estimation error is normal*

$$\hat{\theta}_{T_n} - \theta \stackrel{d}{=} \mathcal{N}\left(0, \frac{1}{(2-2H_1)\delta_{H_1}^2 \int_0^{T_n} h_{T_n}(s)s^{1-2H_1}\,ds}\right).$$

Proof The mean of the estimator (5.108) equals

$$\begin{aligned}
\mathsf{E}_\theta \hat{\theta}_{T_n} &= \frac{1}{\delta_{H_1}} \mathsf{E}_\theta \frac{N(T_n)}{\langle N\rangle(T_n)} = \frac{1}{\delta_{H_1}} \mathsf{E}_0 \left[\frac{N(T_n)}{\langle N\rangle(T_n)} L_{T_n}(X,\theta)\right] \\
&= \frac{1}{\delta_{H_1}^2 \langle N\rangle(T_n)} \exp\left\{-\frac{1}{2}\theta^2\delta_{H_1}^2 \langle N\rangle(T_n)\right\} \frac{d}{d\theta} \mathsf{E}_0 \exp\{\theta\delta_{H_1} N(T_n)\} \\
&= \frac{1}{\delta_{H_1}^2 \langle N\rangle(T_n)} \exp\left\{-\frac{1}{2}\theta^2\delta_{H_1}^2 \langle N\rangle(T_n)\right\} \frac{d}{d\theta} \exp\left\{\frac{1}{2}\theta^2\delta_{H_1}^2 \langle N\rangle(T_n)\right\} = \theta,
\end{aligned}$$

that is, the estimator is unbiased. Similarly, we have

$$\begin{aligned}
\mathsf{E}_\theta \left(\hat{\theta}_{T_n} - \theta\right)^2 &= \mathsf{E}_\theta \hat{\theta}_{T_n}^2 - \theta^2 = \frac{1}{\delta_{H_1}^2} \mathsf{E}_\theta \frac{N^2(T_n)}{\langle N\rangle^2(T_n)} - \theta^2 \\
&= \frac{1}{\delta_{H_1}^2} \exp\left\{-\frac{1}{2}\theta^2\delta_{H_1}^2 \langle N\rangle(T_n)\right\} \mathsf{E}_0 \left[\exp\{\theta\delta_{H_1} N(T_n)\} \frac{N^2(T_n)}{\langle N\rangle^2(T_n)}\right] - \theta^2 \\
&= \frac{1}{\delta_{H_1}^4 \langle N\rangle^2(T_n)} \exp\left\{-\frac{1}{2}\theta^2\delta_{H_1}^2 \langle N\rangle(T_n)\right\} \frac{d^2}{d\theta^2} \mathsf{E}_0 \exp\{\theta\delta_{H_1} N(T_n)\} - \theta^2 \\
&= \frac{1}{\delta_{H_1}^4 \langle N\rangle^2(T_n)} \left(\theta^2\delta_{H_1}^4 \langle N\rangle^2(T_n) + \delta_{H_1}^2 \langle N\rangle(T_n)\right) - \theta^2 = \frac{1}{\delta_{H_1}^2 \langle N\rangle(T_n)}.
\end{aligned} \tag{5.110}$$

By Theorem 5.84, the process N is Gaussian and $\langle N\rangle$ admits the representation (5.109). This concludes the proof. □

Now we establish the asymptotic behavior of the estimator.

Theorem 5.86 *The estimator $\hat{\theta}_{T_n}$ is strongly consistent and*

$$\lim_{T_n\to\infty} T_n^{2-2H_2} \mathsf{E}_\theta \left(\hat{\theta}_{T_n} - \theta\right)^2 = \frac{1}{(2-2H_1)\delta_{H_1}^2 \int_0^1 h_0(u) u^{\frac{1}{2}-H_1}\,du},$$

where the function $h_0(u)$ is the solution of the integral equation of the first kind

$$(K_1 h)(u) = (2-2H_1)u^{1/2-H_1}, \tag{5.111}$$

where

$$(K_1 f)(u) = \int_0^1 f(s) k_1(s,u)\, ds, \tag{5.112}$$

$$k_1(s,u) = s^{H_1-\frac{1}{2}} u^{H_1-\frac{1}{2}} k(s,u) = s^{H_1-\frac{1}{2}} u^{\frac{1}{2}-H_1} \kappa(s,u).$$

We start with solving the integral equation (5.111).

Lemma 5.87 *For any constant $C > 0$, the integral equation of the first kind*

$$u^{1/2-H_1} = C(K_1 h)(u), \quad u \in (0,1] \tag{5.113}$$

has a unique solution.

Proof We can present Eq. (5.113) in equivalent form

$$u^{1/2-H_1} = C \int_0^1 h(s) k_1(s,u)\, ds, \quad u \in (0,1],$$

or, by (5.86),

$$u^{1-2H_1} = C \int_0^1 \breve{h}(s) \int_0^{s \wedge u} \partial_s K_{H_1,H_2}(s,v) \partial_u K_{H_1,H_2}(u,v)\, dv\, ds, \quad u \in (0,1],$$

where $\breve{h}(s) = h(s) s^{1/2-H_1}$, or, at last,

$$u^{1-2H_1} = C \int_0^u \left(\int_v^1 \breve{h}(s) \partial_s K_{H_1,H_2}(s,v) ds \right) \partial_u K_{H_1,H_2}(u,v)\, dv. \tag{5.114}$$

Now, taking into account the transition from Eq. (5.84) to (5.85) with the help of differentiation, we can perform the inverse operation and get from (5.114) the following equivalent equation

$$u^{2-2H_1} = C(2-2H_1) \int_0^u K_{H_1,H_2}(u,v) \left(\int_v^1 \breve{h}(s) \partial_s K_{H_1,H_2}(s,v)\, ds \right) dv, \quad u \in [0,1]. \tag{5.115}$$

The right-hand side of Eq. (5.115) can be rewritten as

$$C(2-2H_1) \int_0^u K_{H_1,H_2}(u,v) q(v)\, dv,$$

where $q(v) = \int_v^1 \breve{h}(s) \partial_s K_{H_1,H_2}(s,v)\, ds$. First, solve the equation

$$u^{2-2H_1} = C_1 \int_0^u K_{H_1,H_2}(u,v) q(v)\, dv,$$

with $C_1 = C(2 - 2H_1)$. Taking into account (5.81), the latter equation can be rewritten in equivalent form

$$u^{2-2H_1} = C_1 c_{H_1} \beta_{H_2} \int_0^u v^{1/2-H_2} \int_v^u (u-z)^{1/2-H_1} z^{H_2-H_1} (z-v)^{H_2-3/2}\, dz\, q(v)\, dv,$$

or

$$u^{2-2H_1} = C_1 c_{H_1} \beta_{H_2} \int_0^u z^{H_2-H_1} (u-z)^{1/2-H_1} \int_0^z v^{1/2-H_2} (z-v)^{H_2-3/2} q(v)\, dv\, dz,$$

or, in terms of fractional integral (1.14),

$$u^{2-2H_1} = C_2 \left(\mathcal{I}_{0+}^{3/2-H_1} p \right)(u),$$

where $C_2 = C_1 c_{H_1} \beta_{H_2} \Gamma(3/2 - H_2)$ and

$$p(z) = z^{H_2-H_1} \int_0^z v^{1/2-H_2} (z-v)^{H_2-3/2} q(v)\, dv. \tag{5.116}$$

This means that

$$\begin{aligned} p(u) &= C_2^{-1} \left(\mathcal{D}_{0+}^{3/2-H_1} \left(\cdot^{2-2H_1} \right) \right)(u) \\ &= \left(C_2 \Gamma(H_1 - 1/2) \right)^{-1} \left(\int_0^u (u-t)^{H_1-3/2} t^{2-2H_1}\, dt \right)'_u = C_3 u^{1/2-H_1}, \end{aligned} \tag{5.117}$$

where $C_3 = \frac{(3/2-H_1)\mathrm{B}(H_1-1/2, 3-2H_1)}{C_2 \Gamma(H_1-1/2)}$. Furthermore, comparing (5.116) and (5.117), we get that

$$\begin{aligned} C_3 z^{1/2-H_2} &= \int_0^z v^{1/2-H_2} (z-v)^{H_2-3/2} q(v)\, dv \\ &= \Gamma(H_2 - 1/2) \left(\mathcal{I}_{0+}^{H_2-1/2} \left(\cdot^{1/2-H_2} q \right) \right)(z), \end{aligned}$$

whence

$$\begin{aligned} v^{1/2-H_2} q(v) &= C_3 \left(\Gamma(H_2 - 1/2) \right)^{-1} \left(\mathcal{D}_{0+}^{H_2-1/2} \cdot^{1/2-H_2} \right)(v) \\ &= C_3 \left(\Gamma(H_2 - 1/2) \Gamma(3/2 - H_2) \right)^{-1} \left(\int_0^v (v-t)^{1/2-H_2} t^{1/2-H_2}\, dt \right)'_v \\ &= C_4 v^{1-2H_2}, \end{aligned}$$

where $C_4 = \frac{C_3(2-2H_2)}{\Gamma(H_2-1/2)\Gamma(3/2-H_2)}$. Obviously, $q(v) = C_4 v^{1/2-H_2}$, and we arrive at the equation

$$C_4 v^{1/2-H_2} = \int_v^1 \breve{h}(s)\partial_s K_{H_1,H_2}(s,v)\,ds. \tag{5.118}$$

Note that

$$\partial_s K_{H_1,H_2}(s,v) = c_{H_1}\beta_{H_2}\Gamma(3/2-H_1)v^{1/2-H_2}\left(\mathcal{D}_{v+}^{H_1-1/2}\left(\cdot^{H_2-H_1}(\cdot - v)^{H_2-3/2}\right)\right)(s),$$

so, with the help of integration by parts formula (1.15), Eq. (5.118) can be rewritten as

$$\begin{aligned} C_5 &= \int_v^1 \breve{h}(s)\left(\mathcal{D}_{v+}^{H_1-1/2}\left(\cdot^{H_2-H_1}(\cdot - v)^{H_2-3/2}\right)\right)(s)\,ds \\ &= \int_v^1 \left(\mathcal{D}_{1-}^{H_1-1/2}\breve{h}\right)(s)s^{H_2-H_1}(s-v)^{H_2-3/2}\,ds \\ &= \left(\Gamma(H_2-1/2)\right)^{-1}\left(\mathcal{I}_{1-}^{H_2-1/2}\left(\mathcal{D}_{1-}^{H_1-1/2}\breve{h}\right)\cdot^{H_2-H_1}\right)(v), \end{aligned} \tag{5.119}$$

where $C_5 = C_4(c_{H_1}\beta_{H_2}\Gamma(3/2-H_1))^{-1}$. The latter equation means that

$$\left(\mathcal{D}_{1-}^{H_1-1/2}\breve{h}\right)(v)v^{H_2-H_1} = C_6(1-v)^{1/2-H_1},$$

$C_6 = \frac{C_5\Gamma(H_1-1/2)}{\Gamma(3/2-H_1)}$. At last, we get that

$$h(v) = v^{H_1-1/2}\breve{h}(v) = C_6 v^{H_1-1/2}\left(\mathcal{I}_{1-}^{H_1-1/2}\left(\cdot^{H_1-H_2}(1-\cdot)^{1/2-H_1}\right)\right)(v),$$

and this solution of Eq. (5.113) is unique. □

Proof of Theorem 5.86 At first we rewrite Eq. (5.85) in the equivalent form

$$(2-2H_1)u^{\frac{1}{2}-H_1} = (2-2H_1)h_T(u)u^{\frac{1}{2}-H_1} + \int_0^T h_T(s)s^{\frac{1}{2}-H_1}s^{H_1-\frac{1}{2}}u^{H_1-\frac{1}{2}}k(s,u)\,ds,$$

or

$$(2-2H_1)u^{\frac{1}{2}-H_1} = (2-2H_1)\widetilde{h}_T(u) + \int_0^T \widetilde{h}_T(s)k_1(s,u)\,ds, \tag{5.120}$$

where $\widetilde{h}_T(u) = h_T(u)u^{\frac{1}{2}-H_1}$.

It is very easy to see that for any $a > 0$

$$\begin{aligned}
K_{H_1,H_2}(ta, sa) &= K_{H_1,H_2}(t, s)a^{\frac{1}{2}+H_2-2H_1},\\
\partial_t K_{H_1,H_2}(ta, sa) &= \partial_t K_{H_1,H_2}(t, s)a^{-\frac{1}{2}+H_2-2H_1},\\
k(ta, sa) &= a^{2H_2-4H_1}k(t, s),
\end{aligned}$$

whence

$$k_1(ta, sa) = k_1(t, s)a^{2H_2-2H_1-1}.$$

Then changing $u = u'T$, $s = s'T$ in (5.120) and omitting superscripts, we get

$$(2-2H_1)u^{\frac{1}{2}-H_1}T^{\frac{1}{2}-H_1} = (2-2H_1)\widetilde{h}_T(uT) + T^{2H_2-2H_1}\int_0^1 \widetilde{h}_T(sT)k_1(s, u)\,ds,$$

or

$$(2-2H_1)u^{\frac{1}{2}-H_1} = (2-2H_1)\widetilde{h}_T(uT)T^{H_1-\frac{1}{2}} + T^{2H_2-2H_1}\int_0^1 \widetilde{h}_T(sT)T^{H_1-\frac{1}{2}}k_1(s, u)\,ds. \tag{5.121}$$

Denote $\mu = T^{2H_2-2H_1}$. Let $h_\mu(u) = \mu\widetilde{h}_T(uT)T^{H_1-\frac{1}{2}}$. Then Eq. (5.121) can be rewritten as

$$(2-2H_1)u^{\frac{1}{2}-H_1} = \frac{1}{\mu}(2-2H_1)h_\mu(u) + \int_0^1 h_\mu(s)k_1(s, u)\,ds,$$

or

$$(2-2H_1)u^{\frac{1}{2}-H_1} = \frac{1}{\mu}(2-2H_1)h_\mu(u) + K_1 h_\mu(u),$$

where K_1 is the operator defined by (5.112).

Note that

$$\begin{aligned}
\langle N\rangle(T) &= (2-2H_1)\int_0^T h_T(s)s^{1-2H_1}\,ds = (2-2H_1)\int_0^T \widetilde{h}_T(s)s^{\frac{1}{2}-H_1}\,ds\\
&= (2-2H_1)T^{2-2H_2}\int_0^1 h_\mu(u)u^{\frac{1}{2}-H_1}\,du.
\end{aligned} \tag{5.122}$$

Define the scalar product $\langle f, g\rangle = \int_0^1 f(s)g(s)\,ds, f, g \in L_2[0,1]$. Note that

$$\begin{aligned}\langle K_1 f, f\rangle &= \int_0^1 (K_1 f)(t) f(t)\,dt = \int_0^1 \left(\int_0^1 f(s)k_1(s,t)\,ds\right) f(t)\,dt \\ &= \int_0^1\int_0^1 f(t)t^{H_1-1/2} f(s)s^{H_1-1/2}\int_0^{s\wedge t} \partial_s K_{H_1,H_2}(s,v)\partial_t K_{H_1,H_2}(t,v)\,dv\,ds\,dt \\ &= \int_0^1 dv \int_v^1 \partial_s K_{H_1,H_2}(s,v) f(s)s^{H_1-1/2} ds \int_v^1 \partial_t K_{H_1,H_2}(t,v) f(t)t^{H_1-1/2} dt \geq 0.\end{aligned} \tag{5.123}$$

Introduce the corresponding first type auxiliary integral equation

$$(2-2H_1)u^{\frac{1}{2}-H_1} = (K_1 h)(u). \tag{5.124}$$

It follows from Lemma 5.87 that (5.124) has a unique solution, say, h_0, obviously, not depending on μ. The function $\delta_\mu = h_\mu - h_0$ satisfies two equations

$$K_1\delta_\mu + \frac{1}{\mu}(2-2H_1)h_\mu = 0 \quad \text{and} \quad K_1\delta_\mu + \frac{1}{\mu}(2-2H_1)\delta_\mu = -\frac{h_0}{\mu}(2-2H_1). \tag{5.125}$$

Multiplying the second equation by δ_μ and integrating, we get

$$\langle K_1\delta_\mu, \delta_\mu\rangle + \frac{1}{\mu}(2-2H_1)\left\|\delta_\mu\right\|^2 = \frac{1}{\mu}(2-2H_1)\left|\langle h_0, \delta_\mu\rangle\right|, \tag{5.126}$$

and it follows from (5.126) and (5.123) that $\left\|\delta_\mu\right\|^2 \leq \left|\langle h_0, \delta_\mu\rangle\right| \leq \|h_0\|\left\|\delta_\mu\right\|$, which implies that $\left\|\delta_\mu\right\| \leq \|h_0\|$. Multiplying the 1st equation in (5.125) by h_0 and integrating we get

$$\langle K_1\delta_\mu, h_0\rangle + \frac{1}{\mu}(2-2H_1)\langle h_\mu, h_0\rangle = 0.$$

Note that the inequality $\left\|\delta_\mu\right\| \leq \|h_0\|$ implies that

$$\left|\langle h_\mu, h_0\rangle\right| \leq \left|\langle \delta_\mu, h_0\rangle\right| + \|h_0\|^2 \leq 2\|h_0\|^2 < \infty,$$

and hence

$$(2-2H_1)\left|\left\langle \delta_\mu, u^{\frac{1}{2}-H_1}\right\rangle\right| = \left|\langle \delta_\mu, K_1 h_0\rangle\right| = \left|\langle K_1\delta_\mu, h_0\rangle\right| = \frac{1}{\mu}(2-2H_1)\left|\langle h_\mu, h_0\rangle\right| \to 0$$

as $T \to \infty$. It means that $\lim_{T\to\infty} \int_0^1 h_\mu(u)u^{\frac{1}{2}-H_1}\,du = \int_0^1 h_0(u)u^{\frac{1}{2}-H_1}\,du$. Therefore, taking into account (5.110) and (5.122), we get

$$T^{2-2H_2}\mathsf{E}_\theta\left(\hat{\theta}_T-\theta\right)^2 = \frac{1}{(2-2H_1)\delta_{H_1}^2\int_0^1 h_\mu(u)u^{\frac{1}{2}-H_1}\,du}$$
$$\to \frac{1}{(2-2H_1)\delta_{H_1}^2\int_0^1 h_0(u)u^{\frac{1}{2}-H_1}\,du},$$

whence the proof follows. □

Remark 5.88 In outline, our method of proof follows the method of the corresponding result from [35], however Lemma 5.87 is specific to our case.

5.6 Drift Parameter Estimation in Models with mfBm

FBm is widely used in the modeling of long-range dependent processes in Internet traffic, stock markets, etc. However the stationarity of increments of fBm means that the behavior of it is the same at each point, and this substantially restricts the area of its application. In particular, it does not allow one to model situations, where the regularity at a point depends on the point. One way to overcome these limitations is to extend the standard fBm to mfBm as described in Sect. 1.5.3. Following this approach, in the present section we investigate two statistical models with mfBm: the linear model and the multifractional Ornstein–Uhlenbeck process. For these models we propose estimators for an unknown drift parameter and prove their strong consistency. The proofs are based on the asymptotic bounds with probability 1 for the rate of the growth of the trajectories of mfBm and of some other functionals of mfBm, including increments and fractional derivatives. These bounds are derived in Appendix B.4.

In Sect. 5.6.1 the linear multifractional model with unknown drift parameter is considered and its strong consistency and normality are shown. In Sect. 5.6.2 we construct the least square estimator for the unknown drift parameter in the multifractional Ornstein–Uhlenbeck model and establish its strong consistency in the non-ergodic case. We consider multifractional Ornstein–Uhlenbeck processes when the index H_t of multifractionality is bounded from below by some constant exceeding $\frac{1}{2}$, and observations are continuous in time. Our results generalize respective theorems supplying conditions of the strong consistency for the least square estimator of the unknown drift parameter in the Ornstein–Uhlenbeck model involving fBm (see Sect. 5.4.2).

5.6.1 *Linear Multifractional Model*

Consider the simplest linear model, namely

$$X_t = \theta t + Y_t, \quad t \geq 0,$$

where $\theta \in \mathbb{R}$ is an unknown parameter, Y_t is an mfBm with the Hurst function H_t satisfying the conditions (C_1)–(C_2) of Sect. 1.5.3. Assume that our aim is to estimate the parameter θ by the observations of X_t. Let us introduce the estimator

$$\hat{\theta}_T = \frac{X_T}{T} = \theta + \frac{Y_T}{T}.$$

Theorem 5.89

1) The estimator $\hat{\theta}_T$ is strongly consistent as $T \to \infty$.
2) For all $T > 0$,

$$\frac{T^{1-H_T}}{C(H_T)}\left(\hat{\theta}_T - \theta\right) \overset{d}{=} \mathcal{N}(0,1),$$

where $C(H) = \left(\frac{\pi}{H\Gamma(2H)\sin(\pi H)}\right)^{1/2}$. Consequently, a confidence interval of level $1-\alpha$ is given by

$$\hat{\theta}_T \pm \frac{C(H_T)}{T^{1-H_T}} z_{1-\alpha/2},$$

where z_p denotes the p-quantile of the standard normal distribution.

Proof

1) By Theorem B.55, for all $T > 1$ and $\delta > 0$

$$|Y_T| \leq T^{h^*+\delta}\xi \quad \text{a. s.},$$

where $\xi = \xi(\delta)$ is some nonnegative random variable, $h^* = \limsup_{t\to\infty} H_t$. Hence, if we choose $\delta < 1 - h^*$, then we get

$$\frac{|Y_T|}{T} \leq \frac{\xi}{T^{1-h^*-\delta}} \to 0, \quad \text{a. s. as } T \to \infty.$$

2) Note that one-dimensional distributions of mfBm Y_t are centered Gaussian with standard deviation $C(H_t)t^{H_t}$, see (1.59). Therefore,

$$\frac{T^{1-H_T}}{C(H_T)}\left(\hat{\theta}_T - \theta\right) = \frac{Y_T}{C(H_T)T^{H_T}} \overset{d}{=} \mathcal{N}(0,1). \qquad \square$$

5.6.2 Multifractional Ornstein–Uhlenbeck Process

Assume that h_3 from (1.61) exceeds $1/2$. In this subsection we consider the estimation of the unknown parameter $\theta \geq 0$ by observations of the process $X = \{X_t, t \geq 0\}$ that is a solution of the SDE of Langevin type,

$$X_t = x_0 + \theta \int_0^t X_s\, ds + Y_t, \tag{5.127}$$

where $x_0 \in \mathbb{R}$ is a known constant, $Y = \{Y_t, t \geq 0\}$ is an mfBm. This solution exists and is unique, see Theorem 1.64.

Note that the trajectories of the processes Y and consequently X are a. s. Hölder continuous up to order h_3. Therefore, path-wise integrals $\int_0^T X_s\, dX_s$ and $\int_0^T X_s\, dY_s$ are well defined. One can verify that the solution of (5.127) can be represented in the following form

$$X_t = e^{\theta t}\left(x_0 + \int_0^t e^{-\theta s}\, dY_s\right).$$

Using the integration-by-parts, this process can be written as follows

$$X_t = x_0 e^{\theta t} + \theta e^{\theta t} \int_0^t e^{-\theta s} Y_s\, ds + Y_t. \tag{5.128}$$

We call the process $X = \{X_t, t \geq 0\}$ a *multifractional Ornstein–Uhlenbeck process*.

Let, more precisely, our goal be to estimate the unknown drift parameter $\theta \in \mathbb{R}$ by the continuous-time observations on the interval $[0, T]$. Consider the estimator

$$\hat{\theta}_T = \frac{\int_0^T X_s\, dX_s}{\int_0^T X_s^2\, ds}. \tag{5.129}$$

Since by (5.127), $dX_s = \theta X_s\, ds + dY_s$, we have that $\hat{\theta}_T$ admits the following stochastic representation

$$\hat{\theta}_T = \theta + \frac{\int_0^T X_s\, dY_s}{\int_0^T X_s^2\, ds}.$$

Lemma 5.90 *Let $\varepsilon > 0$, $T > 1$, $\theta > 0$. Then*

$$\left|\int_0^T X_s\, dY_s\right| \leq \zeta^2 T^{h_2+\varepsilon+1} e^{\theta T}, \tag{5.130}$$

where $\zeta \geq 0$ is a random variable with the following property: there exist positive constants C_1 and C_2 not depending on T such that for all $u > 0$

$$\mathsf{P}(\zeta > u) \leq C_1 e^{-C_2 u^2}.$$

Proof By (5.128),

$$\sup_{0\leq s\leq t} |X_s| \leq |x_0| \, e^{\theta t} + \theta e^{\theta t} \int_0^t e^{-\theta s} \sup_{0\leq u\leq s} |Y_u| \, ds + \sup_{0\leq s\leq t} |Y_s|. \tag{5.131}$$

Then (5.127) implies that for $t_1 > t_2 \geq 0$

$$\begin{aligned} |X_{t_1} - X_{t_2}| \leq \theta \int_{t_2}^{t_1} \left(|x_0| \, e^{\theta s} + \theta e^{\theta s} \int_0^s e^{-\theta v} \sup_{0\leq u\leq v} |Y_u| \, dv + \sup_{0\leq u\leq s} |Y_u| \right) ds \\ + |Y_{t_1} - Y_{t_2}|. \end{aligned} \tag{5.132}$$

Furthermore, using Theorems B.55 and B.59 we get for $t \geq 0$ and $\delta > 0$

$$\sup_{0\leq s\leq t} |Y_s| \leq \left(t^{h_2+\delta} + 1\right) \xi \qquad \text{a. s.,} \tag{5.133}$$

and for $0 \leq t_2 < t_1 \leq t_2 + 1$

$$\begin{aligned} |Y_{t_1} - Y_{t_2}| &\leq \left(t_1^{h_2+\varepsilon} + 1\right) (t_1 - t_2)^{h_3} \left(|\log(t_1 - t_2)|^p + 1\right) \eta \\ &= \left(t_1^{h_2+\varepsilon} + 1\right) \left((t_1 - t_2)^{h_3} |\log(t_1 - t_2)|^p + (t_1 - t_2)^{h_3}\right) \eta \\ &\leq C \left(t_1^{h_2+\varepsilon} + 1\right) (t_1 - t_2)^{h_3 - r} \eta \qquad \text{a. s.,} \end{aligned} \tag{5.134}$$

where $0 < r < h_3 - 1/2$. Then by (5.133),

$$\int_0^t e^{-\theta s} \sup_{0\leq u\leq s} |Y_u| \, ds \leq \xi \int_0^t e^{-\theta s} \left(s^{h_2+\delta} + 1\right) ds \leq C\xi.$$

Therefore, from (5.131) we obtain

$$\sup_{0\leq s\leq t} |X_s| \leq |x_0| \, e^{\theta t} + \theta e^{\theta t} C\xi + \left(t^{h_2+\delta} + 1\right) \xi.$$

It follows from (5.132) and (5.134) that

$$\begin{aligned} |X_{t_1} - X_{t_2}| \leq \theta \int_{t_2}^{t_1} \left(|x_0| \, e^{\theta s} + \theta e^{\theta s} C\xi + \left(s^{h_2+\delta} + 1\right) \xi\right) ds \\ + C \left(t_1^{h_2+\varepsilon} + 1\right) (t_1 - t_2)^{h_3 - r} \eta. \end{aligned}$$

These formulas can be rewritten using simplified notation as follows:

$$\sup_{0\le s\le t} |X_s| \le \left(e^{\theta t} + t^{h_2+\delta}\right)\zeta, \tag{5.135}$$

$$|X_{t_1} - X_{t_2}| \le \zeta\left(e^{\theta t_1} + t_1^{h_2+\delta}\right)(t_1 - t_2) + \zeta\left(t_1^{h_2+\varepsilon} + 1\right)(t_1 - t_2)^{h_3-r}. \tag{5.136}$$

In order to bound $\int_0^T X_s\, dY_s$ we write

$$\begin{aligned}\left|\int_0^T X_s\, dY_s\right| &\le \sum_{k=0}^{[T]+1}\left|\int_k^{k+1} X_s\, dY_s\right| \\ &\le \sum_{k=0}^{[T]+1} \sup_{k\le s\le k+1}\left|\left(\mathcal{D}_{k+1-}^{1-\alpha} Y_{k+1-}\right)(s)\right| \left|\int_k^{k+1}\left(\mathcal{D}_{k+}^{\alpha} X_{k+}\right)(s)\, ds\right|,\end{aligned} \tag{5.137}$$

where $1 - h_3 + r < \alpha < h_3 - r$, see (1.62). Now we need to bound the fractional derivatives. By (5.134),

$$\begin{aligned}&\left|\left(\mathcal{D}_{k+1-}^{1-\alpha} Y_{k+1-}\right)(s)\right| \\ &\le \frac{1}{\Gamma(\alpha)}\left(\frac{|Y_{k+1} - Y_s|}{(k+1-s)^{1-\alpha}} + (1-\alpha)\int_s^{k+1}\frac{|Y_u - Y_s|}{(u-s)^{2-\alpha}}\, du\right) \\ &\le \zeta\left(\left((k+1)^{h_2+\varepsilon} + 1\right)(k+1-s)^{h_3-r-1+\alpha}\right. \\ &\quad \left. + \int_s^{k+1}\left(u^{h_2 u+\varepsilon} + 1\right)(u-s)^{h_3-r-2+\alpha}\, du\right) \\ &\le \zeta\left((k+1)^{h_2+\varepsilon} + 1\right)\left((k+1-s)^{h_3-r-1+\alpha} + \int_s^{k+1}(u-s)^{h_3-r-2+\alpha}\, du\right) \\ &\le \zeta\left((k+1)^{h_2+\varepsilon} + 1\right)(k+1-s)^{h_3-r-1+\alpha} \\ &\le \zeta\left((k+1)^{h_2+\varepsilon} + 1\right).\end{aligned} \tag{5.138}$$

Applying (5.135)–(5.136), we get for $s \in (k, k+1]$

$$\begin{aligned}\left|\left(\mathcal{D}_{k+}^{\alpha} X\right)(s)\right| &\le \frac{1}{\Gamma(1-\alpha)}\left(\frac{|X_s|}{(s-k)^{\alpha}} + \alpha\int_k^s \frac{|X_s - X_u|}{(s-u)^{\alpha+1}}\, du\right) \\ &\le \zeta\left(\left(e^{\theta s} + s^{h_2+\delta}\right)(s-k)^{-\alpha}\right. \\ &\quad \left. + \int_k^s\left(\left(e^{\theta s} + s^{h_2+\delta}\right)(s-u)^{-\alpha} + \left(s^{h_2+\varepsilon} + 1\right)(s-u)^{h_3-r-\alpha-1}\right) du\right)\end{aligned}$$

$$\leq \zeta\left(\left(e^{\theta s}+s^{h_2+\delta}\right)\left((s-k)^{-\alpha}+(s-k)^{1-\alpha}\right)+\left(s^{h_2+\varepsilon}+1\right)(s-k)^{h_3-r-\alpha}\right).$$

Then

$$\int_k^{k+1}\left|\left(\mathcal{D}_{k+}^{\alpha}X\right)(s)\right|ds \leq \zeta\left(e^{\theta(k+1)}+(k+1)^{h_2+\delta}+(k+1)^{h_2+\varepsilon}+1\right). \tag{5.139}$$

Combining (5.137)–(5.139), we get

$$\left|\int_0^T X_s\,dY_s\right| \leq \zeta^2\sum_{k=0}^{[T]+1}\left((k+1)^{h_2+\varepsilon}+1\right) \times\left(e^{\theta(k+1)}+(k+1)^{h_2+\delta}+(k+1)^{h_2+\varepsilon}+1\right).$$

Now, each summand in the right-hand side of the latter inequality can be bounded by $CT^{h_2+\varepsilon}e^{\theta T}$, whence (5.130) follows. □

Theorem 5.91 *Let $\theta > 0$. Then the estimator $\hat{\theta}_T$ is strongly consistent as $T \to \infty$.*

Proof By (5.130),

$$\left|\hat{\theta}_T-\theta\right| = \frac{\left|\int_0^T X_s\,dY_s\right|}{\int_0^T X_s^2\,ds} \leq \zeta^2\frac{T^{h_2+\varepsilon+1}e^{\theta T}}{\int_0^T X_s^2\,ds}.$$

Applying L'Hôpital's rule and (5.128), we get

$$\begin{aligned}\lim_{T\to\infty}\frac{T^{h_2+\varepsilon+1}e^{\theta T}}{\int_0^T X_s^2\,ds} &= \lim_{T\to\infty}\frac{\left((h_2+\varepsilon+1)T^{h_2+\varepsilon}+T^{h_2+\varepsilon+1}\theta\right)e^{\theta T}}{X_T^2}\\ &= \lim_{T\to\infty}\frac{\left((h_2+\varepsilon+1)T^{h_2+\varepsilon}+T^{h_2+\varepsilon+1}\theta\right)}{e^{\theta T}\left(x_0+\theta\int_0^T e^{-\theta s}Y_s\,ds+e^{-\theta T}Y_T\right)^2}.\end{aligned} \tag{5.140}$$

Using the bound (B.66), we obtain that $e^{-\theta T}Y_T \to 0$ a. s. as $T \to \infty$. Moreover, with probability 1 there exists the limit $\lim_{T\to\infty}\int_0^T e^{-\theta s}Y_s\,ds = \int_0^\infty e^{-\theta s}Y_s\,ds$. Obviously, this limit is a Gaussian random variable. This implies that

$$\lim_{T\to\infty}\left(x_0+\theta\int_0^T e^{-\theta s}Y_s\,ds+e^{-\theta T}Y_T\right)^2=\left(x_0+\theta\int_0^\infty e^{-\theta s}Y_s\,ds\right)^2>0 \quad \text{a. s.}$$

Therefore, it follows from (5.140) that

$$\lim_{T\to\infty} \frac{T^{h_2+\varepsilon+1} e^{\theta T}}{\int_0^T X_s^2\, ds} = 0 \quad \text{a. s.}$$

This completes the proof. □

Theorem 5.92 *Let $\theta = 0$. Then the estimator $\hat{\theta}_T$ is consistent as $T \to \infty$.*

Proof In this case $X_t = x_0 + Y_t$. Hence,

$$\int_0^T X_s\, dY_s = x_0 Y_T + \int_0^T Y_s\, dY_s = x_0 Y_T + \frac{1}{2} Y_T^2,$$

see Proposition 1.15. Then for $T > 1$,

$$\left|\hat{\theta}_T - \theta\right| = \frac{\left|\int_0^T X_s\, dY_s\right|}{\int_0^T X_s^2\, ds} \le \frac{x_0 \,|\, Y_T| + \frac{1}{2} Y_T^2}{\int_0^T X_s^2\, ds} \le \zeta^2 \frac{T^{2h_2+2\delta}}{\int_0^T X_s^2\, ds},$$

by Theorem B.55. It follows from the Cauchy–Schwarz inequality that

$$\begin{aligned}\int_0^T X_s^2\, ds &\ge \frac{1}{T}\left(\int_0^T X_s\, ds\right)^2 = \frac{1}{T}\left(\int_0^T (x_0 + Y_s)\, ds\right)^2 = T\left(x_0 + \frac{1}{T}\int_0^T Y_s\, ds\right)^2 \\ &= T\left(x_0 + \sigma_T \mathcal{N}(0,1)\right)^2,\end{aligned}$$

where σ_T^2 denotes the variance of the centered Gaussian random variable $\frac{1}{T}\int_0^T Y_s\, ds$, $\mathcal{N}(0, 1)$ is the standard normal random variable. Therefore, it suffices to show that

$$\frac{T^{2h_2+2\delta-1}}{\left(x_0 + \sigma_T \mathcal{N}(0,1)\right)^2} \xrightarrow{\mathsf{P}} 0 \quad \text{as } T \to \infty.$$

In order to establish this convergence, we will bound σ_T^2 from below. We have

$$\begin{aligned}\sigma_T^2 &= \frac{1}{T^2}\mathsf{E}\left(\int_0^T Y_s\, ds\right)^2 = \frac{1}{T^2}\int_0^T \int_0^T \mathsf{E} Y_s Y_u\, du\, ds \\ &= \frac{1}{T^2}\int_0^T \int_0^T D(H_s, H_u)\left(s^{H_s+H_u} + u^{H_s+H_u} - |s-u|^{H_s+H_u}\right) du\, ds,\end{aligned}$$

where $D(x,y) = \frac{\pi}{\Gamma(x+y+1)\sin(\pi(x+y)/2)}$, see [6] or [159, p. 213]. Further,

$$\sigma_T^2 = \frac{1}{T^2}\int_0^T\int_0^s D(H_s,H_u)\left(s^{H_s+H_u}+u^{H_s+H_u}-(s-u)^{H_s+H_u}\right)du\,ds$$
$$+\frac{1}{T^2}\int_0^T\int_s^T D(H_s,H_u)\left(s^{H_s+H_u}+u^{H_s+H_u}-(u-s)^{H_s+H_u}\right)du\,ds$$
$$\geq \frac{1}{T^2}\left(\int_0^T\int_0^s D(H_s,H_u)u^{H_s+H_u}\,du\,ds+\int_0^T\int_s^T D(H_s,H_u)s^{H_s+H_u}\,du\,ds\right),$$

Since $D(x,y)$ is positive and stays bounded away from 0 for $x,y\in[h_1,h_2]$, we have

$$\sigma_T^2 \geq \frac{C}{T^2}\left(\int_0^T\int_0^s u^{H_s+H_u}\,du\,ds+\int_0^T\int_s^T s^{H_s+H_u}\,du\,ds\right)$$
$$=\frac{2C}{T^2}\int_0^T\int_0^s u^{H_s+H_u}\,du\,ds=\frac{2C}{T^2}\int_0^T\int_0^s T^{H_s+H_u}\left(\frac{u}{T}\right)^{H_s+H_u}du\,ds$$
$$\geq \frac{2C}{T^2}\int_0^T\int_0^s T^{2h_1}\left(\frac{u}{T}\right)^{2h_2}du\,ds=\frac{2CT^{2h_1}}{(2h_2+1)(2h_2+2)}=C_1T^{2h_1}$$

for $T>1$.

Thus, for any $\varepsilon>0$

$$\mathsf{P}\left\{\frac{T^{2h_2+2\delta-1}}{(x_0+\sigma_T\mathcal{N}(0,1))^2}>\varepsilon^2\right\}\leq \mathsf{P}\left\{\left|\frac{x_0}{\sigma_T}+\mathcal{N}(0,1)\right|<\frac{T^{h_2+\delta-1/2}}{\varepsilon\sigma_T}\right\}$$
$$\leq \mathsf{P}\left\{\left|\frac{x_0}{\sigma_T}+\mathcal{N}(0,1)\right|<\frac{T^{h_2-h_1+\delta-1/2}}{\varepsilon C_1^{1/2}}\right\}\to 0$$

as $T\to\infty$ for $0<\delta<1/2-h_2+h_1$. □

Chapter 6
The Extended Orey Index for Gaussian Processes

Stationarity of the increments of fBm is a useful feature in certain applications. However, there are cases when this property is undesirable. In order to enlarge the variety of models to choose from, extensions of fBm have been introduced recently by Houdré and Villa [70] (bifractional Brownian motion) and Bojdecki et al. [27] (sub-fractional Brownian motion). These processes share with fBm such properties as self-similarity, Gaussian property and others, however they do not have stationary increments and possess some new features. Immediately the question arises about the estimation of the parameters of such processes. On tools for statistical estimation, we hope that the reader learned from the Chaps. 2–4 that one of the most important tools is quadratic variation. Except other applications, the asymptotic behavior of the quadratic variation leads to good results in estimation theory. As it was already mentioned in Chap. 2, the problem of the almost sure convergence of a quadratic variation has been solved for a wide class of processes by Baxter [10], Gladyshev [63], Klein and Giné [89], Bégyn [11], Malukas [114] etc. Again, a fBm from this point of view was considered by Gladyshev [63]. Using result of convergence of the quadratic variation, Gladyshev constructed a strongly consistent estimator which is equal H for fBm. Norvaiša [132] extended GladyshevŠs theorem to a class of Gaussian processes that includes bifractional and sub-fractional Brownian motions. Following Klein and Giné [89], Malukas [114] extended the results in [132] to general partitions and proved a central limit theorem in his setting. Bégyn [11, 13] considered second order quadratic variations along general subdivisions for processes with Gaussian increments and asymptotic expansion and central limit theorem for quadratic variations of Gaussian processes. A more complete survey on asymptotic behavior of quadratic variations for Gaussian processes can be found in the thesis of Bégyn [12]. Furthermore, Bégyn considered strongly consistent and asymptotically normal estimator.

In addition, there are other parameters, besides the Hurst index, which are inherent in different classes of processes and which generalize the notion of the Hurst index. One such parameter is Orey index. The present chapter is devoted to

© Springer International Publishing AG 2017

K. Kubilius et al., *Parameter Estimation in Fractional Diffusion Models*,
B&SS – Bocconi & Springer Series 8,
https://doi.org/10.1007/978-3-319-71030-3_6

the introduction of the Orey index concept for the second order stochastic process, construction of the estimators and consideration of the asymptotic properties of these estimators. This index can be determined for a class of Gaussian processes which may not have stationary increments but they have the common property that their trajectories are Hölder continuous up to the Orey index. The fBm is an example of this class of processes. The Orey index of a fBm is equal to H.

In Sect. 6.1, we formulate the definition of the extended Orey index and give the sufficient conditions for the second order stochastic process to possess this index. Moreover, we find the Orey index for some well-known processes. In Sect. 6.2 we consider an almost sure convergence of the quadratic variations of a Gaussian process with the Orey index along arbitrary partition and verify the obtained conditions for some well-known Gaussian processes. General conditions about the asymptotic behavior of the second-order quadratic variations of Gaussian processes were obtained in Bégyn [11]. We formulate the conditions for almost sure convergence of the quadratic variations of a Gaussian process with the Orey index that differ from conditions used by Bégyn. In Sect. 6.3 we apply results obtained to prove the strong consistency of the Orey index estimator based on discrete observations of a Gaussian process along arbitrary partition. In Sect. 6.4 exact (non-asymptotic) confidence intervals for the Orey index of a Gaussian processes are obtained using concentration inequalities for Gaussian quadratic forms and discrete observations of the underlying process. This construction extends the respective results of Breton et al. [31] as well as Breton and Coeurjolly [30], where the exact confidence intervals for the Hurst index of a fBm were derived.

6.1 Gaussian Processes with the Orey Index

For a real-valued zero-mean Gaussian process with stationary increments, Orey suggested the following definition of the index.

Definition 6.1 (See [134, 141]) Let X be a real-valued zero-mean Gaussian stochastic process with stationary increments and continuous in the mean-square sense. Let σ_X be the incremental variance of X given by $\sigma_X^2(h) = \mathsf{E}[X(t+h)-X(t)]^2$ for $t, h \geq 0$. Define

$$\widehat{\beta}_* := \inf\left\{\beta > 0\colon \lim_{h\downarrow 0} \frac{h^\beta}{\sigma_X(h)} = 0\right\} = \limsup_{h\downarrow 0} \frac{\log \sigma_X(h)}{\log h} \tag{6.1}$$

and

$$\widehat{\beta}^* := \sup\left\{\beta > 0\colon \lim_{h\downarrow 0} \frac{h^\beta}{\sigma_X(h)} = +\infty\right\} = \liminf_{h\downarrow 0} \frac{\log \sigma_X(h)}{\log h}\,. \tag{6.2}$$

If $\widehat{\beta}_* = \widehat{\beta}^*$ then X has the *Orey index* $\widehat{\beta}_* = \widehat{\beta}^* = \beta_X$.

If a Gaussian process with stationary increments has Orey index then almost all sample paths satisfy the Hölder condition of order γ for each $\gamma \in (0, \beta_X)$ (see Section 9.4 of Cramèr and Leadbetter [42]). For a fBm B^H with the Hurst index $0 < H < 1$ the Orey index $\beta_{B^H} = H$. It immediately follows from the Definition 6.1 since $\log \sigma_{B^H}(h) = H \log h$. So the class of Gaussian processes with stationary increments that have Orey index β_X is at least not empty. And now our goal is to generalize the definition of this index and significantly expand the class of processes that have Orey index.

Recall that in the previous chapters we have substantially concentrated on fractional Brownian motion. Indeed, it is a popular model in financial mathematics, economics and natural sciences. It is the only continuous Gaussian process which is self-similar with stationary increments and it is completely determined by its index $0 < H < 1$. Moreover, a fBm with Hurst index H is Hölder up to order H. Relatively recently were introduced two extensions of a fBm which preserve some properties of a fBm, but have no stationary increments except for particular parameter values. One of them is a so called sub-fractional Brownian motion (sfBm) (see Sect. 1.5.1 and [27]) and another one is a bifractional Brownian motion (bifBm) (see [70, 152]). All of them are Gaussian processes. Thus it is very natural to extend the definition of the Orey index for Gaussian processes so that it would be possible to consider processes which may not have stationary increments but at the same time have the Orey index.

To provide such extension of the Orey index, we apply the definition of Kubilius [95] which is suitable for the purposes just specified. Moreover, this definition makes it possible to calculate Orey index explicitly. As will be shown later, processes sfBm and bifBm satisfy this extended definition of the Orey index and are Hölder up to their Orey index. Moreover, for fBm, sfBm, and bifBm, the Orey index coincides with their self-similarity parameter (for the definition of the self-similarity parameter see Sect. 1.1.1). Therefore it is reasonable to construct an estimator of the Orey index and study its asymptotic behavior for the classes of processes under consideration. Note that another extension of the definition of the Orey index (more implicit, in a sense) was considered by Norvaiša [133] for the second order stochastic processes which may not have stationary increments. He also showed that sfBm and bifBm satisfy this extended definition of the Orey index.

6.1.1 The Extended Definition of the Orey Index

As usual, we call a process X the second order process if $\mathsf{E}|X(t)|^2 < \infty$ for any time parameter t. Let $X = \{X(t) \colon t \in [0, T]\}$ be a real-valued zero-mean second order stochastic process with the incremental variance function σ_X^2 defined on $[0, T]^2 := [0, T] \times [0, T]$ as

$$\sigma_X^2(s, t) := \mathsf{E}[X(t) - X(s)]^2, \quad (s, t) \in [0, T]^2.$$

In what follows we work with regularly varying at zero functions (definition and properties of such functions are contained in Sect. A.2). However, we restrict the class of regularly varying functions to the following one.

Let Ψ be a class of regularly varying functions at zero $\varphi\colon (0,T) \to [0,\infty)$ with index 1. Every $f \in \Psi$ has representation $f(h) = hL(h)$, where L is some slowly varying function. Let Φ be a subset of Ψ such that

$$\Phi = \{\varphi \in \Psi \,|\varphi(h) \text{ is increasing function in the neighbourhood of zero and } \lim_{h\downarrow 0} L(h) = \infty\}.$$

For any $\varphi \in \Phi$ set

$$\gamma_*(\varphi) := \inf\left\{\gamma > 0\colon \lim_{h\downarrow 0} \sup_{\varphi(h)\le s\le T-h} \frac{h^\gamma}{\sigma_X(s,s+h)} = 0\right\}, \tag{6.3}$$

$$\widetilde{\gamma}_* := \inf\left\{\gamma > 0\colon \lim_{h\downarrow 0} \frac{h^\gamma}{\sigma_X(0,h)} = 0\right\} \tag{6.4}$$

and

$$\gamma^*(\varphi) := \sup\left\{\gamma > 0\colon \lim_{h\downarrow 0} \inf_{\varphi(h)\le s\le T-h} \frac{h^\gamma}{\sigma_X(s,s+h)} = +\infty\right\}, \tag{6.5}$$

$$\widetilde{\gamma}^* := \sup\left\{\gamma > 0\colon \lim_{h\downarrow 0} \frac{h^\gamma}{\sigma_X(0,h)} = +\infty\right\}, \tag{6.6}$$

where $\varphi \in \Phi$. Note that $0 \le \widetilde{\gamma}^* \le \widetilde{\gamma}_* \le +\infty$ and $0 \le \gamma^*(\varphi) \le \gamma_*(\varphi) \le +\infty$.

We give the following extension of the Orey index.

Definition 6.2 Let $X = \{X(t)\colon t \in [0,T]\}$ be a zero-mean second order stochastic process with the incremental variance function σ_X^2 such that $\sup_{0\le s\le T-h} \sigma_X(s,s+h) \to 0$ as $h \to 0$. If $\gamma_*(\varphi) = \widetilde{\gamma}_* = \gamma^*(\varphi) = \widetilde{\gamma}^*$ for any function $\varphi \in \Phi$, then we say that the process X has the Orey index $\gamma_X = \gamma_*(\varphi) = \widetilde{\gamma}_* = \gamma^*(\varphi) = \widetilde{\gamma}^*$.

Remark 6.3 If we consider a real-valued zero-mean Gaussian stochastic process with stationary increments and continuous in the mean square sense then the Orey indices in Definitions 6.1 and 6.2 coincide.

Let us introduce the notions

$$\widehat{\gamma}_*(\varphi) := \limsup_{h\downarrow 0} \sup_{\varphi(h)\le s\le T-h} \frac{\log \sigma_X(s,s+h)}{\log h} \quad \text{and} \quad \overline{\gamma}_* := \limsup_{h\downarrow 0} \frac{\log \sigma_X(0,h)}{\log h}, \tag{6.7}$$

$$\widehat{\gamma}^*(\varphi) := \liminf_{h\downarrow 0} \inf_{\varphi(h)\le s\le T-h} \frac{\log \sigma_X(s,s+h)}{\log h} \quad \text{and} \quad \overline{\gamma}^* := \liminf_{h\downarrow 0} \frac{\log \sigma_X(0,h)}{\log h}. \tag{6.8}$$

It follows from Remark 6.3 and (6.1), (6.2) that $\widetilde{\gamma}_* = \overline{\gamma}_*$ and $\widetilde{\gamma}^* = \overline{\gamma}^*$. Now we compare the values of $\widehat{\gamma}^*(\varphi)$ and $\widehat{\gamma}_*(\varphi)$ with $\gamma^*(\varphi)$ and $\gamma_*(\varphi)$, respectively, for a second order stochastic process X.

Lemma 6.4 *Let $X = \{X(t) \colon t \in [0, T]\}$ be a zero-mean second order stochastic process with the incremental variance function σ_X^2 such that for any $\varphi \in \Phi$*

$$\sup_{\varphi(h) \le s \le T-h} \sigma_X(s, s+h) \longrightarrow 0 \qquad \text{as } h \downarrow 0. \tag{6.9}$$

Then $\widehat{\gamma}^*(\varphi) = \gamma^*(\varphi)$ and $\widehat{\gamma}_*(\varphi) = \gamma_*(\varphi)$ for any $\varphi \in \Phi$.

Proof The proof of the lemma follows the outlines of calculation of limits of the logarithmic ratios (see Annex A.4 in [164]). Fix $\varphi \in \Phi$. For the sake of simplicity we will omit φ in notations $\widehat{\gamma}^*(\varphi)$, $\gamma^*(\varphi)$, $\widehat{\gamma}_*(\varphi)$, $\gamma_*(\varphi)$.

Assume, without lost of generality, that $0 < h < 1$. We first prove that $\widehat{\gamma}_* \le \gamma_*$, where

$$\widehat{\gamma}_* := \limsup_{h \downarrow 0} \sup_{\varphi(h) \le s \le T-h} \frac{\log \sigma_X(s, s+h)}{\log h},$$

$$\gamma_* := \inf \left\{ \gamma > 0 \colon \lim_{h \downarrow 0} \sup_{\varphi(h) \le s \le T-h} \frac{h^\gamma}{\sigma_X(s, s+h)} = 0 \right\}.$$

Let $\gamma > \gamma_*$. It suffices to show that $\gamma \ge \widehat{\gamma}_*$. By definition of the infimum, there exists a real number α such that $\gamma > \alpha > \gamma_*$, and

$$\sup_{\varphi(h) \le s \le T-h} \frac{h^\alpha}{\sigma_X(s, s+h)} \longrightarrow 0 \quad \text{as } h \downarrow 0.$$

But

$$\sup_{\varphi(h) \le s \le T-h} \frac{h^\gamma}{\sigma_X(s, s+h)} = h^{\gamma-\alpha} \sup_{\varphi(h) \le s \le T-h} \frac{h^\alpha}{\sigma_X(s, s+h)} \longrightarrow 0 \quad \text{as } h \downarrow 0 \tag{6.10}$$

as the product of two functions tends to 0. Under the statement

$$\sup_{\varphi(h) \le s \le T-h} \sigma_X(s, s+h) \longrightarrow 0 \qquad \text{as } h \downarrow 0 \tag{6.11}$$

and relation (6.10) there exists h_0 such that for all $h \le h_0 < 1$

$$\sup_{\varphi(h) \le s \le T-h} \frac{h^\gamma}{\sigma_X(s, s+h)} = \frac{h^\gamma}{\inf_{\varphi(h) \le s \le T-h} \sigma_X(s, s+h)} < 1$$

and $\sup_{0\le s\le T-h}\sigma_X(s,s+h)<1$. Moreover,

$$h^\gamma < \inf_{\varphi(h)\le s\le T-h}\sigma_X(s,s+h)$$

for all $h\le h_0<1$. So

$$\log h^\gamma < \log\Big(\inf_{\varphi(h)\le s\le T-h}\sigma_X(s,s+h)\Big)\le \log\Big(\sup_{\varphi(h)\le s\le T-h}\sigma_X(s,s+h)\Big)$$

and

$$\begin{aligned}\gamma &> \frac{\log\big(\sup_{\varphi(h)\le s\le T-h}\sigma_X(s,s+h)\big)}{\log h} = \sup_{\varphi(h)\le s\le T-h}\frac{\log\sigma_X(s,s+h)}{\log h}\\ &\ge \limsup_{h\downarrow 0}\sup_{\varphi(h)\le s\le T-h}\frac{\log\sigma_X(s,s+h)}{\log h} = \widehat{\gamma}_*\,.\end{aligned}$$

Thus $\widehat{\gamma}_*\le\gamma_*$.

Next we prove that $\widehat{\gamma}_*\ge\gamma_*$. Let $\gamma>\alpha>\widehat{\gamma}_*$. It suffices to show that $\gamma\ge\gamma_*$. Under the condition $\alpha>\widehat{\gamma}_*$ and statement (6.11) there exists h_0 such that for $h\le h_0<1$

$$\inf_{\varphi(h)\le s\le T-h}\frac{\log\sigma_X(s,s+h)}{\log h}<\alpha,\qquad \sup_{0\le s\le T-h}\sigma_X(s,s+h)<1.$$

This implies the inequality

$$\log\Big(\inf_{\varphi(h)\le s\le T-h}\sigma_X(s,s+h)\Big)>\log h^\alpha$$

and

$$\inf_{\varphi(h)\le s\le T-h}\sigma_X(s,s+h)>h^\alpha.$$

Thus

$$\sup_{\varphi(h)\le s\le T-h}\frac{h^\alpha}{\sigma_X(s,s+h)}<1.$$

So

$$\sup_{\varphi(h)\le s\le T-h}\frac{h^\gamma}{\sigma_X(s,s+h)}<h^{\gamma-\alpha}\longrightarrow 0\quad\text{as } h\to 0.$$

Therefore $\gamma\ge\gamma_*$.

Now we prove that $\widehat{\gamma}^* = \gamma^*$, where

$$\widehat{\gamma}^* := \liminf_{h\downarrow 0} \inf_{\varphi(h)\le s\le T-h} \frac{\log \sigma_X(s,s+h)}{\log h},$$

$$\gamma^* := \sup\left\{\gamma > 0\colon \lim_{h\downarrow 0} \inf_{\varphi(h)\le s\le T-h} \frac{h^\gamma}{\sigma_X(s,s+h)} = +\infty\right\}.$$

We first prove $\widehat{\gamma}^* \ge \gamma^*$. By definition of supremum, there exists a real number γ such that $\gamma^* > \gamma$, and

$$\lim_{h\downarrow 0} \inf_{\varphi(h)\le s\le T-h} \frac{h^\gamma}{\sigma_X(s,s+h)} = +\infty \tag{6.12}$$

It suffices to show that $\widehat{\gamma}^* \ge \gamma$. Under the condition $\gamma^* > \gamma$ and statements (6.11)–(6.12) there exists h_0 such that for $h \le h_0 < 1$

$$\inf_{\varphi(h)\le s\le T-h} \frac{h^\gamma}{\sigma_X(s,s+h)} > 1, \qquad \sup_{0\le s\le T-h} \sigma_X(s,s+h) < 1.$$

Moreover,

$$h^\gamma > \sup_{\varphi(h)\le s\le T-h} \sigma_X(s,s+h) \ge \inf_{\varphi(h)\le s\le T-h} \sigma_X(s,s+h)$$

and

$$\gamma \log h > \log \inf_{\varphi(h)\le s\le T-h} \sigma_X(s,s+h), \qquad \inf_{\varphi(h)\le s\le T-h} \frac{\log \sigma_X(s,s+h)}{\log h} > \gamma.$$

So $\widehat{\gamma}^* \ge \gamma$.

We show that $\gamma^* \ge \widehat{\gamma}^*$. Assume that $\widehat{\gamma}^* > \alpha > \gamma$. It sufficient to show that $\gamma^* > \gamma$. Under the condition $\widehat{\gamma}^* > \alpha$ and statement (6.11) there exists h_0 such that for $h \le h_0 < 1$

$$\inf_{\varphi(h)\le s\le T-h} \frac{\log \sigma_X(s,s+h)}{\log h} > \alpha, \qquad \sup_{0\le s\le T-h} \sigma_X(s,s+h) < 1.$$

Moreover,

$$\sup_{\varphi(h)\le s\le T-h} \frac{\log \sigma_X(s,s+h)}{\log h} > \alpha$$

and

$$\log\Big(\sup_{\varphi(h)\le s\le T-h}\sigma_X(s,s+h)\Big) < \log h^{\alpha}.$$

Thus

$$\sup_{\varphi(h)\le s\le T-h}\sigma_X(s,s+h) < h^{\alpha} \quad \text{and} \quad \inf_{\varphi(h)\le s\le T-h}\frac{h^{\alpha}}{\sigma_X(s,s+h)} > 1.$$

Then

$$\inf_{\varphi(h)\le s\le T-h}\frac{h^{\gamma}}{\sigma_X(s,s+h)} > h^{\gamma-\alpha}\to\infty$$

and $\gamma^* > \gamma$. □

Assume that for some $\gamma\in(0,1)$ the zero-mean second order stochastic process X satisfies conditions:

(H$_1$) $\sigma_X(0,\delta)\asymp\delta^{\gamma}$, i. e., $\sigma_X(0,\delta)$ and δ^{γ} is of the same order as $\delta\downarrow 0$;
(H$_2$) there exists a constant $\kappa>0$ such that

$$\Lambda(\delta) := \sup_{\varphi(\delta)\le t\le T-\delta}\ \sup_{0<h\le\delta}\left|\frac{\sigma_X(t,t+h)}{\kappa h^{\gamma}}-1\right| \longrightarrow 0 \qquad \text{as } \delta\downarrow 0 \tag{6.13}$$

for every function $\varphi\in\Phi$.

For $(s,t)\in[0,T]^2$, $s\ne t$, set

$$c(s,t) := \frac{\sigma_X^2(s,t)}{\kappa^2|t-s|^{2\gamma}} - 1. \tag{6.14}$$

It follows from (H$_1$) and (H$_2$) that for any $\varphi\in\Phi$

$$\begin{aligned}
&\sup_{0\le s\le T-h}\sigma_X^2(s,s+h)\\
&\quad\le \sup_{0\le s\le\varphi(h)}\sigma_X^2(s,s+h) + \sup_{\varphi(h)\le s\le T-h}\sigma^2(s,s+h)\\
&\quad\le 4\sup_{0\le\delta\le\varphi(h)+h}\sigma_X^2(0,\delta) + \kappa^2h^{2\gamma}\Big(\sup_{\varphi(h)\le s\le T-h}|c(s,s+h)|+1\Big)\\
&\quad\le O\big((\varphi(h))^{2\gamma}\big) + \kappa^2h^{2\gamma}\big[\Lambda^2(h)+2\Lambda(h)+1\big]\longrightarrow 0 \quad \text{as } h\downarrow 0.
\end{aligned} \tag{6.15}$$

Thus the process X is continuous in quadratic mean for all $s\in[0,T-h]$.

Theorem 6.5 *Assume that for some constant $\gamma \in (0,1)$ the zero-mean second order stochastic process X satisfies conditions* (H_1) *and* (H_2)*. Then it has the Orey index that equals γ.*

Proof Due to Lemma 6.4 it is sufficient to show that $\widehat{\gamma}_* = \widehat{\gamma}^* = \gamma$ and $\overline{\gamma}_* = \overline{\gamma}^* = \gamma$.

We first note that condition (H_1) implies $\overline{\gamma}_* = \overline{\gamma}^* = \gamma$. Since $\sigma_X(0,\delta) \asymp \delta^\gamma$, it follows that there exist constants $0 < m < M < \infty$ and δ_0 such that

$$m < \frac{\sigma_X(0,\delta)}{\delta^\gamma} < M \qquad \text{for all } \delta < \delta_0.$$

Thus

$$\gamma + \frac{\log M}{\log \delta} = \frac{\log(M\delta^\gamma)}{\log \delta} < \frac{\log \sigma_X(0,\delta)}{\log \delta} < \frac{\log(m\delta^\gamma)}{\log \delta} = \gamma + \frac{\log m}{\log \delta} \qquad \text{as } \delta < 1,$$

and we obtain equality $\overline{\gamma}_* = \overline{\gamma}^* = \gamma$.

It remains to prove that $\widehat{\gamma}^* = \widehat{\gamma}_*$. We fix some function $\varphi \in \Phi$. It follows from conditions (H_1) and (H_2) that there exists δ_0 such that for $\delta \le \delta_0 < 1$ inequalities $\sigma_X(s, s+\delta) \le 1/2$ for all $0 \le s \le T - \delta_0$ and $\Lambda(\delta) < 1/2$ hold. In what follows we suppose that inequalities hold for $\delta \le \delta_0 < 1$.

Assume that $-1/2 < b(s_0, s_0+\delta_0) \le 0$ for some fixed $s_0 \in [\varphi(\delta_0), T-\delta_0]$, where

$$b(s,t) := \frac{\sigma_X(s,t)}{\kappa|t-s|^\gamma} - 1.$$

Since $-2x \le \log(1-x) \le -x$ for $0 \le x \le 1/2$, it follows that

$$\begin{aligned}
\log \sigma_X(s_0, s_0+\delta_0) &= \log(\kappa\delta_0^\gamma) + \log(1 + b(s_0, s_0+\delta_0)) \\
&= \log(\kappa\delta_0^\gamma) + \log(1 - (-b(s_0, s_0+\delta_0))) \\
&\le \log(\kappa\delta_0^\gamma) + b(s_0, s_0+\delta_0) \le \log(\kappa\delta_0^\gamma) + \Lambda(\delta_0)
\end{aligned}$$

and

$$\begin{aligned}
\log \sigma_X(s_0, s_0+\delta_0) &\ge \log(\kappa\delta_0^\gamma) + 2b(s_0, s_0+\delta_0) = \log(\kappa\delta_0^\gamma) - 2|b(s_0, s_0+\delta_0)| \\
&\ge \log(\kappa\delta_0^\gamma) - 2\Lambda(\delta_0).
\end{aligned}$$

It is known that $|\log(1+x)| \le x$ for $x \ge 0$. Assume that $0 \le b(s_0, s_0+\delta_0) < 1/2$ for some fixed $s_0 \in [\varphi(\delta_0), T-\delta_0]$, then

$$\begin{aligned}
\log \sigma_X(s_0, s_0+\delta_0) &= \log(\kappa\delta_0^\gamma) + \log(1 + b(s_0, s_0+\delta_0)) \le \log(\kappa\delta_0^\gamma) + b(s_0, s_0+\delta_0) \\
&\le \log(\kappa\delta_0^\gamma) + \Lambda(\delta_0)
\end{aligned}$$

and

$$\log \sigma_X(s_0, s_0 + \delta_0) = \log(\kappa \delta_0^\gamma) + \log(1 + b(s_0, s_0 + \delta_0))$$
$$\geq \log(\kappa \delta_0^\gamma) - 2|b(s_0, s_0 + \delta_0)| \geq \log(\kappa \delta_0^\gamma) - 2\Lambda(\delta_0).$$

Thus, for every $s \in [\varphi(\delta_0), T - \delta_0]$ we obtain

$$\log(\kappa \delta_0^\gamma) - 2\Lambda(\delta_0) \leq \log \sigma_X(s, s + \delta_0) \leq \log(\kappa \delta_0^\gamma) + \Lambda(\delta_0).$$

Consequently,

$$\gamma + \frac{\log \kappa}{\log \delta_0} - \frac{\Lambda(\delta_0)}{|\log \delta_0|} \leq \inf_{\varphi(\delta_0) \leq s \leq T - \delta_0} \frac{\log \sigma_X(s, s + \delta_0)}{\log \delta_0}$$
$$\leq \sup_{\varphi(\delta_0) \leq s \leq T - \delta_0} \frac{\log \sigma_X(s, s + \delta_0)}{\log \delta_0} \leq \gamma + \frac{\log \kappa}{\log \delta_0} + 2\frac{\Lambda(\delta_0)}{|\log \delta_0|}.$$

and both sides of the above inequality approach γ as $\delta_0 \to 0$. Thus, $\widehat{\gamma}_* = \widehat{\gamma}^* = \gamma$. □

Proposition 6.6 *Assume that Gaussian process X has the Orey index γ_X. Then almost all sample paths of X are Hölder continuous up to γ_X.*

Proof According to Theorem 1.5, it suffices to show that there exists a constant c_ε such that

$$\sigma_X(s, t) \leq c_\varepsilon |t - s|^{\gamma_X - \varepsilon} \qquad \text{for all } \varepsilon \in (0, \gamma_X), \tag{6.16}$$

where $s, t \in [0, T]$. We prove inequality (6.16) under the condition that $|t - s| \leq h_0$ for some $h_0 < 1$. The proof is easy to complete without this restriction.

Fix $\delta < \varepsilon < \gamma_X$ and fix any $\varphi \in \Phi$. From definition of the Orey index it follows that for any constant $C > 0$ there exists $0 < h_1 < 1$ such that

$$\inf_{\varphi(h) \leq s \leq T - h} \frac{h^{\gamma_X - \delta}}{\sigma_X(s, s + h)} > C \quad \text{and} \quad \frac{h^{\gamma_X - \delta}}{\sigma_X(0, h)} > C \qquad \text{for} \quad h \leq h_1.$$

Suppose that $\varphi(h) + h < h_1$ for $h < h_0 \leq h_1$. Then

$$\sup_{0 \leq s \leq T - h} \sigma_X(s, s + h) \leq \sup_{0 \leq s \leq \varphi(h)} \sigma_X(s, s + h) + \sup_{\varphi(h) \leq s \leq T - h} \sigma_X(s, s + h)$$
$$\leq 2 \sup_{0 \leq s \leq \varphi(h) + h} \sigma_X(0, s) + \sup_{\varphi(h) \leq s \leq T - h} \sigma(s, s + h)$$
$$< \frac{2}{C} (\varphi(h) + h)^{\gamma_X - \delta} + \frac{1}{C} h^{\gamma_X - \delta}$$

$$
\begin{aligned}
&= \frac{1}{C}\left[2h^{\varepsilon-\delta}(1+L(h))^{\gamma_X-\delta} + h^{\varepsilon-\delta}\right]h^{\gamma_X-\varepsilon} \\
&\le \frac{1}{C}\left[2h^{\varepsilon-\delta}(1+L(h))^{\gamma_X} + 1\right]h^{\gamma_X-\varepsilon} \le \widehat{C}_\varepsilon h^{\gamma_X-\varepsilon}
\end{aligned}
$$

for $h < h_0$, since $h^{\varepsilon-\delta}(1+L(h))^{\gamma_X} \to 0$. Consequently, there exists a constant c_ε such that

$$
\sigma_X(s,t) \le c_\varepsilon |t-s|^{\gamma_X-\varepsilon}
$$

for $|t-s| \le h_0 < 1$. □

6.1.2 Examples of Processes with the Orey Index

6.1.2.1 Fractional Brownian Bridge

The fractional Brownian bridge is defined in $[0,T]$ as

$$
X_t^H = B_t^H - \frac{t^{2H} + T^{2H} - |t-T|^{2H}}{2T^{2H}} B_T^H, \tag{6.17}
$$

where B^H, $0 < H < 1$, is a fBm on the interval $[0,T]$ (see [58]).

First, let us check condition (H_1). The incremental variance function of X^H has the following form

$$
\sigma_{X^H}^2(t,t+h) = h^{2H} - \frac{1}{4T^{2H}} f_t^2(h),
$$

where

$$
f_t^2(h) := \left[(t+h)^{2H} - t^{2H} - |t+h-T|^{2H} + |t-T|^{2H}\right]^2.
$$

It is easy to verify that

$$
\frac{f_0^2(h)}{h^{2H}} = \frac{[h^{2H} - |T-h|^{2H} + T^{2H}]^2}{h^{2H}} \longrightarrow 0 \qquad \text{as } h \downarrow 0.
$$

Thus

$$
\sigma_{X^H}^2(t,t+h) \asymp h^{2H}.
$$

So condition (H_1) is satisfied.

Second, check condition (H_2). Assume that $H < 1/2$. Since

$$\left|(t+h)^{2H} - t^{2H}\right| \le h^{2H} \quad \text{and} \quad \left|(T-t-h)^{2H} - (T-t)^{2H}\right| \le h^{2H},$$

we have that for every $\varphi \in \Phi$

$$\sup_{\varphi(\delta)\le t\le T-\delta}\ \sup_{0<h\le\delta}\left|\frac{\sigma^2_{X^H}(t,t+h)}{h^{2H}} - 1\right| = \frac{1}{4T^{2H}}\sup_{\varphi(\delta)\le t\le T-\delta}\ \sup_{0<h\le\delta}\frac{f_t^2(h)}{h^{2H}} \le T^{-2H}\delta^{2H}.$$

Assume that $H \ge 1/2$. Then $f_t(0) = 0$ and we obtain from Taylor's formula that

$$\frac{\sigma^2_{X^H}(t,t+h)}{h^{2H}} - 1 = -\frac{1}{4T^{2H}h^{2H}}\left(\int_0^h f_t'(x)\,dx\right)^2,$$

where

$$f_t'(x) = 2H\left[(t+x)^{2H-1} - (t+x-T)^{2H-1}\right].$$

Thus for every $\varphi \in \Phi$ and $H \ge 1/2$ we get

$$\sup_{\varphi(\delta)\le t\le T-\delta}\ \sup_{0<h\le\delta}\left|\frac{\sigma^2_{X^H}(t,t+h)}{h^{2H}} - 1\right| \le \frac{\delta^{2-2H}}{4T^{2H}}\cdot 4H^2T^{4H-2} = H^2T^{2H-2}\delta^{2-2H}.$$

It follows from Theorem 6.5 that the Orey index of fractional Brownian bridge is equal to H and moreover $\kappa = 1$.

6.1.2.2 Fractional Ornstein–Uhlenbeck Process

The fO-U process

$$X_t = x_0e^{-\theta t} + \sigma\int_0^t e^{-\theta(t-u)}dB_u^H, \qquad t \in [0,T],$$

is the unique solution of the following SDE

$$X_t = x_0 - \theta\int_0^t X_s\,ds + \sigma B_t^H, \qquad t \in [0,T], \tag{6.18}$$

where $x_0 \in \mathbb{R}$, $\theta, \sigma > 0$ are real parameters, and B^H is an fBm with the Hurst index H, $H \in (0,1)$. The integral exists as a Riemann-Stieltjes integral for all $t \in [0,T]$. More detailed information one can find in Sect. 1.3.2.

We defined the incremental variance function for a zero-mean second order stochastic process. The mean of fO-U process is not zero. If we assume that $x_0 = 0$

then it will be centered process. Since

$$|\mathsf{E}X_{t+h} - \mathsf{E}X_t|^2 \le x_0^2 e^{-2\theta T} h^2,$$

it follows that the asymptotic of the incremental variance function of centered or non-centered fO-U process is the same as $h \to 0$.

As well as for the previous example we again check conditions (H_1) and (H_2) for fO-U process. Conclusion about the Orey index will follow from Theorem 6.5.

First check condition (H_1) for this process. Since

$$\int_0^t e^{\theta u} dB_u^H = e^{\theta t} B_t^H - \theta \int_0^t e^{\theta u} B_u^H du,$$

it follows that

$$\begin{aligned} X_t^2 =& \left(x_0 e^{-\theta t} + \sigma \int_0^t e^{-\theta(t-u)} dB_u^H \right)^2 = \left(x_0 e^{-\theta t} + \sigma B_t^H - \sigma \theta e^{-\theta t} \int_0^t e^{\theta u} B_u^H du \right)^2 \\ \le& 3x_0^2 + 3\sigma^2 (B_t^H)^2 + 3\sigma^2 e^{-2\theta t} \left(\int_0^t B_u^H de^{\theta u} \right)^2 \\ \le& 3x_0^2 + 3\sigma^2 (B_t^H)^2 + 3\sigma^2 e^{-\theta t} \int_0^t (B_u^H)^2 de^{\theta u} \end{aligned}$$

and

$$\sup_{t \le T} \mathsf{E}X_t^2 \le 3x_0^2 + 6\sigma^2 T^2. \tag{6.19}$$

The incremental variance function of X has the following form

$$\begin{aligned} \sigma_X^2(t, t+h) =& \theta^2 \mathsf{E} \left(\int_t^{t+h} X_s\, ds \right)^2 - 2\theta\sigma \mathsf{E} \left([B^H(t+h) - B^H(t)] \int_t^{t+h} X_s\, ds \right) \\ &+ \sigma^2 \sigma_{B^H}^2(t, t+h). \end{aligned}$$

The Cauchy-Schwarz inequality yields

$$\mathsf{E} \left(\int_t^{t+h} X_s\, ds \right)^2 \le h^2 \sup_{t \le s \le t+h} \mathsf{E}X_s^2$$

and

$$\begin{aligned} \mathsf{E} \left([B^H(t+h) - B^H(t)] \int_t^{t+h} X_s\, ds \right) \le& \mathsf{E}^{1/2} [B^H(t+h) - B^H(t)]^2 \left(h \int_t^{t+h} \mathsf{E}X_s^2\, ds \right)^{\frac{1}{2}} \\ \le& h^{H+1} \left(\sup_{t \le s \le t+h} \mathsf{E}X_s^2 \right)^{\frac{1}{2}}. \end{aligned}$$

Note that

$$\left|\sigma_X^2(0,h)-\sigma^2 h^{2H}\right| = \left|\sigma_X^2(0,h)-\sigma^2\mathsf{E}(B_h^H)^2\right|$$
$$\le \theta^2 h^2 \sup_{t\le h}\mathsf{E}X_t^2 + 2\theta\sigma h^{1+H}\sqrt{\sup_{t\le h}\mathsf{E}X_t^2}\,.$$

Condition (H_1) follows from these calculations since $\sigma_X^2(0,h)\asymp h^{2H}$ is equivalent to the requirement that $\sigma_X^2(0,h)/h^{2H}$ converges to some finite non-zero limit as $h\downarrow 0$.

Next, for every $\varphi\in\Phi$

$$\sup_{\varphi(\delta)\le t\le T-\delta}\ \sup_{0<h\le\delta}\left|\frac{\sigma_X^2(t,t+h)}{\sigma^2 h^{2H}}-1\right|$$
$$\le \sigma^{-2}\delta^{1-H}\left[\delta^{1-H}\theta^2\sup_{t\le T}\mathsf{E}X_t^2 + 2\theta\sigma\left(\sup_{t\le T}\mathsf{E}X_t^2\right)^{1/2}\right]\longrightarrow 0$$

as $\delta\downarrow 0$. It follows from the inequality

$$\left|\frac{\sigma_X(t,t+h)}{\sigma h^H}-1\right| = \left|\frac{\sigma_X^2(t,t+h)}{\sigma^2 h^{2H}}-1\right|\Big/\left|\frac{\sigma_X(t,t+h)}{\sigma h^H}+1\right| \le \left|\frac{\sigma_X^2(t,t+h)}{\sigma^2 h^{2H}}-1\right|$$

that condition (H_2) is satisfied with $\kappa=\sigma$.

It follows from Theorem 6.5 that the Orey index of fO-U is equal to H.

6.1.2.3 Sub-fractional Brownian Motion

We shall prove that sfBm satisfies conditions (H_1) and (H_2).

A sub-fractional Brownian motion (see Sect. 1.5.1) with index H, $H\in(0,1)$, is a zero-mean Gaussian stochastic process $S^H=(S_t^H, t\ge 0)$ with covariance function

$$G_H(s,t) := s^{2H}+t^{2H}-\frac{1}{2}\left[(s+t)^{2H}+|s-t|^{2H}\right].$$

The incremental variance function of sfBm is of the following form

$$\sigma_{S^H}^2(s,t)=\mathsf{E}|S_t^H-S_s^H|^2 = |t-s|^{2H}+(s+t)^{2H}-2^{2H-1}(t^{2H}+s^{2H}). \tag{6.20}$$

Since for any $0\le s\le t\le T$ inequalities

$$(t-s)^{2H}\le\sigma_{S^H}^2(s,t)\le(2-2^{2H-1})(t-s)^{2H},\qquad \text{if}\quad 0<H<1/2,$$
$$(2-2^{2H-1})(t-s)^{2H}\le\sigma_{S^H}^2(s,t)\le(t-s)^{2H},\qquad \text{if}\quad 1/2<H<1$$

hold, it follows that condition (H_1) is satisfied.

We get from (6.20) that

$$\sigma^2_{SH}(s, s+h) = h^{2H} + f_s(h),$$

where

$$f_s(h) := (2s+h)^{2H} - 2^{2H-1}\big[s^{2H} + (s+h)^{2H}\big].$$

Note that

$$f_s(0) = f_s'(0) = 0.$$

By Taylor's formula we obtain

$$\begin{aligned} f_s(h) =& f_s(0) + f_s'(0)h + \int_0^h f_s''(x)(h-x)\,dx = \int_0^h f_s''(x)(h-x)\,dx \\ =& 2H(2H-1)\int_0^h \big[(2s+x)^{2H-2} - 2^{2H-1}(s+x)^{2H-2}\big](h-x)\,dx. \end{aligned}$$

From inequality

$$\begin{aligned} \big[2^{2H-1}(s+x)^{2H-2} - (2s+x)^{2H-2}\big] &= \frac{1}{(s+x)^{2-2H}}\left[2^{2H-1} - \left(\frac{s+x}{2s+x}\right)^{2-2H}\right] \\ &= \frac{1}{(s+x)^{2-2H}}\left[2^{2H-1} - \left(1 - \frac{s}{2s+x}\right)^{2-2H}\right] \le \frac{1}{(s+x)^{2-2H}}\big[2^{2H-1} - 2^{-1}\big], \end{aligned}$$

it follows that for $s > 0$

$$|f_s(h)| \le (2^{2H} - 1)\int_0^h \frac{h-x}{(s+x)^{2-2H}}\,dx \le \frac{1}{2}(2^{2H} - 1)s^{2H-2}h^2$$

and

$$\begin{aligned} \sup_{\varphi(\delta)\le s\le T-\delta}\ \sup_{0<h\le\delta}\left|\frac{\sigma^2_{SH}(s,s+h)}{h^{2H}} - 1\right| &= \sup_{\varphi(\delta)\le s\le T-\delta}\ \sup_{0<h\le\delta}\frac{|f_s(h)|}{h^{2H}} \\ &\le \sup_{\varphi(\delta)\le s\le T-\delta}\frac{2^{2H-1}\delta^{2-2H}}{s^{2-2H}} \le \frac{2^{2H-1}}{(L(\delta))^{2-2H}} \end{aligned}$$

for every $\varphi \in \Phi$, where $L(h) = \varphi(h)/h$. So we get condition (H_2) with $\kappa = 1$. The Orey index is equal to H.

Remark 6.7 The function $\varphi(\delta)$ could not be replaced by δ or 0 in condition (H_2). Indeed, e. g., $H > 1/2$. Then

$$\sup_{0\le s\le T-\delta}\sup_{0\le h\le\delta}|h^{-2H}f_s(h)| \ge \sup_{\delta\le s\le T-\delta}\sup_{0\le h\le\delta}|h^{-2H}f_s(h)|$$

$$= 2H(2H-1)\sup_{\delta\le s\le T-\delta}\sup_{0\le h\le\delta}\int_0^h\Big[\frac{2^{2H-1}}{h^{2H}(s+x)^{2-2H}}-\frac{1}{h^{2H}(2s+x)^{2-2H}}\Big](h-x)\,dx$$

$$\ge 2H(2H-1)\sup_{\delta\le s\le T-\delta}\sup_{0\le h\le\delta}\int_0^h\frac{2^{2H-1}-1}{h^{2H}(2s+x)^{2-2H}}(h-x)\,dx$$

$$\ge H(2H-1)\sup_{\delta\le s\le T-\delta}\sup_{0\le h\le\delta}\frac{(2^{2H-1}-1)h^{2-2H}}{(2s+h)^{2-2H}}$$

$$= H(2H-1)(2^{2H-1}-1)\sup_{\delta\le s\le T-\delta}\frac{\delta^{2-2H}}{(2s+\delta)^{2-2H}}$$

$$= H(2H-1)(2^{2H-1}-1)3^{2H-2}.$$

6.1.2.4 Bifractional Brownian Motion

A bifractional Brownian motion (see Sect. 1.5.2) $B^{HK} = (B_t^{HK}, t \ge 0)$ with parameters $H \in (0,1)$ and $K \in (0,1]$ is a centered Gaussian process with covariance function

$$R_{HK}(t,s) = 2^{-K}\big((t^{2H}+s^{2H})^K - |t-s|^{2HK}\big), \qquad s,t \ge 0.$$

The incremental variance function of bifBm is of the following form

$$\sigma^2_{BH,K}(s,t) = 2^{1-K}\big[|t-s|^{2HK} - (t^{2H}+s^{2H})^K\big] + t^{2HK} + s^{2HK}.$$

Let $H \in (0,1)$ and $K \in (0,1]$. Then

$$2^{-K}|t-s|^{2HK} \le \sigma^2_{BH,K}(s,t) \le 2^{1-K}|t-s|^{2HK}$$

for all $s,t \in [0,\infty)$. Thus condition (H_1) holds.

Since

$$\sigma^2_{BH,K}(s,s+h) = 2^{1-K}(h^{2HK} - f_s(h))$$

with

$$f_s(h) := \big[s^{2H} + (s+h)^{2H}\big]^K - 2^{K-1}\big[s^{2HK} + (s+h)^{2HK}\big],$$

we get $f_s(0) = f_s'(0) = 0$ and by Taylor's formula we obtain

$$\frac{\sigma^2_{BHK}(s, s+h)}{2^{1-K} h^{2HK}} - 1 = -h^{-2HK} \int_0^h f_s''(x)(h-x)\, dx,$$

where

$$\begin{aligned} f_s''(x) =& 4K(K-1)H^2\big[s^{2H} + (s+x)^{2H}\big]^{K-2}(s+x)^{2(2H-1)} \\ &+ 2HK(2H-1)\big[s^{2H} + (s+x)^{2H}\big]^{K-1}(s+x)^{2H-2} \\ &- 2^K HK(2HK-1)(s+x)^{2HK-2}. \end{aligned}$$

Note that for $H \geq 1/2$

$$\begin{aligned} \frac{(s+x)^{2(2H-1)}}{[s^{2H} + (s+x)^{2H}]^{2-K}} =& \left[\frac{(s+x)^{2H}}{s^{2H} + (s+x)^{2H}}\right]^{2-K} (s+x)^{2HK-2} \\ \leq& (s+x)^{2HK-2}. \end{aligned}$$

Thus for $s > 0$

$$\begin{aligned} \sup_{0\leq x\leq h} |f_s''(x)| \leq& \frac{4}{s^{2-2HK}} \mathbf{1}_{\{H\geq 1/2\}} + \frac{4}{(2s^{2H})^{2-K} s^{2(1-2H)}} \mathbf{1}_{\{H<1/2\}} \\ &+ \frac{2}{(2s^{2H})^{1-K} s^{2-2H}} + \frac{2}{s^{2-2HK}} \leq \frac{8}{s^{2-2HK}} \end{aligned}$$

and

$$\sup_{\varphi(\delta)\leq s\leq T-\delta} \sup_{0<h\leq\delta} \left| \frac{\sigma^2_{B_{HK}}(s, s+h)}{2^{1-K} h^{2H}} - 1 \right| \leq \sup_{\varphi(\delta)\leq s\leq T-\delta} \frac{8\delta^{2-2HK}}{s^{2-2HK}} \leq \frac{8}{(L(\delta))^{2-2HK}}$$

for every $\varphi \in \Phi$. So condition (H_2) holds with $\kappa = 2^{1-K}$. The Orey index of a bifBm is equal to KH.

6.2 The Convergence of the Quadratic Variation of Gaussian Process with the Orey Index

As it was mentioned in the introduction to the chapter, general conditions of the convergence a.s. of the second-order quadratic variations of Gaussian processes along arbitrary partition were obtained in Bégyn [11]. By using conditions (H_1) and (H_2) we suggest another type of conditions of the convergence a.s. of the second-order quadratic variations for a Gaussian process with the Orey index along arbitrary partition. We show that the class Φ of functions plays an important role in asymptotic behavior of the second-order quadratic variations.

Let $\pi_n = \{0 = t_0^n < t_1^n < \cdots < t_{N_n}^n = T\}$, $T > 0$, be a sequence of partitions of the interval $[0, T]$, where (N_n) is an increasing sequence of natural numbers. Recall some definitions from Sect. 3.4.1

$$m_n = \max_{1\le k\le N_n} \Delta_k^n t, \qquad p_n = \min_{1\le k\le N_n} \Delta_k^n t, \qquad \Delta_{n,k} t = t_k^n - t_{k-1}^n,$$

$$\Delta_{\mathrm{ir},k}^{(2)n} X = \Delta_{n,k} t X(t_{k+1}^n) + \Delta_{n,k+1} t X(t_{k-1}^n) - (\Delta_{n,k} t + \Delta_{n,k+1} t) X(t_k^n),$$

$$V_{\pi_n,T}^{(2)X} = 2\sum_{k=1}^{N_n-1} \frac{\Delta_{n,k+1} t (\Delta_{\mathrm{ir},k}^{(2)n} X)^2}{(\Delta_{n,k} t)^{1/2+H} (\Delta_{n,k+1} t)^{1/2+H} [\Delta_{n,k} t + \Delta_{n,k+1} t]}.$$

Further in this section we will use the notations and definitions introduced in Sect. 3.4.1.

Let X be a real-valued zero-mean Gaussian stochastic process with the Orey index γ_X. Define its covariance function

$$R(s,t) = \mathsf{E} X_t X_s, \qquad s, t \in [0, T]$$

and the covariance of the second order increments

$$d_{jk}^{(2)n} := \mathsf{E} \Delta_{\mathrm{ir},j}^{(2)n} X \Delta_{\mathrm{ir},k}^{(2)n} X, \qquad 1 \le j, k \le N_n - 1.$$

If there exists the fourth order mixed partial derivative of the covariance function $R(s, t)$, then

$$d_{jk}^{(2)n} = \int_{t_j^n}^{t_{j+1}^n} du \int_{t_{j-1}^n}^{t_j^n} dv \int_v^u dw \int_{t_k^n}^{t_{k+1}^n} dx \int_{t_{k-1}^n}^{t_k^n} dy \int_y^x \frac{\partial^4 R}{\partial s^2 \partial t^2}(w, z)\, dz$$

holds for $j \ne 1$ and $k \ne 1$.

In the following proposition we will use the definitions from Sect. 3.4.1.

Proposition 6.8 *Let $X = \{X(t) : t \in [0, T]\}$, $T > 0$, be a zero-mean Gaussian process satisfying conditions* (H$_1$) *and* (H$_2$). *Let $(\pi_n)_{n\in\mathbb{N}}$ be a sequence of partitions with asymptotic ratios $(\ell_k)_{k\ge 1}$ and range of the asymptotic ratios $\mathcal{L}$. If the function*

$$h(\lambda) = \frac{1 + \lambda^{2\gamma-1} - (1+\lambda)^{2\gamma-1}}{\lambda^{\gamma-1/2}}$$

is invariant on $\mathcal{L}$ or if the sequence of step functions $\ell_n(t)$, i.e. $\ell_n(t) = \ell_k$ on (t_k^n, t_{k+1}^n), $0 \le k \le N_n - 1$ and $\ell_0 = \ell_1$, converges to a regulated function $\ell(t)$

on the interval $[0, T]$, *then*

$$\mathsf{E}V^{(2)X}_{\pi_n,T} \longrightarrow \begin{cases} 2\kappa^2 h(\ell_1)T, & \text{if } h \text{ is invariant on } \mathcal{L}, \\ 2\kappa^2 \int_0^T h(\ell(s))\,\mathrm{d}s, & \text{if } \ell(s) \text{ is a regulated function.} \end{cases} \tag{6.21}$$

Proof For the sake of simplicity we will omit the index X in notation γ_X. Rewrite the expectation of each increment from the second order variation in the following way

$$\begin{aligned}
\mathsf{E}(\Delta^{(2)n}_{\mathrm{ir},k}X)^2 =&(\Delta_{n,k}t)^2\sigma_X^2(t_k^n, t_{k+1}^n) + (\Delta_{n,k+1}t)^2\sigma_X^2(t_{k-1}^n, t_k^n)\\
&+ \Delta_{n,k}t \cdot \Delta_{n,k+1}t\big[\sigma_X^2(t_k^n, t_{k+1}^n) - \sigma_X^2(t_{k-1}^n, t_{k+1}^n) + \sigma_X^2(t_{k-1}^n, t_k^n)\big]\\
=&[\Delta_{n,k}t + \Delta_{n,k+1}t]\big[\Delta_{n,k}t \cdot \sigma_X^2(t_k^n, t_k^n + \Delta_{n,k+1}t)\\
&+ \Delta_{n,k+1}t \cdot \sigma_X^2(t_{k-1}^n, t_{k-1}^n + \Delta_{n,k}t)\big]\\
&- \Delta_{n,k}t \cdot \Delta_{n,k+1}t \cdot \sigma_X^2(t_{k-1}^n, t_{k-1}^n + \Delta_{n,k}t + \Delta_{n,k+1}t)\\
=&I_k^{(1)} - I_k^{(2)} + I_k^{(3)},
\end{aligned}$$

where $\sigma_X(\cdot,\cdot)$ is incremental variance function of X,

$$\begin{aligned}
I_k^{(1)} :=&[\Delta_{n,k}t + \Delta_{n,k+1}t]\big\{\Delta_{n,k}t\big[\sigma_X^2(t_k^n, t_{k+1}^n) - \kappa^2(\Delta_{n,k+1}t)^{2\gamma}\big]\\
&+ \Delta_{n,k+1}t\big[\sigma_X^2(t_{k-1}^n, t_k^n) - \kappa^2(\Delta_{n,k}t)^{2\gamma}\big]\big\},\\
I_k^{(2)} :=&\Delta_{n,k}t \cdot \Delta_{n,k+1}t\big[\sigma_X^2(t_{k-1}^n, t_{k+1}^n) - \kappa^2(\Delta_{n,k}t + \Delta_{n,k+1}t)^{2\gamma}\big],\\
I_k^{(3)} :=&\kappa^2[\Delta_{n,k}t + \Delta_{n,k+1}t]\Delta_{n,k}t \cdot \Delta_{n,k+1}t\big\{(\Delta_{n,k+1}t)^{2\gamma-1}\\
&+ (\Delta_{n,k}t)^{2\gamma-1} - (\Delta_{n,k}t + \Delta_{n,k+1}t)^{2\gamma-1}\big\}.
\end{aligned}$$

Set

$$\mu_k^n = (\Delta_{n,k+1}t)^{1/2+\gamma}(\Delta_{n,k}t)^{1/2+\gamma}(\Delta_{n,k}t + \Delta_{n,k+1}t) \quad \text{and} \quad \ell_k^n = \frac{\Delta_{n,k}t}{\Delta_{n,k+1}t}.$$

Then

$$\begin{aligned}
I_k^{(1)} =&\kappa^2[\Delta_{n,k}t + \Delta_{n,k+1}t]\Delta_{n,k}t\\
&\times \Delta_{n,k+1}t\big[(\Delta_{n,k+1}t)^{2\gamma-1}c(t_k^n, t_{k+1}^n) + (\Delta_{n,k}t)^{2\gamma-1}c(t_{k-1}^n, t_k^n)\big]\\
=&\kappa^2\mu_k^n\big[(\ell_k^n)^{1/2-\gamma}c(t_k^n, t_{k+1}^n) + (\ell_k^n)^{\gamma-1/2}c(t_{k-1}^n, t_k^n)\big],\\
I_k^{(2)} =&\kappa^2\mu_k^n(\Delta_{n,k}t)^{1/2-\gamma}(\Delta_{n,k+1}t)^{1/2-\gamma}(\Delta_{n,k}t + \Delta_{n,k+1}t)^{2\gamma-1}c(t_{k-1}^n, t_{k+1}^n)\\
=&\kappa^2\mu_k^n(\ell_k^n)^{1/2-\gamma}(1 + \ell_k^n)^{2\gamma-1}c(t_{k-1}^n, t_{k+1}^n)
\end{aligned}$$

and

$$I_k^{(3)} = \kappa^2 \mu_k^n \big[(\ell_k^n)^{1/2-\gamma} + (\ell_k^n)^{\gamma-1/2} - (\ell_k^n)^{1/2-\gamma}(1+\ell_k^n)^{2\gamma-1}\big] = \kappa^2 \mu_k^n h(\ell_k^n),$$

where the function $c(s,t)$ is defined in (6.14). Further, we note that

$$\begin{aligned} \mathsf{E}V_{\pi_n,T}^{(2)X} =& 2\sum_{k=1}^{\tau_n+1} \frac{\Delta_{k+1}^n t \cdot \mathsf{E}(\Delta_{ir,k}^{(2)n} X)^2}{\mu_k^n} + 2\sum_{k=\tau_n+2}^{N_n-1} \frac{\Delta_{k+1}^n t \cdot \mathsf{E}(\Delta_{ir,k}^{(2)n} X)^2}{\mu_k^n} \\ =& 2\sum_{k=1}^{\tau_n+1} \frac{\Delta_{k+1}^n t \cdot \mathsf{E}(\Delta_{\mathrm{ir},k}^{(2)n} X)^2}{\mu_k^n} + 2\kappa^2 \sum_{k=\tau_n+2}^{N_n-1} \Delta_{k+1}^n t \cdot J_k \\ &+ 2\kappa^2 \sum_{k=\tau_n+2}^{N_n-1} \Delta_{k+1}^n t \cdot h(\ell_k^n), \end{aligned} \tag{6.22}$$

where $\tau_n = \lfloor L(2m_n) \rfloor$, $\lfloor a \rfloor$ is an integer part of a real number a,

$$J_k = (\ell_k^n)^{1/2-\gamma}\big[c(t_k^n, t_{k+1}^n) + (\ell_k^n)^{2\gamma-1} c(t_{k-1}^n, t_k^n) - (1+\ell_k^n)^{2\gamma-1} c(t_{k-1}^n, t_{k+1}^n)\big].$$

Assume that $L(2m_n) \geq 4$ and $\varphi(2m_n) \leq T \wedge 1$. It should be noted that

$$\begin{aligned} N_n - 1 \geq& \left\lfloor \frac{T}{m_n} \right\rfloor - 1 \geq \frac{2(T\wedge 1)L(2m_n)}{\varphi(2m_n)} - 2 \geq 2L(2m_n) - 2 \\ \geq& L(2m_n) + 2 \geq \tau_n + 2. \end{aligned}$$

Now we estimate the first term in the right-hand side of (6.22). Note that

$$\tau_n \leq L(2m_n), \qquad 2p_n^{2\gamma+2} \leq \mu_k^n \leq 2m_n^{2\gamma+2}, \tag{6.23}$$

$$\sum_{k=1}^{\tau_n+1} \Delta_{k+1}^n t \leq \varphi(2m_n) + m_n \leq 2\varphi(2m_n). \tag{6.24}$$

From (6.23)-(6.24) and Definition 3.25 we may further conclude that

$$\begin{aligned} 2\sum_{k=1}^{\tau_n+1} \frac{\Delta_{k+1}^n t \cdot \mathsf{E}(\Delta_{\mathrm{ir},k}^{(2)n} X)^2}{\mu_k^n} \leq& \frac{8c^2\varphi(2m_n)}{p_n^{2\gamma}} \max_{1\leq k\leq \tau_n+2} \sigma_X^2(t_{k-1}^n, t_k^n) \\ \leq& \frac{32c^2\varphi(2m_n)}{p_n^{2\gamma}} \sup_{1\leq k\leq \tau_n+2} \sigma_X^2(0, t_k^n). \end{aligned}$$

Since

$$t_k^n \le (\tau_n + 2)m_n \le 3\,\varphi(2m_n)$$

for $L(2m_n) \ge 1$, it follows that condition (H_1) implies that

$$\frac{32c^2\varphi(2m_n)}{p_n^{2\gamma}} \sup_{1\le k\le \tau_n+2} \sigma_X^2(0,t_k^n) = \frac{\varphi(2m_n)}{p_n^{2\gamma}} O\big((\varphi(2m_n))^{2\gamma}\big) = \varphi(2m_n)\,O\big(L^{2\gamma}(2m_n)\big)$$

as $m_n \downarrow 0$. We obtain from the properties of the function φ that the right hand side of the above inequality tends to zero as $m_n \downarrow 0$.

Next, since the function φ is increasing and $(\ell_k)_{k\ge1} \subset [c^{-1}, c]$, $c \ge 1$, (see assumption 1 of Definition 3.25), we get that the second term of equality (6.22) can be estimated as follows

$$\sum_{k=\tau_n+2}^{N_n-1} \Delta_{n,k+1}t \cdot J_k$$

$$\le \max_{\tau_n+1\le k\le N_n-1} \left|c(t_k^n, t_{k+1}^n)\right| \sum_{k=\tau_n+2}^{N_n-1} \Delta_{n,k+1}t[(\ell_k^n)^{1/2-\gamma} + (\ell_k^n)^{\gamma-1/2}]$$

$$+ \max_{\tau_n+2\le k\le N_n-1} \left|c(t_{k-1}^n, t_{k+1}^n)\right| \sum_{k=\tau_n+2}^{N_n-1} \Delta_{n,k+1}t(\ell_k^n)^{1/2-\gamma}(1+\ell_k^n)^{2\gamma-1}$$

$$\le T \sup_{\varphi(m_n)\le s\le T-m_n} \sup_{0<h\le m_n} \left|c(s,s+h)\right| \max_{1\le k\le N_n-1} [(\ell_k^n)^{1/2-\gamma} + (\ell_k^n)^{\gamma-1/2}]$$

$$+ T \sup_{\varphi(2m_n)\le s\le T-2m_n} \sup_{0<h\le 2m_n} \left|c(s,s+h)\right| \max_{1\le k\le N_n-1} (\ell_k^n)^{1/2-\gamma}(1+\ell_k^n)^{2\gamma-1}$$

$$\le Tc^{1/2}\big[\Lambda^2(m_n) + 2\Lambda(m_n)\big] + 2Tc^{3/2}\big[\Lambda(2m_n) + 2\Lambda(2m_n)\big].$$

Thus, the second term of equality (6.22) tends to zero as $n \to \infty$.

It remains to investigate asymptotic behavior of the third term of equality (6.22). If the function h is invariant on $\mathcal{L}$, then

$$2\kappa^2 \sum_{k=\tau_n+2}^{N_n-1} \Delta_{k+1}^n t \cdot h(\ell_k^n) = 2\kappa^2 \sum_{k=\tau_n+2}^{N_n-1} \Delta_{k+1}^n t \cdot [h(\ell_k^n) - h(\ell_k)]$$

$$+ 2\kappa^2 Th(\ell_1) - 2\kappa^2 h(\ell_1) \sum_{k=0}^{\tau_n+1} \Delta_{k+1}^n t.$$

Since the derivative of the function h is bounded on $[c^{-1}, c]$ by $c^{3/2}(1+c)$, it follows that

$$\sum_{k=\tau_n+2}^{N_n-1} \Delta_{k+1}^n t \cdot |h(\ell_k^n) - h(\ell_k)| \le c^{3/2}(1+c)T \max_{1\le k\le N_n-1} |\ell_k^n - \ell_k| .$$

Thus,

$$2\kappa^2 \sum_{k=\tau_n+2}^{N_n-1} \Delta_{k+1}^n t \cdot h(\ell_k^n) \longrightarrow 2\kappa^2 h(\ell_1)T \quad \text{as } n\to\infty$$

by assumption 2 of Definition 3.25 and the inequality (6.24).

Assume that the sequence of step functions $\ell_n(t)$ converges uniformly to $\ell(t)$ on the interval $[0, T]$. Then $\ell_n(t), \ell(t) \in [c^{-1}, c]$ and

$$|h(\ell_n(t)) - h(\ell(t))| \le c^{3/2}(1+c) \sup_{0\le t\le T} |\ell_n(t) - \ell(t)| \longrightarrow 0 \quad \text{as } n\to\infty,$$

i.e. the sequence $h(\ell_n(t))$ converges uniformly to $h(\ell(t))$ on $[0, T]$ and $h(\ell(t))$ is a regulated function. Thus,

$$\begin{aligned} 2\kappa^2 \sum_{k=\tau_n+2}^{N_n-1} \Delta_{k+1}^n t \cdot h(\ell_k^n) =& 2\kappa^2 \sum_{k=\tau_n+2}^{N_n-1} \Delta_{k+1}^n t \cdot [h(\ell_k^n) - h(\ell_k)] + 2\kappa^2 \int_0^T h(\ell_n(s))\, ds \\ & - 2\kappa^2 \sum_{k=0}^{\tau_n+1} \Delta_{k+1}^n t \cdot h(\ell_k) \longrightarrow 2\kappa^2 \int_0^T h(\ell(s))\, ds \end{aligned}$$

since regulated functions are Riemann integrable and $h(\ell_k) \le c^{1/2}(1+c)$ for all $k \ge 1$.

Consequently, we obtain that

$$\mathsf{E}V_{\pi_n,T}^{(2)X} \longrightarrow \begin{cases} 2\kappa^2 h(\ell_1)T, & \text{if } h \text{ is invariant on } \mathcal{L}, \\ 2\kappa^2 \int_0^T h(\ell(s))\, \mathrm{d}s, & \text{if } \ell(s) \text{ is a regulated function.} \end{cases}$$

□

Corollary 6.9 *Let* $(\pi_n)_{n\in\mathbb{N}}$ *be a sequence of regular partitions of the interval* $[0, T]$, $T > 0$, *and let* $X = \{X(t) : t \in [0, T]\}$, $T > 0$, *be a zero-mean Gaussian process satisfying conditions* (H_1) *and* (H_2). *Then*

$$\mathsf{E}V_{N_n,T}^{(2)X} = (T^{-1}N_n)^{2\gamma-1}\mathsf{E} \sum_{k=1}^{N_n-1} \left(\Delta_{n,k}^{(2)} X\right)^2 \longrightarrow \kappa^2(4 - 2^{2\gamma})T \qquad \text{as } n\to\infty.$$

Proof For regular subdivision we have that $\ell_k = 1$. Thus $h(\ell_1) = 2 - 2^{2\gamma-1}$ and the statement of the corollary follows immediately from Proposition 6.8. □

Now we formulate a slightly more general version of Corollary 6.9.

Proposition 6.10 *Let $(\pi_n)_{n\in\mathbb{N}}$ be a sequence of regular partitions of the interval $[0,T]$, $T>0$, and let $X = \{X(t) : t \in [0,T]\}$, $T>0$, be a zero-mean Gaussian process. Assume that condition* (H$_1$) *is fulfilled for some constant $\gamma \in (0,1)$ and there exists a continuous bounded function $g_0 : (0,T) \to \mathbb{R}$ such that*

$$\lim_{h\to 0+} \sup_{\varphi(h)\le t\le T-h} \left| \frac{\mathsf{E}\big(X_{t+h} - 2X_t + X_{t-h}\big)^2}{h^{2\gamma}} - g_0(t) \right| = 0. \tag{6.25}$$

Then

$$\mathsf{E} V^{(2)X}_{N_n,T} \longrightarrow \int_0^T g_0(s)\,ds \qquad \text{as } n \to \infty.$$

Proof Note that

$$\begin{aligned}
&\left| \mathsf{E} V^{(2)X}_{N_n,T} - \int_0^T g_0(s)\,ds \right| \\
&\le \left(\frac{T}{N_n}\right)^{1-2\gamma} \sum_{k=1}^{\tau_n+1} \mathsf{E}(\Delta^{(2)}_{n,k}X)^2 + \frac{T}{N_n} \sum_{k=\tau_n+2}^{N_n-1} \left| \frac{\mathsf{E}(\Delta^{(2)}_{n,k}X)^2}{T^{2\gamma}N_n^{-2\gamma}} - g_0\left(\frac{kT}{N_n}\right) \right| \\
&\quad + \left| \int_0^t g_0(s)\,ds - \frac{T}{N_n} \sum_{k=\tau_n+2}^{N_n-1} g_0\left(\frac{kT}{N_n}\right) \right|,
\end{aligned} \tag{6.26}$$

where $\tau_n = \lfloor L(TN_n^{-1}) \rfloor$. From condition (H$_1$) it follows that

$$\begin{aligned}
\left(\frac{T}{N_n}\right)^{1-2\gamma} \sum_{k=1}^{\tau_n+1} \mathsf{E}(\Delta^{(2)}_{n,k}X)^2 &\le 4T\left(\frac{T}{N_n}\right)^{-2\gamma} \varphi\left(\frac{T}{N_n}\right) \max_{1\le k\le \tau_n+1} \sigma^2(t^n_{k-1}, t^n_k) \\
&= \left(\frac{T}{N_n}\right)^{-2\gamma} \varphi\left(\frac{T}{N_n}\right) O\left(\left(\varphi\left(\frac{T}{N_n}\right)\right)^{2\gamma}\right)
\end{aligned}$$

and the first term in inequality (6.26) tends to zero as $n \to \infty$.

Assumption (6.25) yields

$$\begin{aligned}
&\max_{\tau_n+2\le k\le N_n-1} \left| \frac{\mathsf{E}(\Delta^{(2)}_{n,k}X)^2}{T^{2\gamma}N_n^{-2\gamma}} - g_0\left(\frac{kT}{N_n}\right) \right| \\
&\le \sup_{\varphi(TN_n^{-1})\le s\le t-TN_n^{-1}} \left| \frac{\mathsf{E}\big(X_{s+TN_n^{-1}} - 2X_s + X_{s-TN_n^{-1}}\big)^2}{(TN_n^{-1})^{2\gamma}} - g_0(t) \right| \xrightarrow[n\to\infty]{} 0.
\end{aligned}$$

The third term of the right hand side of (6.26) also converges towards 0 as $n \to \infty$ that is a consequence of classical results for Riemann sums and inequality

$$\frac{T}{N_n}\sum_{k=1}^{N_n-1}\left|g_0\left(\frac{kT}{N_n}\right)\right| \le T \sup_{0\le t\le T}|g_0(t)|\,\varphi\left(\frac{T}{N_n}\right). \qquad \square$$

Remark 6.11 Generally speaking, the function $\varphi(h)$ could not be replaced by h in assumption (6.25). Indeed, let us consider sfBm X with $H \neq 1/2$. Observe that following equality

$$\begin{aligned}\mathsf{E}\big(X_{t+h} - 2X_t + X_{t-h}\big)^2 =& (4 - 2^{2H})h^{2H} - 2^{2H-1}(t+h)^{2H} - 3\cdot 2^{2H}t^{2H}\\ &- 2^{2H-1}(t-h)^{2H} + 2(2t+h)^{2H} + 2(2t-h)^{2H}\end{aligned}$$

holds. Set

$$\begin{aligned}\lambda_t(h) := \mathsf{E}\big(X_{t+h} - 2X_t + X_{t-h}\big)^2 - (4-2^{2H})h^{2H} &= -2^{2H-1}(t+h)^{2H} - 3\cdot 2^{2H}t^{2H}\\ - 2^{2H-1}(t-h)^{2H} + 2(2t+h)^{2H} + 2(2t-h)^{2H}.\end{aligned}$$

Then

$$\begin{aligned}\sup_{h\le t\le T-h}\left|\frac{\mathsf{E}\big(X_{t+h} - 2X_t + X_{t-h}\big)^2}{h^{2H}} - (4-2^{2H})\right| &= \sup_{h\le t\le T-h}\left|\frac{\lambda_t(h)}{h^{2H}}\right| \ge \left|\frac{\lambda_h(h)}{h^{2H}}\right|\\ = \left|\frac{-2^{2H-1}(2h)^{2H} - 3\cdot 2^{2H}h^{2H} + 2(3h)^{2H} + 2h^{2H}}{h^{2H}}\right|&\\ = \left|-2^{4H-1} - 3\cdot 2^{2H} + 2\cdot 3^{2H} + 2\right| > 0.&\end{aligned}$$

On the other hand, assumption (6.25) is satisfied for sfBm. Note that $\lambda_t(0) = \lambda_t'(0) = \lambda_t''(0) = \lambda_t^{(3)}(0) = 0$. The Taylor formula yields

$$\lambda_t(h) = \int_0^h \frac{(h-x)^3}{3!}\,\lambda_t^{(4)}(x)\,dx, \qquad \forall\, h \le t \le T-h,$$

where

$$\begin{aligned}\lambda_t^{(4)}(x) =& C_H\Big(2\big[(2t+x)^{2H-4} + (2t-x)^{2H-4}\big] - 2^{2H-1}\big[(t+x)^{2H-4} + (t-x)^{2H-4}\big]\Big),\\ C_H =& 2H(2H-1)(2H-2)(2H-3).\end{aligned}$$

From inequality

$$\sup_{\varphi(h)\le t\le T-h}\left|\frac{\mathsf{E}\big(X_{t+h}-2X_t+X_{t-h}\big)^2}{h^{2H}}-(4-2^{2H})\right|$$

$$\le h^{-2H}\sup_{\varphi(h)\le t\le T-h}\sup_{0\le x\le h}|\lambda_t^{(4)}(x)|\int_0^h (h-x)^3\,dx$$

$$\le |C_H|\cdot h^{4-2H}\sup_{\varphi(h)\le t\le T-h}\left(\frac{2}{(2t)^{4-2H}}+\frac{2}{(2t-h)^{4-2H}}+\frac{2^{2H-1}}{t^{4-2H}}+\frac{2^{2H-1}}{(t-h)^{4-2H}}\right)$$

$$\le |C_H|\cdot h^{4-2H}\left[\frac{2}{(2\varphi(h))^{4-2H}}+\frac{2}{(2\varphi(h)-h)^{4-2H}}+\frac{2^{2H-1}}{\varphi(h)^{4-2H}}+\frac{2^{2H-1}}{(\varphi(h)-h)^{4-2H}}\right]$$

$$\le |C_H|\left[\left(\frac{h}{\varphi(h)}\right)^{4-2H}+\frac{2}{(2L(h)-1)^{4-2H}}+2^{2H-1}\left(\frac{h}{\varphi(h)}\right)^{4-2H}\right.$$

$$\left.+\frac{2^{2H-1}}{(L(h)-1)^{4-2H}}\right]$$

we obtain the required assertion.

Theorem 6.12 *Assume that conditions of Proposition 6.8 are satisfied and the partition π_n is such that $p_n = o(\log^{-1} n)$. Moreover assume that X is a Gaussian process with the Orey index γ and*

$$\max_{1\le k\le N_n-1}\sum_{j=1}^{N_n-1}|d_{jk}^{(2)n}|\le Cp_n^{2+2\gamma},\tag{6.27}$$

for some constant C and any sequence of partitions (π_n) of the interval $[0,T]$, where $d_{jk}^{(2)n} = \mathsf{E}(\Delta_{\mathrm{ir},j}^{(2)n}X\Delta_{\mathrm{ir},k}^{(2)n}X)$, $1\le j,k\le n$. Then

$$V_{\pi_n,T}^{(2)X}\xrightarrow{a.s.}\begin{cases}2\kappa^2 h(\ell_1)T, & \text{if } h \text{ is invariant on } \mathcal{L},\\ 2\kappa^2\int_0^T h(\ell(s))\,\mathrm{d}s, & \text{if } \ell(s) \text{ is a regulated function.}\end{cases}$$

Proof The proof of the theorem follows the outlines of the proof of Theorem 4 in [11].

We can consider $V_{\pi_n,T}^{(2)X}$ as the square of the Euclidean norm of the (N_n-1)-dimensional Gaussian vector X_n with the components

$$\sqrt{\frac{2\Delta_{n,k+1}t}{\mu_k}}\,\Delta_{\mathrm{ir},k}^{(2)n}X,\qquad 1\le k\le N_n-1.$$

So by the classical Cochran theorem, one can find r_n nonnegative real numbers $(\lambda_{1,n}, \dots, \lambda_{r_n,n})$ and r_n-dimensional Gaussian vector Y_n with independent Gaussian $\mathcal{N}(0,1)$-components, satisfying the equality

$$V^{(2)X}_{\pi_n,T} = \sum_{j=1}^{r_n} \lambda_{j,n} \big(Y_n^{(j)} \big)^2.$$

The numbers $(\lambda_{1,n}, \dots, \lambda_{r_n,n})$ are the eigenvalues of the symmetric $(N_n - 1) \times (N_n - 1)$-matrix

$$\left(2 \sqrt{\frac{\Delta_{n,j+1}t \Delta_{n,k+1}t}{\mu_j \mu_k}} \, \mathsf{E}\big(\Delta^{(2)n}_{\mathrm{ir},j} X \Delta^{(2)n}_{\mathrm{ir},k} X \big) \right)_{1 \le j,k \le N_n - 1}.$$

Now we can apply Hanson and Wright's inequality (see Hanson et al. [68] or Bégyn [11]), and it yields that for $\varepsilon > 0$

$$\mathsf{P}\left(\left| \sum_{j=1}^{r_n} \lambda_{j,n} \big[\big(Y_n^{(j)} \big)^2 - 1 \big] \right| \ge \varepsilon \right) \le 2 \exp\left(-\min\left[\frac{C_1 \varepsilon}{\lambda_n^*}, \frac{C_2 \varepsilon^2}{\sum_{j=1}^{r_n} \lambda_{j,n}^2} \right] \right), \tag{6.28}$$

where C_1, C_2 are nonnegative constants, $\lambda_n^* = \max_{1 \le j \le r_n} \lambda_{j,n}$.

Evidently,

$$\sum_{j=1}^{r_n} \lambda_{j,n} = \mathsf{E} V^{(2)X}_{\pi_n,T}.$$

Furthermore, it follows from (6.21) that the sequence $(\mathsf{E} V^{(2)X}_{\pi_n,T})$ is bounded. So, the sums $\sum_{j=1}^{r_n} \lambda_{j,n}$ are bounded as well. It is easy to check that

$$\sum_{j=1}^{r_n} (\lambda_{j,n})^2 \le \lambda_n^* \sum_{j=1}^{r_n} \lambda_{j,n}.$$

Therefore for any $0 < \varepsilon \le 1$ inequality (6.28) can be rewritten as

$$\mathsf{P}\big(\big| V^{(2)X}_{\pi_n,T} - \mathsf{E} V^{(2)X}_{\pi_n,T} \big| \ge \varepsilon \big) \le 2 \exp\left(-\frac{K \varepsilon^2}{\lambda_n^*} \right), \tag{6.29}$$

where K is a positive constant.

We use the following elementary result of linear algebra. Let $S = (s_{ij})_{1 \le i,j \le n}$ be a $n \times n$ symmetric matrix. We denote by $\lambda_{\max}$ its highest eigenvalue. Then one has

$$\lambda_{\max} \le \max_{1 \le j \le n} \sum_{i=1}^{n} |s_{ij}|.$$

This fact with inequality (6.23) yield

$$\lambda_n^* \le 2 \max_{1\le k\le N_n-1} \sum_{j=1}^{N_n-1} \sqrt{\frac{\Delta_{n,j+1}t\Delta_{n,k+1}t}{\mu_j\mu_k}} \left|\mathsf{E}(\Delta_{\mathrm{ir},j}^{(2)n}X\Delta_{\mathrm{ir},k}^{(2)n}X)\right|$$

$$\le \frac{m_n}{p_n^{2\gamma+2}} \max_{1\le k\le N_n-1} \sum_{j=1}^{N_n-1} |d_{jk}^{(2)n}| \le Cm_n.$$

Set

$$\varepsilon_n^2 = \frac{2C}{K} m_n \log n.$$

From (6.29) we conclude

$$\mathsf{P}\left(|V_{\pi_n,T}^{(2)X} - \mathsf{E}V_{\pi_n,T}^{(2)X}| \ge \varepsilon_n\right) \le 2\exp\left(-2\log n\right) = \frac{2}{n^2}.$$

It means that

$$\sum_{n=1}^{\infty} \mathsf{P}\left(|V_{\pi_n,T}^{(2)X} - \mathsf{E}V_{\pi_n,T}^{(2)X}| \ge \varepsilon_n\right) < \infty.$$

So, we get the statement of the theorem from the Borel–Cantelli lemma. □

Corollary 6.13 *Let $(\pi_n)_{n\in\mathbb{N}}$ be a sequence of regular partitions of the interval $[0,T]$, $T>0$. Assume that X is a Gaussian process satisfying conditions* (H_1) *and* (H_2) *and having the Orey index γ. Moreover, assume that*

$$\max_{1\le k\le N_n-1} \sum_{j=1}^{N_n-1} |d_{jk}^{(2)n}| \le C\left(\frac{T}{N_n}\right)^{2\gamma} \tag{6.30}$$

for some constant C, and any sequence of partitions (π_n) of the interval $[0,T]$, where $d_{jk}^{(2)n} = \mathsf{E}(\Delta_{n,j}^{(2)}X\Delta_{n,k}^{(2)}X)$, $1\le j,k\le N_n-1$. Then

$$V_{N_n,T}^{(2)X} \xrightarrow{a.s.} \kappa^2(4-2^{2\gamma})T \qquad \text{as } n\to\infty.$$

Proof For a regular partition π_n condition (6.27) transforms to (6.30). □

Theorem 6.14 *Assume that conditions of Proposition 6.10 are satisfied. Moreover, assume that inequality (6.30) holds, then*

$$V_{N_n,T}^{(2)X} \xrightarrow{a.s.} \int_0^T g_0(t)\,dt \qquad \text{as } n\to\infty.$$

Proof The proof of the theorem evidently follows from Proposition 6.10 and arguments used to prove Theorem 6.12. □

6.2.1 Examples

6.2.1.1 Fractional Brownian Bridge

We recall that conditions (H_1) and (H_2) are satisfied for fractional Brownian bridge (see Sect. 6.1.2). So the statement of Theorem 6.12 is satisfied if inequality (6.27) holds.

First of all we prove inequality (6.27) for fBm B^H. Following the outlines of the proof of Theorem 4 of Bégyn [11], we divide the study of the asymptotic properties of $d_{jk}^{(2)n}$ into three steps, according to the value of $k-j$.

If $j = k$ then

$$\begin{aligned} d_{kk}^{(2)n} &\leq 2\big[(\Delta_{n,k}t)^2 \mathsf{E}(\Delta_{n,k+1}B^H)^2 + (\Delta_{n,k+1}t)^2 \mathsf{E}(\Delta_{n,k}B^H)^2\big] \\ &= 2\big[(\Delta_{n,k}t)^2 |t_{k+1}^n - t_k^n|^{2H} + (\Delta_{n,k+1}t)^2 |t_k^n - t_{k-1}^n|^{2H}\big] = 4m_n^{2+2H}. \end{aligned} \tag{6.31}$$

By using the Cauchy-Schwarz inequality we get

$$\left|d_{jk}^{(2)n}\right| \leq \mathsf{E}^{1/2}\left|(\Delta_{\mathrm{ir},j}^{(2)n} B^H)\right|^2 \cdot \mathsf{E}^{1/2}\left|(\Delta_{\mathrm{ir},k}^{(2)n} B^H)\right|^2 \leq 4m_n^{2+2H} \tag{6.32}$$

for $1 \leq k-j \leq 2$ and

$$d_{j1}^{(2)n} \leq 4m_n^{2+2H} \quad \text{for } 1 \leq j \leq N_n - 1, \tag{6.33}$$

$$d_{1k}^{(2)n} \leq 4m_n^{2+2H} \quad \text{for } 1 \leq k \leq N_n - 1. \tag{6.34}$$

Now consider the case $|j-k| \geq 3$. By symmetry of $d_{jk}^{(2)n}$ one can take $j-k \geq 3$. Note that for $j \neq 1$ and $k \neq 1$ equality

$$d_{jk}^{(2)n} = \int_{t_j^n}^{t_{j+1}^n} du \int_{t_{j-1}^n}^{t_j^n} dv \int_v^u dw \int_{t_k^n}^{t_{k+1}^n} dx \int_{t_{k-1}^n}^{t_k^n} dy \int_y^x \frac{\partial^4 R_H}{\partial s^2 \partial t^2}(w,z)\, dz$$

holds, where $R_H(s,t)$ is covariance function of fBm. The fourth order mixed partial derivative of the covariance function $R_H(s,t)$ is of the following form

$$\frac{\partial^4 R_H}{\partial s^2 \partial t^2}(s,t) = -\frac{C_H}{|s-t|^{2(2-H)}} \qquad \forall s,t \in [0,T]^2 \backslash \{s=t\},$$

where $C_H = H(2H-1)(2H-2)(2H-3)$. By inequality

$$|w-z| \geq t_{j-1}^n - t_{k+1}^n = \sum_{i=k+2}^{j-1} \Delta_{n,i}t \geq (j-k-2)p_n$$

on the integration set implies

$$|d_{jk}^{(2)n}| \leq \frac{4|C_H|m_n^6}{(j-k-2)^{2(2-H)}p_n^{2(2-H)}} \leq \frac{4|C_H|c^6 p_n^{2+2H}}{(j-k-2)^{2(2-H)}},$$

where c is a constant from Definition 3.25. Moreover,

$$\sum_{j-k\geq 3}^{n-1} \frac{1}{(j-k-2)^{2(2-H)}} \leq \sum_{j=1}^{\infty} \frac{1}{j^{2(2-H)}} < \infty.$$

Thus, we have inequality (6.27) for fBm and the statement of Theorem 6.12 is satisfied for fBm.

Now we prove the statement of Theorem 6.12 for the fractional Brownian bridge. For brevity, we rewrite the fractional Brownian bridge X_t^H given by (6.17) as follows:

$$X_t^H = B_t^H - g(t,T),$$

where

$$g(t,T) = \frac{t^{2H} + T^{2H} - |t-T|^{2H}}{2T^{2H}} B_T^H.$$

It is evident that

$$\Delta_{\text{ir},k}^{(2)n} X^H = \Delta_{\text{ir},k}^{(2)n} B^H - \Delta_{\text{ir},k}^{(2)n} g(\cdot,T),$$

where

$$\begin{aligned}\Delta_{\text{ir},k}^{(2)n} g(\cdot,T) =& \Delta_{n,k}t \frac{((t_{k+1}^n)^{2H} - (t_k^n)^{2H}) - (|T-t_{k+1}^n|^{2H} - |T-t_k^n|^{2H})}{2T^{2H}} B_T^H \\ &- \Delta_{n,k+1}t \frac{((t_k^n)^{2H} - (t_{k-1}^n)^{2H}) - (|T-t_k^n|^{2H} - |T-t_{k-1}^n|^{2H})}{2T^{2H}} B_T^H.\end{aligned}$$

Since

$$|\Delta_{\text{ir},k}^{(2)n} g(\cdot,T)| \leq \frac{4m_n^{1+2H}}{2T^{2H}} 2^{(2H-1)\vee 0} |B_T^H| \leq \frac{4m_n^{1+2H}}{T^{2H}} |B_T^H|,$$

we have

$$V_{\pi_n,T}^{(2)g(\cdot,T)} \le 2T \frac{16m_n^{2+4H}}{T^{4H} 2p_n^{2+2H}} |B_T^H| = 16T^{1-4H} c^{2+2H} m_n^{2H} |B_T^H| .$$

From the obtained inequalities and definition of $V_{\pi_n,T}^{(2)\bullet}$ we conclude that

$$\begin{aligned}
&\left|V_{\pi_n,T}^{(2)X^H} - V_{\pi_n,T}^{(2)B^H}\right| \\
&= \left|V_{\pi_n,T}^{(2)g(\cdot,T)} - 4\sum_{k=1}^{N_n-1} \frac{\Delta_{n,k+1}t\Delta_{\mathrm{ir},k}^{(2)n}B^H\Delta_{\mathrm{ir},k}^{(2)n}g(\cdot,T)}{(\Delta_{n,k}t)^{H+1/2}(\Delta_{n,k+1}t)^{H+1/2}[\Delta_{n,k}t+\Delta_{n,k+1}t]}\right| \\
&\le V_{\pi_n,T}^{(2)g(\cdot,T)} + 2\frac{4TG_{\varepsilon,T}m_n^{1+H-\varepsilon}m_n^{1+2H}}{T^{2H}p_n^{2+2H}}|B_T^H| \\
&\le V_{\pi_n,T}^{(2)g(\cdot,T)} + 8c^{2+2H}T^{1-2H}G_{\varepsilon,T}m_n^{H-\varepsilon}|B_T^H| = O_\omega(m_n^{H-\varepsilon})
\end{aligned}$$

for $0 < \varepsilon < H$.

Thus, the limits of $V_{\pi_n,T}^{(2)X^H}$ and $V_{\pi_n,T}^{(2)B^H}$ are the same, where X^H is the fractional Brownian bridge.

6.2.1.2 Fractional Ornstein-Uhlenbeck Process

We shall prove that the conditions of Theorem 6.12 are satisfied for fO-U process X which is the unique solution of the following SDE

$$X_t = x_0 - \theta \int_0^t X_s\,ds + \sigma B_t^H, \qquad t \in [0,T], \tag{6.35}$$

where $x_0 \in \mathbb{R}$, $\theta, \sigma > 0$ are real parameters, and B^H is an fBm with the Hurst index H, $H \in (0,1)$. The fO-U process satisfies conditions (H_1) and (H_2). So it suffices to verify the inequality (6.27).

The outlines of the proof is similar as for the fractional Brownian bridge. It suffices to show that the limits of $V_{\pi_n,T}^{(2)X}$ and $V_{\pi_n,T}^{(2)B^H}$ are the same, where X is the fO-U process, B^H is a fBm.

Lemma 6.15 *Let X be the solution of Eq. (6.35). Then*

$$\left|V_{\pi_n,T}^{(2)X} - \sigma^2 V_{\pi_n,T}^{(2)B^H}\right| = O_\omega(m_n^{1-2\varepsilon})$$

for every $0 < \varepsilon < 1/2 \wedge H$.

Proof It is evident that

$$\Delta_{\mathrm{ir},k}^{(2)n} X = -\theta \left(\Delta_k^n t \int_{t_k^n}^{t_{k+1}^n} X_s \, ds - \Delta_{k+1}^n t \int_{t_{k-1}^n}^{t_k^n} X_s \, ds \right) + \sigma \Delta_{\mathrm{ir},k}^{(2)n} B^H.$$

For simplicity, we denote $X_k = X(t_k^n)$ and $B_k^H = B^H(t_k^n)$. After simple calculations we get the estimate

$$\begin{aligned} \sup_{t_k^n \le s \le t_{k+1}^n} |X_s - X_k| &\le \theta(\Delta_{k+1}^n t) \sup_{t \le T} |X_t| + \sigma \sup_{t_k^n \le s \le t_{k+1}^n} |B_s^H - B_k^H| \\ &\le \theta \, m_n \sup_{t \le T} |X_t| + \sigma G_{\varepsilon,T} m_n^{H-\varepsilon}, \end{aligned}$$

where $G_{\varepsilon,T}$ is defined in Sect. 1.1. Thus,

$$\begin{aligned} &\left(\Delta_{n,k} t \int_{t_k^n}^{t_{k+1}^n} (X_s - X_k) \, ds - \Delta_{n,k+1} t \int_{t_{k-1}^n}^{t_k^n} (X_s - X_k) \, ds \right)^2 \\ &\le 2m_n^3 \int_{t_k^n}^{t_{k+1}^n} (X_s - X_k)^2 \, ds + 2m_n^3 \int_{t_{k-1}^n}^{t_k^n} (X_s - X_k)^2 \, ds \\ &\le 2m_n^4 \Big(\sup_{t_k^n \le s \le t_{k+1}^n} (X_s - X_k)^2 + \sup_{t_{k-1}^n \le s \le t_k^n} (X_k - X_s)^2 \Big) \\ &\le 8m_n^{4+2H-2\varepsilon} \Big(\theta^2 m_n^{2-2H+2\varepsilon} \sup_{t \le T} X_t^2 + \sigma^2 G_{\varepsilon,T}^2 \Big) \end{aligned}$$

and

$$\begin{aligned} &\left| \left(\Delta_{n,k} t \int_{t_k^n}^{t_{k+1}^n} X_s \, ds - \Delta_{n,k+1} t \int_{t_{k-1}^n}^{t_k^n} X_s \, ds \right) \Delta_{\mathrm{ir},k}^{(2)n} B^H \right| \\ &= \left| \left(\Delta_{n,k} t \int_{t_k^n}^{t_{k+1}^n} (X_s - X_k) \, ds - \Delta_{n,k+1} t \int_{t_{k-1}^n}^{t_k^n} (X_s - X_k) \, ds \right) \Delta_{\mathrm{ir},k}^{(2)n} B^H \right| \\ &\le 2m_n^{2+H-\varepsilon} \Big(\theta \, m_n^{1-H+\varepsilon} \sup_{t \le T} |X_t| + \sigma G_T^{H,H-\varepsilon} \Big) \cdot 2m_n G_{\varepsilon,T} m_n^{H-\varepsilon} \\ &= 4m_n^{3+2H-2\varepsilon} \Big(\theta \, m_n^{1-H+\varepsilon} \sup_{t \le T} |X_t| + \sigma G_T^{H,H-\varepsilon} \Big) \cdot G_{\varepsilon,T}. \end{aligned}$$

We get from the obtained inequalities and definition of $V^{(2)\bullet}_{\pi_n,T}$ that

$$\begin{aligned}\left|V^{(2)X}_{\pi_n,T} - \sigma^2 V^{(2)B^H}_{\pi_n,T}\right| \leq & 8c^{2+2H} m_n^{2-2\varepsilon}\left(\theta^2 m_n^{2-2H+2\varepsilon} \sup_{t\leq T} X_t^2 + 2\sigma^2 G^2_{\varepsilon,T}\right)T \\ & + 4c^{2+2H} m_n^{1-2\varepsilon}\left(\theta m_n^{1-H+\varepsilon} \sup_{t\leq T} |X_t| + \sigma G_{\varepsilon,T}\right)\cdot G_{\varepsilon,T}T \\ = & O_\omega(m_n^{1-2\varepsilon}).\end{aligned}$$

□

6.2.1.3 Sub-fractional Brownian Motion

We recall that conditions (H_1) and (H_2) are satisfied for a sfBm (see Sect. 6.1.2.3). So the statement of Theorem 6.12 is satisfied if a inequality (6.27) holds. In order to prove inequality (6.27), we apply similar arguments as for a fBm.

If $j = k$ or $1 \leq j - k \leq 2$ then (1.55) and (1.56) yields

$$d^{(2)n}_{jk} \leq 8m_n^{2+2H}.$$

The same inequality holds for $d^{(2)n}_{j1}$, $1 \leq j \leq N_n - 1$ and $d^{(2)n}_{1k}$, $1 \leq k \leq N_n - 1$.

The fourth order mixed partial derivative of the covariance function $G_H(s,t)$ is of the following form

$$\frac{\partial^4 G_H}{\partial s^2 \partial t^2}(s,t) = -H(2H-1)(2H-2)(2H-3)\left[\frac{1}{|s-t|^{2(2-H)}} + \frac{1}{(s+t)^{2(2-H)}}\right].$$

for each $s, t > 0$ such that $s \neq t$. Note that $(s+t)^{2(2-H)} \geq |s-t|^{2(2-H)}$ if $s \neq t$.

Thus,

$$\left|\frac{\partial^4 G_H}{\partial s^2 \partial t^2}(s,t)\right| \leq \frac{|C_H|}{|s-t|^{2(2-H)}}$$

and

$$|d^{(2)n}_{jk}| \leq \frac{4|C_H| m_n^6}{(j-k-2)^{2(2-H)} p_n^{2(2-H)}} \leq \frac{4|C_H| c^6 p_n^{2+2H}}{(j-k-2)^{2(2-H)}}$$

for $j - k \geq 3$, $2 \leq k \leq N_n - 1$, where $C_H = 2H(2H-1)(2H-2)(2H-3)$, c is a constant from Definition 3.25. So, we have

$$\max_{2\leq k\leq N_n-1} \sum_{j-k\geq 3} |d_{jk}^{(2)n}| \leq 4|C_H|c^6 p_n^{2+2H} \max_{2\leq k\leq N_n-1} \sum_{j-k\geq 3} \frac{1}{(j-k-2)^{2(2-H)}}$$

$$\leq 4|C_H|c^6 p_n^{2+2H} \sum_{j=1}^{\infty} \frac{1}{j^{2(2-H)}} \leq C p_n^{2+2H} \qquad (6.36)$$

for some constant C. This proves (6.27).

6.2.1.4 Bifractional Brownian Motion

We shall prove that the conditions of Theorem 6.12 are satisfied for a bifBm. A bifBm satisfies conditions (H_1) and (H_2). So it suffices to verify the inequality (6.27). To prove inequality (6.27), we apply similar arguments as for fBm.

If $j = k$ or $1 \leq j - k \leq 2$ then (1.57) yields

$$d_{kk}^{(2)n} \leq 2^{3-K} m_n^{2+2HK}.$$

The same inequality holds for $d_{j1}^{(2)n}$, $1 \leq j \leq N_n - 1$ and $d_{1k}^{(2)n}$, $1 \leq k \leq N_n - 1$.

Now consider the case $|j-k| \geq 3$. By symmetry of $d_{jk}^{(2)n}$ one can take $j - k \geq 3$. The fourth order mixed partial derivative of the covariance function $R_{HK}(s,t)$ is of the following form

$$\begin{aligned}
\frac{\partial^4 R_{HK}}{\partial s^2 \partial t^2}(s,t) &= -\frac{2HK(2H-1)(2HK-2)(2HK-3)}{2^K |s-t|^{2(2-KH)}} \\
&+ \frac{K(K-1)(2H)^2(2H-1)^2}{2^K} (st)^{2H-2} \left(s^{2H} + t^{2H}\right)^{K-2} \\
&+ \frac{K(K-1)(K-2)(2H)^3(2H-1)}{2^K} (st)^{2H-2} t^{2H} \left(s^{2H} + t^{2H}\right)^{K-3} \\
&+ \frac{K(K-1)(K-2)(2H)^3(2H-1)}{2^K} (st)^{2H-2} s^{2H} \left(s^{2H} + t^{2H}\right)^{K-3} \\
&+ \frac{K(K-1)(K-2)(K-3)(2H)^4}{2^K} (st)^{4H-2} \left(s^{2H} + t^{2H}\right)^{K-4} \\
&= -\frac{2HK(2H-1)(2HK-2)(2HK-3)}{2^K |s-t|^{2(2-KH)}} \\
&+ \frac{K(K-1)(K-2)(K-3)(2H)^4}{2^K} (st)^{4H-2} \left(s^{2H} + t^{2H}\right)^{K-4}
\end{aligned}$$

$$+\frac{K(K-1)(2H)^2(2H-1)}{2^K}\big[(K-2)(2H)+(2H-1)\big]$$
$$\times (st)^{2H-2}\big(s^{2H}+t^{2H}\big)^{K-2}$$

for each $s,t>0$ such that $s\neq t$. Since $2s^Ht^H\leq s^{2H}+t^{2H}$ and $K-2<0$, $K-4<0$, it follows that

$$(st)^{2H-2}\big(s^{2H}+t^{2H}\big)^{K-2}\leq 2^{K-2}(st)^{KH-2},$$
$$(st)^{4H-2}\big(s^{2H}+t^{2H}\big)^{K-4}\leq 2^{K-4}(st)^{KH-2}.$$

Thus,

$$\left|\frac{\partial^4 R^{HK}}{\partial s^2\partial t^2}(s,t)\right|\leq\frac{C_1}{|s-t|^{2(2-KH)}}+\frac{C_2}{(st)^{2-KH}}$$

and

$$|d_{jk}^{(2)n}|\leq\int_{t_j^n}^{t_{j+1}^n}du\int_{t_{j-1}^n}^{t_j^n}dv\int_v^u dw\int_{t_k^n}^{t_{k+1}^n}dx\int_{t_{k-1}^n}^{t_k^n}dy\int_y^x\frac{C_1}{|w-z|^{2(2-KH)}}dz$$
$$+\int_{t_j^n}^{t_{j+1}^n}du\int_{t_{j-1}^n}^{t_j^n}dv\int_v^u dw\int_{t_k^n}^{t_{k+1}^n}dx\int_{t_{k-1}^n}^{t_k^n}dy\int_y^x\frac{C_2}{(wz)^{2-KH}}dz$$
$$=:I_{jk}^{n,1}+I_{jk}^{n,2},\tag{6.37}$$

where constants C_1 and C_2 depend on H and K. Inequality

$$|w-z|\geq t_{j-1}^n-t_{k+1}^n=\sum_{i=k+2}^{j-1}\Delta_{n,i}t\geq(j-k-2)p_n$$

on the integration set implies

$$I_{jk}^{n,1}\leq\frac{4C_1m_n^6}{(j-k-2)^{2(2-HK)}p_n^{2(2-HK)}}\leq\frac{4C_1c^6p_n^{2+2HK}}{(j-k-2)^{2(2-HK)}},\tag{6.38}$$

where c is a constant from Definition 3.25. Moreover,

$$\sum_{j-k\geq 3}^{n-1}\frac{1}{(j-k-2)^{2(2-HK)}}\leq\sum_{j=1}^{\infty}\frac{1}{j^{2(2-KH)}}<\infty.\tag{6.39}$$

Now we estimate $I_{jk}^{n,2}$. By modifying the computations above we similarly find that

$$
\begin{aligned}
I_{jk}^{n,2} \leq & \frac{4C_2 m_n^6}{(t_{j-1}t_{k-1})^{2-KH}} = \frac{4C_2 m_n^6}{(t_{k-1}\sum_{i=k}^{j-1}\Delta_i t + t_{k-1}^2)^{2-KH}} \\
\leq & \frac{4C_2 m_n^6}{p_n^{2-KH}((t_{j-1}-t_{k-1})+t_{k-1})^{2-KH}} \leq \frac{4C_2 c^6 p_n^{4+KH}}{(t_{j-1}-t_{k-1})^{2-KH}} \\
\leq & 4C_2 c^6 \frac{p_n^{2+2KH}}{(j-k)^{2-KH}}. \qquad (6.40)
\end{aligned}
$$

Note that

$$
\sum_{j-k\geq 3}^{N_n-1} \frac{1}{(j-k)^{2-KH}} \leq \sum_{j=1}^{\infty} \frac{1}{j^{2-KH}} < \infty. \qquad (6.41)
$$

The inequality (6.27) follows from inequalities (6.37)–(6.41).

6.3 On the Estimation of the Orey Index for Arbitrary Partition

Let $(\pi_n)_{n\geq 1}$ be a sequence of partitions of $[0, T]$ such that $0 = t_0^n < t_1^n < \cdots < t_{m(n)}^n = T$ for all $n \geq 1$. Assume that we have two sequences of partitions $(\pi_{i(n)})_{n\geq 1}$ and $(\pi_{j(n)})_{n\geq 1}$ of $[0, T]$ such that $\pi_{i(n)} \subset \pi_{j(n)} \subseteq \pi_n$, $i(n) < j(n) \leq m(n)$, for all $n \in \mathbb{N}$, where $\pi_{i(n)} = \{0 = t_0^n < t_{i(1)}^n < t_{i(2)}^n < \cdots < t_{i(n)}^n = T\}$ and $\pi_{j(n)} = \{0 = t_0^n < t_{j(1)}^n < t_{j(2)}^n < \cdots < t_{j(n)}^n = T\}$. Set

$$
\Delta_{n,i(k)}t = t_{i(k)}^n - t_{i(k-1)}^n, \qquad m_{i(n)} = \max_{1\leq k\leq i(n)} \Delta_{n,i(k)}t, \qquad p_{i(n)} = \min_{1\leq k\leq i(n)} \Delta_{n,i(k)}t.
$$

Moreover, assume that $p_{j(n)} \neq m_{i(n)}$ and $m_{i(n)} \leq c p_{i(n)}$, for all $i(n)$, $n \geq 1$, $c \geq 1$. Note that $p_{j(n)} \leq p_{i(n)}$.

Let X be a Gaussian process with the Orey index $\gamma \in (0, 1)$. Set

$$
V_{\pi_{i(n)},T}^{(2)X} = 2\sum_{k=1}^{i(n)-1} \frac{\Delta_{n,i(k+1)}t(\Delta_{\mathrm{ir},i(k)}^{(2)n}X)^2}{(\Delta_{n,i(k)}t)^{\gamma+1/2}(\Delta_{n,i(k+1)}t)^{\gamma+1/2}[\Delta_{n,i(k)}t + \Delta_{n,i(k+1)}t]},
$$

where

$$
\Delta_{\mathrm{ir},i(k)}^{(2)n}X = \Delta_{n,i(k)}t \cdot X(t_{i(k+1)}^n) + \Delta_{n,i(k+1)}t \cdot X(t_{i(k-1)}^n) - (\Delta_{n,i(k)}t + \Delta_{n,i(k+1)}t)X(t_{i(k)}^n).
$$

Denote

$$V_{i(n)}^{(2)X} = \sum_{k=1}^{i(n)-1} (\Delta_{\mathrm{ir},i(k)}^{(2)n} X)^2 \text{ and } \mu_k^n = (\Delta_{n,i(k)}t)^{\gamma+1/2}(\Delta_{n,i(k)}t)^{\gamma+1/2}[\Delta_{n,i(k)}t + \Delta_{n,i(k+1)}t].$$

Define

$$\widehat{\gamma}_n = -\frac{1}{2} + \frac{1}{2\log(p_{j(n)}/m_{i(n)})} \log \frac{V_{j(n)}^{(2)X}}{V_{i(n)}^{(2)X}}.$$

Theorem 6.16 *Assume that the conditions of Theorem 6.12 are satisfied for two sequences of partitions* $(\pi_{i(n)})_{n\geq 1}$ *and* $(\pi_{j(n)})_{n\geq 1}$ *of* $[0, T]$ *with the properties mentioned above. Then*

$$V_{\pi_{k(n)},T}^{(2)X} \xrightarrow{a.s.} \begin{cases} 2\kappa^2 h(\ell_1)T, & \text{if } h \text{ is invariant on } \mathcal{L}, \\ 2\kappa^2 \int_0^T h(\ell(s))\,\mathrm{d}s, & \text{if } \ell(s) \text{ is a regulated function} \end{cases} \tag{6.42}$$

for $k(n) = i(n)$ *and for* $k(n) = j(n)$. *If sequences of partitions* $\{\pi_{i(n)}\}$ *and* $\{\pi_{j(n)}\}$, $i(n) < j(n)$, *are such that* $p_{j(n)}/p_{i(n)} \to 0$ *as* $n \to \infty$, *then*

$$\widehat{\gamma}_n \xrightarrow{a.s.} \gamma.$$

Proof of the theorem is the same as for Theorem 3.30.

6.4 Exact Confidence Intervals of the Extended Orey Index for Gaussian Processes

In the papers of Breton et al. [31] and Breton and Coeurjolly [30], an exact (non-asymptotic) confidence interval for the Hurst index of a fBm was derived with the aid of concentration inequalities for quadratic forms of a Gaussian process. The obtained confidence intervals for the Hurst parameter were based on a single observation of a discretized sample path of the interval $[0, 1]$ of a fBm. Exact confidence intervals for a sub-fractional Brownian motion were considered in [113] but were not sufficiently precise.

In this section exact confidence intervals for the Orey index of Gaussian processes are obtained using concentration inequalities for Gaussian quadratic forms and discrete observations of the underlying process. The obtained result is applied for fBm, sfBm and the fractional Ornstein-Uhlenbeck process.

6.4.1 Confidence Intervals

Breton and Coeurjolly [30] slightly improve the bounds of the concentration inequality obtained by Nourdin and Viens [137] (see Theorem 2.4). This allows them to improve the result of Breton et al. [31].

Recall the definitions used in Sect. 2.1.1. Consider a finite centered Gaussian family $X = \{X_k : k = 1, \ldots, M\}$, and write $d_{kj} = \mathsf{E}(X_k X_j)$. Define two quadratic forms associated with X and with some real coefficient c:

$$Q_1(c, X) = c \sum_{k=1}^{M} (X_k^2 - d_{kk}), \qquad Q_2(c, X) = 2c^2 \sum_{k,j=1}^{M} X_k X_j d_{kj}. \tag{6.43}$$

The following statement characterizes the tail behavior of $Q_1(c, X)$.

Theorem 6.17 ([30, 31]) *Suppose that $Q_1(c, X)$ is not a.s. zero and fix $\alpha \geq 0$ and $\beta > 0$. Assume that $Q_2(c, X) \leq \alpha Q_1(c, X) + \beta$, a.s.-P. Then, for all $z > 0$, we have*

$$\mathsf{P}(Q_1(c, X) \geq z) \leq \varphi_r(z; \alpha, \beta) := e^{-z/\alpha}\left(1 + \frac{\alpha z}{\beta}\right)^{\beta/\alpha^2}$$

$$\mathsf{P}(Q_1(c, X) \leq -z) \leq \varphi_l(z; \alpha, \beta) := e^{z/\alpha}\left(1 - \frac{\alpha z}{\beta}\right)^{\beta/\alpha^2} \mathbf{1}_{[0, \beta/\alpha]}(z).$$

Remark 6.18 ([30]) Note that $\varphi_r(\cdot; \alpha, \beta)$ (resp. $\varphi_l(\cdot; \alpha, \beta)$) is a bijective function from $(0, +\infty)$ (resp. $(0, \beta/\alpha)$) to $(0, 1)$.

Next, we apply the obtained concentration inequality to second order quadratic variations.

Let X be a centered Gaussian process satisfying conditions (H_1) and (H_2) with the Orey index $\gamma_X \in (0, 1)$. For the sake of simplicity we will omit the index X in notation γ_X. Denote $Y_n = (Y_{1,n}, \ldots, Y_{n-1,n})$,

$$Y_{k,n} = \frac{n^\gamma}{T^\gamma \varkappa \sqrt{4 - 2^{2\gamma}}} \Delta_{n,k}^{(2)} X, \qquad \Delta_{n,k}^{(2)} X = X(t_{k+1}^n) - 2X(t_k^n) + X(t_{k-1}^n),$$

$t_k^n = \frac{kT}{n}$ and κ is a constant defined in condition (H_2). Set $d_{jk}^Y = \mathsf{E} Y_{j,n} Y_{k,n}$.

Proposition 6.19 *Assume that*

$$\varepsilon_n := \left| \frac{1}{n-1} \sum_{k=1}^{n-1} d_{kk}^Y - 1 \right| \longrightarrow 0 \qquad as\ n \to \infty.$$

Then for all $z > 0$ we have

$$\mathsf{P}\left(\frac{1}{\sqrt{n-1}}\sum_{k=1}^{n-1}\left[\left(Y_{k,n}\right)^2 - d_{kk}^Y\right] \geq z\right) \leq \varphi_{r,n}(z; \nu_n, \varepsilon_n),$$

$$\mathsf{P}\left(\frac{1}{\sqrt{n-1}}\sum_{k=1}^{n-1}\left[\left(Y_{k,n}\right)^2 - d_{kk}^Y\right] \leq -z\right) \leq \varphi_{l,n}(z; \nu_n, \varepsilon_n),$$

where

$$\varphi_{r,n}(z; \nu_n, \varepsilon_n) := \exp\left\{-\frac{z\sqrt{n-1}}{2\nu_n}\right\}\left(1 + \frac{z}{(\varepsilon_n+1)\sqrt{n-1}}\right)^{\frac{(\varepsilon_n+1)(n-1)}{2\nu_n}} \tag{6.44}$$

$$\begin{aligned}\varphi_{l,n}(z; \nu_n, \varepsilon_n) := \exp\left\{\frac{z\sqrt{n-1}}{2\nu_n}\right\}\left(1 - \frac{z}{(\varepsilon_n+1)\sqrt{n-1}}\right)^{\frac{(\varepsilon_n+1)(n-1)}{2\nu_n}} \\ \times \mathbf{1}_{[0,(\varepsilon_n+1)\sqrt{n-1}]}(z),\end{aligned} \tag{6.45}$$

$$\nu_n = \max_{1\leq k\leq n-1}\sum_{j=1}^{n-1}|d_{kj}^Y|.$$

Proof of lemma immediately follows from Lemma 2.5 and Theorem 6.17.

Remark 6.20 Since $\varphi_{l,n}(z; \cdot, \varepsilon_n)$ and $\varphi_{r,n}(z; \cdot, \varepsilon_n)$ are non-decreasing, inequalities (6.44) and (6.45) remain true with ν_n replacing υ_n, where $\nu_n \leq \upsilon_n$.

For any $\alpha \in (0, 1)$ and υ_n ($\nu_n \leq \upsilon_n$), denote by $q_{l,n}(\alpha) := \varphi_{l,n}^{-1}(\alpha; \upsilon_n, \varepsilon_n)$ and $q_{r,n}(\alpha) := \varphi_{r,n}^{-1}(\alpha; \upsilon_n, \varepsilon_n)$. For convenience we define

$$x_{l,n-1}(\alpha) := 1 - \frac{q_{l,n-1}(\alpha)}{\sqrt{n-1}} \quad \text{and} \quad x_{r,n-1}(\alpha) := 1 + \frac{q_{r,n-1}(\alpha)}{\sqrt{n-1}}.$$

Note that Remark 6.18 above ensures that for any $\alpha \in (0, 1)$ and for all $n > 1$, we have $x_{l,n-1}(\alpha) > 0$. Set

$$S_n := \frac{1}{n-1}\sum_{k=1}^{n-1}\left(\Delta_{n,k}^{(2)}X\right)^2, \qquad g_{n,T}(x) := 2x\log(n/T) - \log(4 - 2^{2x}), \qquad x \in (0, 1),$$

and

$$\log^* x = \begin{cases} -\infty & \text{if} \quad x = 0, \\ \log x & \text{if} \quad x > 0. \end{cases}$$

The function $g_{n,T}(x)$ is a strictly increasing bijection from $(0, 1)$ to $(-\log 3, +\infty)$ if $n > T$.

Theorem 6.21 *Let $\alpha \in (0, 1)$. Assume that conditions of Proposition 6.19 are satisfied and there are constants υ_n such that $v_n \le \upsilon_n$. Then*

$$\mathsf{P}(\gamma \in [\gamma_n^{inf}(\alpha), \gamma_n^{sup}(\alpha)]) \ge 1 - \alpha,$$

where

$$\gamma_n^{inf}(\alpha) := \max\left(0, g_{n,T}^{-1}\left[\max\left(\log^*\left(\left(\frac{x_{l,n-1}(\alpha/2) - \varepsilon_n}{S_n}\kappa^2\right) \vee 0\right), -\log 3\right)\right]\right),$$

$$\gamma_n^{sup}(\alpha) := \min\left(1, g_{n,T}^{-1}\left[\log\left(\frac{x_{r,n-1}(\alpha/2) + \varepsilon_n}{S_n}\kappa^2\right)\right]\right).$$

Proof Denote

$$Z_n = (n-1)^{-1/2} V_{n,T}^{(2)Y} - \sqrt{n-1} = \sqrt{n-1}\left[(n-1)^{-1} V_{n,T}^{(2)Y} - 1\right],$$

where

$$V_{n,T}^{(2)Y} = \sum_{i=1}^{n-1} (\Delta_{n,k}^{(2)} Y)^2.$$

Then

$$\left\{ -q_{l,n-1}(\alpha/2) \le (n-1)^{-1/2} \sum_{k=1}^{n-1} \left[(Y_{k,n})^2 - d_{kk}^Y \right] \le q_{r,n-1}(\alpha/2) \right\}$$

$$= \left\{ -q_{l,n-1}(\alpha/2) + (n-1)^{-1/2} \sum_{k=1}^{n-1} [d_{kk}^Y - 1] \le Z_n \right.$$

$$\left. \le q_{r,n-1}(\alpha/2) + (n-1)^{-1/2} \sum_{k=1}^{n-1} [d_{kk}^Y - 1] \right\}$$

$$= \left\{ x_{l,n-1}(\alpha/2) + (n-1)^{-1} \sum_{k=1}^{n-1} [d_{kk}^Y - 1] \le \frac{n^{2\gamma}}{T^{2\gamma}\kappa^2(4 - 4^\gamma)} S_n \right.$$

$$\left. \le x_{r,n-1}(\alpha/2) + (n-1)^{-1} \sum_{k=1}^{n-1} [d_{kk}^Y - 1] \right\}$$

$$\subset \left\{ x_{l,n-1}(\alpha/2) - \varepsilon_n \le \frac{n^{2\gamma}}{T^{2\gamma}\kappa^2(4 - 2^{2\gamma})} S_n \le x_{r,n-1}(\alpha/2) + \varepsilon_n \right\}$$

$$
\begin{aligned}
&= \left\{ \left(x_{l,n-1}(\alpha/2) - \varepsilon_n\right) \vee 0 \leq \frac{n^{2\gamma}}{T^{2\gamma}\kappa^2(4-2^{2\gamma})} S_n \leq x_{r,n-1}(\alpha/2) + \varepsilon_n \right\} \\
&= \left\{ \log^* \left(\left(x_{l,n-1}(\alpha/2) - \varepsilon_n\right) \vee 0 \right) - \log S_n + \log \kappa^2 \leq g_{n,T}(\gamma) \right. \\
&\qquad \left. \leq \log(x_{r,n-1}(\alpha/2) + \varepsilon_n) - \log S_n + \log \kappa^2 \right\} \\
&= \left\{ \left[\log^* \left(\frac{(x_{l,n-1}(\alpha/2) - \varepsilon_n) \vee 0}{S_n} \kappa^2 \right) \right] \vee (-\log 3) \leq g_{n,T}(\gamma) \right. \\
&\qquad \left. \leq \log \left(\frac{x_{r,n-1}(\alpha/2) + \varepsilon_n}{S_n} \kappa^2 \right) \right\}.
\end{aligned}
$$

Note that

$$
\mathsf{P}\Big(- q_{l,n-1}(\alpha/2) \leq \frac{1}{\sqrt{n-1}} \sum_{k=1}^{n-1} \big[\left(Y_{k,n} \right)^2 - d_{kk}^{Y} \big] \leq q_{r,n-1}(\alpha/2) \Big) \geq 1 - \alpha.
$$

Thus

$$
\mathsf{P}\big(\gamma \in \left[\gamma_n^{inf}(\alpha), \gamma_n^{sup}(\alpha)\right]\big) \geq 1 - \alpha.
$$

The proof is completed. □

6.4.2 *Examples*

6.4.2.1 Fractional Brownian Motion

The Orey index of a fBm is equal to the Hurst index H. We apply Theorem 6.21 to construct a confidence interval for the Hurst index H. In order to apply Theorem 6.21, it suffices to observe that $d_{kk}^{\widehat{B^H}} = 1$, $\varepsilon_n = 0$, $\kappa = 1$, and (see (2.5))

$$
\nu_n \leq \sup_{H \in (0,1)} \sum_{j \in \mathbb{Z}} |\rho_H^{(2)}(j)| = \frac{8}{3}.
$$

Then

$$
\mathsf{P}\big(H \in \left[H_n^{inf}(\alpha), H_n^{sup}(\alpha)\right]\big) \geq 1 - \alpha,
$$

where

$$
\begin{aligned}
H_n^{inf}(\alpha) &:= \max\left(0, g_{n,T}^{-1}\big(\log\left(x_{l,n-1}(\alpha/2)\right) - \log S_n\big)\right), \\
H_n^{sup}(\alpha) &:= \min\left(1, g_{n,T}^{-1}\big(\log\left(x_{l,n-1}(\alpha/2)\right) - \log S_n\big)\right),
\end{aligned}
$$

$$x_{l,n-1}(\alpha) = 1 - \frac{\varphi_{l,n}^{-1}(\alpha; 8/3)}{\sqrt{n-1}}, \quad x_{r,n-1}(\alpha) = 1 - \frac{\varphi_{r,n}^{-1}(\alpha; 8/3)}{\sqrt{n-1}}, \tag{6.46}$$

$$\varphi_{r,n}(z; 8/3) := \exp\left\{ -\frac{3z\sqrt{n-1}}{16} \right\} \left(1 + \frac{z}{\sqrt{n-1}} \right)^{\frac{3(n-1)}{16}},$$

$$\varphi_{l,n}(z; 8/3) := \exp\left\{ \frac{3z\sqrt{n-1}}{16} \right\} \left(1 - \frac{z}{\sqrt{n-1}} \right)^{\frac{3(n-1)}{16}} \mathbf{1}_{[0,\sqrt{n-1}]}(z).$$

The Brownian motion B appears as the special case among the considered cases. Since

$$\frac{n}{2T} d_{kk}^B = 1, \qquad \frac{n}{2T} \max_{1 \le k \le n-1} \sum_{j=1}^{n-1} |d_{jk}^B| = 2,$$

then in (6.46) the number $8/3$ may be replaced by 2.

6.4.2.2 Fractional Ornstein–Uhlenbeck Process

Let us consider a fO-U process which satisfies SDE

$$X_t = x_0 - \theta \int_0^t X_s \, ds + \sigma B_t^H, \qquad t \in [0, T], \tag{6.47}$$

with $\theta, \sigma > 0$, where $B^H, 0 < H < 1$, is a fBm. It is known that for the fO-U process the Orey index is equal to $H, 0 < H < 1$, (see Sect. 6.1.2.2).

First we show the following lemma holds true.

Lemma 6.22 *Let X be the solution of Eq. (6.47). Assume that $B^H = \{B^H(t) : t \in [0, T]\}$ is a fBm with $H \in (0, H^*]$, where a real number $0 < H^* < 1$ is known. Then for $H \ne 1/2$*

$$\max_{1 \le k \le n-1} \sum_{j=1}^{n-1} |d_{jk}^{\overline{X}}| \le \frac{4\theta T}{4 - 2^{2H^*}} \left(2\theta \frac{T}{n} \left[C \left(\frac{T}{n} \right)^{2-2H^*} + 1 \right] \right.$$
$$\left. + \sqrt{4 - 2^{2H^*}} \left[C \left(\frac{T}{n} \right)^{2-2H^*} + 1 \right]^{1/2} \right) + \frac{8}{3},$$

where $C = \theta^2 [3x_0^2 \sigma^{-2} + 6T^2]$ and

$$d_{kk}^{\overline{X}} = \frac{n^{2H}}{T^{2H} \sigma^2 (4 - 2^{2H})} \mathsf{E}(\Delta_{k,n}^{(2)} X)^2.$$

Proof It is evident that

$$\Delta^{(2)}_{n,k}X = -\theta\left(\int_{t^n_k}^{t^n_{k+1}} X_s\,ds - \int_{t^n_{k-1}}^{t^n_k} X_s\,ds\right) + \sigma\Delta^{(2)}_{n,k}B^H,$$

where $t^n_k = \frac{k}{n}T$, and

$$\begin{aligned}\left|d^X_{j,k} - \sigma^2 d^{B^H}_{j,k}\right| \le& \theta^2 \mathsf{E}\left(\int_{t^n_j}^{t^n_{j+1}} X_s\,ds - \int_{t^n_{j-1}}^{t^n_j} X_s\,ds\right)\left(\int_{t^n_k}^{t^n_{k+1}} X_s\,ds - \int_{t^n_{k-1}}^{t^n_k} X_s\,ds\right)\\ &+ \theta\sigma\,\mathsf{E}\left|\int_{t^n_j}^{t^n_{j+1}} X_s\,ds - \int_{t^n_{j-1}}^{t^n_j} X_s\,ds\right| \cdot \left|\Delta^{(2)}_{n,k}B^H\right|\\ &+ \theta\sigma\,\mathsf{E}\left|\int_{t^n_k}^{t^n_{k+1}} X_s\,ds - \int_{t^n_{k-1}}^{t^n_k} X_s\,ds\right| \cdot \left|\Delta^{(2)}_{n,j}B^H\right| := \sum_{i=1}^{3} I^{(i)}_{j,k}.\end{aligned}$$

Moreover, for $s < t$

$$\begin{aligned}\mathsf{E}|X_t - X_s|^2 \le& 2\theta^2(t-s)\int_s^t \mathsf{E}X_u^2\,du + 2\sigma^2\,\mathsf{E}|B^H_t - B^H_s|^2\\ \le& 2\theta^2(t-s)^2 \sup_{s\le u\le t} \mathsf{E}X_u^2 + 2\sigma^2(t-s)^{2H}\\ =& 2(t-s)^{2H}\left[\theta^2(t-s)^{2-2H} \sup_{s\le u\le t} \mathsf{E}X_u^2 + \sigma^2\right]\end{aligned}$$

and (see (6.19))

$$\sup_{t\le T} \mathsf{E}X_t^2 \le 3x_0^2 + 6\sigma^2T^2.$$

From above we have

$$\begin{aligned}|I^{(1)}_{j,k}| \le& \theta^2\left(\frac{2T}{n}\int_{t^n_j}^{t^n_{j+1}} \mathsf{E}(X_s - X_{t^n_j})^2\,ds + \frac{2T}{n}\int_{t^n_{j-1}}^{t^n_j} \mathsf{E}(X_s - X_{t^n_j})^2\,ds\right)^{1/2}\\ &\times\left(\frac{2T}{n}\int_{t^n_k}^{t^n_{k+1}} \mathsf{E}(X_s - X_{t^n_k})^2\,ds + \frac{2T}{n}\int_{t^n_{k-1}}^{t^n_k} \mathsf{E}(X_s - X_{t^{2n}_{k+m}})^2\,ds\right)^{1/2}\\ \le& 8\theta^2\left(\frac{T}{n}\right)^{2+2H}\left[\theta^2\left(\frac{T}{n}\right)^{2-2H} \sup_{t\le T} \mathsf{E}X_t^2 + \sigma^2\right]\end{aligned}$$

and

$$|I_{j,k}^{(2)}| \leq \theta\sigma \left(\frac{2T}{n} \int_{t_j^n}^{t_{j+1}^n} \mathsf{E}(X_s - X_{t_j^n})^2 \, ds + \frac{2T}{n} \int_{t_{j-1}^n}^{t_j^n} \mathsf{E}(X_s - X_{t_j^n})^2 \, ds \right)^{1/2}$$

$$\times \sqrt{4 - 2^{2H}} \left(\frac{T}{n} \right)^H$$

$$\leq 2\sqrt{2(4 - 2^{2H})}\, \theta\sigma \left(\frac{T}{n} \right)^{1+2H} \left[\theta^2 \left(\frac{T}{n} \right)^{2-2H} \sup_{t \leq T} \mathsf{E} X_t^2 + \sigma^2 \right]^{1/2},$$

$$|I_{j,k}^{(3)}| \leq 2\sqrt{2(4 - 2^{2H})}\, \theta\sigma \left(\frac{T}{n} \right)^{1+2H} \left[\theta^2 \left(\frac{T}{n} \right)^{2-2H} \sup_{t \leq T} \mathsf{E} X_t^2 + \sigma^2 \right]^{1/2}.$$

Recall that (see Lemma 2.7)

$$\max_{1 \leq k \leq n-1} \sum_{j=1}^{n-1} |d_{jk}^{\widehat{B^H},n}| \leq \frac{8}{3}.$$

Consequently,

$$\max_{1 \leq k \leq n-1} \sum_{j=1}^{n-1} |d_{jk}^{\overline{X},n}| \leq \max_{1 \leq k \leq n-1} \sum_{j=1}^{n-1} |d_{jk}^{\overline{X},n} - \sigma^2 d_{jk}^{\overline{B}^H,n}| + \sigma^2 \max_{1 \leq k \leq n-1} \sum_{j=1}^{n-1} |d_{jk}^{\overline{B}^H,n}|$$

$$\leq \frac{8\theta^2}{4 - 2^{2H^*}} \frac{T^2}{n} \left[C \left(\frac{T}{n} \right)^{2-2H^*} + 1 \right] + \frac{4\sqrt{2}\theta T}{\sqrt{4 - 2^{2H^*}}} \left[C \left(\frac{T}{n} \right)^{2-2H^*} + 1 \right]^{1/2} + \frac{8}{3}$$

$$= \frac{4\theta T}{4 - 2^{2H^*}} \left(2\theta \frac{T}{n} \left[C \left(\frac{T}{n} \right)^{2-2H^*} + 1 \right] + \sqrt{2(4 - 2^{2H^*})} \left[C \left(\frac{T}{n} \right)^{2-2H^*} + 1 \right]^{1/2} \right)$$

$$+ \frac{8}{3},$$

where $C = \theta^2[3x_0^2\sigma^{-2} + 6T^2]$. □

Lemma 6.23 *Let X be the solution of the Eq. (6.47). Then*

$$\left| \frac{1}{n-1} \sum_{k=1}^{n-1} d_{kk}^{\overline{X}} - 1 \right| \leq \frac{4\theta}{4 - 2^{2H^*}} \frac{T}{n} \left(2\theta \frac{T}{n} \left[C \left(\frac{T}{n} \right)^{2-2H^*} + 1 \right] \right.$$

$$\left. + \sqrt{2(4 - 2^{2H^*})} \left[C \left(\frac{T}{n} \right)^{2-2H^*} + 1 \right]^{1/2} \right),$$

where $C = \theta^2[3x_0^2\sigma^{-2} + 6T^2]$.

Proof In order to prove this lemma, we observe that

$$\left|\frac{1}{n-1}\sum_{k=1}^{n-1} d_{kk}^{\overline{X}} - 1\right| \le \max_{1\le k\le n-1} \left|d_{kk}^{\overline{X}} - 1\right| = \max_{1\le k\le n-1} \left|d_{kk}^{\overline{X}} - \theta^2 d_{kk}^{\overline{B}^H}\right|.$$

The statement of the lemma directly follows from results of Lemma 6.22. □

6.4.2.3 Sub-fractional Brownian Motion

It is known that for a sfBm the Orey index is equal to H, $0 < H < 1$, (see Sect. 6.1.2.3). Now we prove the following lemma.

Lemma 6.24 *Assume that* $S^H = \{S^H(t) : t \in [0, T]\}$ *is a sfBm. If* $H \neq 1/2$ *then*

$$\left|\frac{1}{n-1}\sum_{k=1}^{n-1} d_{kk}^{\overline{S}^H} - 1\right| \le \left(\frac{T}{n}\right)^{2/3}\left(\frac{1}{6}\frac{n}{n-1} + \frac{33}{9\log 4}\right),$$

where

$$d_{kk}^{\overline{S}^H} = \frac{n^{2H}}{T^{2H}(4-2^{2H})}\,\mathsf{E}\big(\Delta_{k,n}^{(2)} S^H\big)^2.$$

Proof Observe that the following equality

$$\begin{aligned}\mathsf{E}\big(S_{t+h}^H - 2S_t^H + S_{t-h}^H\big)^2 =&(4-2^{2H})h^{2H} - 2^{2H-1}(t+h)^{2H} - 3\cdot 2^{2H}t^{2H}\\ &- 2^{2H-1}(t-h)^{2H} + 2(2t+h)^{2H} + 2(2t-h)^{2H}\end{aligned}$$

holds. Thus

$$\begin{aligned}d_{kk}^{\overline{S}^H,n} &= 1 - \big[2^{2H-1}(k+1)^{2H} + 3\cdot 2^{2H}k^{2H} + 2^{2H-1}(k-1)^{2H}\\ &\quad - 2(2k+1)^{2H} - 2(2k-1)^{2H}\big] / \big(4-2^{2H}\big)\\ &= 1 - \frac{b(k,H)}{4-2^{2H}}.\end{aligned}$$

For simplicity we shall omit the index n for $d_{kk}^{\overline{S}^H,n}$. In what follows we need estimates of $d_{kk}^{\overline{S}^H}$ and $\frac{|b(k,H)|}{4-2^{2H}}$. The simplest way to obtain them is by plotting these functions. The function $\frac{b(k,H)}{4-2^{2H}}$ graphs for $k = 1, 2, \ldots, 5$ are illustrated in Fig. 6.1.

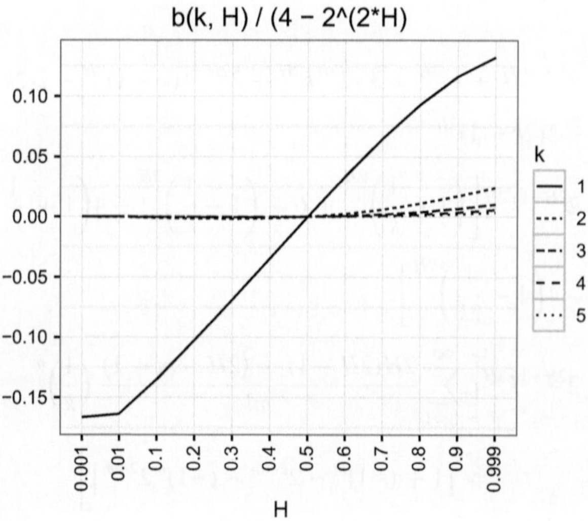

Fig. 6.1 Function $\frac{b(k,H)}{4-2^{2H}}$

Then it is easy to check that the inequalities

$$\max_{H\in(0,1)} d_{kk}^{\overline{S}^H} \le \max_{H\in(0,1)} d_{11}^{\overline{S}^H} = \frac{7}{6}, \quad \max_{H\in(0,1)} \frac{|b(k,H)|}{4-2^{2H}} \le \max_{H\in(0,1)} \frac{|b(1,H)|}{4-2^{2H}} \le \frac{1}{6} \tag{6.48}$$

hold.

Denote $\tau_n = \lfloor (nT^{-1})^{1/3} \rfloor$, where $\lfloor a \rfloor$ is an integer part of a. Then

$$\begin{aligned}\left|\frac{1}{n-1}\sum_{k=1}^{n-1} d_{kk}^{\overline{S}^H} - 1\right| \le &\frac{1}{n-1}\sum_{k=1}^{\tau_n} \frac{|b(k,H)|}{4-2^{2H}} + \max_{\tau_n+1\le k\le n-1} \frac{|b(k,H)|}{4-2^{2H}} \\ =&I_1 + I_2.\end{aligned}$$

It is clear that

$$I_1 \le \frac{\tau_n}{n-1} \max_{1\le k\le \tau_n} \frac{|b(k,H)|}{4-2^{2H}} \le \frac{\tau_n}{n-1}\frac{|b(1,H)|}{4-2^{2H}} \le \frac{1}{6}\left(\frac{T}{n}\right)^{2/3} \frac{n}{T(n-1)}.$$

Now we estimate $b(k,H)$ for $k \ge 2$. Using the formula

$$(1+x)^\alpha = 1 + \sum_{k=1}^{\infty} \frac{\alpha(\alpha-1)\cdots(\alpha-k+1)}{k!} x^k \qquad \text{for } -1 < x < 1,$$

we obtain

$$\begin{aligned}
b(k,H) &= 2^{2H-1}(k+1)^{2H} + 3\cdot 2^{2H}k^{2H} + 2^{2H-1}(k-1)^{2H} - 2(2k+1)^{2H} \\
&\quad - 2(2k-1)^{2H} \\
&= 2^{2H-1}k^{2H}\left[\left(1+\frac{1}{k}\right)^{2H} + 6 + \left(1-\frac{1}{k}\right)^{2H} - 4\left(1+\frac{1}{2k}\right)^{2H}\right. \\
&\quad \left. - 4\left(1-\frac{1}{2k}\right)^{2H}\right] \\
&= 2^{2H-1}k^{2H}\left[\sum_{m=1}^{\infty} \frac{2H(2H-1)\cdots(2H-m+1)}{m!}\left(\frac{1}{k}\right)^m\right. \\
&\qquad \left. \times\left[1+(-1)^m - 2^{2-m} - (-1)^m 2^{2-m}\right]\right] \\
&= 2^{2H}k^{2H}\left[\sum_{m=2}^{\infty} \frac{2H(2H-1)\cdots(2H-2m+1)}{(2m)!}\left(\frac{1}{k}\right)^{2m}\left(1-2^{2-2m}\right)\right].
\end{aligned}$$

Note that the sign of $2H(2H-1)(2H-2)\cdots(2H-2m+1)$ is the same as that of $2H-1$. Thus,

$$\begin{aligned}
\frac{|b(k,H)|}{4-2^{2H}} &\le \frac{2^{2H}k^{2H}}{4-2^{2H}} \sum_{m=2}^{\infty} \frac{2H|2H-1|(2-2H)\cdots(2m-1-2H)}{(2m)!}\left(\frac{1}{k}\right)^{2m}\left[1-2^{2-2m}\right] \\
&\le \frac{3\cdot 2^{2H}k^{2H}}{4\log 4} \sum_{m=2}^{\infty} \frac{2(2m-1)!}{(2m)!}\left(\frac{1}{k}\right)^{2m} \\
&\le \frac{3\cdot 2^{2H}}{4\log 4} k^{2H} \sum_{m=2}^{\infty} \frac{1}{m}\left(\frac{1}{k^2}\right)^m \le \frac{3\cdot 2^{2H}}{4\log 4} k^{2H-2} \sum_{m=2}^{\infty} \frac{1}{m-1}\left(\frac{1}{k^2}\right)^{m-1} \\
&= -\frac{3\cdot 2^{2H}}{4\log 4} k^{2H-2}\log\left(1-\frac{1}{k^2}\right) \le \frac{33\cdot 2^{2H}}{36\log 4} k^{2H-4} \le \frac{33}{9\log 4} k^{-2}
\end{aligned}$$

since

$$2H|1-2H|(2-2H)\cdots(2m-1-2H) \le 2(1-H)(2m-1)!\,,$$

$$\log(1-u) = -\sum_{k=1}^{\infty} \frac{u^k}{k} \quad \text{if } 0 \le u < 1 \quad \text{and} \quad -\log(1-u) \le \frac{11}{9}u \quad \text{if } 0 \le u \le 1/4,$$

$$(1-H)\log 4 < 4 - 2^{2H}.$$

Therefore,

$$\left|\frac{1}{n-1}\sum_{k=1}^{n-1} d_{kk}^{\overline{S}^H} - 1\right| \le \frac{1}{6}\left(\frac{T}{n}\right)^{2/3} \frac{n}{T(n-1)} + \frac{33}{9\log 4}\frac{1}{(\tau_n+1)^2}$$
$$\le \frac{1}{6}\left(\frac{T}{n}\right)^{2/3} \frac{n}{T(n-1)} + \frac{33}{9\log 4}\left(\frac{T}{n}\right)^{2/3}$$
$$= \left(\frac{T}{n}\right)^{2/3}\left(\frac{1}{6}\frac{n}{n-1} + \frac{33}{9\log 4}\right).$$

□

Lemma 6.25 *Assume that $S^H = \{S_t^H : t \in [0,T]\}$ is a sfBm. If $H \neq 1/2$ then*

$$\max_{1\le k\le n-1}\sum_{j=1}^{n-1} |d_{jk}^{\overline{S}^H}| \le \frac{9}{2}. \tag{6.49}$$

Proof The fourth order mixed partial derivative of the covariance function $G_H(s,t)$ is of the following form

$$\frac{\partial^4 G_H}{\partial s^2 \partial t^2}(s,t) = -C_H\left[\frac{1}{|s-t|^{2(2-H)}} + \frac{1}{(s+t)^{2(2-H)}}\right]$$

for each $s, t > 0$ such that $s \neq t$, where $C_H = H(2H-1)(2H-2)(2H-3)$. Since the covariance function $G_H(s,t)$ is continuous in $[0,T]^2$ and the derivative $\frac{\partial^4 R}{\partial s^2 \partial t^2}$ is continuous in $(0,T]^2/\{s=t\}$ then for $H \neq 1/2$ and $j \neq 1$ or $k \neq 1$

$$d_{jk}^{S^H} = \mathsf{E}\Delta_{j,n}^{(2)} S^H \Delta_{k,n}^{(2)} S^H = \int_{t_j^n}^{t_{j+1}^n} du \int_{u-T/n}^{u} dv \int_{t_k^n}^{t_{k+1}^n} dx \int_{x-T/n}^{x} \frac{\partial^4 G_H}{\partial s^2 \partial t^2}(v,y)\, dy.$$

Assume that R_H is the covariance function of the fBm B^H. Then the derivative

$$\frac{\partial^4 R_H}{\partial s^2 \partial t^2}(s,t) = -\frac{C_H}{|s-t|^{4-2H}}$$

of the covariance function R_H is continuous in $(0,T]^2/\{s=t\}$ and

$$d_{jk}^{B^H} = -\int_{t_j^n}^{t_{j+1}^n} du \int_{u-T/n}^{u} dv \int_{t_k^n}^{t_{k+1}^n} dx \int_{x-T/n}^{x} \frac{C_H}{(v-y)^{4-2H}}\, dy.$$

First of all by simple calculations we obtain

$$\int_t^{t+h} du \int_{u-h}^{u} dv \int_s^{s+h} dx \int_{x-h}^{x} \frac{dy}{(v-y)^{4-2H}}$$
$$= (2C_H)^{-1}\{6(t-s)^{2H} - 4(t-s-h)^{2H} + (t-s-2h)^{2H}$$
$$- 4(t-s+h)^{2H} + (t-s+2h)^{2H}\}. \tag{6.50}$$

In view of (6.50) and (2.4) we get

$$\left|d_{j,k}^{S^H} - d_{j,k}^{B^H}\right| = \left|\int_{\frac{j}{n}T}^{\frac{j+1}{n}T} du \int_{u-T/n}^{u} dv \int_{\frac{k}{n}T}^{\frac{k+1}{n}T} dx \int_{x-T/n}^{x} \frac{C_H}{(v+y)^{2(2-H)}} dy\right|$$
$$\leq |C_H| \int_{\frac{j}{n}T}^{\frac{j+1}{n}T} du \int_{u-T/n}^{u} dv \int_{\frac{k}{n}T}^{\frac{k+1}{n}T} dx \int_{x-T/n}^{x} \frac{dy}{|v-y|^{2(2-H)}}$$
$$= -\operatorname{sgn}(2H-1)\widehat{\rho}_H(j-k)(\tfrac{T}{n})^{2H} = |\widehat{\rho}_H(j-k)|(\tfrac{T}{n})^{2H}$$

for $|j-k| \geq 1$ and $j \neq 1, k \neq 1$.

It remains to us to prove the cases when $j = 1$ and $k > 1$ or $k = 1$ and $j > 1$. Let $k = 1$ and $\varepsilon \in (0, 1)$. Denote

$$d_{j,1}^{S^H(\varepsilon)} = \int_{\frac{j}{n}T}^{\frac{j+1-\varepsilon}{n}T} du \int_{u-\frac{1-\varepsilon}{n}T}^{u} dv \int_{\frac{1}{n}T}^{\frac{2-\varepsilon}{n}T} dx \int_{x-\frac{1-\varepsilon}{n}T}^{x} \frac{\partial^4 G_H}{\partial s^2 \partial t^2}(v, y)\, dy$$

and

$$d_{j,1}^{B^H(\varepsilon)} = -\int_{\frac{j}{n}T}^{\frac{j+1-\varepsilon}{n}T} du \int_{u-\frac{1-\varepsilon}{n}T}^{u} dv \int_{\frac{1}{n}T}^{\frac{2-\varepsilon}{n}T} dx \int_{x-\frac{1-\varepsilon}{n}T}^{x} \frac{C_H}{|v-y|^{2(2-H)}} dy.$$

Then

$$\left|d_{j,1}^{S^H(\varepsilon)} - d_{j,1}^{B^H(\varepsilon)}\right|$$
$$\leq |C_H| \int_{\frac{j}{n}T}^{\frac{j+1-\varepsilon}{n}T} du \int_{u-\frac{1-\varepsilon}{n}T}^{u} dv \int_{\frac{1}{n}T}^{\frac{2-\varepsilon}{n}T} dx \int_{x-\frac{1-\varepsilon}{n}T}^{x} \frac{dy}{|v-y|^{2(2-H)}}$$
$$= \frac{1}{2}\left|6(j-1)^{2H} - 4(j-2+\varepsilon)^{2H} + (j-3+2\varepsilon)^{2H}\right.$$
$$\left.- 4(j-\varepsilon)^{2H} + (j+1-2\varepsilon)^{2H}\right|(\tfrac{T}{n})^{2H}.$$

In view of continuity of the functions G_H and R_H

$$\lim_{\varepsilon\to 0} d_{j,1}^{S^H(\varepsilon)} = d_{j,1}^{S^H}, \qquad \lim_{\varepsilon\to 0} d_{j,1}^{B^H(\varepsilon)} = d_{j,1}^{B^H}.$$

Thus,

$$\left|d_{j,1}^{S^H} - d_{j,1}^{B^H}\right| \leq \left|\widehat{\rho}_H(j-1)\right|\left(\tfrac{T}{n}\right)^{2H}.$$

A similar argument yields

$$\left|d_{1,k}^{S^H} - d_{1,k}^{B^H}\right| \leq \left|\widehat{\rho}_H(1-k)\right|\left(\tfrac{T}{n}\right)^{2H}.$$

Thus,

$$\left|d_{j,k}^{S^H}\right| \leq 2\left|\widehat{\rho}_H(j-k)\right|(Tn^{-1})^{2H} \qquad \text{for } |j-k| \geq 1.$$

Now we will prove the statement of the lemma. We will use the estimate (6.48) and the equalities $\widehat{\rho}_H(-r) = \widehat{\rho}_H(r)$, and $\widehat{\rho}_H(1) = -\frac{1}{2}(7-4\cdot 2^{2H}+3^{2H})$. Note that for $H \neq 1/2$

$$\max_{1\leq k\leq n-1} \sum_{j=1}^{n-1} \left|d_{jk}^{\overline{S}^H}\right| \leq \max_{1\leq k\leq n-1} d_{kk}^{\overline{S}^H} + 2 \max_{1\leq k\leq n-1} \sum_{\substack{j=1\\ j\neq k}}^{n-1} \frac{|\widehat{\rho}_H(j-k)|}{4-2^{2H}}$$

$$\leq \frac{7}{6} + 4\sum_{j=1}^{\infty} \frac{|\widehat{\rho}_H(j)|}{4-2^{2H}} = \frac{7}{6} + 2\,\frac{7-4\cdot 2^{2H}+3^{2H}}{4-4^{H}} + 4\sum_{j=2}^{\infty} \frac{|\widehat{\rho}_H(j)|}{4-2^{2H}}.$$

Since (see (2.3))

$$\sum_{j=2}^{\infty} \frac{|\widehat{\rho}_H(j)|}{4-2^{2H}} = -\frac{1}{2}\cdot \operatorname{sgn}(2H-1)\,\frac{3-3\cdot 2^{2H}+3^{2H}}{4-2^{2H}},$$

we obtain

$$\max_{1\leq k\leq n-1} \sum_{j=1}^{n-1} \left|d_{jk}^{\overline{S}^H}\right| \leq \frac{7}{6} + 2\,\frac{7-4\cdot 2^{2H}+3^{2H}}{4-2^{2H}} - 2\cdot \operatorname{sgn}(2H-1)\,\frac{3-3\cdot 2^{2H}+3^{2H}}{4-2^{2H}}$$

$$= \begin{cases} \frac{7}{6} + 2\,\frac{10-7\cdot 4^{H}+2\cdot 3^{2H}}{4-2^{2H}} & \text{for} \quad H < 1/2, \\ \frac{7}{6} + 2 & \text{for} \quad 1/2 < H < 1 \end{cases}$$

$$= \begin{cases} \frac{9}{2} & \text{for} \quad H < 1/2, \\ 2\frac{7}{6} & \text{for} \quad 1/2 < H < 1. \end{cases}$$

□

6.4.3 Simulations

The simulations of the obtained confidence intervals presented below were performed using the R software environment [147]. Sample paths of fBm were generated using the circulant matrix embedding method and were further used to simulate the sample paths of the fO-U process (6.47). The constants for the latter were (arbitrarily) chosen as $x_0 = 0$ and $\mu = 0.5$. Sample paths of the sub-fractional Brownian motion were simulated using the Cholesky method. Due to the notable computational requirements of this method the maximum sample path length considered was $n = 1600$. Figures presented below correspond to the case of the confidence level $1-\alpha$, $\alpha = 0.1$. The observed coverage percentages in all cases were at least as good as claimed in Theorem 6.21.

Figures 6.2, 6.3, 6.4 present the confidence interval (CI) lengths for all the process types considered in this paper. Figure 6.5 shows the median ratios of the confidence intervals lengths, where the CI lengths of the sfBm and the fO-U processes were divided by the corresponding CI lengths of fBm. It can be seen that in almost all cases the confidence intervals behave in a similar way, one notable exception being the case of fO-U as the value of H approaches 1. This is hardly unexpected given the normalization used in Lemma 6.22, and in this scenario the CI covers the whole interval of possible parameter values $0 < H < 1$.

Fig. 6.2 $\left|H_n^{sup}(\alpha) - H_n^{inf}(\alpha)\right|$

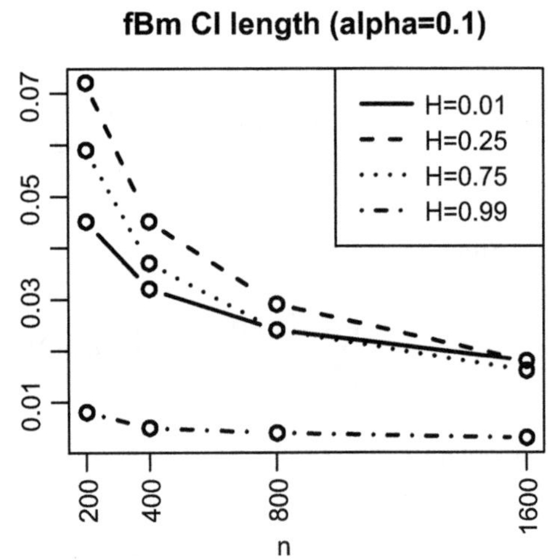

Fig. 6.3 $\left|H_n^{sup}(\alpha) - H_n^{inf}(\alpha)\right|$

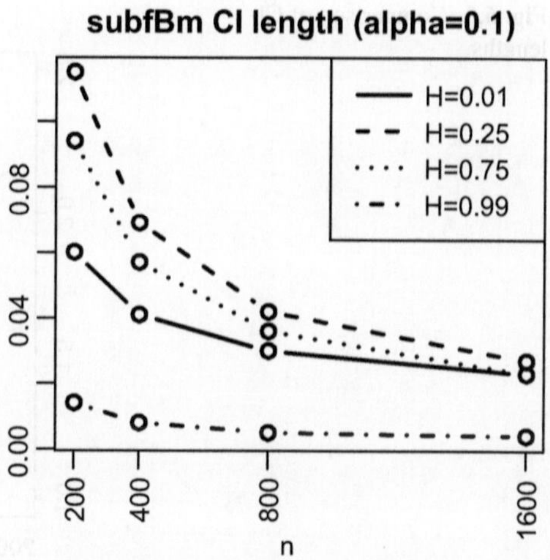

Fig. 6.4 $\left|H_n^{sup}(\alpha) - H_n^{inf}(\alpha)\right|$

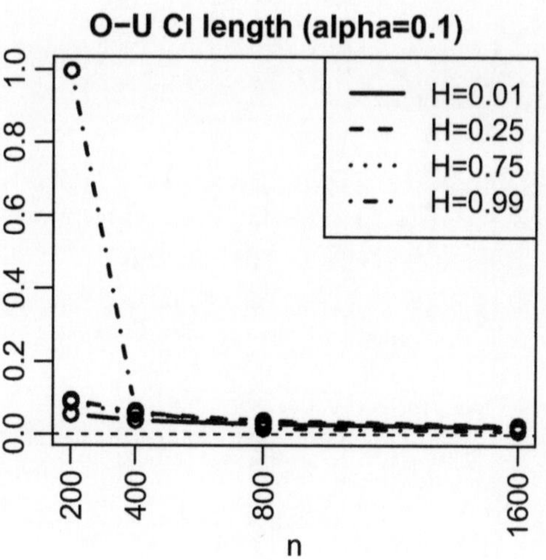

Fig. 6.5 Comparison of CI lengths

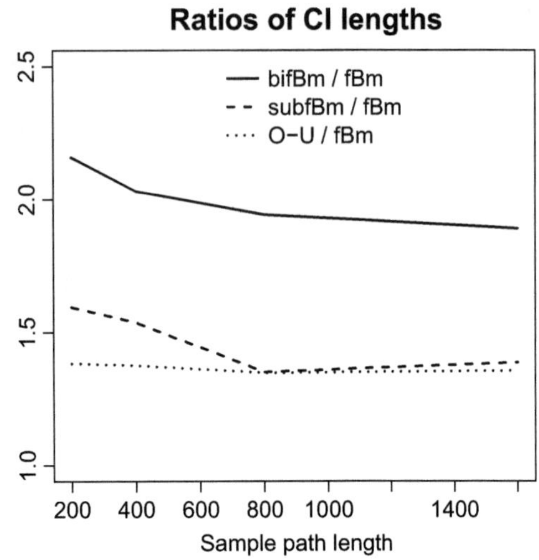

Appendix A
Selected Facts from Mathematical and Functional Analysis

A.1 Some Properties of the Hypergeometric Function

Recall the integral representation of the Gauss hypergeometric function and some of its properties.

For $c > b > 0$ and $x < 1$, the *Gauss hypergeometric function* is defined as the integral (see [1], formula 15.3.1)

$$F(a, b, c; x) = {}_2F_1(a, b, c; x) = \frac{1}{\mathrm{B}(b, c-b)} \int_0^1 t^{b-1}(1-t)^{c-b-1}(1-xt)^{-a}\, dt. \tag{A.1}$$

For the same values of parameters the following equality holds (see [1], 15.3.4):

$$F(a, b, c; x) = (1-x)^{-a} F\left(a, c-b, c; \frac{x}{x-1}\right), \tag{A.2}$$

Evidently, $F(a, b, c; x)$ at $x = 1$ is correctly defined for $c - a - b > 1$ and in this case equals

$$F(a, b, c; 1) = \frac{\Gamma(c)\Gamma(c-a-b)}{\Gamma(c-a)\Gamma(c-b)}. \tag{A.3}$$

Finally, it is easy to check with the help of (A.1) that

$$F(a, b, c; 0) = F(0, b, c; x) = 1. \tag{A.4}$$

The following result gives upper bounds for the hypergeometric function (see [84], Theorems 4 and 5, respectively).

© Springer International Publishing AG 2017

K. Kubilius et al., *Parameter Estimation in Fractional Diffusion Models*,
B&SS – Bocconi & Springer Series 8,
https://doi.org/10.1007/978-3-319-71030-3

Proposition A.1

(i) For $c > b > 1, x > 0$, and $0 < a \le 1$, we have the inequality

$$F(a,b,c;-x) < \frac{1}{\left(1 + x(b-1)/(c-1)\right)^a}.$$

(ii) For $0 < a \le 1$, $b > 0$, $c - b > 1$, and $x \in (0,1)$, we have the inequality

$$F(a,b,c;x) < \frac{1}{\left(1 - \frac{b}{c-1}x\right)^a}.$$

A.2 Regularly Varying Functions

The following facts about regularly varying functions are taken from Bingham et al. [27] (see also Marcus and Rosen [117], Resnick [150]).

Definition A.2 A Borel function $f\colon (0,a) \to \mathbb{R}$, $a > 0$, is *regularly varying* at zero with index $\rho \in \mathbb{R}$ if for all $\lambda > 0$

$$\lim_{x \downarrow 0} \frac{f(\lambda x)}{f(x)} = \lambda^\rho.$$

If $\rho = 0, f$ is said to be *slowly varying* at zero.

A regularly varying function f at zero with index ρ can be written in the form

$$f(x) = x^\rho \beta(x) \exp\left(\int_1^x \frac{\epsilon(u)}{u}\, du\right),$$

where $\beta(u)$ is measurable function, $\lim_{u\to 0} \beta(u) = C \in (0,\infty)$ and $\lim_{u \to 0} \epsilon(u) = 0$.

A function f is regularly varying with index ρ if and only if it can be written in the form $f(x) = x^\rho L(x)$, where L is some slowly varying function. From the representation theorem we may conclude that for a regularly varying f with index $\rho, f(x) \to 0$ as $x \to 0$, if $\rho > 0$.

Typical examples of slowly varying functions are positive constants or functions converging at zero to a positive constant, logarithms and iterated logarithms. Moreover, for all $\rho \in \mathbb{R}$ the functions

$$x^\rho, \quad x^\rho |\log x|, \quad (x|\log x|)^\rho, \quad x^\rho \log|\log x|$$

are regularly varying at zero with index ρ.

Assume that L is slowly varying function at zero and $\alpha, \beta > 0$. Then $L^\alpha(x)$ is a slowly varying function while $x^\beta L(x) \to 0$.

A.3 *p*-Variation Calculus

The following facts about p-variation functions are taken from Dudley and Norvaiša [51, 52].

Fix $p > 0$ and $-\infty < a < b < \infty$. Let

$$\varkappa = \{\{x_0, \ldots, x_n\} \mid a = x_0 < \cdots < x_n = b, n \geq 1\}$$

denotes a set of all possible partitions of $[a, b]$. For any $f: [a, b] \to \mathbb{R}$ define

$$v_p(f; [a, b]) = \sup_{\varkappa} \sum_{k=1}^{n} |f(x_k) - f(x_{k-1})|^p, \qquad V_p(f; [a, b]) = v_p^{1/p}(f; [a, b]),$$

$$\begin{aligned} \mathcal{W}_p([a, b]) &= \{f : [a, b] \to \mathbb{R} \mid v_p(f; [a, b]) < \infty\}, \\ C\mathcal{W}_p([a, b]) &= \{f \in \mathcal{W}_p([a, b]) \mid f \text{ is continuous}\}. \end{aligned}$$

Recall that v_p is called p-variation of f on $[a, b]$ and any f in $\mathcal{W}_p([a, b])$ is said to have bounded p-variation on $[a, b]$. Below we list several facts used further on.

- $f \mapsto V_p(f; [a, b])$ is a seminorm on $\mathcal{W}_p([a, b])$; $V_p(f; [a, b]) = 0$ if and only if f is a constant.
- For each f, $V_p(f; [a, b])$ is a non-increasing function of p, i. e., if $q < p$ then $V_p(f; [a, b]) \leq V_q(g; [a, b])$. Thus $\mathcal{W}_q([a, b]) \subseteq \mathcal{W}_p([a, b])$ if $1 \leq q < p < \infty$.
- $V_{p,\infty}(f; [a, b]) = V_p(f; [a, b]) + \sup_{x \in [a,b]} |f(x)|$ is a norm on $\mathcal{W}_p([a, b])$, $p \geq 1$.
- $f \in \mathcal{W}_p([a, b]) \Rightarrow \sup_{x \in [a,b]} |f(x)| < \infty$.
- $f, g \in \mathcal{W}_p([a, b]) \Rightarrow fg \in \mathcal{W}_p([a, b])$.
- $q > p \geq 1 \Rightarrow \mathcal{W}_p([a, b]) \subset \mathcal{W}_q([a, b])$.
- If $p \geq 1$ and $f, g \in \mathcal{W}_p([a, b])$, then $V_p(f+g; [a, b]) \leq V_p(f; [a, b]) + V_p(g; [a, b])$.
- Let $a < c < b$ and let $f \in \mathcal{W}_p([a, b])$ with $0 < p < \infty$. Then $v_p\big(f; [a, c]\big) + v_p\big(f; [c, b]\big) \leq v_p\big(f; [a, b]\big)$ and $V_p(f; [a, b]) \leq V_p(f; [a, c]) + V_p(f; [c, b])$.
- Let $f \in C\mathcal{W}_p([a, b])$ with $1 \leq p < \infty$. Then, for every $\varepsilon > 0$, there exists a partition $\varkappa = \{a = t_0 < \cdots < t_n = b\}$ of $[a, b]$ such that $\max_{1 \leq i \leq n} v_p\big(f; [t_{i-1}, t_i]\big) < \varepsilon$.
- Let $f \in \mathcal{W}_q([a, b])$ and $h \in \mathcal{W}_p([a, b])$ with $p > 0$, $q > 0$, $1/p + 1/q > 1$. If f and h have no common discontinuities then the extended RS integral $\int_a^b f\, dh$ exists and the Love–Young inequality

$$\left| \int_a^b f\, dh - f(y)\big[h(b) - h(a)\big] \right| \leq C_{p,q} V_q(f; [a, b]) V_p(h; [a, b]), \tag{A.5}$$

holds for any $y \in [a, b]$, where $C_{p,q} = \zeta(p^{-1} + q^{-1})$, $\zeta(s)$ denotes the Riemann zeta function, i.e., $\zeta(s) = \sum_{n \geq 1} n^{-s}$.

- If the function $h \in C\mathcal{W}_p([a, b]), f \in \mathcal{W}_q$, $1/p + 1/q > 1$, then the indefinite integral $\int_a^y f\,dh$, $y \in [a, b]$, is a continuous function.
- Let $f \in \mathcal{W}_q([a, b])$ and $h \in C\mathcal{W}_p([a, b])$. From (A.5) it follows that

$$V_p\left(\int_a^{\cdot} f\,dh; [a,b]\right) \leq \left[C_{p,q}V_q(f;[a,b]) + |f|_{\infty,[a,b]}\right]V_p(h;[a,b])$$
$$\leq C_{p,q}V_{q,\infty}(f;[a,b])V_p(h;[a,b]). \tag{A.6}$$

Lemma A.3 ([51, 154]) *Let $f \in C\mathcal{W}_p([a, b])$ with $1 \leq p < \infty$. Then, for every $\varepsilon > 0$, there exists a partition $\varkappa = \{a = t_0 < \cdots < t_n = b\}$ of $[a, b]$ such that*

$$\max_{1 \leq i \leq n} v_p\big(f; [t_{i-1}, t_i]\big) < \varepsilon.$$

Let $v_p(B^H; [s, t])$ denote the p-variation of B^H and $V_p(B^H; [s, t]) = v_p^{1/p}(B^H; [s, t])$. The following estimate for the p-variation, $p = \frac{1}{H-\varepsilon}$, $0 < \varepsilon < H$, of an fBm is evident:

$$V_p\big(B^H; [s,t]\big) \leq G_{\varepsilon,T}(t-s)^{H-\varepsilon},$$

where $s < t \leq T$, $G_{\varepsilon,T}$ is defined in (1.3).

A.4 The Wasserstein Distance and Kantorovich–Rubinstein Duality Theorem

The following facts are taken from Dudley [50] and Deza [44].

Given a metric space $(\mathcal{S}, d)$, the *Wasserstein distance* is the metric defined by

$$W(P,Q) := \inf_{Y \sim P, Z \sim Q} \mathsf{E}d(Y,Z) = \inf\left\{\int d(x,y)d\mu(x,y) : \mu \in M(P,Q)\right\},$$

where $M(P, Q)$ denotes the set of all laws on $\mathcal{S} \times \mathcal{S}$ with marginals P and Q. The infimum is taken over all distributions μ of pairs (Y, Z) of random variables Y, Z such that marginal distributions of Y and Z are P and Q.

$\mathcal{P}_1(\mathcal{S})$ will denote the collection of all laws P on $\mathcal{S}$ for which $\int d(x, z)dP(x) < \infty$ for every z in $\mathcal{S}$. For example, if $S = \mathbb{R}$ and X is a real-valued random variable, we have $\mathcal{L}(X) \in \mathcal{P}_1(S)$ if and only if $\mathsf{E}|X| < \infty$. $W(P, Q)$ is finite for any P and Q in $\mathcal{P}_1(\mathcal{S})$.

Now recall the Lipschitz seminorm for suitable real-valued functions f on $\mathcal{S}$:

$$\|f\|_L := \sup\big\{|f(x) - f(y)|/d(x,y) : x \neq y \text{ in } \mathcal{S}\big\}.$$

For two laws P and Q on $\mathcal{S}$, let

$$\gamma(P,Q) := \|P-Q\|_L^* := \sup\left\{\left|\int f\, d(P-Q)\right| : \|f\|_L \leq 1\right\}.$$

Theorem A.4 (Kantorovich–Rubinstein) *For any separable metric space $(\mathcal{S}, d)$ and any two laws P and Q in $\mathcal{P}_1(\mathcal{S})$,*

$$W(P,Q) = \gamma(P,Q).$$

If P and Q are tight, for example if $\mathcal{S}$ is complete, then there exists a law Pr in $M(P,Q)$ such that $\int d(x,y)dPr(x,y) = W(P,Q)$, so that the infimum in the definition of $W(P,Q)$ is attained.
Assume that $F_P(x) = P(Y \leq x)$ and $F_Q(x) = Q(Y \leq x)$. Then

$$W(P,Q) = \int_{\mathbb{R}} |F_P(x) - F_Q(x)|\, dx.$$

Appendix B
Selected Facts from Probability, Stochastic Processes and Stochastic Calculus

B.1 Weak and Strong Limit Theorems

B.1.1 Ergodic Theorem

Theorem B.1 ([5, Corollary 8.6.3]) *Let $(X_i)_{i\geq 1}$ be a stationary ergodic sequence of $\mathbb{R}^k$ valued random variables on some probability space $\{\Omega, \mathfrak{F}, \mathbb{F}, \mathsf{P}\}$. Let $h\colon \mathbb{R}^k \to \mathbb{R}$ be Borel measurable and let $\mathsf{E}|h(X_1, X_2, \ldots, X_k)| < \infty$. Then*

$$\frac{1}{n}\sum_{i=1}^{n} h(X_i, X_{i+1}, \ldots, X_{i+k-1}) \xrightarrow{a.s.} \mathsf{E}h(X_1, X_2, \ldots, X_k).$$

Corollary B.2 *Let $(X_i)_{i\geq 1}$ be a real-valued stationary zero-mean Gaussian sequence with $\mathsf{E}(X_1 X_n) \to 0$ as $n \to \infty$. Then it is ergodic and consequently, Theorem B.1 holds.*

Originally this statement was proved in [60], Theorem 5, page 144, see also Theorem A2.15 from [127].

B.1.2 The Strong Law of Large Numbers for Martingales

Various strong law results for martingales can be found in [67, Ch. 2] and [111, Ch. 2, § 6]. The following theorem is widely used in this book. It is taken from [111, Ch. 2, § 6, Th. 10, Cor. 1].

Theorem B.3 (SLLN for Martingales) *Let $M = \{M_t, t \geq 0\}$ be a locally square integrable martingale on some probability space $\{\Omega, \mathfrak{F}, \mathbb{F}, \mathsf{P}\}$. that is right*

© Springer International Publishing AG 2017
K. Kubilius et al., *Parameter Estimation in Fractional Diffusion Models*,
B&SS – Bocconi & Springer Series 8,
https://doi.org/10.1007/978-3-319-71030-3

continuous with limits from the left (càdlàg) such that $\langle M\rangle_t \xrightarrow{a.s.} \infty$, *as* $t \to \infty$. *Then* $M_t/\langle M\rangle_t \xrightarrow{a.s.} 0$, *as* $t \to \infty$.

B.1.3 Classical and Non-classical Limit Theorems

Theorem B.4 *[Helly–Bray Theorem [85]] Let* $F_n(x)$ *be a sequence of nondecreasing functions in* $[a,b]$, $-\infty < a < b < \infty$. *Suppose that* $F_n(x)$ *converges weakly to* $F(x)$ *and*

$$F_n(a) \to F(a), \qquad F_n(b) \to F(b).$$

If $f(x)$ *is a continuous function on* $[a,b]$, *then*

$$\lim_{n\to\infty} \int_a^b f(x)\, dF_n(x) = \int_a^b f(x)\, dF(x).$$

Theorem B.5 (Slutsky Theorem) *Let* $\{\mathbf{X}_n, n \geq 1\}$ *and* $\{\mathbf{Y}_n, n \geq 1\}$ *be sequences of random vectors in* $\mathbb{R}^d$ *and* $\mathbb{R}^k$ *respectively such that* $\mathbf{X}_n \xrightarrow{d} \mathbf{X}$, *and* $\mathbf{Y}_n \xrightarrow{\mathsf{P}} \mathbf{C}$, $n \to \infty$, *where* $\mathbf{C}$ *is a constant vector, then*

$$(\mathbf{X}_n, \mathbf{Y}_n) \xrightarrow{d} (\mathbf{X}, \mathbf{C}),\ n \to \infty.$$

In particular, for any continuous function $f\colon \mathbb{R}^d \times \mathbb{R}^k \to \mathbb{R}^m$,

$$f(\mathbf{X}_n, \mathbf{Y}_n) \xrightarrow{d} f(\mathbf{X}, \mathbf{C}),\ n \to \infty.$$

Theorem B.6 (The Delta Method) *Let* $g\colon \mathbb{R}^d \to \mathbb{R}^k$ *be a function continuously differentiable in a neighborhood of* $\boldsymbol{\theta} \in \mathbb{R}^d$. *If* $\mathbf{T_n}$ *is a sequence of d-dimensional random vectors such that*

$$\sqrt{n}(\mathbf{T_n} - \boldsymbol{\theta}) \xrightarrow{d} \mathbf{T},$$

then

$$\sqrt{n}\big(g(\mathbf{T_n}) - g(\boldsymbol{\theta})\big) \xrightarrow{d} g'(\boldsymbol{\theta})\mathbf{T}.$$

In particular, if

$$\sqrt{n}(\mathbf{T_n} - \boldsymbol{\theta}) \xrightarrow{d} \mathbf{T} \sim \mathcal{N}(\mathbf{0}, \boldsymbol{\Sigma}),$$

then

$$\sqrt{n}\big(g(\mathbf{T_n}) - g(\boldsymbol{\theta})\big) \xrightarrow{d} \mathbf{Y} \sim \mathcal{N}(\mathbf{0}, g'(\boldsymbol{\theta})\boldsymbol{\Sigma}(g'(\boldsymbol{\theta}))^T),$$

where $(g'(\boldsymbol{\theta}))^T$ *is the transpose of the matrix* $g'(\boldsymbol{\theta})$.

Example B.7 Assume that $g(x,y) = \frac{x}{y}$ or $g(x,y) = \log \frac{x}{y}$ and

$$\sqrt{n}(\mathbf{T_n} - \mathbf{1}) \xrightarrow{d} \mathcal{N}(\mathbf{0}, \boldsymbol{\Sigma}).$$

Then

$$\sqrt{n}\left(\frac{T_n^{(1)}}{T_n^{(2)}} - 1\right) \xrightarrow{d} \mathcal{N}(0,\sigma^2) \quad \text{and} \quad \sqrt{n}\left(\log \frac{T_n^{(1)}}{T_n^{(2)}} - \log 1\right) \xrightarrow{d} \mathcal{N}(0,\sigma^2),$$

where $\sigma^2 = \Sigma_{11} - 2\Sigma_{12} + \Sigma_{22}$.

B.1.4 Central and Non-central Limit Theorems for Stationary Gaussian Sequences

Let $H_0(x) = 1$ and

$$H_n(x) = (-1)^n e^{x^2/2} \frac{d^n}{dx^n} e^{-x^2/2},\ n \geq 1,$$

be the nth Hermite polynomial. The Hermite polynomials form an orthogonal base in $L_2(\mathbb{R}, \gamma)$, where γ is the probability distribution of a standard Gaussian random variable on $\mathbb{R}$. In other words, any function $G \in L_2(\mathbb{R}, \gamma)$ admits the Hermite expansion

$$G(x) = \sum_{k=0}^{\infty} a_k H_k(x), \tag{B.1}$$

convergent in $L_2(\mathbb{R}, \gamma)$. Equivalently, if ζ is a standard Gaussian variable and $G\colon \mathbb{R} \to \mathbb{R}$ is such that $\mathsf{E}G^2(\zeta) < \infty$, then

$$G(\zeta) = \sum_{k=0}^{\infty} a_k H_k(\zeta),$$

and the series converges in $L_2(\Omega)$. The smallest k such that $a_k \neq 0$ is called the Hermite rank of G.

The following result [32, Th. 1] is a CLT for stationary sequences with short-range dependence.

Theorem B.8 (Breuer–Major) *Let $\{\xi_n, n \geq 1\}$ be a stationary Gaussian sequence with $\mathsf{E}\xi_1 = 0$, $\mathsf{E}\xi_1^2 = 1$, $\mathsf{E}\xi_n\xi_{n+k} = r(k)$, $k \geq 1$. Consider the sum*

$$S_n = \sum_{k=1}^{n} G(\xi_k)$$

with $G \in L_2(\mathbb{R}^2, \gamma)$ having Hermite rank $m \geq 1$. Assume that

$$\sum_{k=-\infty}^{\infty} |r(k)|^m < \infty. \tag{B.2}$$

Then

$$\sigma_l^2 = \lim_{n\to\infty} n^{-1} \sum_{i,j=0}^{n} r(i-j)^l$$

exists for all $l \geq m$,

$$\sigma^2 = \sum_{l=m}^{\infty} a_l^2 l! \sigma_l^2 < \infty,$$

and

$$\frac{S_n}{\sqrt{n}} \xrightarrow{d} \mathcal{N}(0, \sigma^2), \quad n \to \infty,$$

where a_l, $l \geq m$, are the coefficients in the decomposition (B.1) *of the function G.*

In the following theorem [32, Th. 1′] condition (B.2) is slightly weakened. Recall that a Borel function $L\colon [a, \infty) \to \mathbb{R}$, $a > 0$, is slowly varying at infinity if for all $\lambda > 0$

$$\lim_{t\to\infty} \frac{L(\lambda t)}{L(t)} = 1.$$

Theorem B.9 *Let $\{\xi_n, n \geq 1\}$ be a stationary Gaussian sequence with $\mathsf{E}\xi_1 = 0$, $\mathsf{E}\xi_1^2 = 1$, $\mathsf{E}\xi_n\xi_{n+k} = r(k)$, $k \geq 1$. Let the function G and the sum S_n be the same as in Theorem B.8. Assume that*

$$\sum_{k=-n}^{n} |r(k)|^m = L(n)$$

and

$$\lim_{n\to\infty} \frac{1}{L(n)} \sum_{k=-n}^{n} |r(k)|^l$$

exists for all $l \geq m$, where L is a slowly varying function at infinity. Then

$$\sigma_l^2 = \lim_{n\to\infty} \frac{1}{nL(n)} \sum_{i,j=0}^{n} r(i-j)^l$$

exists for all $l \geq m$,

$$\sigma^2 = \sum_{l=m}^{\infty} a_l^2 l! \sigma_l^2 < \infty,$$

and

$$\frac{S_n}{\sqrt{nL(n)}} \xrightarrow{d} \mathcal{N}(0,\sigma^2), \quad n \to \infty,$$

where a_l, $l \geq m$, are the coefficients in the decomposition (B.1) *of the function G.*

In the case of long-range dependence (when the condition (B.2) does not hold), we have the following limit theorem. It follows from Corollary 5.1 and Theorem 6.1 from [163] (where a functional analogue of the below results is proved).

Theorem B.10 (Taqqu) *Let $\{\xi_n, n \geq 1\}$ be a stationary Gaussian sequence with $\mathsf{E}\xi_1 = 0$, $\mathsf{E}\xi_1^2 = 1$, $\mathsf{E}\xi_n\xi_{n+k} = r(k)$, $k \geq 1$.*

Assume that for some integer $m \geq 1$ and $H \in (1 - 1/(2m), 1)$, the covariance r satisfies $r(n) \sim n^{2H-2}L(n)$, $n \to \infty$, where L is some slowly varying function at infinity.

Consider the sum

$$S_n = \sum_{k=1}^{n} G(\xi_k)$$

with $G \in L_2(\mathbb{R}^2, \gamma)$ having Hermite rank m.

1. If $m = 1$ (and hence $H \in (1/2, 1)$), then

$$\frac{S_n}{n^H\sqrt{L(n)}} \xrightarrow{d} \frac{a_1}{\sqrt{H(2H-1)}} \mathcal{N}(0,1), \ n \to \infty,$$

where $a_1 = \mathsf{E}\big(\zeta G(\zeta)\big)$.

2. *If* $m = 2$ *(and hence* $H \in (3/4, 1)$*), then*

$$\frac{S_n}{n^{2H-1}L(n)} \xrightarrow{d} \frac{b_2}{\sqrt{2(2H-1)(4H-3)}}\zeta_{2H-1},\ n \to \infty,$$

where $b_2 = \mathsf{E}\big(\zeta^2 G(\zeta)\big)$*, and* ζ_{2H-1} *has the Rosenblatt distribution with the Hurst parameter* $2H-1$*, considered in Sect. B.2.1.*

Theorem B.8 can be generalized to the multivariate case. Let $(Z_i)_{i\geq 1}$, $Z_i = (Z_{i,1}, \ldots, Z_{i,d})^T$ be a stationary sequence of centered $\mathbb{R}^d$-valued Gaussian random vectors. For $k \in \mathbb{Z}$ and $p, q \in \{1, \ldots, d\}$ denote

$$r^{(p,q)}(k) = \mathsf{E}Z_{m,p}Z_{m+k,q},$$

where m is any positive integer satisfying $m + k \geq 1$. Note that

$$r^{(p,q)}(0) = \mathsf{E}Z_{m,p}Z_{m,q} = \mathsf{E}Z_{1,p}Z_{1,q} \tag{B.3}$$

and

$$r^{(p,q)}(-k) = \mathsf{E}Z_{m,p}Z_{m-k,q} = \mathsf{E}Z_{k+1,p}Z_{(k+1)-k,q} = \mathsf{E}Z_{k+1,p}Z_{1,q} = r^{(q,p)}(k) \tag{B.4}$$

for $k \geq 1$.

We make use of the following result given in Arcones [4, Theorem 2].

Theorem B.11 ([4, Theorem 2]) *Assume that the function* $f\colon \mathbb{R}^d \to \mathbb{R}$ *is measurable,* $\mathsf{E}f^2(Z_1) < \infty$*, and for each* $(p, q) \in \{(i,j) \mid i,j = 1, \ldots, d\}$ *there exist finite limits*

$$\lim_{n\to\infty} \frac{1}{n} \sum_{j,k=1}^{n} r^{(p,q)}(j-k), \qquad \lim_{n\to\infty} \frac{1}{n} \sum_{j,k=1}^{n} \left(r^{(p,q)}(j-k)\right)^2. \tag{B.5}$$

Then

$$\frac{1}{\sqrt{n}} \sum_{i=1}^{n} \big(f(Z_i) - \mathsf{E}f(Z_1)\big) \xrightarrow{d} \mathcal{N}\left(0, \sigma_f^2\right), \tag{B.6}$$

where

$$\sigma_f^2 = \mathrm{Var}(f(Z_1)) + 2\sum_{k=1}^{\infty} \mathrm{cov}(f(Z_1), f(Z_{1+k})).$$

B.2 Wiener Chaos, Multiple Integrals, and Rosenblatt Distribution

In this section we briefly recall the notion of Wiener chaos. More information may be found in [81, 139].

Let $\mathcal{H}$ be a real separable Hilbert space with scalar product $\langle\cdot,\cdot\rangle_{\mathcal{H}}$ and the corresponding norm $\|\cdot\|_{\mathcal{H}}$.

Definition B.12 A collection of centered Gaussian random variables r. vs. $\{W(h), h \in \mathcal{H}\}$ is called an *isonormal Gaussian process* in $\mathcal{H}$, if for any $h_1, h_2 \in \mathcal{H}$,

$$\mathsf{E}W(h_1)W(h_2) = \langle h_1, h_2\rangle_{\mathcal{H}}.$$

Remark B.13 An isonormal Gaussian process exists for any Hilbert space. Indeed, let $\{\xi_n, n \geq 1\}$ be independent standard Gaussian random variables, $\{e_n, n \geq 1\}$ be an orthonormal base of $\mathcal{H}$. The series

$$W(h) = \sum_{n\geq 1}\langle h, e_n\rangle \xi_n,$$

converges a.s. thanks to the Kolmogorov theorem, and we can set $W(h) = 0$ for $\omega \in \Omega$ such that the series diverges. Then $W(h)$ is an isonormal Gaussian process. Indeed,

$$\mathsf{E}W(h_1)W(h_2) = \mathsf{E}\sum_{n=1}^{\infty}\langle h_1, e_n\rangle \xi_n \sum_{m=1}^{\infty}\langle h_2, e_m\rangle \xi_m = \sum_{n,m=1}^{\infty}\langle h_1, e_n\rangle\langle h_2, e_m\rangle \mathsf{E}\xi_n\xi_m$$
$$= \sum_{n=1}^{\infty}\langle h_1, e_n\rangle\langle h_2, e_n\rangle = \langle h_1, h_2\rangle_{\mathcal{H}}.$$

Remark B.14 By definition, an isonormal Gaussian process is a map from $\mathcal{H}$ into a subspace of $L^2(\Omega)$, which preserves the scalar product. Then this map is linear. Indeed, for any $h_1, h_2 \in \mathcal{H}$, $a_1, a_2 \in \mathbb{R}$,

$$\mathsf{E}\big(W(a_1h_1 + a_2h_2) - a_1W(h_1) - a_2W(h_2)\big)^2_{\mathcal{H}}$$
$$= \|a_1h_1 + a_2h_2\|^2_{\mathcal{H}} + \|a_1h_1\|^2_{\mathcal{H}} + \|a_2h_2\|^2_{\mathcal{H}}$$
$$-2a_1\langle a_1h_1 + a_2h_2, h_1\rangle_{\mathcal{H}} - 2a_2\langle a_1h_1 + a_2h_2, h_2\rangle_{\mathcal{H}} + 2a_1a_2\langle h_1, h_2\rangle_{\mathcal{H}} = 0,$$

so $W(a_1h_1 + a_2h_2) = a_1W(h_1) + a_2W(h_2)$ a.s. This also implies that all finite linear combinations of the values of isonormal Gaussian process have Gaussian distribution, meaning that W is a Gaussian process indexed by $\mathcal{H}$.

Let further $\mathcal{L}^2$ denote the space of square integrable random variables, measurable with respect to the sigma-algebra $\mathcal{F}^W$ generated by the isonormal Gaussian process W. Further we will establish a decomposition of $\mathcal{L}^2$ into orthogonal sum of special subspaces, the so-called chaos decomposition.

For $n \geq 1$, consider the nth tensor power $\mathcal{H}_n$ of $\mathcal{H}$, i.e. the set of linear combinations of expressions of the form

$$h_1 \otimes \cdots \otimes h_n,\ h_1, \dots, h_n \in \mathcal{H}.$$

Define the scalar product

$$\langle h_1 \otimes \cdots \otimes h_n, g_1 \otimes \cdots \otimes g_n \rangle_{\mathcal{H}_n} = \prod_{k=1}^{n} \langle h_i, g_i \rangle_{\mathcal{H}}$$

and extend it by linearity to $\mathcal{H}_n$. The completion of $\mathcal{H}_n$ with respect to this norm will be denoted by $\mathcal{H}^{\otimes n}$.

Further, define the symmetrization operator $f \in \mathcal{H}^{\otimes n} \mapsto \tilde{f} \in \mathcal{H}^{\otimes n}$, extending the map

$$h_1 \otimes \cdots \otimes h_n \mapsto \frac{1}{n!} \sum_{\sigma \in S_n} h_{\sigma(1)} \otimes \cdots \otimes h_{\sigma(n)},$$

where the sum runs over permutations of the set $\{1, \dots, n\}$, by linearity and continuity to $\mathcal{H}_n$. Call an element $f \in \mathcal{H}^{\otimes n}$ symmetric if $\tilde{f} = f$ and denote by $\mathcal{H}^{\odot n}$ be space of such elements (it is called the n-fold symmetric tensor power of $\mathcal{H}$). This is a Hilbert space with the scalar product inherited from $\mathcal{H}^{\otimes n}$.

Let

$$H_n(x) = (-1)^n e^{x^2/2} \frac{d^n}{dx^n} e^{-x^2/2},\ n \geq 1,$$

be the nth Hermite polynomial. For $h \in \mathcal{H}$ with $\|h\| = 1$ define

$$I_n(h^{\otimes n}) := I_n(h \otimes \cdots \otimes h) = H_n(W(h)).$$

This map can be extended by linearity and continuity to the whole $\mathcal{H}^{\odot n}$.

For $f \in \mathcal{H}^{\odot n}$, the corresponding value $I_n(f)$ may be called the multiple Wiener–Itô integral of f. However, this name is more appropriate when $\mathcal{H} = L^2(X, \mathcal{B}, \mu)$, where W generates a measure on X by $W(B) = W(\mathbf{1}_B)$, $B \in \mathcal{B}$. In this case $\mathcal{H}^{\odot n}$ can be identified with the space of symmetric functions $f: X^n \to \mathbb{R}$, measurable with respect to $\mathcal{B}^{\otimes n}$ and square integrable with respect to the product measure $\mu^{\otimes n}$, and one usually writes

$$I_n(f) = \int_{X^n} f(x_1, \dots, x_n) W(dx_1) \cdots W(dx_n).$$

For general case, there is no measure, so it is not clear what the word "integral" refers to, but this name still seems appropriate.

By construction, multiple Wiener-Itô integrals are linear in the argument. They are also centered:

$$\mathsf{E} I_n(f) = 0$$

and have the following property of orthogonality and isomorphism: for $f \in \mathcal{H}^{\odot n}$, $g \in \mathcal{H}^{\odot m}$,

$$\mathsf{E} I_n(f) I_m(g) = \begin{cases} n!\langle f, g\rangle_{\mathcal{H}_n}, & n = m, \\ 0, & n \neq m; \end{cases}$$

in particular, there is an isometry property

$$\mathsf{E}(I_n(f)^2) = n!\|f\|^2_{\mathcal{H}_n}.$$

For a non-symmetric $f \in \mathcal{H}^{\otimes n}$, the multiple Wiener-Itô integral may be defined as an integral of symmetrization $I_n(f) = I_n(\tilde{f})$. With such definition, the isometry fails in general, but it is still true that $\left(\mathsf{E} I_n(f)^2\right) \le n!\|f\|^2_{\mathcal{H}_n}$.

Let $\mathcal{I}_n \subset \mathcal{L}^2$ be the space of multiple stochastic integrals of order n, i.e. $\mathcal{I}_n = I_n(\mathcal{H}^{\odot n})$. By definition, it is equivalent to say that $\mathcal{I}_n$ is a closed linear span of r. vs. of the form $H_n(W(h))$. Thanks to the orthogonality property, the spaces $\mathcal{I}_n$ are orthogonal for different values of n. Define also $\mathcal{I}_0$ the space of constant random variables. Since multiple Wiener-Itô integrals are centered, this subspace is orthogonal to all $\mathcal{I}_n$ with $n \ge 1$. For convenience define $\mathcal{H}^{\odot 0} = \mathbb{R}$ and $I_0(f) = f$ for $f \in \mathcal{H}^{\odot 0}$.

Theorem B.15 *The space $\mathcal{L}^2$ is the direct sum of spaces $\mathcal{I}_n$:*

$$\mathcal{L}^2 = \bigotimes_{n=0}^{\infty} \mathcal{I}_n.$$

In other words, any square integrable r.v. ξ, measurable with respect to $\mathcal{F}^W$, has a unique orthogonal decomposition

$$\xi = \sum_{n=0}^{\infty} I_n(f_n)$$

with $f_n \in \mathcal{H}^{\odot n}$, where the series converges in the mean-square sense.

Remark B.16 Since the multiple Wiener-Itô integrals have mean zero, in this decomposition $I_0(f_0) = f_0 = \mathsf{E}\xi$.

We will need the following property of hypercontractivity of multiple Wiener-Itô integrals.

Theorem B.17 ([81, Corollary 7.36]) *For each $n, m \geq 1$, there is a universal (independent of the Hilbert space $\mathcal{H}$) constant $C_{n,m} > 0$ such that for any $\xi \in \mathcal{I}_n$,*

$$\mathsf{E}\xi^{2m} \leq C_{n,m} \left(\mathsf{E}\xi^2\right)^m.$$

A similar estimate then holds for r. vs. having finite chaos expansion.

Corollary B.18 *For each $n, m \geq 1$, there is a universal constant $K_{n,m} > 0$ such that for any $\xi \in \mathcal{L}^2$ of the form*

$$\xi = \sum_{k=0}^{n} I_k(f_k)$$

with $f_k \in \mathcal{H}^{\odot k}$, $k = 0, 1, \ldots, n$, the following estimate

$$\mathsf{E}\xi^{2m} \leq K_{n,m} \left(\mathsf{E}\xi^2\right)^m.$$

Proof By Jensen's inequality,

$$\xi^{2m} \leq n^{2m-1} \sum_{k=0}^{n} I_k(f_k)^{2m}.$$

Therefore, using Theorem B.17,

$$\mathsf{E}\xi^{2m} \leq n^{2m-1} C_{n,m} \sum_{k=0}^{n} \left(\mathsf{E}I_k(f_k)^2\right)^m. \tag{B.7}$$

Thanks to orthogonality,

$$\mathsf{E}\xi^2 = \sum_{k=0}^{n} \mathsf{E}I_k(f_k)^2,$$

in particular, $\mathsf{E}I_k(f_k)^2 \leq \mathsf{E}\xi^2$, $k = 0, \ldots, n$. Plugging this into (B.7), we get

$$\mathsf{E}\xi^{2m} \leq n^{2m} C_{n,m} \left(\mathsf{E}\xi^2\right)^m.$$

□

B.2.1 Rosenblatt Random Variable

Let $W = \{W_x, x \in \mathbb{R}\}$ be a two-sided standard Wiener process, $\mathcal{H} = L^2(\mathbb{R})$. Then, by the properties of Wiener integral, $W(h) = \int_{\mathbb{R}} h(x)dW_x$, $h \in \mathcal{H}$, is an isonormal Gaussian process. Define for $H \in (1/2, 1)$ the kernel

$$f_H(x_1, x_2) = \int_0^1 (s - x_1)_+^{H/2-1} (s - x_2)_+^{H/2-1} ds.$$

Then $f_H \in L^2(\mathbb{R}^2) = \mathcal{H}^{\otimes 2}$. Indeed,

$$\int_{\mathbb{R}^2} \left(\int_0^1 (s - x_1)_+^{H/2-1} (s - x_2)_+^{H/2-1} ds \right)^2 dx_1\, dx_2 = \int_0^1 \int_0^1 g(s_1, s_2)^2 ds_1\, ds_2,$$

where

$$\begin{aligned} g(s_1, s_2) &= \int_{\mathbb{R}} (s_1 - x)_+^{H/2-1} (s_2 - x)_+^{H/2-1} dx \\ &= |s_1 - s_2|^{H-1} \int_0^\infty z^{H/2-1} (1 + z)^{H/2-1} dz = |s_1 - s_2|^{H-1} \mathrm{B}(H/2, 1 - H), \end{aligned}$$

so

$$\|f_H\|_{L^2(\mathbb{R}^2)}^2 = \mathrm{B}(H/2, 1 - H)^2 \int_0^1 \int_0^1 |s_1 - s_2|^{2H-2} ds_1 ds_2 = \frac{\mathrm{B}(H/2, 1 - H)^2}{H(2H - 1)}.$$

Moreover, f_H is symmetric, so $f_H \in \mathcal{H}^{\odot 2}$. In view of this, we can define the double integral

$$I_2(f_H) = \int_{\mathbb{R}^2} f_H(x_1, x_2) dW_{x_1} dW_{x_2}.$$

Definition B.19 A random variable ζ has a *standard Rosenblatt distribution* with Hurst parameter $H \in (1/2, 1)$, if

$$\zeta_H \stackrel{d}{=} \frac{I_2(f_H)}{\sqrt{2}\, \|f_H\|_{L^2(\mathbb{R}^2)}} = \frac{\sqrt{H(H - 1/2)}}{\mathrm{B}(H/2, 1 - H)} I_2(f_H).$$

Remark B.20 From the properties of multiple integrals, we have $\mathsf{E}\zeta_H = 0$ and

$$\mathsf{E}\zeta_H^2 = \frac{\mathsf{E}\big(I_2(f_H)^2\big)}{2\, \|f_H\|_{L^2(\mathbb{R}^2)}^2} = 1,$$

so ζ_H indeed has a standardized distribution.

B.3 Exponential Maximal Bounds and Asymptotic Growth of Trajectories of Gaussian Processes

In the present section we obtain exponential maximal bounds for the trajectories of Gaussian processes. These results enable us to derive bounds for the asymptotic growth of an fBm and its fractional derivative, as well as of an mfBm and its increments (an mfBm is studied in the next section).

Let $\mathbf{T}$ be a parameter set and $X = \{X(t), t \in \mathbf{T}\}$ be a centered Gaussian process. Introduce the notation

$$\rho_X(t,s) = \left(\mathsf{E}\big(X(t) - X(s)\big)^2\right)^{1/2}, \quad s,t \in \mathbf{T}.$$

Evidently, ρ_X is a pseudometric on $\mathbf{T}$. Also, denote

$$m(\mathbf{T}) = \sup_{t \in \mathbf{T}} \left(\mathsf{E}\,|X(t)|^2\right)^{1/2}.$$

Throughout the section we assume that the following conditions hold.

($\mathbf{I}_1$) $m(\mathbf{T}) < \infty$.
($\mathbf{I}_2$) The space $(\mathbf{T}, \rho_X)$ is separable and the process X is separable on this space.

The section is organized as follows. In Sect. B.3.1 we present exponential maximal upper bounds in terms of metric massiveness for a Gaussian processes defined on an arbitrary parameter set $\mathbf{T}$. Further, we apply these general results to several particular cases related to fBm and mfBm, their increments and fractional derivatives. In Sect. B.3.2 we consider a Gaussian process defined on the positive half-axis, i. e., on $\mathbf{T} = \mathbb{R}_+$. We obtain exponential maximal bounds and the asymptotic growth with probability 1 of this process. These results are needed to study the asymptotic growth of an mfBm (see Sect. B.4.2). In order to obtain upper bounds with probability 1 for the increments of an mfBm, in Sect. B.3.3 we consider a two-parameter Gaussian process defined on the set $\mathbf{T} = \left\{\mathbf{t} = (t_1, t_2) \in \mathbb{R}_+^2 : t_1 - \Delta \le t_2 \le t_1\right\}$, $\Delta > 0$ (see Sect. B.4.3 for application of these results to increments of an mfBm). In Sect. B.3.4 we investigate Gaussian processes defined on the set $\mathbf{T} = \{\mathbf{t} = (t_1, t_2) \in \mathbb{R}_+^2 : 0 \le t_2 \le t_1\}$. We derive the asymptotic maximal bounds for such processes. This allows us to obtain similar bounds for a fractional derivative of an fBm. This is done in Sect. B.3.5.

B.3.1 Exponential Maximal Upper Bound for Gaussian Process in Terms of Metric Massiveness

In this subsection we present the general results concerning exponential maximal upper bound for a Gaussian process defined on an arbitrary parameter set, in terms of metric massiveness. Let $N(u), u > 0$ be the metric massiveness of the space $(\mathbf{T}, \rho_X)$, that is, $N(u)$ is the number of open balls in the minimal u-covering of $(\mathbf{T}, \rho_X)$.

We start with a particular case of the general theorem proved in [34, p. 100].

Lemma B.21 *Let*

$$\int_0^{m(\mathbf{T})} (\log N(u))^{\frac{1}{2}}\, du < \infty.$$

Then for any $\lambda > 0$ and any $\theta \in (0,1)$ the following inequality holds:

$$\mathsf{E}\exp\left\{\lambda \sup_{t\in\mathbf{T}} |X(t)|\right\} \le 2A_0(\lambda,\theta),$$

where

$$A_0(\lambda,\theta) = \exp\left\{\frac{\lambda^2 m(\mathbf{T})^2}{2(1-\theta)^2} + \frac{2\lambda}{\theta(1-\theta)}\int_0^{\theta m(\mathbf{T})} (\log N(u))^{\frac{1}{2}}\, du\right\}.$$

In order to establish a more general result, we consider the function $r(x), x \ge 1$ satisfying the following properties:

1. r is non-negative and nondecreasing;
2. $r(e^y), y \ge 0$ is a convex function.

Introduce one more notation: let $I_r(x) = \int_0^x r(N(u))\, du, x > 0$.

Theorem B.22 *Let $I_r(m(\mathbf{T})) < \infty$. Then the following bounds hold:*

(i) For any $\theta \in (0,1)$ and any $\lambda > 0$

$$\mathsf{E}\exp\left\{\lambda \sup_{t\in\mathbf{T}} |X(t)|\right\} \le 2A_1(\lambda,\theta), \tag{B.8}$$

where

$$A_1(\lambda,\theta) = \exp\left\{\frac{\lambda^2 m^2(\mathbf{T})}{2(1-\theta)^2}\right\} r^{(-1)}\left(\frac{I_r(\theta m(\mathbf{T}))}{\theta m(\mathbf{T})}\right),$$

$r^{(-1)}(t)$ is the generalized inverse function of $r(t)$ that is

$$r^{(-1)}(t) = \sup\{u \ge 0 : r(u) \le t\}.$$

(ii) For any $\theta \in (0,1)$ and any $\mu > 0$

$$\mathsf{P}\left\{\sup_{t\in\mathbf{T}} |X(t)| \ge \mu\right\} \le 2A_2(\mu,\theta), \tag{B.9}$$

where

$$A_2(\mu, \theta) = \exp\left\{-\frac{\mu^2(1-\theta)^2}{2m^2(\mathbf{T})}\right\} r^{(-1)}\left(\frac{I_r(\theta m(\mathbf{T}))}{\theta m(\mathbf{T})}\right).$$

Proof (*i*) First, we simplify the notation: let $m := m(\mathbf{T})$. Now our goal is to establish the following bound: for arbitrary $\theta \in (0, 1)$ and any sequence $r_n > 0$ such that $\sum_{n=1}^{\infty} \frac{1}{r_n} = 1$,

$$\mathsf{E} \exp\left\{\lambda \sup_{t\in\mathbf{T}} |X(t)|\right\} \le \prod_{n=1}^{\infty} \left[2N(m\theta^n) \exp\left\{\tfrac{1}{2}\lambda^2 r_n^2 \theta^{2(n-1)} m^2\right\}\right]^{\frac{1}{r_n}}. \tag{B.10}$$

Let $\theta \in (0, 1)$ and put $u_n = m\theta^n$, $n \ge 0$. Denote by S_n a minimal u_n-net in the set $\mathbf{T}$ with respect to the pseudometric ρ_X and put $S = \bigcup_{n=0}^{\infty} S_n$. According to condition (I_2), the set S is countable and everywhere dense in $\mathbf{T}$ with respect to the pseudometric ρ_X, and the process X is continuous in probability in $(\mathbf{T}, \rho_X)$. Therefore the set S is a ρ_X-separability set for the process X and moreover

$$\sup_{t\in\mathbf{T}} |X(t)| = \sup_{t\in S} |X(t)|$$

with probability 1.

Suppose that $t \in S$. Then there exists a number $n(t)$ such that $t \in S_{n(t)}$. Define a function $\alpha_k : S \to S_k$, $k \ge 0$ as $\alpha_k(x) = x$ if $x \in S_k$ and $\alpha_k(x)$ is the point of S_k closest to x if $x \notin S_k$. If there is more than one closest point then we may choose any of these points. The family of maps $\{\alpha_k, k \ge 0\}$ is called the α-procedure for choosing points in S. Using the α-procedure we can choose a sequence of points $t_{n(t)} = t$, $t_{n(t)-1} = \alpha_{n(t)-1}\left(t_{n(t)}\right), \dots, t_1 = \alpha_1(t_2)$ such that $t_k \in S_k$, $k = 1, \dots, n(t)$ and $\rho_X(t_k, \alpha_{k-1}(t_k)) \le u_{k-1}$.

Evidently,

$$X(t) = X(t_1) + \sum_{k=2}^{n(t)} (X(t_k) - X(t_{k-1})).$$

Therefore we have an upper bound

$$\sup_{t\in S} |X(t)| \le \max_{s\in S_1} |X(s)| + \sum_{n=2}^{\infty} \max_{s\in S_n} |X(s) - X(\alpha_{n-1}(s))|.$$

Take any sequence of numbers $r_n > 0, n \geq 1$ such that $\sum_{n=1}^{\infty} r_n^{-1} = 1$. It follows from the Hölder inequality that for any $\lambda > 0$

$$\mathsf{E}\exp\left\{\lambda \sup_{t\in S} |X(t)|\right\} \leq \mathsf{E}\exp\left\{\lambda\left(\max_{s\in S_1}|X(s)| + \sum_{n=2}^{\infty}\max_{s\in S_n}|X(s) - X(\alpha_{n-1}(s))|\right)\right\}$$

$$\leq \left[\mathsf{E}\exp\left\{\lambda r_1 \max_{s\in S_1}|X(s)|\right\}\right]^{\frac{1}{r_1}} \prod_{n=2}^{\infty}\left[\mathsf{E}\exp\left\{\lambda r_n \max_{s\in S_n}|X(s) - X(\alpha_{n-1}(s))|\right\}\right]^{\frac{1}{r_n}}. \tag{B.11}$$

Furthermore, all the multipliers in the right-hand side of (B.11), except the 1st one, can be bounded as

$$\mathsf{E}\exp\left\{\lambda r_n \max_{s\in S_n}|X(s) - X(\alpha_{n-1}(s))|\right\}$$
$$\leq N(u_n) \max_{s\in S_n} \mathsf{E}\exp\{\lambda r_n |X(s) - X(\alpha_{n-1}(s))|\}, \tag{B.12}$$

and, in addition,

$$\left(\mathsf{E}\,|X(s) - X(\alpha_{n-1}(s))|^2\right)^{\frac{1}{2}} = \rho_X(s, \alpha_{n-1}(s)) \leq u_{n-1} = \theta^{n-1}u_0$$

for $s \in S_n$. Therefore, for any $n \geq 2$

$$\max_{s\in S_n} \mathsf{E}\exp\{\lambda r_n |X(s) - X(\alpha_{n-1}(s))|\} \leq \max_{s\in S_n}\Big(\mathsf{E}\exp\{\lambda r_n (X(s) - X(\alpha_{n-1}(s)))\}$$
$$+\mathsf{E}\exp\{-\lambda r_n (X(s) - X(\alpha_{n-1}(s)))\}\Big)$$
$$= \max_{s\in S_n} 2\exp\left\{\frac{\lambda^2}{2} r_n^2 \mathsf{E}(X(s) - X(\alpha_{n-1}(s)))^2\right\} \leq 2\exp\left\{\frac{\lambda^2}{2} r_n^2 (\theta^{n-1}u_0)^2\right\}. \tag{B.13}$$

Now we estimate the first multiplier in the right-hand side of (B.11):

$$\mathsf{E}\exp\left\{\lambda r_1 \max_{s\in S_1}|X(s)|\right\} \leq N(u_1)\max_{s\in S_1}\mathsf{E}\exp\{\lambda r_1 |X(s)|\}$$
$$\leq 2N(u_1)\exp\left\{\frac{\lambda^2 r_1^2}{2}\max_{s\in S_1}\mathsf{E}\,|X(s)|^2\right\} \leq 2N(u_1)\exp\left\{\frac{\lambda^2 r_1^2}{2}u_0^2\right\}. \tag{B.14}$$

Taking into account that it follows from the separability of X that

$$\mathsf{E}\exp\left\{\lambda \sup_{t\in\mathbf{T}} |X(t)|\right\} = \mathsf{E}\exp\left\{\lambda \sup_{t\in S} |X(t)|\right\},$$

we get inequality (B.10) from (B.11)–(B.14). It follows from (B.10) that

$$\mathsf{E}\exp\left\{\lambda \sup_{t\in\mathbf{T}} |X(t)|\right\} \le 2\exp\left\{\sum_{n=1}^{\infty}\frac{1}{r_n}H(m\theta^n) + \sum_{n=1}^{\infty}\frac{\lambda^2 m^2}{2}\theta^{2(n-1)}r_n\right\},$$

where $H(u) = \log N(u)$ is the metric entropy. Now, choose $r_n = \frac{1}{\theta^{n-1}(1-\theta)}$. Then

$$\mathsf{E}\exp\left\{\lambda \sup_{t\in\mathbf{T}} |X(t)|\right\} \le 2\exp\left\{(1-\theta)\sum_{n=1}^{\infty}\theta^{n-1}H(m\theta^n) + \frac{\lambda^2 m^2}{2(1-\theta)^2}\right\}. \tag{B.15}$$

Since $r(e^x)$ is a convex function, we have that

$$\begin{aligned} & r^{(-1)}\left(r\left(\exp\left\{\sum_{n=1}^{\infty}(1-\theta)\theta^{n-1}H(\theta^n m)\right\}\right)\right) \\ & \le r^{(-1)}\left(\sum_{n=1}^{\infty}(1-\theta)\theta^{n-1}r\left(\exp\{H(\theta^n m)\}\right)\right) \\ & = r^{(-1)}\left(\sum_{n=1}^{\infty}(1-\theta)\theta^{n-1}r\left(N(\theta^n m)\right)\right) \\ & \le r^{(-1)}\left(\sum_{n=1}^{\infty}(1-\theta)\theta^{n-1}\frac{1}{m\theta^n(1-\theta)}\int_{\theta^{n+1}m}^{\theta^n m} r(N(u))du\right) \\ & = r^{(-1)}\left(\frac{1}{m\theta}\int_0^{m\theta} r(N(u))du\right). \end{aligned} \tag{B.16}$$

Now, inequality (B.8) follows from (B.15) and (B.16).

(*ii*) Now we are in position to establish inequality (B.9). Let $u > 0$, $0 < \theta < 1$, $\lambda > 0$. Then Chebyshev's inequality and (B.8) yield that

$$\begin{aligned} \mathsf{P}\left\{\sup_{t\in\mathbf{T}} |X(t)| \ge u\right\} & \le \mathsf{E}\exp\left\{\lambda \sup_{t\in\mathbf{T}} |X(t)|\right\}\exp\{-\lambda u\} \\ & \le 2r^{(-1)}\left(\frac{I_r(\theta m)}{\theta m}\right)\exp\left\{\frac{\lambda^2 m^2}{2(1-\theta)^2} - \lambda u\right\}. \end{aligned}$$

Minimizing $\frac{\lambda^2 m^2}{2(1-\theta)^2} - \lambda u$ with respect to $\lambda > 0$, we note that minimum is achieved at the point $\lambda = \frac{u(1-\theta)^2}{m^2}$, whence (B.9) immediately follows. □

Applying this result to the parameter set $\mathbf{T} = [a, b]$, we get the following result.

Corollary B.23 *Let* $\mathbf{T} = [a, b]$, $X = \{X(t), t \in [a, b]\}$ *be a centered separable Gaussian process and* $m := m([a, b]) = \sup_{t\in[a,b]} \left(\mathsf{E}|X(t)|^2\right)^{1/2} < \infty$. *Assume that there exists a strictly increasing function* $\sigma = \{\sigma(h), h > 0\}$ *such that* $\sigma(h) > 0$, $h > 0$, $\sigma(h) \downarrow 0$ *as* $h \downarrow 0$, *and*

$$\sup_{|t-s|<h} \left(\mathsf{E}|X(t) - X(s)|^2\right)^{1/2} \le \sigma(h).$$

Then for any $\theta \in (0, 1)$ *and any* $\lambda > 0$

$$\mathsf{E}\exp\left\{\lambda \sup_{t\in[a,b]} |X(t)|\right\} \le 2A_3(\lambda, \theta), \tag{B.17}$$

where

$$A_3(\lambda, \theta) = \exp\left\{\frac{\lambda^2 m^2}{2(1-\theta)^2}\right\} r^{(-1)}\left(\frac{\widetilde{I}_r(\theta m)}{\theta m}\right), \tag{B.18}$$

and

$$\widetilde{I}_r(x) = \int_0^x r\left(\frac{b-a}{2\sigma^{(-1)}(v)} + 1\right) dv.$$

Proof In this case condition (I_2) holds and $N(v) \le \frac{b-a}{2\sigma^{(-1)}(v)} + 1$, whence (B.17)–(B.18) immediately follow. □

Corollary B.24 *Put* $\sigma(h) = ch^\beta$ *with* $c > 0$, $0 < \beta \le 1$ *in Corollary B.23. Then for any* $\theta \in (0, 1)$ *and any* $\lambda > 0$

$$\mathsf{E}\exp\left\{\lambda \sup_{t\in[a,b]} |X(t)|\right\} \le 2^{\frac{2}{\beta}} \exp\left\{\frac{\lambda^2 m^2}{2(1-\theta)^2}\right\} \left(\frac{2^{2/\beta-1}(b-a)c^{1/\beta}}{(\theta m)^{1/\beta}} + 1\right). \tag{B.19}$$

Proof Consider $r(x) = x^\alpha - 1$, $x \ge 1$, where $0 < \alpha < \beta$. Since $\sigma^{(-1)}(s) = \left(\frac{s}{c}\right)^{1/\beta}$, we have

$$\begin{aligned}\widetilde{I}_r(\theta m) &= \int_0^{\theta m} \left(\left(\frac{b-a}{2\sigma^{(-1)}(s)} + 1\right)^\alpha - 1\right) ds \le \int_0^{\theta m} \left(\frac{b-a}{2\sigma^{(-1)}(s)}\right)^\alpha ds \\ &= \int_0^{\theta m} \left(\frac{(b-a)c^{1/\beta}}{2s^{1/\beta}}\right)^\alpha ds = \left(\frac{(b-a)c^{1/\beta}}{2}\right)^\alpha \frac{(\theta m)^{1-\alpha/\beta}}{1-\frac{\alpha}{\beta}}.\end{aligned}$$

Therefore in this case

$$A_3(\lambda,\theta) \le \exp\left\{\frac{\lambda^2 m^2}{2(1-\theta)^2}\right\}\left(\left(\frac{(b-a)c^{1/\beta}}{2}\right)^{\alpha}\frac{1}{\left(1-\frac{\alpha}{\beta}\right)(\theta m)^{\alpha/\beta}}+1\right)^{\frac{1}{\alpha}}.$$

Applying the elementary inequality $(a+b)^p \le 2^{p-1}(a^p+b^p), p \ge 1$, we get

$$A_3(\lambda,\theta) \le \exp\left\{\frac{\lambda^2 m^2}{2(1-\theta)^2}\right\}\cdot 2^{\frac{1}{\alpha}-1}\left(\frac{(b-a)c^{1/\beta}}{2\left(1-\frac{\alpha}{\beta}\right)^{1/\alpha}(\theta m)^{1/\beta}}+1\right). \tag{B.20}$$

Now (B.19) follows from (B.20) if we put $\alpha = \frac{\beta}{2}$. □

Corollary B.25 *Under the assumptions of Corollary B.24, for $x > m$*

$$\mathsf{P}\left\{\sup_{t\in[a,b]}|X(t)| > x\right\} \le \exp\left\{-\frac{x^2}{2m^2}\right\} 2^{\frac{2}{\beta}} e^{\frac{1}{2}}\left((b-a)\frac{x^{\frac{2}{\beta}}2^{\frac{3}{\beta}-1}c^{\frac{1}{\beta}}}{m^{\frac{3}{\beta}}}+1\right). \tag{B.21}$$

Proof We get from Chebyshev's inequality combined with (B.19) that

$$\begin{aligned}\mathsf{P}\left\{\sup_{t\in[a,b]}|X(t)| > x\right\} &\le \inf_{\lambda>0}\mathsf{E}\exp\{\lambda\,|X(t)|\}\,e^{-\lambda x}\\ &\le 2^{\frac{2}{\beta}}\inf_{\lambda>0}\left(\exp\left\{\frac{\lambda^2 m^2}{2(1-\theta)^2}-\lambda x\right\}\right)\left(2^{\frac{2}{\beta}-1}(b-a)c^{\frac{1}{\beta}}(\theta m)^{-\frac{1}{\beta}}+1\right)\\ &= 2^{\frac{2}{\beta}}\exp\left\{-\frac{x^2(1-\theta)^2}{2m^2}\right\}\left(2^{\frac{2}{\beta}-1}(b-a)c^{\frac{1}{\beta}}(\theta m)^{-\frac{1}{\beta}}+1\right).\end{aligned} \tag{B.22}$$

Let $x > m$. Put in (B.22) $\theta = 1-\sqrt{1-\frac{m^2}{x^2}}$ and get that

$$\begin{aligned}\mathsf{P}\left\{\sup_{t\in[a,b]}|X(t)| > x\right\} &\le 2^{\frac{2}{\beta}} e^{\frac{1}{2}}\exp\left\{-\frac{x^2}{2m^2}\right\}\\ &\times\left(2^{\frac{2}{\beta}-1}(b-a)c^{\frac{1}{\beta}}\left(\left(1-\sqrt{1-\frac{m^2}{x^2}}\right)m\right)^{-\frac{1}{\beta}}+1\right).\end{aligned} \tag{B.23}$$

Using the inequality $1-\sqrt{1-z} \ge z/2$ for $0 < z < 1$, from (B.23) we get (B.21). □

B.3.2 Exponential Maximal Upper Bound for the Weighted Gaussian Process Defined on the Half-Axis

Now, let $X = \{X(t), t \geq 0\}$ be a centered Gaussian process and $a(t) > 0$ be a continuous strictly increasing function such that $a(t) \to \infty$ as $t \to \infty$. Introduce the sequence $b_0 = 0$, $b_{k+1} > b_k$, $b_k \to \infty$ as $k \to \infty$. Denote $a_k = a(b_k)$ and $m_k = m([b_k, b_{k+1}]) = \sup_{t\in[b_k,b_{k+1}]} \left(\mathsf{E}\,|X(t)|^2\right)^{1/2}$. Our goal is to get exponential maximal upper bound for the weighted Gaussian process $\frac{X(t)}{a(t)}$, applying the above results, in particular, Corollary B.24.

Theorem B.26 *Let the following conditions hold:*

(i) There exist $c_k > 0$ *and* $0 < \beta < 1$ *such that*

$$\sup_{\substack{t,s\in[b_k,b_{k+1}]\\ |t-s|\leq h}} \left(\mathsf{E}\,|X(t) - X(s)|^2\right)^{1/2} \leq c_k h^\beta;$$

(ii) $0 < m_k < \infty$ *and* $A = \sum_{k=0}^{\infty} \frac{m_k}{a_k} < \infty;$

(iii) There exists $0 < \gamma \leq 1$ *such that*

$$\sum_{k=0}^{\infty} \frac{m_k^{1-\gamma/\beta}(b_{k+1} - b_k)^\gamma c_k^{\gamma/\beta}}{a_k} < \infty.$$

Then for any $\theta \in (0, 1)$ *and any* $\lambda > 0$

$$I(\lambda) = \mathsf{E}\exp\left\{\lambda \sup_{t>0} \frac{|X(t)|}{a(t)}\right\} \leq 2^{\frac{2}{\beta}-1} \exp\left\{\frac{\lambda^2 A^2}{2(1-\theta)^2}\right\} A_4(\theta, \gamma),$$

where

$$A_4(\theta, \gamma) = \exp\left\{\frac{1}{\gamma A}\left(\sum_{k=0}^{\infty} \frac{m_k^{1-\gamma/\beta}}{a_k}(b_{k+1} - b_k)^\gamma c_k^{\gamma/\beta}\right)\left(\frac{2^{2/\beta-1}}{\theta^{1/\beta}}\right)^\gamma\right\}.$$

Proof Let $r_k > 0$, $k = 0, 1, 2\ldots$ and $\sum_{k=0}^{\infty} \frac{1}{r_k} = 1$. Then for any $\lambda > 0$

$$I(\lambda) \leq \mathsf{E}\exp\left\{\lambda \sum_{k=0}^{\infty} \sup_{t\in[b_k,b_{k+1}]} \frac{|X(t)|}{a(t)}\right\} \leq \prod_{k=0}^{\infty}\left(\mathsf{E}\exp\left\{\lambda r_k \sup_{t\in[b_k,b_{k+1}]} \frac{|X(t)|}{a(t)}\right\}\right)^{\frac{1}{r_k}}.$$

It follows from Corollary B.24 that for any $\theta \in (0,1)$

$$\mathsf{E}\exp\left\{\lambda r_k \sup_{t\in[b_k,b_{k+1}]}\frac{|X(t)|}{a(t)}\right\} \le \mathsf{E}\exp\left\{\lambda r_k \sup_{t\in[b_k,b_{k+1}]}\frac{|X(t)|}{a_k}\right\}$$
$$\le 2^{\frac{2}{\beta}-1}\exp\left\{\frac{\lambda^2 r_k^2 m_k^2}{2(1-\theta)^2 a_k^2}\right\}\left(1+\frac{b_{k+1}-b_k}{\theta^{1/\beta}}2^{2/\beta-1}\left(\frac{c_k}{m_k}\right)^{1/\beta}\right).$$

Therefore,

$$I(\lambda) \le 2^{\frac{2}{\beta}-1}\exp\left\{\frac{\lambda^2}{2(1-\theta)^2}\sum_{k=0}^{\infty}\frac{r_k m_k^2}{a_k^2}\right\}\prod_{k=0}^{\infty}\left[1+\frac{b_{k+1}-b_k}{\theta^{1/\beta}}2^{2/\beta-1}\left(\frac{c_k}{m_k}\right)^{1/\beta}\right]^{\frac{1}{r_k}}$$
$$= 2^{\frac{2}{\beta}-1}\exp\left\{\frac{\lambda^2}{2(1-\theta)^2}\sum_{k=0}^{\infty}\frac{r_k m_k^2}{a_k^2}\right\}$$
$$\times\exp\left\{\sum_{k=0}^{\infty}\frac{1}{r_k}\log\left[1+\frac{b_{k+1}-b_k}{\theta^{1/\beta}}2^{2/\beta-1}\left(\frac{c_k}{m_k}\right)^{1/\beta}\right]\right\}.$$

Recall the elementary inequality: for $0<\gamma\le 1$ and $x\ge 0$,

$$\log(1+x) = \frac{1}{\gamma}\log(1+x)^{\gamma} \le \frac{x^{\gamma}}{\gamma}. \tag{B.24}$$

Taking this into account, we continue with the upper bound for $I(\lambda)$:

$$I(\lambda) \le 2^{\frac{2}{\beta}-1}\exp\left\{\frac{\lambda^2}{2(1-\theta)^2}\sum_{k=0}^{\infty}\frac{r_k m_k^2}{a_k^2}\right\}$$
$$\times\exp\left\{\frac{1}{\gamma}\sum_{k=0}^{\infty}\frac{1}{r_k}\left(\frac{b_{k+1}-b_k}{\theta^{1/\beta}}2^{2/\beta-1}\left(\frac{c_k}{m_k}\right)^{1/\beta}\right)^{\gamma}\right\}.$$

Let $r_k=\frac{Aa_k}{m_k}$. Then we get immediately the claimed upper bound:

$$I(\lambda) \le 2^{\frac{2}{\beta}-1}\exp\left\{\frac{\lambda^2A^2}{2(1-\theta)^2}\right\}\exp\left\{\frac{1}{\gamma A}\sum_{k=0}^{\infty}\frac{m_k}{a_k}\left(\frac{b_{k+1}-b_k}{\theta^{1/\beta}}2^{\frac{2}{\beta}-1}\left(\frac{c_k}{m_k}\right)^{1/\beta}\right)^{\gamma}\right\}$$
$$= 2^{\frac{2}{\beta}-1}\exp\left\{\frac{\lambda^2A^2}{2(1-\theta)^2}\right\}$$
$$\times\exp\left\{\frac{1}{\gamma A}\left(\sum_{k=0}^{\infty}\frac{m_k^{1-\gamma/\beta}}{a_k}(b_{k+1}-b_k)^{\gamma}c_k^{\gamma/\beta}\right)\left(\frac{2^{2/\beta-1}}{\theta^{1/\beta}}\right)^{\gamma}\right\}.$$

□

Remark B.27 Theorem B.26 and all subsequent results are formulated in order to finally derive the asymptotic rate of growth with probability 1 of the mfBm. In the process of obtaining this result we shall specify the choice of parameters, optimizing them in such a way, in a sense. However, for the reader's better understanding we can consider one example of "optimal" choice. We put quotation marks because the true optimization, as just mentioned, is produced in the process of further proofs, and this example is illustrative. So, let $b_k = k$, $a_k = k^{\varrho+1+\varepsilon}$, $\varrho > 0, \varepsilon > 0$, $m_k = k^{\varrho}$, $c_k = 1$. Then conditions *(ii)* and *(iii)* hold for any $\gamma \in (0, \beta]$. Put $\gamma = \beta$ (it is an optimal choice in (B.24) when the argument does not exceed e, otherwise the choice of optimal γ depends on x). Then we get

$$I(\lambda) \le 2^{\frac{2}{\beta}-1} \exp\left\{\frac{\lambda^2 A^2}{2(1-\theta)^2}\right\} \exp\left\{\frac{1}{\beta A}\left(\sum_{k=0}^{\infty} k^{-1-\varrho-\varepsilon}\right)\left(\frac{2^{2-\beta}}{\theta}\right)\right\}.$$

Obviously, analytical optimization of the latter value in θ, on the one hand is difficult, on the other hand, it will not lead to a new significant results, better than if we just put $\theta = \frac{1}{2}$.

Corollary B.28 *Let the assumptions of Theorem B.26 hold. Then for any* $\theta \in (0, 1)$ *and any* $u > 0$ *the following inequality holds:*

$$\mathsf{P}\left\{\sup_{t>0} \frac{|X(t)|}{a(t)} > u\right\} \le 2^{\frac{2}{\beta}-1} \exp\left\{-\frac{u^2(1-\theta)^2}{2A^2}\right\} A_4(\theta, \gamma). \tag{B.25}$$

Proof From Chebyshev's inequality we get that

$$\begin{aligned}\mathsf{P}\left\{\sup_{t>0} \frac{|X(t)|}{a(t)} > u\right\} &\le \frac{\mathsf{E}\exp\left\{\lambda \sup_{t>0} \frac{|X(t)|}{a(t)}\right\}}{\exp\{\lambda u\}} \\ &\le 2^{\frac{2}{\beta}-1} \exp\left\{\frac{\lambda^2 A^2}{2(1-\theta)^2} - \lambda u\right\} A_4(\theta, \gamma).\end{aligned} \tag{B.26}$$

The inequality (B.25) follows from (B.26) if we put $\lambda = \frac{u(1-\theta)^2}{A^2}$. □

Corollary B.29 *Let the assumptions of Theorem B.26 hold. Then for any* $u > A$ *we can get the following bound:*

$$\mathsf{P}\left\{\sup_{t>0} \frac{|X(t)|}{a(t)} > u\right\} \le 2^{\frac{2}{\beta}-1} e^{\frac{1}{2}} \exp\left\{-\frac{u^2}{2A^2}\right\} A_4\left(1 - \sqrt{1 - \frac{A^2}{u^2}}, \gamma\right). \tag{B.27}$$

Proof The inequality (B.27) follows from (B.25) if we put $\theta = 1 - \sqrt{1 - \frac{A^2}{u^2}}$. □

Corollary B.30 *Let the assumptions of Theorem B.26 hold. Then for all $t > 0$ we have with probability 1 the following bound,*

$$|X(t)| \le a(t)\xi,$$

where ξ is non-negative random variable whose distribution has the tail admitting the following upper bound: for any $u > A$

$$\mathsf{P}\{\xi > u\} \le 2^{\frac{2}{\beta}-1}\sqrt{e}\exp\left\{-\frac{u^2}{2A^2}\right\} A_4\left(1-\sqrt{1-\frac{A^2}{u^2}}, \gamma\right).$$

Remark B.31 Theorem B.26 remains true if the process $X(t)$ is considered on the domain $t > s$, where s is an arbitrary number. In this case we put $b_0 = s$.

B.3.3 Exponential Maximal Upper Bound for Gaussian Process in the Bounded Strip on the Plane

Additionally, we need in exponential maximal upper bound for Gaussian process defined in the bounded strip on the plane. We get it, applying Theorem B.22. So, let $0 \le a < b < \infty$, $\Delta > 0$,

$$\mathbf{T}_{a,b,\Delta} = \left\{\mathbf{t} = (t_1, t_2) \in \mathbb{R}_+^2 : a \le t_1 \le b, t_1 - \Delta \le t_2 \le t_1\right\},$$
$$d(\mathbf{t}, \mathbf{s}) = \max\left\{|t_1 - s_1|, |t_2 - s_2|\right\} \text{ for } \mathbf{t}, \mathbf{s} \in \mathbf{T}_{a,b,\Delta}.$$

Let $r(x)$, $x \ge 1$, be a non-negative nondecreasing function such that $r(e^y)$, $y \ge 0$ is a convex function.

Theorem B.32 *Assume that $X = \{X(\mathbf{t}), \mathbf{t} \in \mathbf{T}_{a,b,\Delta}\}$ is a centered separable Gaussian process satisfying the following conditions:*

$$m(\mathbf{T}_{a,b,\Delta}) = \sup_{\mathbf{t}\in\mathbf{T}_{a,b,\Delta}} \left(\mathsf{E}(X(\mathbf{t}))^2\right)^{1/2} < \infty;$$

$$\sup_{\substack{d(\mathbf{t},\mathbf{s})\le h \\ \mathbf{t},\mathbf{s}\in\mathbf{T}_{a,b,\Delta}}} \left(\mathsf{E}(X(\mathbf{t}) - X(\mathbf{s}))^2\right)^{1/2} \le \sigma(h),$$

where $\sigma = \{\sigma(h), h > 0\}$ is an increasing continuous function, $\sigma(h) \ge 0$, $\sigma(0) = 0$. If

$$\widehat{I}_r(m(\mathbf{T}_{a,b,\Delta})) = \int_0^{m(\mathbf{T}_{a,b,\Delta})} r\left(\frac{(b-a)\Delta}{4}\left(\frac{1}{\sigma^{(-1)}(v)} + d\right)^2\right) dv < \infty,$$

where $d = \max\left\{\frac{2}{b-a}, \frac{4}{\Delta}\right\}$, *then for any* $\theta \in (0,1)$ *and any* $\lambda > 0$

$$\mathsf{E}\exp\left\{\lambda \sup_{\mathbf{t}\in\mathbf{T}_{a,b,\Delta}} |X(\mathbf{t})|\right\} \le 2A_5(\lambda,\theta),$$

where

$$A_5(\lambda,\theta) = \exp\left\{\frac{\lambda^2 m^2(\mathbf{T}_{a,b,\Delta})}{2(1-\theta)^2}\right\} r^{(-1)}\left(\frac{\widehat{I}_r(\theta m(\mathbf{T}_{a,b,\Delta}))}{\theta m(\mathbf{T}_{a,b,\Delta})}\right).$$

Proof The statement follows from Theorem B.22, since in this case

$$\begin{aligned} N(v) &\le \left(\frac{b-a}{2\sigma^{(-1)}(v)} + 1\right)\left(\frac{\Delta + 2\sigma^{(-1)}(v)}{2\sigma^{(-1)}(v)} + 1\right) \\ &= \left(\frac{b-a}{2\sigma^{(-1)}(v)} + 1\right)\left(\frac{\Delta}{2\sigma^{(-1)}(v)} + 2\right) \\ &= \frac{b-a}{2}\cdot\frac{\Delta}{2}\left(\frac{1}{\sigma^{(-1)}(v)} + \frac{2}{b-a}\right)\left(\frac{1}{\sigma^{(-1)}(v)} + \frac{4}{\Delta}\right) \\ &\le \frac{(b-a)\Delta}{4}\left(\frac{1}{\sigma^{(-1)}(v)} + d\right)^2. \end{aligned}$$

□

Corollary B.33 *Let in Theorem B.32* $\sigma(h) = ch^\beta$ *for some* $c > 0$ *and* $\beta \in (0,1]$, $m = m(\mathbf{T}_{a,b,\Delta})$. *Then for all* $\lambda > 0$, $0 < \varepsilon < \beta$, $0 < \theta < 1$

$$\mathsf{E}\exp\left\{\lambda \sup_{\mathbf{t}\in\mathbf{T}_{a,b,\Delta}} |X(\mathbf{t})|\right\} \le 2^{\frac{2}{\varepsilon}-2}(b-a)\Delta \exp\left\{\frac{\lambda^2 m^2}{2(1-\theta)^2}\right\} \times \left(\frac{c^{2/\beta}}{\left(1-\frac{\varepsilon}{\beta}\right)^{2/\varepsilon}(\theta m)^{2/\beta}} + d^2\right),$$

where $d = \max\left\{\frac{2}{b-a}, \frac{4}{\Delta}\right\}$.

Proof We can choose

$$r(s) = \begin{cases} 0, & 1 \le s \le \frac{d^2(b-a)\Delta}{4}, \\ s^{\varepsilon/2} - \left(\frac{d^2(b-a)\Delta}{4}\right)^{\varepsilon/2}, & s > \frac{d^2(b-a)\Delta}{4}. \end{cases}$$

Then $r^{(-1)}(t) = \left(t + \left(\frac{d^2(b-a)\Delta}{4}\right)^{\varepsilon/2}\right)^{\frac{2}{\varepsilon}}$, and

$$\begin{aligned}
\widehat{I}_r(\theta m) &= \int_0^{\theta m} r\left(\frac{(b-a)\Delta}{4}\left(\frac{c^{1/\beta}}{v^{1/\beta}} + d\right)^2\right) dv \\
&= \int_0^{\theta m}\left(\left(\frac{(b-a)\Delta}{4}\left(\frac{c^{1/\beta}}{v^{1/\beta}} + d\right)^2\right)^{\varepsilon/2} - \left(\frac{d^2(b-a)\Delta}{4}\right)^{\varepsilon/2}\right) dv \\
&= \left(\frac{(b-a)\Delta}{4}\right)^{\frac{\varepsilon}{2}} \int_0^{\theta m}\left(\left(\frac{c^{1/\beta}}{v^{1/\beta}} + d\right)^{\varepsilon} - d^{\varepsilon}\right) dv \\
&\le \left(\frac{(b-a)\Delta}{4}\right)^{\frac{\varepsilon}{2}} \int_0^{\theta m} \frac{c^{\varepsilon/\beta}}{v^{\varepsilon/\beta}} dv = \left(\frac{(b-a)\Delta}{4}\right)^{\frac{\varepsilon}{2}} \frac{c^{\varepsilon/\beta}(\theta m)^{1-\varepsilon/\beta}}{1-\frac{\varepsilon}{\beta}}.
\end{aligned}$$

Hence,

$$\begin{aligned}
A_5(\lambda,\theta) &= \exp\left\{\frac{\lambda^2 m^2}{2(1-\theta)^2}\right\}\left(\frac{\widehat{I}_r(\theta m)}{\theta m} + \left(\frac{d^2(b-a)\Delta}{4}\right)^{\frac{\varepsilon}{2}}\right)^{\frac{2}{\varepsilon}} \\
&\le \exp\left\{\frac{\lambda^2 m^2}{2(1-\theta)^2}\right\}\frac{(b-a)\Delta}{4}\left(\frac{c^{\varepsilon/\beta}}{\left(1-\frac{\varepsilon}{\beta}\right)(\theta m)^{\varepsilon/\beta}} + d^{\varepsilon}\right)^{\frac{2}{\varepsilon}} \\
&\le \exp\left\{\frac{\lambda^2 m^2}{2(1-\theta)^2}\right\}\frac{(b-a)\Delta}{4} 2^{\frac{2}{\varepsilon}-1}\left(\frac{c^{2/\beta}}{\left(1-\frac{\varepsilon}{\beta}\right)^{2/\varepsilon}(\theta m)^{2/\beta}} + d^2\right).
\end{aligned}$$

□

Now we get the upper bound for the weighted Gaussian process defined on the bounded strip on the plane, similarly to getting Theorem B.26 from Theorem B.22. Let $\mathbf{T}_\Delta = \left\{\mathbf{t} = (t_1,t_2) \in \mathbb{R}_+^2 : t_1 - \Delta \le t_2 \le t_1\right\}$, $\Delta > 0$,

$$d(\mathbf{t},\mathbf{s}) = \max\left\{|t_1 - s_1|, |t_2 - s_2|\right\}, \quad \mathbf{s},\mathbf{t} \in \mathbf{T}_\Delta.$$

Also, let b_l be an increasing sequence such that $b_0 = 0, b_{l+1} - b_l \ge 1$, $b_l \to \infty$, $l \to \infty$, and $a(t) > 0$ is a continuous increasing function and denote $a_l = a(b_l)$,

$$\mathbf{T}_{b_l,b_{l+1},\Delta} = \left\{\mathbf{t} = (t_1,t_2) \in \mathbb{R}_+^2 : b_l \le t_1 \le b_{l+1}, t_1 - \Delta \le t_2 \le t_1\right\}.$$

Theorem B.34 *Let $0 < \Delta \le 2(b_{l+1} - b_l)$, $l \ge 0$, $X = \{X(\mathbf{t}), \mathbf{t} \in \mathbf{T}_\Delta\}$ be a centered Gaussian process satisfying following conditions:*

(i) $m_l = m(\mathbf{T}_{b_l, b_{l+1}, \Delta}) = \sup_{\mathbf{t} \in \mathbf{T}_{b_l, b_{l+1}, \Delta}} \left(\mathsf{E}(X(\mathbf{t}))^2\right)^{\frac{1}{2}} < \infty;$

(ii) there exist $\beta \in (0, 1]$ and constants $c_l > 0$ such that

$$\sup_{\substack{d(\mathbf{t},\mathbf{s}) \le h, \\ \mathbf{t},\mathbf{s} \in \mathbf{T}_{b_l, b_{l+1}, \Delta}}} \left(\mathsf{E}(X(\mathbf{t}) - X(\mathbf{s}))^2\right)^{\frac{1}{2}} \le c_l h^\beta.$$

(iii) $A = \sum_{l=0}^{\infty} \frac{m_l}{a_l} < \infty, \quad \sum_{l=0}^{\infty} \frac{m_l \log(b_{l+1} - b_l)}{a_l} < \infty,$ *and for some $\gamma \in (0, 1]$*

$$\sum_{l=0}^{\infty} \frac{m_l^{1-2\gamma/\beta} c_l^{2\gamma/\beta}}{a_l} < \infty.$$

Then for any $\theta \in (0, 1)$, $\varepsilon \in (0, \beta)$ and $\lambda > 0$

$$I(\lambda) = \mathsf{E} \exp\left\{\lambda \sup_{\mathbf{t} \in \mathbf{T}_\Delta} \frac{|X(\mathbf{t})|}{a(t_1)}\right\} \le \exp\left\{\frac{\lambda^2 A^2}{2(1-\theta)^2}\right\} A_6(\theta, \gamma, \varepsilon),$$

where

$$A_6(\theta, \gamma, \varepsilon) = \frac{2^{\frac{2}{\varepsilon}+2}}{\Delta} \exp\left\{\frac{1}{A} \sum_{l=0}^{\infty} \frac{m_l \log(b_{l+1} - b_l)}{a_l}\right\}$$
$$\times \exp\left\{\frac{\Delta^{2\gamma}}{\gamma A \left(1 - \frac{\varepsilon}{\beta}\right)^{2\gamma/\varepsilon} \theta^{2\gamma/\beta} 4^{2\gamma}} \sum_{l=0}^{\infty} \frac{c_l^{2\gamma/\beta} m_l^{1-2\gamma/\beta}}{a_l}\right\}.$$

Proof The theorem follows from Corollary B.33. Indeed, let $r_l > 0$, $\sum_{l=0}^{\infty} \frac{1}{r_l} = 1$. Then we easily get the following upper bounds

$$I(\lambda) \le \mathsf{E} \exp\left\{\lambda \sum_{l=0}^{\infty} \frac{1}{a_l} \sup_{\mathbf{t} \in \mathbf{T}_{b_l, b_{l+1}, \Delta}} |X(\mathbf{t})|\right\}$$
$$\le \prod_{l=0}^{\infty} \left(\mathsf{E} \exp\left\{\lambda \frac{r_l}{a_l} \sup_{\mathbf{t} \in \mathbf{T}_{b_l, b_{l+1}, \Delta}} |X(\mathbf{t})|\right\}\right)^{1/r_l}$$

$$\leq 2 \prod_{l=0}^{\infty} \exp\left\{ \frac{\lambda^2 m_l^2 r_l}{2a_l^2(1-\theta)^2} \right\}$$

$$\times \left((b_{l+1} - b_l) \Delta 2^{\frac{2}{\varepsilon}-3} \left(\frac{c_l^{2/\beta}}{\left(1 - \frac{\varepsilon}{\beta}\right)^{2/\varepsilon} (\theta m_l)^{2/\beta}} + \left(\frac{4}{\Delta}\right)^2 \right) \right)^{1/r_l}$$

$$= 2^{\frac{2}{\varepsilon}} \prod_{l=0}^{\infty} \exp\left\{ \frac{\lambda^2 m_l^2 r_l}{2a_l^2(1-\theta)^2} \right\} (b_{l+1} - b_l)^{1/r_l} \left(\frac{4}{\Delta}\right)^{1/r_l}$$

$$\times \left(\frac{c_l^{2/\beta}}{\left(1 - \frac{\varepsilon}{\beta}\right)^{2/\varepsilon} (\theta m_l)^{2/\beta}} \left(\frac{\Delta}{4}\right)^2 + 1 \right)^{1/r_l}$$

$$= \frac{2^{\frac{2}{\varepsilon}+2}}{\Delta} \exp\left\{ \frac{\lambda^2}{2(1-\theta)^2} \sum_{l=0}^{\infty} \frac{r_l m_l^2}{a_l^2} \right\} \exp\left\{ \sum_{l=0}^{\infty} \frac{\log(b_{l+1} - b_l)}{r_l} \right\}$$

$$\times \exp\left\{ \sum_{l=0}^{\infty} \frac{1}{r_l} \log\left(\frac{c_l^{2/\beta} \Delta^2}{\left(1 - \frac{\varepsilon}{\beta}\right)^{2/\varepsilon} (\theta m_l)^{2/\beta} 4^2} + 1 \right) \right\}.$$

Applying (B.24), we get

$$I(\lambda) \leq \frac{2^{\frac{2}{\varepsilon}+2}}{\Delta} \exp\left\{ \frac{\lambda^2}{2(1-\theta)^2} \sum_{l=0}^{\infty} \frac{r_l m_l^2}{a_l^2} \right\} \exp\left\{ \sum_{l=0}^{\infty} \frac{\log(b_{l+1} - b_l)}{r_l} \right\}$$

$$\times \exp\left\{ \sum_{l=0}^{\infty} \frac{c_l^{2\gamma/\beta} \Delta^{2\gamma}}{\gamma r_l \left(1 - \frac{\varepsilon}{\beta}\right)^{2\gamma/\varepsilon} (\theta m_l)^{2\gamma/\beta} 4^{2\gamma}} \right\}.$$

Now we choose $r_l = \frac{A a_l}{m_l}$. Then

$$I(\lambda) \leq \frac{2^{\frac{2}{\varepsilon}+2}}{\Delta} \exp\left\{ \frac{\lambda^2 A^2}{2(1-\theta)^2} \right\} \exp\left\{ \frac{1}{A} \sum_{l=0}^{\infty} \frac{m_l \log(b_{l+1} - b_l)}{a_l} \right\}$$

$$\times \exp\left\{ \frac{\Delta^{2\gamma}}{\gamma A \left(1 - \frac{\varepsilon}{\beta}\right)^{2\gamma/\varepsilon} \theta^{2\gamma/\beta} 4^{2\gamma}} \sum_{l=0}^{\infty} \frac{c_l^{2\gamma/\beta} m_l^{1-2\gamma/\beta}}{a_l} \right\}.$$

□

Corollary B.35 *Let the assumptions of Theorem B.34 hold. Then for all* $\theta \in (0, 1)$, $\varepsilon \in (0, \beta)$ *and* $u > 0$,

$$\mathsf{P}\left\{\sup_{\mathbf{t}\in\mathbf{T}_\Delta} \frac{|X(\mathbf{t})|}{a(t_1)} > u\right\} \leq \exp\left\{-\frac{u^2(1-\theta)^2}{2A^2}\right\} A_6(\theta, \gamma, \varepsilon).$$

Corollary B.36 *Let the assumptions of Theorem B.34 hold. Then for any* $\varepsilon \in (0, \beta)$ *and* $u > A$ *we have that*

$$\mathsf{P}\left\{\sup_{\mathbf{t}\in\mathbf{T}_\Delta} \frac{|X(\mathbf{t})|}{a(t_1)} > u\right\} \leq \sqrt{e} \exp\left\{-\frac{u^2}{2A^2}\right\} A_6\left(1 - \sqrt{1 - \frac{A^2}{u^2}}, \gamma, \varepsilon\right).$$

Corollary B.37 *Let the assumptions of Theorem B.34 hold. Then for all* $\mathbf{t} \in \mathbf{T}_\Delta$

$$|X(\mathbf{t})| \leq a(t_1)\xi \quad a.\,s.,$$

where ξ *is a non-negative random variable such that*

$$\mathsf{P}\{\xi > u\} \leq \exp\left\{-\frac{u^2(1-\theta)^2}{2A^2}\right\} A_6(\theta, \gamma, \varepsilon),$$

and for $u > A$

$$\mathsf{P}\{\xi > u\} \leq \sqrt{e} \exp\left\{-\frac{u^2}{2A^2}\right\} A_6\left(1 - \sqrt{1 - \frac{A^2}{u^2}}, \gamma, \varepsilon\right).$$

Remark B.38 Theorem B.34 remains true if the set $\mathbf{T}_\Delta$ is replaced with the set

$$\mathbf{T}_{\Delta,R} = \left\{\mathbf{t} = (t_1, t_2) \in \mathbb{R}_+^2 : t_1 - \Delta \leq t_2 \leq t_1, t_1 > R\right\},$$

and $b_0 = R$, where R is an arbitrary number. Corollaries B.35–B.37 are also true if supremum is considered over the domain $\mathbf{t} \in \mathbf{T}_{\Delta,R}$.

B.3.4 Exponential Maximal Upper Bound for Gaussian Process Defined in the Angle of the Plane

Consider set $\mathbf{T} = \{\mathbf{t} = (t_1, t_2) \in \mathbb{R}_+^2 : 0 \leq t_2 \leq t_1\}$ supplied with the distance

$$d\,(\mathbf{t}, \mathbf{s}) = \max\left\{|t_1 - s_1|, |t_2 - s_2|\right\}.$$

Assume that random process $X = \{X(\mathbf{t}), \mathbf{t} \in \mathbf{T}\}$ satisfies the following conditions.

($\mathbf{J}_1$) Process X is a centered Gaussian process on $\mathbf{T}$, separable on metric space $(\mathbf{T}, d)$.

(J_2) There exist $\beta > 0, \gamma > 0$ and a constant $C(\beta, \gamma)$ independent of X, $\mathbf{t}$ and $\mathbf{s}$ such that for any $\mathbf{t}, \mathbf{s} \in \mathbf{T}$

$$\left(\mathsf{E}(X(\mathbf{t}) - X(\mathbf{s}))^2\right)^{\frac{1}{2}} \leq C(\beta, \gamma)\,(t_1 \vee s_1)^{\beta}\,(d\,(\mathbf{t}, \mathbf{s}))^{\gamma}. \tag{B.28}$$

(J_3) There exist $\delta > 0$ and a constant $C(\delta)$ independent of X and $\mathbf{t}$ such that for any $\mathbf{t} \in \mathbf{T}$

$$\left(\mathsf{E}(X(\mathbf{t}))^2\right)^{\frac{1}{2}} \leq C(\delta) t_1^{\delta}. \tag{B.29}$$

Let us introduce the following notations. Let $A(t) \geq 1, t \geq 0$ be an increasing function such that $A(t) \to \infty$, $t \to \infty$. Consider an increasing sequence $b_0 = 0$, $b_l < b_{l+1}, l \geq 1$ and suppose that $b_l \to \infty$, $l \to \infty$. For $\delta_l = A(b_l)$ and $\kappa > 0$ we denote

$$S(\delta) = \sum_{l=0}^{\infty} b_{l+1}^{\delta} \delta_l^{-1},\ \kappa_1 = \frac{\kappa}{2}\left(1 + \frac{\beta}{\gamma} - \frac{\delta}{\gamma}\right),\ B_1 = C(\delta)S(\delta),$$

$$C_1 = C_2 \kappa^{-\frac{1}{2}} S(\delta + \kappa_1) \text{ and } C_2 = \frac{2^{\frac{1-\kappa}{2}}}{1 - \frac{\kappa}{2\gamma}} (C(\delta))^{1-\frac{\kappa}{2\gamma}} (C(\beta, \gamma))^{\frac{\kappa}{2\gamma}}.$$

Now we shall present the auxiliary exponential maximal bound for a Gaussian process defined on $(\mathbf{T}, d)$.

Theorem B.39 *Let $\{X(\mathbf{t}), \mathbf{t} \in \mathbf{T}\}$ be a random process satisfying assumptions* (J_1)–(J_3). *Let $0 \leq a < b$, set $\mathbf{T}_{a,b} = \{\mathbf{t} = (t_1, t_2) \in \mathbf{T} : a \leq t_1 \leq b,\ 0 \leq t_2 \leq t_1\}$. Then for any $0 < \theta < 1,\ \lambda > 0$ and $0 < \kappa < 1 \wedge 2\gamma$ the following inequality holds:*

$$\mathsf{E} \exp\left\{\lambda \sup_{\mathbf{t} \in \mathbf{T}_{a,b}} |X(\mathbf{t})|\right\} \leq 2A_7(\lambda, \theta),$$

where

$$A_7(\lambda, \theta) = \exp\left\{\frac{\lambda^2 (b^{\delta} C(\delta))^2}{2(1-\theta)^2} + \frac{2\lambda}{1-\theta} b^{\delta+\kappa_1} \frac{C_2}{\theta^{\frac{\kappa}{2\gamma}} \kappa^{\frac{1}{2}}}\right\}.$$

Proof It follows from (B.28) and (B.29) that

$$m := \sup_{\mathbf{t} \in \mathbf{T}_{a,b}} \left(\mathsf{E}|X(\mathbf{t})|^2\right)^{\frac{1}{2}} \leq C(\delta) b^{\delta}, \tag{B.30}$$

and

$$\sup_{d(\mathbf{t},\mathbf{s}) \leq h, \mathbf{t},\mathbf{s} \in \mathbf{T}_{a,b}} \left(\mathsf{E}(X(\mathbf{t}) - X(\mathbf{s}))^2\right)^{\frac{1}{2}} \leq \sigma(h) := C(\beta, \gamma) b^{\beta} h^{\gamma}. \tag{B.31}$$

In turn, it follows from (B.31) that

$$N_{\mathbf{T}_{a,b}}(v) \le \left(\frac{b-a}{2\sigma^{(-1)}(v)}+1\right)\left(\frac{b}{2\sigma^{(-1)}(v)}+1\right) \le \left(\frac{(C(\beta,\gamma))^{\frac{1}{\gamma}} b^{1+\frac{\beta}{\gamma}}}{2v^{\frac{1}{\gamma}}}+1\right)^2. \tag{B.32}$$

Define $J(\theta m) := \int_0^{\theta m} \left(\log N_{\mathbf{T}_{a,b}}(u)\right)^{\frac{1}{2}} du$. It follows from (B.32) that

$$J(\theta m) \le \int_0^{\theta m} \sqrt{2}\left[\log\left(\frac{(C(\beta,\gamma))^{\frac{1}{\gamma}} b^{1+\frac{\beta}{\gamma}}}{2v^{\frac{1}{\gamma}}}+1\right)\right]^{\frac{1}{2}} dv. \tag{B.33}$$

For any $0 < \kappa \le 1$,

$$\log(1+x) = \frac{1}{\kappa}\log(1+x)^\kappa \le \frac{x^\kappa}{\kappa}.$$

Now, let $\kappa \in (0, 1 \wedge 2\gamma)$. Then it follows from (B.30) and (B.33) that

$$\begin{aligned} J(\theta m) &\le \frac{\sqrt{2}}{\kappa^{\frac{1}{2}}}\int_0^{\theta m} \frac{\left((C(\beta,\gamma))^{\frac{1}{\gamma}} b^{1+\frac{\beta}{\gamma}}\right)^{\frac{\kappa}{2}}}{(2v^{\frac{1}{\gamma}})^{\frac{\kappa}{2}}} dv \\ &= \frac{\sqrt{2}}{\kappa^{\frac{1}{2}}(1-\frac{\kappa}{2\gamma})}\left(\frac{(C(\beta,\gamma))^{\frac{1}{\gamma}} b^{1+\frac{\beta}{\gamma}}}{2}\right)^{\frac{\kappa}{2}} (\theta m)^{1-\frac{\kappa}{2\gamma}} \le b^{\delta+\kappa_1}\frac{\theta^{1-\frac{\kappa}{2\gamma}}}{\kappa^{\frac{1}{2}}} C_2. \end{aligned}$$

Separability of X on $(\mathbf{T}, d)$ and relation (B.31) ensure separability of X on $(\mathbf{T}, \rho)$ with $\rho(\mathbf{t},\mathbf{s}) = \left(\mathsf{E}(X(\mathbf{t}) - X(\mathbf{s}))^2\right)^{\frac{1}{2}}$. Hence the statement of the theorem follows from Lemma B.21. □

Now we are ready to state the general result concerning the asymptotic maximal growth of a Gaussian process defined on $(\mathbf{T}, d)$.

Theorem B.40 *Let $X = \{X(\mathbf{t}), \mathbf{t} \in \mathbf{T}\}$ satisfy assumptions* (J_1)–(J_3). *Suppose that function $A(t)$ is chosen in such a way that series $S(\delta)$ converges. In case when $1 + \frac{\beta}{\gamma} - \frac{\delta}{\gamma} > 0$, assume additionally that there exists $0 < \kappa < 1$ such that series $S(\delta+\kappa_1)$ converges with $\kappa_1 = \frac{\kappa}{2}\left(1 + \frac{\beta}{\gamma} - \frac{\delta}{\gamma}\right)$.*

Then there exists a r. v. $\xi > 0$ such that on any $\omega \in \Omega$ and for any $\mathbf{t} \in \mathbf{T}$

$$|X(\mathbf{t})| \le A(t_1)\xi,$$

and ξ satisfies the following assumption: for any $\varepsilon > (2C_1+1)^{\frac{2\gamma}{2\gamma+\kappa}}$

$$\mathsf{P}\{\xi > \varepsilon\} \le 2\exp\left\{-\frac{\left(\varepsilon - \varepsilon^{\frac{\kappa}{2\gamma+\kappa}}(2C_1+1)\right)^2}{2B_1^2}\right\}. \tag{B.34}$$

Here the value of $\kappa < 2\gamma$ is chosen to ensure the convergence of series $S(\delta+\kappa_1)$ in case when $1+\frac{\beta}{\gamma}-\frac{\delta}{\gamma} > 0$, and we set $\kappa = \frac{1}{2}\wedge\gamma$ in case when $1+\frac{\beta}{\gamma}-\frac{\delta}{\gamma} \le 0$.

Proof It is easy to check that

$$I := \mathsf{E}\exp\left\{\lambda \sup_{\mathbf{t}\in\mathbf{T}} \frac{|X(\mathbf{t})|}{A(t_1)}\right\} \le \mathsf{E}\exp\left\{\lambda\sum_{l=0}^{\infty}(\delta_l)^{-1}\sup_{t_1\in(b_l,b_{l+1})}|X(\mathbf{t})|\right\}. \tag{B.35}$$

Let $l \ge 0$, $r_l > 1$ be integers such that $\sum_{l=0}^{\infty}\frac{1}{r_l} = 1$. Then it follows from (B.35), Theorem B.39 and the Hölder inequality that for any $\theta \in (0,1)$ and $0 < \kappa < 1\wedge 2\gamma$

$$I \le \prod_{l=0}^{\infty}\left(\mathsf{E}\exp\left\{\lambda\frac{r_l}{\delta_l}\sup_{t_1\in(b_l,b_{l+1})}|X(\mathbf{t})|\right\}\right)^{\frac{1}{r_l}} \le \prod_{l=0}^{\infty}(2Q_l(\lambda,\theta))^{\frac{1}{r_l}} = 2\prod_{l=0}^{\infty}(Q_l(\lambda,\theta))^{\frac{1}{r_l}},$$

where

$$Q_l(\lambda,\theta) = \exp\left\{\frac{\lambda^2 r_l^2}{2\delta_l^2}\frac{(b_{l+1}^{\delta}C(\delta))^2}{(1-\theta)^2} + \frac{2\lambda r_l}{(1-\theta)\delta_l}b_{l+1}^{\delta+\kappa_1}\frac{C_2}{\theta^{\frac{\kappa}{2\gamma}}\kappa^{\frac{1}{2}}}\right\}.$$

Therefore, if we take such value of $\kappa < 2\gamma$ that series $S(\delta+\kappa_1)$ converges in case when $1+\frac{\beta}{\gamma}-\frac{\delta}{\gamma} > 0$ and set $\kappa = \frac{1}{2}\wedge\gamma$ in case when $1+\frac{\beta}{\gamma}-\frac{\delta}{\gamma} \le 0$, we obtain

$$I \le 2\exp\left\{\frac{\lambda^2(C(\delta))^2}{2(1-\theta)^2}\sum_{l=0}^{\infty}\frac{r_l(b_{l+1}^{\delta})^2}{\delta_l^2} + \frac{2\lambda C_2\kappa^{-\frac{1}{2}}S(\delta+\kappa_1)}{(1-\theta)\theta^{\frac{\kappa}{2\gamma}}}\right\}. \tag{B.36}$$

Now we can substitute $r_l = S(\delta)b_{l+1}^{-\delta}\delta_l$ into (B.36):

$$I \le 2\exp\left\{\frac{\lambda^2(S(\delta)C(\delta))^2}{2(1-\theta)^2} + \frac{2\lambda C_2\kappa^{-\frac{1}{2}}S(\delta+\kappa_1)}{(1-\theta)\theta^{\frac{\kappa}{2\gamma}}}\right\}.$$

Therefore,

$$\mathsf{E}\exp\left\{\lambda\sup_{\mathbf{t}\in\mathbf{T}}\frac{|X(\mathbf{t})|}{A(t_1)}\right\} \le 2\exp\left\{\frac{\lambda^2}{2}\hat{B}^2 + 2\lambda\hat{C}\right\}, \tag{B.37}$$

where

$$\hat{B} = \frac{S(\delta)C(\delta)}{1-\theta} \quad \text{and} \quad \hat{C} = \frac{C_2\kappa^{-\frac{1}{2}}S(\delta+\kappa_1)}{(1-\theta)\theta^{\frac{\kappa}{2\gamma}}}.$$

It follows immediately from (B.37) that for any $\lambda > 0$, $\varepsilon > 0$

$$\mathsf{P}\left\{\sup_{\mathbf{t}\in\mathbf{T}} \frac{|X(\mathbf{t})|}{A(t_1)} > \varepsilon\right\} \le \exp\{-\lambda\varepsilon\}\mathsf{E}\exp\left\{\lambda \sup_{\mathbf{t}\in\mathbf{T}} \frac{|X(\mathbf{t})|}{A(t_1)}\right\} \le$$

$$\le 2\exp\left\{\frac{\lambda^2}{2}\hat{B}^2 + 2\lambda\hat{C} - \lambda\varepsilon\right\}. \tag{B.38}$$

If we minimize the right-hand side of (B.38) w.r.t. λ then we obtain that for any $\varepsilon > 2\hat{C}$

$$\mathsf{P}\left\{\sup_{\mathbf{t}\in\mathbf{T}} \frac{|X(\mathbf{t})|}{A(t_1)} > \varepsilon\right\} \le 2\exp\left\{-\frac{(\varepsilon-2\hat{C})^2}{2\hat{B}^2}\right\}$$

$$= 2\exp\left\{-\frac{(\varepsilon(1-\theta)-2\theta^{-\frac{\kappa}{2\gamma}}C_1)^2}{2B_1^2}\right\}. \tag{B.39}$$

Finally, we can insert $\theta = \varepsilon^{-\frac{2\gamma}{2\gamma+\kappa}}$ into (B.39) and derive that for $\varepsilon > (2C_1+1)^{\frac{2\gamma}{2\gamma+\kappa}}$

$$\mathsf{P}\left\{\sup_{t\in\mathbf{T}} \frac{|X(\mathbf{t})|}{A(t_1)} > \varepsilon\right\} \le 2\exp\left\{-\frac{(\varepsilon-\varepsilon^{\frac{\kappa}{\kappa+2\gamma}}(1+2C_1))^2}{2B_1^2}\right\}.$$

Denote $\xi := \sup_{\mathbf{t}\in\mathbf{T}} \frac{|X(\mathbf{t})|}{A(t_1)}$. Then ξ satisfies assumption (B.34), and on any $\omega \in \Omega$

$$X(\mathbf{t}) \le A(t_1)\xi,$$

which concludes the proof. □

B.3.5 *Asymptotic Growth with Probability 1 of Fractional Derivative of fBm*

In order to find upper bounds for integrals with respect to an fBm, we need to bound from above the fractional derivative of B^H.

Theorem B.41 *Let* $0 < H < 1, 1 - H < \alpha < 1$, $\mathbf{T} = \{\mathbf{t} = (t_1, t_2), 0 \le t_2 < t_1\}$,

$$X(\mathbf{t}) = \frac{B_{t_1}^H - B_{t_2}^H}{(t_1 - t_2)^{1-\alpha}} + \int_{t_2}^{t_1} \frac{B_u^H - B_{t_2}^H}{(u - t_2)^{2-\alpha}} du.$$

Then for any $p > 1$ *there exists random variable* $\xi = \xi(p)$ *such that for any* $\mathbf{t} \in \mathbf{T}$

$$|X(\mathbf{t})| \le \left(1 \vee \left(t_1^{H+\alpha-1} (\log^+ t_1)^p\right)\right) \xi(p),$$

where $\xi(p)$ *satisfies assumption* (B.34) *with some constants* B_1 *and* C_1, *and with* $\kappa = \gamma = 1/2$ *in the case when* $\alpha + H > 3/2$ *and* $\kappa = \gamma = \alpha + H - 1 - \nu$ *with any* $0 < \nu < (\alpha + H - 1) \wedge 1/2$ *in the case when* $\alpha + H \le 3/2$.

Remark B.42 Equality $\gamma = \kappa$ means that condition (B.34) can be rewritten as

$$\mathsf{P}\{\xi > \varepsilon\} \le 2 \exp\left\{ - \frac{\left(\varepsilon - \varepsilon^{\frac{1}{3}}(2C_1 + 1)\right)^2}{2B_1^2} \right\}$$

for any $\varepsilon > (2C_1 + 1)^{\frac{2}{3}}$.

To apply Theorem B.40 to the fractional derivative of the fBm and to prove Theorem B.41, we need an auxiliary result. In what follows we denote by $C(H, \alpha)$ a constant depending only on H and α and not on other parameters.

Lemma B.43 *Let* $z_i > 0$ *for* $i = 1, 2$. *In addition, let* $0 < H < 1$, $1 - H < \alpha < 1$ *and*

$$I = {z_2}^{2(H+\alpha-1)} + {z_1}^{2(H+\alpha-1)} + \frac{|z_2 - z_1|^{2H} - z_1^{2H} - z_2^{2H}}{(z_1 z_2)^{1-\alpha}}.$$

Then $I \le C(H, \alpha)|z_2 - z_1|^{2(H+\alpha-1)}$.

Proof Let $z_2 > z_1 > 0$ (the case $z_1 > z_2 > 0$ can be dealt with in a similar way). We can rewrite I as

$$\begin{aligned} I &= (z_2^{H+\alpha-1} - z_1^{H+\alpha-1})^2 + 2(z_1 z_2)^{H+\alpha-1} \\ &\quad + ((z_2 - z_1)^{2H} - (z_2^H - z_1^H)^2 - 2(z_1 z_2)^H)(z_1 z_2)^{\alpha-1} \\ &= (z_2^{H+\alpha-1} - z_1^{H+\alpha-1})^2 + \frac{(z_2 - z_1)^{2H} - (z_2^H - z_1^H)^2}{(z_1 z_2)^{1-\alpha}} = I_1 + I_2. \end{aligned}$$

Recall a simple inequality $b^r - a^r \le (b - a)^r$ for $b > a$, $0 < r \le 1$. Since $0 < H + \alpha - 1 < 1$, we can estimate I_1 by $(z_2 - z_1)^{2(H+\alpha-1)}$. Furthermore, I_2 can be rewritten as

$$I_2 = (z_2 - z_1)^{2(H+\alpha-1)} \frac{|z_2 - z_1|^{2H} - (z_2^H - z_1^H)^2}{(z_1 z_2)^{1-\alpha}(z_2 - z_1)^{2(H+\alpha-1)}} = (z_2 - z_1)^{2(H+\alpha-1)} f(u),$$

where $u = \frac{z_2}{z_1} > 1, f(u) = \frac{(u-1)^{2H} - (u^H - 1)^2}{u^{1-\alpha}(u-1)^{2(H+\alpha-1)}} \geq 0$.

Calculate the limit of function f at 1:

$$\lim_{u \to 1} f(u) = \lim_{u \to 1} \frac{(u-1)^{2H} - (u^H - 1)^2}{(u-1)^{2(H+\alpha-1)}}.$$

Here

$$\lim_{u \to 1} \frac{(u-1)^{2H}}{(u-1)^{2(H+\alpha-1)}} = \lim_{u \to 1} (u-1)^{2-2\alpha} = 0,$$

and

$$\lim_{u \to 1} \frac{(u^H - 1)^2}{(u-1)^{2(H+\alpha-1)}} = H^2 \lim_{u \to 1} (u-1)^{4-2H-2\alpha} = 0,$$

since $\lim_{u \to 1} \frac{u^H - 1}{u - 1} = H$. Calculate the limit of the function f at infinity:

$$\begin{aligned} 0 \leq \lim_{u \to \infty} f(u) &= \lim_{u \to \infty} \frac{(u-1)^{2H} - (u^H - 1)^2}{u^{1-\alpha}(u-1)^{2(H+\alpha-1)}} \\ &\leq \lim_{u \to \infty} \frac{u^{2H} - (u^H - 1)^2}{u^{2H+\alpha-1}} = \lim_{u \to \infty} \frac{2u^H - 1}{u^{2H+\alpha-1}} = 0. \end{aligned}$$

This implies that the function f is bounded, i. e., there exists $C(H, \alpha) > 0$ such that

$$I_2 \leq C(H, \alpha)(z_2 - z_1)^{2(H+\alpha-1)},$$

and the proof follows if we combine the bounds for I_1 and I_2. □

We are now ready to check conditions (J_2) and (J_3) of Sect. B.3.4 for the fractional derivative of an fBm.

Lemma B.44 *Let*

$$X(\mathbf{t}) = \frac{B_{t_1}^H - B_{t_2}^H}{(t_1 - t_2)^{1-\alpha}} + \int_{t_2}^{t_1} \frac{B_u^H - B_{t_2}^H}{(u - t_2)^{2-\alpha}} du,$$

where $0 \leq t_2 < t_1,\ 0 < H < 1, 1 - H < \alpha < 1$.

Then the following bounds hold:

1) for any $0 \le t_2 < t_1$

$$\left(\mathsf{E}(X(\mathbf{t}))^2\right)^{\frac{1}{2}} \le C(H,\alpha)(t_1 - t_2)^{H+\alpha-1};$$

2) (a) Let $H + \alpha \le \frac{3}{2}$. *Then for any* $0 \le t_2 < t_1$, $0 \le s_2 < s_1$ *and any* $0 < \nu < (H + \alpha - 1) \wedge \frac{1}{2}$

$$\begin{aligned}&(\mathsf{E}|X(\mathbf{t}) - X(\mathbf{s})|^2)^{\frac{1}{2}}\\ &\le C(H,\alpha)\big(1 + \nu^{-1}\big)(|t_1 - s_1| \vee |t_2 - s_2|)^{H+\alpha-1-\nu}(t_1 \vee s_1)^{\nu}\end{aligned}$$

with $C(H,\alpha)$ *not depending on* X, *its arguments and* ν.

(b) Let $H + \alpha > \frac{3}{2}$. *Then for any* $0 \le t_2 < t_1$, $0 \le s_2 < s_1$

$$(\mathsf{E}|X(\mathbf{t}) - X(\mathbf{s})|^2)^{\frac{1}{2}} \le C(H,\alpha)(|t_1 - s_1| \vee |t_2 - s_2|)^{\frac{1}{2}}(t_1 \vee s_1)^{H+\alpha-\frac{3}{2}}.$$

Proof The first statement follows immediately from Minkowski's integral inequality:

$$\begin{aligned}\left(\mathsf{E}(X(\mathbf{t}))^2\right)^{\frac{1}{2}} &\le \left(\mathsf{E}\left(\frac{B^H_{t_1} - B^H_{t_2}}{(t_1 - t_2)^{1-\alpha}}\right)^2\right)^{\frac{1}{2}} + \left(\mathsf{E}\left(\int_{t_2}^{t_1} \frac{B^H_u - B^H_{t_2}}{(u - t_2)^{2-\alpha}} du\right)^2\right)^{\frac{1}{2}}\\ &\le \left(\frac{(t_1 - t_2)^{2H}}{(t_1 - t_2)^{2(1-\alpha)}}\right)^{\frac{1}{2}} + \int_{t_2}^{t_1} \left(\mathsf{E}\left(\frac{B^H_u - B^H_{t_2}}{(u - t_2)^{2-\alpha}}\right)^2\right)^{\frac{1}{2}} du\\ &= (t_1 - t_2)^{H+\alpha-1} + \int_{t_2}^{t_1} \left(\frac{(u - t_2)^{2H}}{(u - t_2)^{2(2-\alpha)}}\right)^{\frac{1}{2}} du\\ &= \frac{\alpha + H}{\alpha + H - 1}(t_1 - t_2)^{H+\alpha-1}.\end{aligned}$$

In order to prove the second statement, denote

$$X_1(\mathbf{t}) = \frac{B^H_{t_1} - B^H_{t_2}}{(t_1 - t_2)^{1-\alpha}} \quad \text{and} \quad X_2(\mathbf{t}) = \int_{t_2}^{t_1} \frac{B^H_u - B^H_{t_2}}{(u - t_2)^{2-\alpha}} du.$$

Evidently,

$$(\mathsf{E}|X(\mathbf{t}) - X(\mathbf{s})|^2)^{\frac{1}{2}} \le (\mathsf{E}|X_1(\mathbf{t}) - X_1(\mathbf{s})|^2)^{\frac{1}{2}} + (\mathsf{E}|X_2(\mathbf{t}) - X_2(\mathbf{s})|^2)^{\frac{1}{2}}. \tag{B.40}$$

Let $t_1 > s_1$, the opposite case can be considered in a similar way. Then

$$
\begin{aligned}
&(\mathsf{E}|X_1(\mathbf{t}) - X_1(\mathbf{s})|^2)^{\frac{1}{2}} \\
&= \left(\mathsf{E}\left(\frac{B^H_{t_1} - B^H_{t_2}}{(t_1 - t_2)^{1-\alpha}} - \frac{B^H_{t_1} - B^H_{s_2}}{(t_1 - s_2)^{1-\alpha}} + \frac{B^H_{t_1} - B^H_{s_2}}{(t_1 - s_2)^{1-\alpha}} - \frac{B^H_{s_1} - B^H_{s_2}}{(s_1 - s_2)^{1-\alpha}}\right)^2\right)^{\frac{1}{2}} \\
&\le \left(\mathsf{E}\left(\frac{B^H_{t_1} - B^H_{t_2}}{(t_1 - t_2)^{1-\alpha}} - \frac{B^H_{t_1} - B^H_{s_2}}{(t_1 - s_2)^{1-\alpha}}\right)^2\right)^{\frac{1}{2}} \\
&\quad + \left(\mathsf{E}\left(\frac{B^H_{t_1} - B^H_{s_2}}{(t_1 - s_2)^{1-\alpha}} - \frac{B^H_{s_1} - B^H_{s_2}}{(s_1 - s_2)^{1-\alpha}}\right)^2\right)^{\frac{1}{2}} =: I_3 + I_4.
\end{aligned} \tag{B.41}
$$

It is more convenient to estimate the squares $(I_3)^2$ and $(I_4)^2$ from (B.41) instead of I_3 and I_4. As for $(I_3)^2$, we can calculate it explicitly and then estimate it with the help of Lemma B.43; $(I_4)^2$ can be evaluated similarly.

$$
\begin{aligned}
(I_3)^2 &= (t_1 - t_2)^{2(H+\alpha-1)} + (t_1 - s_2)^{2(H+\alpha-1)} - 2\frac{\mathsf{E}(B^H_{t_1} - B^H_{t_2})(B^H_{t_1} - B^H_{s_2})}{(t_1 - t_2)^{1-\alpha}(t_1 - s_2)^{1-\alpha}} \\
&= (t_1 - t_2)^{2(H+\alpha-1)} + (t_1 - s_2)^{2(H+\alpha-1)} - \frac{2}{(t_1 - t_2)^{1-\alpha}(t_1 - s_2)^{1-\alpha}} \\
&\quad \times \left[t_1^{2H} - \frac{1}{2}\left(t_2^{2H} + t_1^{2H} - (t_1 - t_2)^{2H}\right) - \frac{1}{2}\left(t_1^{2H} + s_2^{2H} - (t_1 - s_2)^{2H}\right)\right. \\
&\quad \left. + \frac{1}{2}\left(t_2^{2H} + s_2^{2H} - |t_2 - s_2|^{2H}\right)\right] \\
&= (t_1 - t_2)^{2(H+\alpha-1)} + (t_1 - s_2)^{2(H+\alpha-1)} \\
&\quad + \frac{|t_2 - s_2|^{2H} - (t_1 - t_2)^{2H} - (t_1 - s_2)^{2H}}{(t_1 - t_2)^{1-\alpha}(t_1 - s_2)^{1-\alpha}} \le C(H,\alpha)|t_2 - s_2|^{2(H+\alpha-1)}.
\end{aligned} \tag{B.42}
$$

We derive from (B.42) that

$$
I_3 \le C(H,\alpha)|t_2 - s_2|^{H+\alpha-1}, \tag{B.43}
$$

and similarly,

$$
I_4 \le C(H,\alpha)|t_1 - s_1|^{H+\alpha-1}. \tag{B.44}
$$

It follows immediately from (B.43) and (B.44) that

$$(\mathsf{E}|X_1(\mathbf{t}) - X_1(\mathbf{s})|^2)^{\frac{1}{2}} \leq C(H,\alpha)\left(|t_1 - s_1| \vee |t_2 - s_2|\right)^{H+\alpha-1}. \tag{B.45}$$

Now estimate

$$\begin{aligned} F(\mathbf{t},\mathbf{s}) &= (\mathsf{E}|X_2(\mathbf{t}) - X_2(\mathbf{s})|^2)^{\frac{1}{2}} \\ &= \left(\mathsf{E}\left(\int_{t_2}^{t_1} \frac{B_u^H - B_{t_2}^H}{(u-t_2)^{2-\alpha}} du - \int_{s_2}^{s_1} \frac{B_u^H - B_{s_2}^H}{(u-s_2)^{2-\alpha}} du\right)^2\right)^{\frac{1}{2}}. \end{aligned}$$

Let, for instance, $0 \leq t_2 < s_2 < s_1 < t_1$ (other types of relation between these points can be handled similarly). Then

$$\begin{aligned} F(\mathbf{t},\mathbf{s}) \leq &\left(\mathsf{E}\left(\int_{t_2}^{s_2} \frac{B_u^H - B_{t_2}^H}{(u-t_2)^{2-\alpha}} du\right)^2\right)^{\frac{1}{2}} \\ &+ \left(\mathsf{E}\left(\int_{s_2}^{s_1} \left(\frac{B_u^H - B_{t_2}^H}{(u-t_2)^{2-\alpha}} - \frac{B_u^H - B_{s_2}^H}{(u-s_2)^{2-\alpha}}\right) du\right)^2\right)^{\frac{1}{2}} \\ &+ \left(\mathsf{E}\left(\int_{s_1}^{t_1} \frac{B_u^H - B_{t_2}^H}{(u-t_2)^{2-\alpha}} du\right)^2\right)^{\frac{1}{2}} =: I_5 + I_6 + I_7. \end{aligned} \tag{B.46}$$

Using Minkowski's integral inequality we immediately obtain

$$\begin{aligned} I_5 &\leq \int_{t_2}^{s_2} \left(\mathsf{E}\left(\frac{B_u^H - B_{t_2}^H}{(u-t_2)^{2-\alpha}}\right)^2\right)^{\frac{1}{2}} du \\ &= \int_{t_2}^{s_2} (u-t_2)^{H+\alpha-2} du = \frac{1}{H+\alpha-1}(s_2-t_2)^{H+\alpha-1}. \end{aligned} \tag{B.47}$$

Similarly,

$$I_7 \leq \frac{1}{H+\alpha-1}(t_1-s_1)^{H+\alpha-1}. \tag{B.48}$$

Again, using Minkowski's integral inequality and Lemma B.43 we conclude that

$$
\begin{aligned}
I_6 &\le \int_{s_2}^{s_1} \left(\mathsf{E} \left(\frac{B_u^H - B_{t_2}^H}{(u-t_2)^{2-\alpha}} - \frac{B_u^H - B_{s_2}^H}{(u-s_2)^{2-\alpha}} \right)^2 \right)^{\frac{1}{2}} du \\
&= \int_{s_2}^{s_1} \Bigg[(u-t_2)^{2(H+\alpha-2)} + (u-s_2)^{2(H+\alpha-2)} \\
&\quad + \frac{(s_2-t_2)^{2H} - (u-t_2)^{2H} - (u-s_2)^{2H}}{(u-t_2)^{2-\alpha}(u-s_2)^{2-\alpha}} \Bigg]^{\frac{1}{2}} du \\
&= \int_{s_2}^{s_1} (u-s_2)^{-\frac{1}{2}} (u-t_2)^{-\frac{1}{2}} \Bigg[(u-t_2)^{2(H+\alpha-2)} (u-s_2)(u-t_2) \\
&\quad + (u-s_2)^{2(H+\alpha-2)} (u-s_2)(u-t_2) \\
&\quad + \frac{(s_2-t_2)^{2H} - (u-t_2)^{2H} - (u-s_2)^{2H}}{(u-t_2)^{1-\alpha}(u-s_2)^{1-\alpha}} \Bigg]^{\frac{1}{2}} du \\
&\le \int_{s_2}^{s_1} (u-s_2)^{-\frac{1}{2}} (u-t_2)^{-\frac{1}{2}} \Bigg[(u-t_2)^{2(H+\alpha-1)} + (u-s_2)^{2(H+\alpha-1)} \\
&\quad + (u-s_2)^{2(H+\alpha-2)+1}(s_2-t_2) + \frac{(s_2-t_2)^{2H} - (u-t_2)^{2H} - (u-s_2)^{2H}}{(u-t_2)^{1-\alpha}(u-s_2)^{1-\alpha}} \Bigg]^{\frac{1}{2}} du \\
&\le C(H,\alpha) \int_{s_2}^{s_1} (u-s_2)^{-\frac{1}{2}} (u-t_2)^{-\frac{1}{2}} (s_2-t_2)^{H+\alpha-1} du \\
&\quad + C(H,\alpha) \int_{s_2}^{s_1} (u-s_2)^{H+\alpha-2} (u-t_2)^{-\frac{1}{2}} (s_2-t_2)^{\frac{1}{2}} du =: I_8 + I_9.
\end{aligned}
\tag{B.49}
$$

Evidently,

$$
I_8 = (s_2-t_2)^{H+\alpha-1} \int_{s_2}^{s_1} (u-s_2)^{-\frac{1}{2}} (u-t_2)^{-\frac{1}{2}} du = (s_2-t_2)^{H+\alpha-1} I_{10}
$$

up to the constant multiplier and for any $0 < \rho < \frac{1}{2}$ and $0 < \nu < \frac{1}{2} - \rho$ integral I_{10} can be rewritten as

$$\begin{aligned}
I_{10} &= \int_{s_2}^{s_1} (u - s_2)^{-\frac{1}{2}} (u - t_2)^{-\frac{1}{2}} du \\
&= \int_0^{\frac{s_1 - s_2}{s_2 - t_2}} (y+1)^{-\frac{1}{2}} y^{-\frac{1}{2}} dy \le \left(\frac{s_1 - s_2}{s_2 - t_2}\right)^{\nu} \int_0^{\frac{s_1 - s_2}{s_2 - t_2}} (y+1)^{-\frac{1}{2}} y^{-\frac{1}{2} - \nu} dy \\
&\le \left(\frac{s_1 - s_2}{s_2 - t_2}\right)^{\nu} \int_0^{\infty} (y+1)^{-\frac{1}{2}} y^{-\frac{1}{2} - \nu} dy \le C\big(1 + \nu^{-1}\big) \left(\frac{s_1 - s_2}{s_2 - t_2}\right)^{\nu}.
\end{aligned}$$

Therefore, for any $0 < \nu < (H + \alpha - 1) \wedge (\frac{1}{2} - \rho)$

$$I_8 \le C(H, \alpha, \nu)\big(1 + \nu^{-1}\big)(s_2 - t_2)^{H+\alpha-1-\nu}(s_1 - s_2)^{\nu}. \tag{B.50}$$

Furthermore,

$$I_9 = (s_2 - t_2)^{\frac{1}{2}} \int_{s_2}^{s_1} (u - s_2)^{H+\alpha-2} (u - t_2)^{-\frac{1}{2}} du = (s_2 - t_2)^{\frac{1}{2}} I_{11}$$

up to a constant multiplier. In the case when $H + \alpha < \frac{3}{2}$ the integral I_{11} can be rewritten as

$$\begin{aligned}
I_{11} &= \int_{s_2}^{s_1} (u - s_2)^{H+\alpha-2} (u - t_2)^{-\frac{1}{2}} du \\
&= \int_0^{\frac{s_1 - s_2}{s_2 - t_2}} y^{H+\alpha-2} (1 + y)^{-\frac{1}{2}} (s_2 - t_2)^{H+\alpha-2+\frac{1}{2}} du \\
&\le (s_2 - t_2)^{H+\alpha-\frac{3}{2}} \int_0^{\infty} y^{H+\alpha-2} (1 + y)^{-\frac{1}{2}} du \le C(H, \alpha)(s_2 - t_2)^{H+\alpha-\frac{3}{2}}.
\end{aligned}$$

In the case when $H + \alpha > \frac{3}{2}$ integral I_{11} admits an obvious bound

$$I_{11} \le \int_{s_2}^{s_1} (u - s_2)^{H+\alpha-2} (u - s_2)^{-\frac{1}{2}} du \le C(H, \alpha)(s_1 - s_2)^{H+\alpha-\frac{3}{2}}.$$

Finally, for $H + \alpha = \frac{3}{2}$ integral I_{11} admits the same bound as I_{10}. Therefore,

$$I_9 \le C(H, \alpha)(s_2 - t_2)^{H+\alpha-1} \tag{B.51}$$

for $H + \alpha < \frac{3}{2}$,

$$I_9 \le C(H, \alpha)(s_2 - t_2)^{\frac{1}{2}} (s_1 - s_2)^{H+\alpha-\frac{3}{2}} \tag{B.52}$$

for $H+\alpha > \frac{3}{2}$, and

$$I_9 \le C(H,\alpha,\nu)(s_2-t_2)^{\frac{1}{2}-\nu}(s_1-s_2)^{\nu} \tag{B.53}$$

for $H+\alpha = \frac{3}{2}$.

This implies that

$$F(\mathbf{t},\mathbf{s}) \le C(H,\alpha)\big(1+\nu^{-1}\big)(|t_1-s_1| \vee |t_2-s_2|)^{H+\alpha-1-\nu}(s_1 \vee t_1)^{\nu} \tag{B.54}$$

for $H+\alpha \le \frac{3}{2}$. In case $H+\alpha > \frac{3}{2}$ we can put $\nu = H+\alpha-\frac{3}{2} \in (0,\frac{1}{2})$ in (B.50) and conclude that

$$F(\mathbf{t},\mathbf{s}) \le C(H,\alpha)(|t_1-s_1| \vee |t_2-s_2|)^{\frac{1}{2}}(s_1 \vee t_1)^{H+\alpha-\frac{3}{2}}. \tag{B.55}$$

The proof follows immediately from (B.40) and (B.45)–(B.55). □

Proof (Proof of Theorem B.41) First of all we should verify conditions (J_1)–(J_3) of Sect. B.3.4. Condition (J_1) is evident, since X is continuous in both variables. According to the 2nd statement of Theorem B.44, condition (J_2) holds with $\beta = \nu$, $0 < \nu < (H+\alpha-1) \wedge \frac{1}{2}$ and $\gamma = H+\alpha-1-\nu$ in case when $\alpha+H \le \frac{3}{2}$, and with $\beta = H+\alpha-\frac{3}{2}$ and $\gamma = \frac{1}{2}$ in case when $\alpha+H > \frac{3}{2}$. According to the first statement of Theorem B.44, condition (J_3) holds with $\delta = H+\alpha-1$.

Let $A(t) = 1 \vee \big(t_1^{H+\alpha-1}(\log^+ t_1)^p\big)$ for some $p>1$ and for any $t>0$ and let $b_l = e^l$, $l \ge 0$. Then $\delta_l = (e^{l(H+\alpha-1)}l^p) \vee 1$ and $b_{l+1}^{\delta} = e^{(l+1)(H+\alpha-1)}$. Therefore, in this case series $S(\delta)$ converges since

$$S(\delta) = e^{H+\alpha-1} + \sum_{l=1}^{\infty} \frac{e^{(l+1)(H+\alpha-1)}}{e^{l(H+\alpha-1)}l^p} = e^{H+\alpha-1}\left(1+\sum_{l=1}^{\infty} l^{-p}\right) < \infty.$$

Moreover, it is easy to check that $1+\frac{\beta}{\gamma}-\frac{\delta}{\gamma} = 0$ for any values of $\alpha+H$, hence $\kappa_1 = 0$, $\kappa = 1/2$ in the case when $\alpha+H > 3/2$ and $\kappa = \alpha+H-1-\nu$ with any $0 < \nu < (\alpha+H-1) \wedge 1/2$ in the case when $\alpha+H \le 3/2$. This implies that all conditions of Theorem B.40 hold true and we can apply the theorem with $A(t) = 1 \vee \big(t_1^{H+\alpha-1}(\log^+ t_1)^p\big)$ which concludes the proof. □

Remark B.45 Instead of the fractional derivative, we can consider the fBm B_t^H itself and apply the same reasoning to it. This case is much simpler and we immediately obtain that

$$\sup_{0\le s\le t} \left|B_s^H\right| \le \big(1 \vee \big(t^H(\log^+ t)^p\big)\big)\,\xi(p) \tag{B.56}$$

for any $p > 1$.

The following result is a generalization of Theorem B.41.

Theorem B.46 *Let $H \in (1/2, 1)$, $\alpha \in (1 - H, 1/2)$. Denote for $t_1 < t_2$*

$$Z(t_1, t_2) = \frac{1}{\Gamma(\alpha)} \left(\frac{B^H_{t_1} - B^H_{t_2}}{(t_2 - t_1)^{1-\alpha}} + (1 - \alpha) \int_{t_1}^{t_2} \frac{B^H_{t_1} - B^H_u}{(u - t_1)^{2-\alpha}} du \right).$$

Then for any $\gamma > 1/2$,

$$\eta_{H,\alpha,\gamma} := \sup_{0 \le t_1 < t_2 \le t_1 + 1} \frac{|Z(t_1, t_2)|}{(t_2 - t_1)^{H+\alpha-1} \left(|\log(t_2 - t_1)|^{1/2} + 1 \right) \left(\log(t_2 + 2) \right)^{\gamma}} \tag{B.57}$$

is finite a. s. Moreover, there exists $c_{H,\alpha,\gamma} > 0$ such that $\mathsf{E}\left[\exp\left\{ x\eta^2_{H,\alpha,\gamma} \right\} \right] < \infty$ for $x < c_{H,\alpha,\gamma}$.

Proof Throughout the proof, the symbol C will denote a generic constant, whose value is not important and may change from one line to another.

Let $h(s) = s^{H+\alpha-1} \left(|\log s|^{1/2} + 1 \right)$, $s > 0$. Define for $T > 0$

$$M_T = \sup_{0 \le t_1 < t_2 \le t_1 + 1 \le T} \frac{|Z(t_1, t_2)|}{h(t_2 - t_1)}.$$

We will first prove that M_T is finite a. s. Since $\mathsf{E}\left[\left(B^H_t - B^H_s \right)^2 \right] = (t - s)^{2H}$, it follows from [116, Theorem 4] that there exists a random variable η_T such that a. s. for all t_1, t_2 with $0 \le t_1 < t_2 \le t_1 + 1$

$$\left| B^H_{t_1} - B^H_{t_2} \right| \le \eta_T (t_2 - t_1)^H \left(|\log(t_2 - t_1)|^{1/2} + 1 \right).$$

Then

$$|Z(t_1, t_2)| \le \frac{\eta_T}{\Gamma(\alpha)} (t_2 - t_1)^{H+\alpha-1} \left(|\log(t_2 - t_1)|^{1/2} + 1 \right) + I,$$

where

$$\begin{aligned} I &= \left| \int_{t_1}^{t_2} \frac{B^H_u - B^H_{t_1}}{(u - t_1)^{2-\alpha}} du \right| \le \frac{\eta_T}{\Gamma(\alpha)} \int_{t_1}^{t_2} (u - t_1)^{H+\alpha-2} \left(|\log(u - t_1)|^{1/2} + 1 \right) du \\ &\le \frac{\eta_T}{\Gamma(\alpha)} (t_2 - t_1)^{H+\alpha-1} \int_0^1 z^{H+\alpha-2} \left(|\log z|^{1/2} + |\log(t_2 - t_1)|^{1/2} + 1 \right) dz \\ &\le C\eta_T (t_2 - t_1)^{H+\alpha-1} \left(|\log(t_2 - t_1)|^{1/2} + 1 \right), \end{aligned}$$

whence finiteness of M_T follows. Since M_T is a supremum of Gaussian family, Fernique's theorem implies that $\mathsf{E}\left[e^{\varepsilon M_T^2}\right] < \infty$ for some $\varepsilon > 0$; in particular, all moments of M_T are finite.

Now observe that from H-selfsimilarity of B^H it follows that for any $a > 0$

$$\{Z(at_1, at_2), 0 \le t_1 < t_2\} \overset{d}{=} \{a^{H+\alpha-1} Z(t_1, t_2), 0 \le t_1 < t_2\}.$$

Therefore, for any $k \ge 1$

$$\begin{aligned} M_1 &\overset{d}{=} \sup_{0\le t_1<t_2\le 1} \frac{2^{-k(H+\alpha-1)}\,|Z(2^k t_1, 2^k t_2)|}{|t_2-t_1|^{H+\alpha-1}\left(|\log(t_2-t_1)|^{1/2}+1\right)} \\ &= \sup_{0\le s_1<s_2\le 2^k} \frac{|Z(s_1,s_2)|}{|s_2-s_1|^{H+\alpha-1}\left(|\log(s_2-s_1)-k\log 2|^{1/2}+1\right)} \\ &\ge \sup_{0\le s_1<s_2\le s_2+1\le 2^k} \frac{|Z(s_1,s_2)|}{|s_2-s_1|^{H+\alpha-1}\left(|\log(s_2-s_1)-k\log 2|^{1/2}+1\right)} \\ &\ge \sup_{0\le s_1<s_2\le s_2+1\le 2^k} \frac{|Z(s_1,s_2)|}{|s_2-s_1|^{H+\alpha-1}\left(|\log(s_2-s_1)|^{1/2}+(k\log 2)^{1/2}+1\right)} \\ &\ge \frac{M_{2^k}}{(k\log 2)^{1/2}+1}. \end{aligned}$$

Hence, for any $q \ge 1$

$$\mathsf{E}\left[M_{2^k}^q\right] \le \mathsf{E}\left[M_1^q\right]\left((k\log 2)^{1/2}+1\right)^q.$$

This implies that for any $p > q/2+1$

$$\mathsf{E}\left[\sum_{k=1}^{\infty}\frac{M_{2^k}^q}{k^p}\right] = \sum_{k=1}^{\infty}\frac{\mathsf{E}\left[M_{2^k}\right]^q}{k^p} \le C\,\mathsf{E}\left[M_1^q\right]\sum_{k=1}^{\infty} k^{q/2-p} < \infty.$$

In particular, the sum $\sum_{k=1}^{\infty}|M_{2^k}|^q k^{-p}$ is finite a. s., so $M_{2^k} = o(k^{p/q})$, $k \to \infty$, a. s. If we choose some $q > (\gamma - 1/2)^{-1}$, then $q/2+1 < \gamma q$. Hence, we can take some $p \in (q/2+1, \gamma q)$ and arrive at $M_{2^k} = o(k^\gamma)$, $k \to \infty$, a. s. Consequently, the random variable $\zeta = \sup_k M_{2^k} k^{-\gamma}$ is finite a. s.

Obviously, for $t_2 \le 2$

$$\frac{|Z(t_1,t_2)|}{h(t_2-t_1)\log(t_2+2)} \le \frac{M_2}{\log 2}.$$

Now let $t_2 \in (2^{k-1}, 2^k]$ for some $k \geq 2$. Then we have for any $t_1 \in [t_2 - 1, t_2)$

$$|Z(t_1, t_2)| \leq M_{2^k} h(t_2 - t_1) \leq \zeta k^\gamma h(t_2 - t_1) \leq \zeta \left(\frac{\log t_2}{\log 2} + 1 \right)^\gamma h(t_2 - t_1)$$
$$\leq 2^\gamma \zeta (\log t_2)^\gamma h(t_2 - t_1) < 2^\gamma \zeta (\log(t_2 + 2))^\gamma h(t_2 - t_1).$$

Consequently, $\eta_{H,\alpha,\gamma} \leq \max\{M_2/\log 2, 2^\gamma \zeta\} < \infty$ a. s.

The second statement follows from Fernique's theorem, since $\eta_{H,\alpha,\gamma}$ is a supremum of absolute values of a centered Gaussian family. □

Corollary B.47 *For any $\gamma > \frac{1}{2}$ and any $H \in (0, 1)$ there exists a nonnegative random variable $\eta_{H,\gamma}$ such that for any $0 < t_1 < t_2 < \infty$*

$$\left| B_{t_2}^H - B_{t_1}^H \right| \leq (t_2 - t_1)^H \left(|\log(t_2 - t_1)|^{1/2} + 1 \right) (\log(t_2 + 2))^\gamma \eta_{H,\gamma}, \qquad \text{(B.58)}$$

and there exists a number $c_{H,\gamma} > 0$ such that for any $0 < y < c_{H,\gamma}$, $\mathsf{E} \exp\{y \eta_{H,\gamma}^2\} < \infty$.

B.4 Asymptotic Growth with Probability 1 of mfBm and Its Increments

In this section we shall apply the results of Sect. B.3 to get the asymptotic growth of the trajectories of an mfBm. Let $H: \mathbb{R}_+ \to (0, 1)$ be a continuous function satisfying the assumptions (C_1)–(C_2) of Sect. 1.5.3. Denote

$$h_3 = \min\{h_1, \kappa\}, \quad h_4 = \max\{h_2, \kappa\}, \quad h_5 = h_4 - h_3, \qquad \text{(B.59)}$$

where h_1, h_2 and κ are the constants from assumptions (C_1)–(C_2). Let $Y(t)$ be an mfBm defined by (1.58).

B.4.1 *Upper Bounds for the Incremental Variances of mfBm*

The first result gives the upper bound for the incremental variance function of mfBm.

Lemma B.48 *Under the assumption* (C_1), *there exists a constant $K_2 > 0$ such that for all $t \geq s \geq 0$*

$$\mathsf{E}(Y(t) - Y(s))^2 \leq K_2 |t - s|^{2H_t} + K_2 (H_t - H_s)^2 z^2(s), \qquad \text{(B.60)}$$

where

$$z(s) = \begin{cases} s^{h_2}\left(\log^2 s + 1\right)^{1/2}, & s \geq 1, \\ 1, & 0 < s < 1. \end{cases}$$

Proof By the isometry property,

$$\begin{aligned}
\mathsf{E}(Y(t) - Y(s))^2 &= \mathsf{E}\left(\int_{\mathbb{R}}\left(\frac{e^{itu}-1}{|u|^{H_t+1/2}} - \frac{e^{isu}-1}{|u|^{H_s+1/2}}\right)\widetilde{W}(du)\right)^2 \\
&= \int_{\mathbb{R}}\left|\frac{e^{itu}-1}{|u|^{H_t+1/2}} - \frac{e^{isu}-1}{|u|^{H_s+1/2}}\right|^2 du \\
&= \int_{\mathbb{R}}\left|\frac{e^{itu}-1}{|u|^{H_t+1/2}} - \frac{e^{isu}-1}{|u|^{H_t+1/2}} + \frac{e^{isu}-1}{|u|^{H_t+1/2}} - \frac{e^{isu}-1}{|u|^{H_s+1/2}}\right|^2 du \\
&\leq 2(I_1 + I_2),
\end{aligned}$$

where

$$I_1 = \int_{\mathbb{R}} \frac{\left|e^{itu} - e^{isu}\right|^2}{|u|^{2H_t+1}} du,$$

$$I_2 = \int_{\mathbb{R}} \left|e^{isu} - 1\right|^2 \left(|u|^{-H_t-1/2} - |u|^{-H_s-1/2}\right)^2 du.$$

Consider now I_1 and I_2 separately. For I_1 we get the following bound

$$I_1 = \int_{\mathbb{R}} \frac{\left|e^{i(t-s)u} - 1\right|^2}{|u|^{2H_t+1}} du = |t-s|^{2H_t} \int_{\mathbb{R}} \frac{\left|e^{iv} - 1\right|^2}{|v|^{2H_t+1}} dv = |t-s|^{2H_t} \int_{\mathbb{R}} \frac{4\sin^2 \frac{v}{2}}{|v|^{2H_t+1}} dv.$$

Hence, $I_1 \leq C_1 \, |t-s|^{2H_t}$, where

$$\begin{aligned}
C_1 &= \int_{|v|<1} \frac{4\sin^2 \frac{v}{2}}{|v|^{2h_2+1}} dv + \int_{|v|>1} \frac{4\sin^2 \frac{v}{2}}{|v|^{2h_1+1}} dv \\
&\leq \int_{|v|<1} \frac{1}{|v|^{2h_2-1}} dv + \int_{|v|>1} \frac{4}{|v|^{2h_1+1}} dv < \infty.
\end{aligned}$$

Consider I_2. By the mean value theorem,

$$|u|^{-H_t-1/2} - |u|^{-H_s-1/2} = -\,|u|^{-h(u)-1/2} \log|u| \, (H_t - H_s),$$

where $h(u) \in [h_1, h_2]$. Therefore,

$$I_2 = (H_t - H_s)^2 \int_{\mathbb{R}} \left|e^{isu} - 1\right|^2 |u|^{-2h(u)-1} \log^2 |u| \, du$$
$$= (H_t - H_s)^2 \int_{\mathbb{R}} \frac{\left|e^{iv} - 1\right|^2}{|v|^{2h(v/s)+1}} s^{2h(v/s)} (\log|v| - \log s)^2 dv$$
$$\leq \begin{cases} s^{2h_1}(H_t - H_s)^2 I_3, & 0 \leq s < 1, \\ s^{2h_2}(H_t - H_s)^2 I_3, & s \geq 1, \end{cases}$$

where

$$I_3 = \int_{\mathbb{R}} \frac{\left|e^{iv} - 1\right|^2}{|v|^{2h(v/s)+1}} (\log|v| - \log s)^2 dv$$
$$\leq 2\left(\int_{\mathbb{R}} \frac{4\sin^2 \frac{v}{2}}{|v|^{2h(v/s)+1}} \log^2 |v| \, dv + \log^2 s \int_{\mathbb{R}} \frac{4\sin^2 \frac{v}{2}}{|v|^{2h(v/s)+1}} dv\right)$$
$$\leq 2\left(C_2 + C_1 \log^2 s\right),$$
$$C_2 = \int_{|v|<1} \frac{4\sin^2 \frac{v}{2}}{|v|^{2h_2+1}} \log^2 |v| \, dv + \int_{|v|>1} \frac{4\sin^2 \frac{v}{2}}{|v|^{2h_1+1}} \log^2 |v| \, dv < \infty.$$

Note that $f(s) = s^{2h_1}\left(C_2 + C_1 \log^2 s\right)$, $s > 0$, is a continuous function and $f(s) \to 0$ as $s \downarrow 0$. Therefore, f is bounded on $[0, 1]$, whence $I_2 \leq C_3(H_t - H_s)^2 z^2(s)$ for some $C_3 > 0$. □

Remark B.49 Bound (B.60) is inconvenient in the sense that it contains two different exponents of $|t - s|$ and therefore one should every time relate corresponding terms depending upon the value of $|t - s|$. To avoid this technical difficulty, we establish the next result.

Lemma B.50 *Assume that the Hurst function H satisfies the conditions* (C_1) *and* (C_2). *Let $a, b \in \mathbb{R}_+$, $b - a \geq 1$. Then*

(a) for all $t, s \in [a, b]$ such that $|t - s| \leq 1$,

$$\left(\mathsf{E}(Y(t) - Y(s))^2\right)^{1/2} \leq K_3 |t - s|^{h_3} z(b),$$

where $K_3 = K_2^{1/2}\left(1 + D^2\right)^{1/2}$, D is the Hölder coefficient from assumption (C_2).

(b) for all $t, s \in [a, b]$

$$\left(\mathsf{E}(Y(t) - Y(s))^2\right)^{1/2} \leq K_3 |t - s|^{h_3} (b - a)^{h_5} z(b);$$

(c) for all $t_1, t_2, s_1, s_2 \in [a, b]$

$$\left(\mathsf{E}\,(Y(t_1) - Y(t_2) - Y(s_1) + Y(s_2))^2\right)^{1/2} \\ \leq 2K_3 \max\left(|t_1 - s_1|, |t_2 - s_2|\right)^{h_3} (b-a)^{h_5} z(b).$$

Proof It follows from the assumptions (C_1)–(C_2) and Lemma B.48 that

$$\mathsf{E}(Y(t) - Y(s))^2 \leq K_2 |t-s|^{2H_t} + K_2 D^2 |t-s|^{2\kappa} z^2(s). \tag{B.61}$$

(a) In the case $|t-s| \leq 1$, we have that

$$\begin{aligned}\mathsf{E}(Y(t) - Y(s))^2 &\leq K_2 |t-s|^{2h_3} + K_2 D^2 |t-s|^{2h_3} z^2(s) \\ &= K_2 |t-s|^{2h_3} \left(1 + D^2 z^2(s)\right) \leq K_3^2 |t-s|^{2h_3} z^2(b).\end{aligned}$$

(b) For arbitrary values of arguments, inequality (B.61) implies that

$$\begin{aligned}&\mathsf{E}(Y(t) - Y(s))^2 \\ &\leq K_2 (b-a)^{2H_t} \left|\frac{t-s}{b-a}\right|^{2H_t} + K_2 D^2 (b-a)^{2\kappa} \left|\frac{t-s}{b-a}\right|^{2\kappa} z^2(s) \\ &\leq K_2 (b-a)^{2h_4} \left|\frac{t-s}{b-a}\right|^{2h_3} \left(1 + D^2 z^2(s)\right) \\ &\leq K_3^2 (b-a)^{2h_4 - 2h_3} |t-s|^{2h_3} z^2(s).\end{aligned}$$

(c) Proof follows immediately from the part (b), since by Minkowski's inequality,

$$\left(\mathsf{E}\,(Y(t_1) - Y(t_2) - Y(s_1) + Y(s_2))^2\right)^{1/2} \\ \leq \left(\mathsf{E}(Y(t_1) - Y(s_1))^2\right)^{1/2} + \left(\mathsf{E}(Y(t_2) - Y(s_2))^2\right)^{1/2}.$$

□

B.4.2 Asymptotic Growth of the Trajectories of mfBm with Probability 1

Now we apply the results of Appendix B.3 to mfBm. The first result gives the maximal exponential bound for the weighted mfBm. In order to get it, introduce the following notations. Let b_k, $k \geq 0$, be a sequence such that $b_0 = 0$, $b_{k+1} - b_k \geq 1$, and let $a(t) > 0$ be an increasing continuous function such that $a(t) \to \infty$ as $t \to \infty$, $a_k = a(b_k)$.

Theorem B.51 *Let the Hurst function H satisfy the conditions* (C_1) *and* (C_2). *Assume that there exists* $0 < \gamma \leq 1$ *such that*

$$\sum_{k=0}^{\infty} \frac{b_{k+1}^{h_2+\frac{\gamma h_4}{h_3}} \left(\log^2 b_{k+1} + 1\right)^{\frac{\gamma}{2h_3}}}{a_k} < \infty. \tag{B.62}$$

Then for all $0 < \theta < 1$, $u > 0$

$$\mathsf{P}\left\{\sup_{t>0} \frac{|Y(t)|}{a(t)} > u\right\} \leq 2^{\frac{2}{h_3}-1} \exp\left\{-\frac{u^2(1-\theta)^2}{2A^2}\right\} A_8(\theta, \gamma), \tag{B.63}$$

where

$$A = K_1 \sum_{k=0}^{\infty} \frac{b_{k+1}^{h_2}}{a_k}, \tag{B.64}$$

$$A_8(\theta, \gamma) = \exp\left\{\frac{K_1^{1-\frac{\gamma}{h_3}} K_3^{\frac{\gamma}{h_3}}}{\gamma A} \sum_{k=0}^{\infty} \frac{b_{k+1}^{h_2+\frac{\gamma h_4}{h_3}} \left(\log^2 b_{k+1} + 1\right)^{\frac{\gamma}{2h_3}}}{a_k} \left(\frac{2^{\frac{2}{h_3}-1}}{\theta^{\frac{1}{h_3}}}\right)^{\gamma}\right\},$$

and additionally for all $u > A$

$$\mathsf{P}\left\{\sup_{t>0} \frac{|Y(t)|}{a(t)} > u\right\} \leq 2^{\frac{2}{h_3}-1} e^{\frac{1}{2}} \exp\left\{-\frac{u^2}{2A^2}\right\} A_8\left(1 - \sqrt{1 - \frac{A^2}{u^2}}, \gamma\right). \tag{B.65}$$

Here K_1 *is the constant from* (1.60).

Proof First, we check the conditions (*i*)–(*iii*) of Theorem B.26. Lemma B.50(b) implies that the assumption (*i*) holds with $c_k = K_3(b_{k+1} - b_k)^{h_5} z(b_{k+1})$ and $\beta = h_3$. According to (1.60), we can choose $m_k = K_1 b_{k+1}^{h_2}$. Then

$$\sum_{k=0}^{\infty} \frac{m_k}{a_k} = K_1 \sum_{k=0}^{\infty} \frac{b_{k+1}^{h_2}}{a_k} < \infty,$$

by (B.62), and the condition (*ii*) is satisfied. Finally, the condition (*iii*) follows from (B.62), because in our case

$$\begin{aligned} m_k^{1-\gamma/\beta} (b_{k+1} - b_k)^{\gamma} c_k^{\gamma/\beta} &= K_1^{1-\frac{\gamma}{h_3}} K_3^{\frac{\gamma}{h_3}} b_{k+1}^{h_2} (b_{k+1} - b_k)^{\gamma\left(1+\frac{h_5}{h_3}\right)} \left(\log^2 b_{k+1} + 1\right)^{\frac{\gamma}{2h_3}} \\ &\leq K_1^{1-\frac{\gamma}{h_3}} K_3^{\frac{\gamma}{h_3}} b_{k+1}^{h_2+\frac{\gamma h_4}{h_3}} \left(\log^2 b_{k+1} + 1\right)^{\frac{\gamma}{2h_3}}, \end{aligned}$$

where we used the equality $h_5 = h_4 - h_3$. Thus, the assumptions of Theorem B.26 are satisfied. Now the statements (B.63) and (B.65) follow from Corollaries B.28 and B.29 respectively. □

Remark B.52 Theorem B.51 remains true if the process $Y(t)$ is considered on the set $t > R$, where $R > 0$ is an arbitrary number. In this case the conditions (C_1) and (C_2) are replaced with the assumptions

(C_{1R}) There exist constants $0 < h_{1R} < h_{2R} \le 1$ such that for any $t > R$

$$h_{1R} \le H_t \le h_{2R}.$$

(One can put $h_{2R} = \sup_{t>R} H_t$.)

(C_{2R}) There exist constants $D_R > 0$ and $\kappa \in (0, 1]$ such that for all $t \ge s > R$

$$|H_t - H_s| \le D_R \, |t-s|^{\kappa}.$$

Now we present the first main result of this appendix, namely, the power upper bound for the asymptotic growth of the trajectories of mfBm with probability 1.

Theorem B.53 *Let the Hurst function H satisfy the conditions (C_{1R}) and (C_{2R}) for some $R > 0$. Then for any $\delta > 0$ there exists a nonnegative random variable $\xi = \xi(\delta)$ such that for all $t > R$*

$$|Y(t)| \le \left(t^{h_{2R}+\delta} \vee 1\right) \xi \quad a.s., \tag{B.66}$$

and there exist positive constants $C_1 = C_1(\delta)$ and $C_2 = C_2(\delta)$ such that for all $u > 0$

$$\mathsf{P}(\xi > u) \le C_1 e^{-C_2 u^2}. \tag{B.67}$$

Let $A = \frac{K_1 (Re)^{h_{2R}}}{1-e^{-\delta}}$. Then for any $\gamma \in \left(0, \frac{\delta h_{3R}}{h_{4R}} \wedge 1\right)$ there exists a constant $C_3 > 0$ such that for all $u > A$

$$\mathsf{P}(\xi > u) \le 2^{\frac{2}{h_{3R}}-1} e^{\frac{1}{2}} \exp\left\{-\frac{u^2}{2A^2}\right\} \exp\left\{C_3 \left(1 - \sqrt{1 - \frac{A^2}{u^2}}\right)^{-\gamma/h_{3R}}\right\}. \tag{B.68}$$

Proof Put in Theorem B.51 $a(t) = t^{h_{2R}+\delta} \vee 1$, $b_0 = R$, $b_k = Re^k$, $k \ge 1$, and arbitrary $\gamma \in \left(0, \frac{\delta h_{3R}}{h_{4R}} \wedge 1\right)$, $\theta \in (0,1)$. Then $a_0 = 1$, $a_k = e^{k(h_{2R}+\delta)}$, $k \ge 1$,

$$\sum_{k=0}^{\infty} \frac{b_{k+1}^{h_{2R}+\frac{\gamma h_{4R}}{h_{3R}}} \left(\log^2 b_{k+1} + 1\right)^{\frac{\gamma}{2h_{3R}}}}{a_k}$$

$$= (Re)^{h_{2R}+\frac{\gamma h_{4R}}{h_{3R}}} \left(\sum_{k=0}^{\infty} e^{k\left(\frac{\gamma h_{4R}}{h_{3R}}-\delta\right)} \left((k+1+\log R)^2+1\right)^{\frac{\gamma}{2h_{3R}}}\right) < \infty,$$

and (B.64) becomes

$$A = K_1 \sum_{k=0}^{\infty} \frac{b_{k+1}^{h_{2R}}}{a_k} = K_1 (Re)^{h_{2R}} \sum_{k=0}^{\infty} e^{-k\delta} = \frac{K_1 (Re)^{h_{2R}}}{1-e^{-\delta}}.$$

Now (B.67) follows from the inequality (B.63) of Theorem B.51, if we additionally put

$$\xi = \sup_{t>R} \frac{|Y(t)|}{t^{h_{2R}+\delta} \vee 1}, \quad C_1 = 2^{\frac{2}{h_{3R}}-1} A_8(\theta,\gamma), \quad C_2 = \frac{(1-\theta)^2}{2A^2}.$$

The bound (B.68) follows from (B.65), since $A_8(\theta,\gamma) = \exp\left\{C_3 \theta^{-\gamma/h_{3R}}\right\}$, where

$$C_3 = \frac{K_1^{1-\frac{\gamma}{h_{3R}}} K_3^{\frac{\gamma}{h_{3R}}}}{\gamma A} \sum_{k=0}^{\infty} \frac{b_{k+1}^{h_{2R}+\frac{\gamma h_{4R}}{h_{3R}}} \left(\log^2 b_{k+1}+1\right)^{\frac{\gamma}{2h_{3R}}}}{a_k} 2^{\left(\frac{2}{h_{3R}}-1\right)\gamma}. \qquad \square$$

Remark B.54 The result of Theorem B.53 cannot be improved substantially. Indeed, $Y(t)$ is a normal random variable, $\mathsf{E}Y(t) = 0$, $\sqrt{\mathsf{E}Y^2(t)} = K_1 t^{H_t}$. Therefore, $Y(t) = \xi K_1 t^{H_t}$, where ξ is a standard random variable. If there exists t_m such that $H(t_m) = \sup H_t$, then $Y(t_m) = \xi K_1 t^{H_{t_m}}$.

Theorem B.55 *Let the Hurst function H satisfy the conditions* (C_1) *and* (C_2). *Then for any $\delta > 0$ there exists t_δ such that for all $t > t_\delta$*

$$|Y(t)| \le \left(t^{h^*+\delta} \vee 1\right) \xi \qquad a.s.,$$

where $h^* = \limsup_{t\to\infty} H_t$, ξ is a nonnegative random variable such that for all $u > 0$

$$\mathsf{P}(\xi > u) \le C_1 e^{-C_2 u^2}$$

for some positive constants $C_1 = C_1(\delta)$ and $C_2 = C_2(\delta)$.

Proof The result follows from Theorem B.53, since for each $\delta > 0$ there exists t_δ such that $H(t) \le h^* + \frac{\delta}{2}$ for $t > t_\delta$. □

B.4.3 Asymptotic Growth with Probability 1 of the Increments of mfBm

Let $\Delta \in (0, 1]$, $\mathbf{T}_\Delta = \{\mathbf{t} = (t_1, t_2) \in \mathbb{R}_+^2 : t_1 - \Delta \le t_2 \le t_1\}$. Consider the increment of mfBm $Z(\mathbf{t}) = Y(t_1) - Y(t_2)$, $\mathbf{t} \in \mathbf{T}_\Delta$. Let b_k, $k \ge 0$, be a sequence such that $b_0 = 0$, $b_{k+1} - b_k \ge 1$, and let $a(t) > 0$ be an increasing continuous function such that $a(t) \to \infty$ as $t \to \infty$, $a_k = a(b_k)$.

Theorem B.56 *Let the Hurst function H satisfy the conditions* (C$_1$) *and* (C$_2$). *Assume that there exists* $0 < \gamma \le 1$ *such that*

$$\sum_{l=0}^{\infty} \frac{b_{l+1}^{\frac{2h_5\gamma}{h_3}} z(b_{l+1})}{a_l} < \infty. \tag{B.69}$$

Then for all $\theta \in (0, 1)$, $\varepsilon \in (0, h_3)$ *and* $\lambda > 0$

$$\mathsf{E} \exp\left\{\lambda \sup_{\mathbf{t} \in \mathbf{T}_\Delta} \frac{|Z(\mathbf{t})|}{a(t_1)}\right\} \le \frac{1}{\Delta} \exp\left\{\frac{\lambda^2 A^2 \Delta^{2h_3}}{2(1-\theta)^2}\right\} A_9(\theta, \gamma, \varepsilon),$$

where $A = K_3 \sum_{l=0}^{\infty} \frac{z(b_{l+1})}{a_l}$,

$$A_9(\theta, \gamma, \varepsilon) = 2^{\frac{2}{\varepsilon}+2} \exp\left\{\frac{K_3}{A} \sum_{l=0}^{\infty} \frac{z(b_{l+1}) \log(b_{l+1} - b_l)}{a_l}\right\}$$

$$\times \exp\left\{\frac{K_3}{\gamma A 4^{2\gamma} \left(1 - \frac{\varepsilon}{h_3}\right)^{\frac{2\gamma}{\varepsilon}} \theta^{\frac{2\gamma}{h_3}}} \sum_{l=0}^{\infty} \frac{z(b_{l+1})(b_{l+1} - b_l)^{\frac{2h_5\gamma}{h_3}}}{a_l}\right\}.$$

Proof We need to verify the assumptions of Theorem B.34 for the process Z. By Lemma B.50(a), for all $\mathbf{t} \in \mathbf{T}_{b_l, b_{l+1}, \Delta}$

$$\left(\mathsf{E}(Z(\mathbf{t}))^2\right)^{\frac{1}{2}} = \left(\mathsf{E}(Y(t_1) - Y(t_2))^2\right)^{\frac{1}{2}} \le K_3 z(b_{l+1}) \Delta^{h_3}.$$

Hence, the condition (i) of Theorem B.34 is satisfied with $m_l = K_3 z(b_{l+1})\Delta^{h_3}$. Further, Lemma B.50(c) implies

$$\sup_{\substack{d(\mathbf{t},\mathbf{s})\le h,\\ \mathbf{t},\mathbf{s}\in \mathbf{T}_{b_l,b_{l+1},\Delta}}} \left(\mathsf{E}(Z(\mathbf{t})-Z(\mathbf{s}))^2\right)^{\frac{1}{2}} \le 2K_3(b_{l+1}-b_l)^{h_5} z(b_{l+1})h^{h_3}.$$

Thus, the condition (ii) of Theorem B.34 holds with $c_l = 2K_3(b_{l+1}-b_l)^{h_5} z(b_{l+1})$, $\beta = h_3$. It is not hard to see that in this case the condition (iii) of Theorem B.34 is equivalent to the condition

$$\sum_{l=0}^{\infty} \frac{z(b_{l+1})}{a_l} < \infty, \quad \sum_{l=0}^{\infty} \frac{z(b_{l+1})\log(b_{l+1}-b_l)}{a_l} < \infty,$$

$$\sum_{l=0}^{\infty} \frac{z(b_{l+1})(b_{l+1}-b_l)^{\frac{2h_5\gamma}{h_3}}}{a_l} < \infty.$$

Obviously, these three series converge when (B.69) holds. Now the result follows from Theorem B.34. □

Let d_k, $k \ge 0$, be a strictly decreasing sequence such that $d_0 = 1$, $d_k \downarrow 0$ as $k \to \infty$. Let $g\colon (0,1] \to (0,\infty)$ be a continuous function and g_k, $k \ge 0$, be a sequence such that $0 < g_k \le \min_{d_{k+1}\le t\le d_k} g(t)$.

Theorem B.57 *Assume that the assumptions of Theorem B.56 hold and*

$$\sum_{k=0}^{\infty} \frac{d_k^{h_3}\,|\log d_k|}{g_k} < \infty.$$

Then for all $\theta \in (0,1)$, $\varepsilon \in (0,h_3)$ *and* $\lambda > 0$

$$I(\lambda) = \mathsf{E}\exp\left\{\lambda \sup_{0\le t_2<t_1\le t_2+1} \frac{|Z(\mathbf{t})|}{a(t_1)g(t_1-t_2)}\right\} \le \exp\left\{\frac{\lambda^2 A^2 B^2}{2(1-\theta)^2}\right\} A_{10}(\theta,\gamma,\varepsilon),$$

where

$$B = \sum_{k=0}^{\infty} \frac{d_k^{h_3}}{g_k}, \qquad A_{10}(\theta,\gamma,\varepsilon) = \exp\left\{\frac{1}{B}\sum_{k=0}^{\infty} \frac{d_k^{h_3}\,|\log d_k|}{g_k}\right\} A_9(\theta,\gamma,\varepsilon).$$

Proof Denote $\mathbf{T}^{(k)} = \{(t_1,t_2)\in\mathbb{R}_+ : d_{k+1} < t_1 - t_2 \le d_k\}$. Then $\mathbf{T}^{(k)} \subset \mathbf{T}_{d_k}$ and

$$\{(t_1,t_2)\in\mathbb{R} : 0 \le t_2 < t_1 \le t_2+1\} = \bigcup_{k=0}^{\infty} \mathbf{T}^{(k)}.$$

Therefore

$$\sup_{0\le t_2<t_1\le t_2+1} \frac{|Z(\mathbf{t})|}{a(t_1)g(t_1-t_2)} \le \sum_{k=0}^{\infty} \sup_{\mathbf{t}\in\mathbf{T}^{(k)}} \frac{|Z(\mathbf{t})|}{a(t_1)g(t_1-t_2)}$$

$$\le \sum_{k=0}^{\infty} \frac{1}{g_k} \sup_{\mathbf{t}\in\mathbf{T}^{(k)}} \frac{|Z(\mathbf{t})|}{a(t_1)} \le \sum_{k=0}^{\infty} \frac{1}{g_k} \sup_{\mathbf{t}\in\mathbf{T}_{d_k}} \frac{|Z(\mathbf{t})|}{a(t_1)}.$$

Let $r_k > 0$, $\sum_{k=0}^{\infty} \frac{1}{r_k} = 1$. Then

$$I(\lambda) \le \mathsf{E}\exp\left\{\sum_{k=0}^{\infty} \frac{\lambda}{g_k} \sup_{\mathbf{t}\in\mathbf{T}_{d_k}} \frac{|Z(\mathbf{t})|}{a(t_1)}\right\} \le \prod_{k=0}^{\infty}\left(\mathsf{E}\exp\left\{\frac{\lambda r_k}{g_k} \sup_{\mathbf{t}\in\mathbf{T}_{d_k}} \frac{|Z(\mathbf{t})|}{a(t_1)}\right\}\right)^{\frac{1}{r_k}}.$$

By Theorem B.56, we get

$$I(\lambda) \le \prod_{k=0}^{\infty}\left(\frac{1}{d_k}\exp\left\{\frac{\lambda^2 r_k^2 A^2 d_k^{2h_3}}{2(1-\theta)^2 g_k^2}\right\} A_9(\theta,\gamma,\varepsilon)\right)^{\frac{1}{r_k}}$$

$$= A_9(\theta,\gamma,\varepsilon)\exp\left\{\frac{\lambda^2 A^2}{2(1-\theta)^2}\sum_{k=0}^{\infty}\frac{r_k d_k^{2h_3}}{g_k^2}\right\}\exp\left\{-\sum_{k=0}^{\infty}\frac{\log d_k}{r_k}\right\}.$$

Put $r_k = \frac{B g_k}{d_k^{h_3}}$. Then

$$I(\lambda) \le A_9(\theta,\gamma,\varepsilon)\exp\left\{\frac{\lambda^2 A^2 B^2}{2(1-\theta)^2}\right\}\exp\left\{-\frac{1}{B}\sum_{k=0}^{\infty}\frac{d_k^{h_3}\log d_k}{g_k}\right\}. \qquad \square$$

Corollary B.58 *Let the assumptions of Theorem B.57 hold. Then for all* $\theta \in (0,1)$, $\varepsilon \in (0, h_3)$ *and* $u > 0$,

$$\mathsf{P}\left\{\sup_{0\le t_2<t_1\le t_2+1} \frac{|Z(\mathbf{t})|}{a(t_1)g(t_1-t_2)} > u\right\} \le \exp\left\{-\frac{u^2(1-\theta)^2}{2A^2B^2}\right\} A_{10}(\theta,\gamma,\varepsilon). \tag{B.70}$$

Proof By Chebyshev's inequality,

$$\begin{aligned}
&\mathsf{P}\left\{\sup_{0\le t_2<t_1\le t_2+1}\frac{|Z(\mathbf{t})|}{a(t_1)g(t_1-t_2)}>u\right\}\\
&\quad\le\frac{1}{e^{\lambda u}}\mathsf{E}\exp\left\{\lambda\sup_{0\le t_2<t_1\le t_2+1}\frac{|Z(\mathbf{t})|}{a(t_1)g(t_1-t_2)}\right\}\\
&\quad\le\exp\left\{\frac{\lambda^2A^2B^2}{2(1-\theta)^2}-\lambda u\right\}A_{10}(\theta,\gamma,\varepsilon).
\end{aligned}$$

If we put $\lambda=\frac{u(1-\theta)^2}{A^2B^2}$, we get (B.70). □

With the help of Corollary B.58, we can now state the second main result of this appendix, which is the following upper bound for the asymptotic growth of the increments of mfBm with probability 1.

Theorem B.59 *For any $\varepsilon>0$ and any $p>2$ there exists a nonnegative random variable $\eta=\eta(\varepsilon,p)$ such that for all $0\le t_2<t_1\le t_2+1$*

$$|Z(\mathbf{t})|\le\left(t_1^{h_2+\varepsilon}\vee 1\right)(t_1-t_2)^{h_3}\left(|\log(t_1-t_2)|^p\vee 1\right)\eta\qquad a.\,s.,$$

and there exist positive constants $C_1=C_1(\varepsilon,p)$ and $C_2=C_2(\varepsilon,p)$ such that for all $u>0$

$$\mathsf{P}(\eta>u)\le C_1e^{-C_2u^2}.$$

Proof Put in Theorem B.56 $a(t)=t^{h_2+\varepsilon}\vee 1$, $b_0=0$, $b_l=e^l$, $l\ge 1$. Then $a_0=1$, $a_l=e^{l(h_2+\varepsilon)}$, $l\ge 1$, and

$$\sum_{l=0}^{\infty}\frac{b_{l+1}^{\frac{2h_5\gamma}{h_3}}z(b_{l+1})}{a_l}=e^{\frac{2h_5\gamma}{h_3}+h_2}\left(\sqrt{2}+\sum_{l=1}^{\infty}e^{l\left(\frac{2h_5\gamma}{h_3}-\varepsilon\right)}\left((l+1)^2+1\right)^{1/2}\right).$$

Therefore, (B.69) holds, if we choose $\gamma\in\left(0,\frac{\varepsilon h_3}{2h_5}\wedge 1\right)$.

Further, put in Theorem B.57 $g(t)=t^{h_3}|\log t|^p$, $d_k=e^{-k}$, $k\ge 0$, $g_0=e^{-h_3}$, $g_k=d_{k+1}^{h_3}|\log d_k|^p=e^{-(k+1)h_3}k^p$, $k\ge 1$. We have

$$\sum_{k=0}^{\infty}\frac{d_k^{h_3}}{g_k}=e^{h_3}\left(1+\sum_{k=1}^{\infty}\frac{1}{k^p}\right)<\infty,$$

$$\sum_{k=0}^{\infty}\frac{d_k^{h_3}|\log d_k|}{g_k}=e^{h_3}\sum_{k=1}^{\infty}\frac{1}{k^{p-1}}<\infty.$$

Thus, the conditions of Theorem B.57 are satisfied. The result follows from Corollary B.58, if we put

$$\eta = \sup_{0\le t_2<t_1\le t_2+1} \frac{|Z(\mathbf{t})|}{\left(t_1^{h_2+\varepsilon} \vee 1\right)(t_1-t_2)^{h_3}\left(|\log(t_1-t_2)|^p \vee 1\right)}. \qquad \square$$

B.5 Boxplot

The lower and upper hinges correspond to the first and third quartiles (the 25th and 75th percentiles), and the band inside the box is the median. The upper whisker extends from the hinge to the largest value no further than 1.5 * IQR (where IQR is the inter-quartile range, or distance between the first and third quartiles). The lower whisker extends from the hinge to the smallest value at most 1.5 * IQR. Data beyond the end of the whiskers are called "outlying" points and are plotted individually (Fig. B.1).

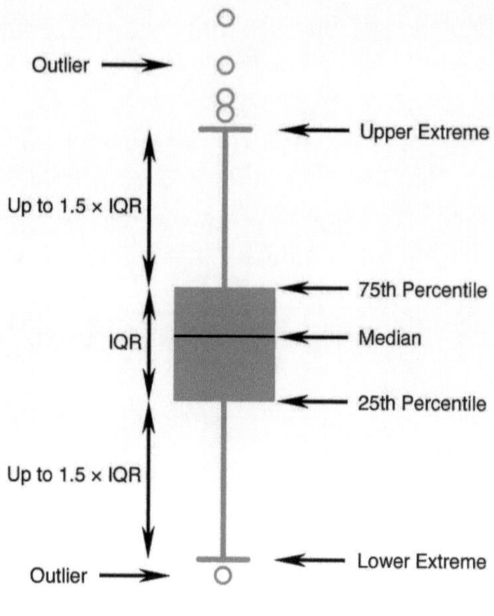

Fig. B.1 Box and whisker

References

1. Abramowitz, M., Stegun, I.A.: Handbook of mathematical functions. Appl. Math. Ser. **55**, 62 (1966)
2. Achard, S., Coeurjolly, J.F.: Discrete variations of the fractional Brownian motion in the presence of outliers and an additive noise. Stat. Surv. **4**, 117–147 (2010)
3. Androshchuk, T., Mishura, Y.: Mixed Brownian-fractional Brownian model: absence of arbitrage and related topics. Stochastics **78**(5), 281–300 (2006)
4. Arcones, M.A.: Limit theorems for nonlinear functionals of a stationary Gaussian sequence of vectors. Ann. Probab. **22**(4), 2242–2274 (1994)
5. Athreya, K.B., Lahiri, S.N.: Measure Theory and Probability Theory. Springer Texts in Statistics. Springer, New York (2006)
6. Ayache, A., Cohen, S., Lévy Véhel, J.: The covariance structure of multifractional Brownian motion, with application to long range dependence. In: 2000 IEEE International Conference on Acoustics, Speech, and Signal Processing – ICASSP'00. Proceedings, vol. 6, pp. 3810–3813 (2000)
7. Azmoodeh, E., Sottinen, T., Viitasaari, L., Yazigi, A.: Necessary and sufficient conditions for Hölder continuity of Gaussian processes. Stat. Probab. Lett. **94**, 230–235 (2014)
8. Bardet, J.M., Surgailis, D.: Measuring the roughness of random paths by increment ratios. Bernoulli **17**(2), 749–780 (2011)
9. Bardina, X., Es-Sebaiy, K.: An extension of bifractional Brownian motion. Commun. Stoch. Anal. **5**(2), 333–340 (2011)
10. Baxter, G.: A strong limit theorem for Gaussian processes. Proc. Am. Math. Soc. **7**, 522–527 (1956)
11. Bégyn, A.: Quadratic variations along irregular subdivisions for Gaussian processes. Electron. J. Probab. **10**, 691–717 (2005)
12. Bégyn, A.: Generalized quadratic variations of Gaussian processes: limit theorems and applications to fractional processes. Ph.D. thesis, Toulouse 3 (2006)
13. Bégyn, A.: Asymptotic expansion and central limit theorem for quadratic variations of Gaussian processes. Bernoulli **13**(3), 712–753 (2007)
14. Bel Hadj Khalifa, M., Mishura, Y., Zili, M.: Asymptotic properties of non-standard drift parameter estimators in the models involving fractional Brownian motion. Theory Probab. Math. Stat. **94**, 73–84 (2016)
15. Belfadli, R., Es-Sebaiy, K., Ouknine, Y.: Parameter estimation for fractional Ornstein–Uhlenbeck processes: non-ergodic case. Front. Sci. Eng. **1**(1), 1–16 (2011)

© Springer International Publishing AG 2017
K. Kubilius et al., *Parameter Estimation in Fractional Diffusion Models*,
B&SS – Bocconi & Springer Series 8,
https://doi.org/10.1007/978-3-319-71030-3

16. Benassi, A., Jaffard, S., Roux, D.: Gaussian processes and pseudodifferential elliptic operators. Rev. Math. Iberoam. **13**(1), 19–89 (1997)
17. Benassi, A., Cohen, S., Istas, J.: Identifying the multifractional function of a Gaussian process. Stat. Probab. Lett. **39**(4), 337–345 (1998)
18. Benassi, A., Cohen, S., Istas, J., Jaffard, S.: Identification of filtered white noises. Stoch. Process. Appl. **75**(1), 31–49 (1998)
19. Beran, J.: Statistics for Long-Memory Processes. Monographs on Statistics and Applied Probability, vol. 61. Chapman and Hall, New York (1994)
20. Beran, J., Feng, Y., Ghosh, S., Kulik, R.: Long-Memory Processes. Probabilistic Properties and Statistical Methods. Springer, Heidelberg (2013)
21. Bercu, B., Coutin, L., Savy, N.: Sharp large deviations for the fractional Ornstein–Uhlenbeck process. Teor. Veroyatn. Primen. **55**(4), 732–771 (2010)
22. Berezansky, Y.M., Sheftel, Z.G., Us, G.F.: Functional Analysis, vol. 1. Birkhäuser, Basel (2012)
23. Bertin, K., Torres, S., Tudor, C.A.: Drift parameter estimation in fractional diffusions driven by perturbed random walks. Stat. Probab. Lett. **81**(2), 243–249 (2011)
24. Berzin, C., Latour, A., León, J.R.: Inference on the Hurst Parameter and the Variance of Diffusions Driven by Fractional Brownian Motion. Lecture Notes in Statistics, vol. 216. Springer, Cham (2014)
25. Biagini, F., Hu, Y., Øksendal, B., Zhang, T.: Stochastic Calculus for Fractional Brownian Motion and Applications. Springer, London (2008)
26. Bishwal, J.P.N.: Parameter Estimation in Stochastic Differential Equations. Springer, Berlin (2008)
27. Bojdecki, T., Gorostiza, L.G., Talarczyk, A.: Sub-fractional Brownian motion and its relation to occupation times. Stat. Probab. Lett. **69**(4), 405–419 (2004)
28. Borodin, A., Salminen, P.: Handbook of Brownian Motion: Facts and Formulae, 2nd edn. Birkhäuser, Basel (2002)
29. Boufoussi, B., Dozzi, M., Marty, R.: Local time and Tanaka formula for a Volterra-type multifractional Gaussian process. Bernoulli **16**(4), 1294–1311 (2010)
30. Breton, J.C., Coeurjolly, J.F.: Confidence intervals for the Hurst parameter of a fractional Brownian motion based on finite sample size. Stat. Infer. Stoch. Process. **15**(1), 1–26 (2012)
31. Breton, J.C., Nourdin, I., Peccati, G.: Exact confidence intervals for the Hurst parameter of a fractional Brownian motion. Electron. J. Stat. **3**, 416–425 (2009)
32. Breuer, P., Major, P.: Central limit theorems for non-linear functionals of Gaussian fields. J. Multivar. Anal. **13**, 425–441 (1983)
33. Brouste, A., Kleptsyna, M.: Asymptotic properties of MLE for partially observed fractional diffusion system. Stat. Infer. Stoch. Process. **13**(1), 1–13 (2010)
34. Buldygin, V., Kozachenko, Y.: Metric Characterization of Random Variables and Random Processes. Translated from the Russian by V. Zaiats. American Mathematical Society, Providence, RI (2000)
35. Cai, C., Chigansky, P., Kleptsyna, M.: Mixed Gaussian processes: a filtering approach. Ann. Probab. **44**(4), 3032–3075 (2016)
36. Cénac, P., Es-Sebaiy, K.: Almost sure central limit theorems for random ratios and applications to LSE for fractional Ornstein–Uhlenbeck processes. Probab. Math. Stat. **35**(2), 285–300 (2015)
37. Cheridito, P.: Mixed fractional Brownian motion. Bernoulli **7**(6), 913–934 (2001)
38. Cheridito, P., Kawaguchi, H., Maejima, M.: Fractional Ornstein-Uhlenbeck processes. Electron. J. Probab. **8**, 1–14 (2003)
39. Cherny, A.S., Engelbert, H.J.: Singular Stochastic Differential Equations. Lecture Notes in Mathematics, vol. 1858. Springer, Berlin (2005)
40. Coeurjolly, J.F.: Estimating the parameters of a fractional Brownian motion by discrete variations of its sample paths. Stat. Inf. Stoch. Process. **4**(2), 199–227 (2001)
41. Coeurjolly, J.F.: Identification of multifractional Brownian motion. Bernoulli **11**(6), 987–1008 (2005)

42. Cramer, H., Leadbetter, M.: Stationary and Related Stochastic Processes. Sample Function Properties and Their Applications, vol. XII, 348 p. Wiley, New York-London-Sydney (1967)
43. Dellacherie, C., Meyer, P.A.: Probabilities and Potential. C. North-Holland Mathematics Studies, vol. 151. North-Holland, Amsterdam (1988)
44. Deza, M., Deza, E.: Encyclopedia of Distances, 3rd edn. Springer, Berlin (2014)
45. Dieudonné, J.: Foundations of Modern Analysis. Enlarged and Corrected Printing, vol. XV, 387 p. Academic, New York-London (1969)
46. Dobrić, V., Ojeda, F.M.: Fractional Brownian fields, duality, and martingales. In: High Dimensional Probability. IMS Lecture Notes Monographs Series, vol. 51, pp. 77–95. Institute of Mathematical Statistics, Beachwood, OH (2006)
47. Dobrushin, R., Major, P.: Non-central limit theorems for non-linear functionals of Gaussian fields. Z. Wahrscheinlichkeitstheor. Verw. Geb. **50**, 27–52 (1979)
48. Dozzi, M., Kozachenko, Y., Mishura, Y., Ralchenko, K.: Asymptotic growth of trajectories of multifractional Brownian motion, with statistical applications to drift parameter estimation. Stat. Inf. Stoch. Process. (2016)
49. Dozzi, M., Mishura, Y., Shevchenko, G.: Asymptotic behavior of mixed power variations and statistical estimation in mixed models. Stat. Infer. Stoch. Process. **18**(2), 151–175 (2015)
50. Dudley, R.: Real Analysis and Probability. Cambridge University Press, Cambridge (2004)
51. Dudley, R.M., Norvaiša, R.: An Introduction to p-Variation and Young Integrals. MaPhySto Lecture Notes, vol. 1. Aarhus, Denmark (1998)
52. Dudley, R.M., Norvaiša, R.: Concrete Functional Calculus. Springer, New York (2011)
53. El Machkouri, M., Es-Sebaiy, K., Ouknine, Y.: Least squares estimator for non-ergodic Ornstein–Uhlenbeck processes driven by Gaussian processes. J. Korean Statist. Soc. **45**(3), 329–341 (2016)
54. Engelbert, H.J., Schmidt, W.: Strong Markov continuous local martingales and solutions of one-dimensional stochastic differential equations, I, II, III. Math. Nachr. **143**(1), 167–184 (1989); **144**(1), 241–281 (1989); **151**(1), 149–197 (1991)
55. Es-Sebaiy, K.: Berry-Esséen bounds for the least squares estimator for discretely observed fractional Ornstein–Uhlenbeck processes. Stat. Probab. Lett. **83**(10), 2372–2385 (2013)
56. Es-sebaiy, K., Ndiaye, D.: On drift estimation for non-ergodic fractional Ornstein–Uhlenbeck process with discrete observations. Afr. Stat. **9**(1), 615–625 (2014)
57. Filatova, D.: Mixed fractional Brownian motion: some related questions for computer network traffic modeling. In: International Conference on Signals and Electronic Systems, Kraków 2008, pp. 393–396 (2008)
58. Gasbarra, D., Sottinen, T., Valkeila, E.: Gaussian bridges. In: Stochastic Analysis and Applications. The Abel Symposium 2005. Proceedings of the Second Abel Symposium, Oslo, Norway, July 29–August 4, 2005, Held in Honor of Kiyosi Itô., pp. 361–382. Springer, Berlin (2007)
59. Gikhman, I.I., Skorokhod, A.V.: Introduction to the Theory of Random Processes. Dover, Mineola, NY (1996)
60. Gikhman, I.I., Skorokhod, A.V.: The Theory of Stochastic Processes. I. Springer, Berlin (2004)
61. Giraitis, L., Surgailis, D.: CLT and other limit theorems for functionals of Gaussian processes. Z. Wahrscheinlichkeitstheor. Verw. Geb. **70**, 191–212 (1985)
62. Giraitis, L., Robinson, P.M., Surgailis, D.: Variance-type estimation of long memory. Stoch. Process. Appl. **80**(1), 1–24 (1999)
63. Gladyshev, E.: A new limit theorem for stochastic processes with Gaussian increments. Theory Probab. Appl. **6**, 52–61 (1962)
64. Gradshteyn, I.S., Ryzhik, I.M.: Table of Integrals, Series, and Products. Academic [Harcourt Brace Jovanovich, Publishers], New York-London-Toronto, Ontario (1980)
65. Grimm, C., Schlüchtermann, G.: IP-Traffic Theory and Performance. Springer Series on Signals and Communication Technology. Springer, Berlin (2008)
66. Guerra, J., Nualart, D.: Stochastic differential equations driven by fractional Brownian motion and standard Brownian motion. Stoch. Anal. Appl. **26**(5), 1053–1075 (2008)

67. Hall, P., Heyde, C.C.: Martingale Limit Theory and Its Application. Academic, New York–London (1980)
68. Hanson, D., Wright, F.: A bound on tail probabilities for quadratic forms in independent random variables. Ann. Math. Stat. **42**, 1079–1083 (1971)
69. Heyde, C.: Quasi-Likelihood and Its Application: a General Approach to Optimal Parameter Estimation. Springer, New York (1997)
70. Houdré, C., Villa, J.: An example of infinite dimensional quasi-helix. In: Stochastic Models. Seventh Symposium on Probability and Stochastic Processes, June 23–28, 2002, Mexico City, Mexico. Selected Papers, pp. 195–201. American Mathematical Society, Providence, RI (2003)
71. Hu, Y., Nualart, D.: Parameter estimation for fractional Ornstein–Uhlenbeck processes. Stat. Probab. Lett. **80**(11–12), 1030–1038 (2010)
72. Hu, Y., Song, J.: Parameter estimation for fractional Ornstein-Uhlenbeck processes with discrete observations. In: Malliavin Calculus and Stochastic Analysis. A Festschrift in Honor of David Nualart, pp. 427–442. Springer, New York (2013)
73. Hu, Y., Nualart, D., Xiao, W., Zhang, W.: Exact maximum likelihood estimators for drift fractional Brownian motion at discrete observation. Acta Math. Sci. Ser. B Engl. Ed. **31**(5), 1851–1859 (2011)
74. Hu, Y., Nualart, D., Zhou, H.: Parameter estimation for fractional Ornstein–Uhlenbeck processes of general Hurst parameter (2017). arXiv preprint arXiv:1703.09372
75. Iacus, S.M.: Simulation and Inference for Stochastic Differential Equations. With R Examples. Springer, New York (2008)
76. Ibe, O.: Elements of Random Walk and Diffusion Processes, 1st edn. Wiley Series in Operations Research and Management Science. Wiley, Hoboken, NJ (2013)
77. Ikeda, N., Watanabe, S.: Stochastic Differential Equations and Diffusion Processes, 2nd edn. North-Holland, Amsterdam (1989)
78. Istas, J., Lang, G.: Quadratic variations and estimation of the local Hölder index of a Gaussian process. Ann. Inst. Henri Poincaré Probab. Stat. **33**(4), 407–436 (1997)
79. Itô, K.: On stochastic differential equations. Mem. Am. Math. Soc. **4**, 1–51 (1951)
80. Itô, K., McKean Jr., H.P.: Diffusion Processes and Their Sample Paths, Second printing, Corrected. Die Grundlehren der mathematischen Wissenschaften, Band 125. Springer, Berlin (1974)
81. Janson, S.: Gaussian Hilbert Spaces. Cambridge University Press, Cambridge (1997)
82. Jost, C.: Transformation formulas for fractional Brownian motion. Stoch. Process. Appl. **116**(10), 1341–1357 (2006)
83. Kahane, J.P.: Hélices et quasi-hélices. In: Mathematical Analysis and Applications, Part B. Advances in Mathematics Supplement Studies, vol. 7, pp. 417–433. Academic, New York-London (1981)
84. Karp, D., Sitnik, S.: Two-sided inequalities for generalized hypergeometric function. Res. Rep. Collect. **10**(2) (2007)
85. Kawata, T.: Fourier Analysis in Probability Theory. Probability and Mathematical Statistics. A Series of Monographs and Textbooks. Academic, New York/London (1972)
86. Kent, J.T., Wood, A.T.: Estimating the fractal dimension of a locally self-similar Gaussian process by using increments. J. R. Stat. Soc. Ser. B **59**(3), 679–699 (1997)
87. Kessler, M., Lindner, A., Sørensen, M. (eds.): Statistical Methods for Stochastic Differential Equations. Selected Papers Based on the Presentations at the 7th Séminaire Européen de Statistiques on "Statistics for stochastic differential equations models", La Manga del Mar Menor, Cartagena, May 7–12, 2007. CRC, Boca Raton, FL (2012)
88. Khoshnevisan, D., Salminen, P., Yor, M.: A note on a.s. finiteness of perpetual integral functionals of diffusions. Electron. Commun. Probab. **11**, 108–117 (electronic) (2006)
89. Klein, R., Gine, E.: On quadratic variations of processes with Gaussian increments. Ann. Probab. **3**(4), 716–721 (1975)
90. Kleptsyna, M.L., Le Breton, A.: Statistical analysis of the fractional Ornstein–Uhlenbeck type process. Stat. Infer. Stoch. Process. **5**, 229–248 (2002)

91. Kolmogoroff, A.N.: Wienersche Spiralen und einige andere interessante Kurven im Hilbertschen Raum. C. R. (Doklady) Acad. Sci. URSS (N.S.) **26**, 115–118 (1940)
92. Kozachenko, Y., Melnikov, A., Mishura, Y.: On drift parameter estimation in models with fractional Brownian motion. Statistics **49**(1), 35–62 (2015)
93. Krylov, N.V.: Itô's stochastic integral equations. Theor. Probab. Appl. **14**(2), 330–336 (1969)
94. Kubilius, K.: The existence and uniqueness of the solution of an integral equation driven by a p-semimartingale of special type. Stoch. Process. Appl. **98**(2), 289–315 (2002)
95. Kubilius, K.: On estimation of the extended Orey index for Gaussian processes. Stochastics **87**(4), 562–591 (2015)
96. Kubilius, K., Melichov, D.: On estimation of the Hurst index of solutions of stochastic integral equations. Liet. Mat. Rink. **48/49**, 401–406 (2008)
97. Kubilius, K., Melichov, D.: Estimating the Hurst index of the solution of a stochastic integral equation. Liet. Mat. Rink. **50**, 24–29 (2009)
98. Kubilius, K., Melichov, D.: Quadratic variations and estimation of the Hurst index of the solution of SDE driven by a fractional Brownian motion. Lith. Math. J. **50**(4), 401–417 (2010)
99. Kubilius, K., Melichov, D.: On comparison of the estimators of the Hurst index of the solutions of stochastic differential equations driven by the fractional Brownian motion. Informatica **22**(1), 97–114 (2011)
100. Kubilius, K., Mishura, Y.: The rate of convergence of Hurst index estimate for the stochastic differential equation. Stoch. Process. Appl. **122**(11), 3718–3739 (2012)
101. Kubilius, K., Skorniakov, V.: On some estimators of the Hurst index of the solution of SDE driven by a fractional Brownian motion. Statist. Probab. Lett. **109**, 159–167 (2016)
102. Kubilius, K., Skorniakov, V., Melichov, D.: Estimation of parameters of SDE driven by fractional Brownian motion with polynomial drift. J. Stat. Comput. Simul. **86**(10), 1954–1969 (2016)
103. Kubilius, K., Mishura, Y., Ralchenko, K., Seleznjev, O.: Consistency of the drift parameter estimator for the discretized fractional Ornstein–Uhlenbeck process with Hurst index $H \in (0, \frac{1}{2})$. Electron. J. Stat. **9**(2), 1799–1825 (2015)
104. Kubilius, K., Skorniakov, V., Ralchenko, K.: The rate of convergence of the Hurst index estimate for a stochastic differential equation. Nonlinear Anal. Model. Control **22**(2), 273–284 (2017)
105. Kukush, A., Mishura, Y., Ralchenko, K.: Hypothesis testing of the drift parameter sign for fractional Ornstein–Uhlenbeck process. Electron. J. Stat. **11**(1), 385–400 (2017)
106. Kutoyants, Y.A.: Statistical Inference for Ergodic Diffusion Processes. Springer, London (2004)
107. Le Breton, A.: Filtering and parameter estimation in a simple linear system driven by a fractional Brownian motion. Stat. Probab. Lett. **38**(3), 263–274 (1998)
108. Lei, P., Nualart, D.: A decomposition of the bifractional Brownian motion and some applications. Stat. Probab. Lett. **79**(5), 619–624 (2009)
109. Lifshits, M., Volkova, K.: Bifractional Brownian motion: existence and border cases. ESAIM Probab. Stat. **19**, 766–781 (2015)
110. Liptser, R., Shiryayev, A.: Statistics of Random Processes. II. Applications. Translated by A. B. Aries. Applications of Mathematics, vol. 6, X, 339 p. Springer, New York, Heidelberg, Berlin (1978)
111. Liptser, R., Shiryayev, A.: Theory of Martingales. Kluwer Academic, Dordrecht (1989)
112. Liptser, R.S., Shiryaev, A.N.: Statistics of Random Processes. II. Applications, Applications of Mathematics, vol. 6. Springer, Berlin (2001)
113. Liu, J., Yan, L., Peng, Z., Wang, D.: Remarks on confidence intervals for self-similarity parameter of a subfractional Brownian motion. Abstr. Appl. Anal. **2012**, 14 (2012). Art. ID 804942
114. Malukas, R.: Limit theorems for a quadratic variation of Gaussian processes. Nonlinear Anal. Model. Control **16**(4), 435–452 (2011)

115. Mandelbrot, B.B., Van Ness, J.W.: Fractional Brownian motions, fractional noises and applications. SIAM Rev. **10**, 422–437 (1968)
116. Marcus, M.B.: Hölder conditions for Gaussian processes with stationary increments. Trans. Am. Math. Soc. **134**, 29–52 (1968)
117. Marcus, M.B., Rosen, J.: Markov Processes, Gaussian Processes, and Local Times. Cambridge University Press, Cambridge (2006)
118. Mémin, J., Mishura, Y., Valkeila, E.: Inequalities for the moments of Wiener integrals with respect to a fractional Brownian motion. Stat. Probab. Lett. **51**(2), 197–206 (2001)
119. Mijatović, A., Urusov, M.: Convergence of integral functionals of one-dimensional diffusions. Electron. Commun. Probab. **17**, 1–13 (2012). Article 61
120. Mishura, Y.: Stochastic Calculus for Fractional Brownian Motion and Related Processes, vol. 1929. Springer Science & Business Media, Berlin (2008)
121. Mishura, Y.: Standard maximum likelihood drift parameter estimator in the homogeneous diffusion model is always strongly consistent. Stat. Probab. Lett. **86**, 24–29 (2014)
122. Mishura, Y.: Maximum likelihood drift estimation for the mixing of two fractional Brownian motions. In: Stochastic and Infinite Dimensional Analysis, pp. 263–280. Springer, Berlin (2016)
123. Mishura, Y., Ralchenko, K.: On drift parameter estimation in models with fractional Brownian motion by discrete observations. Austrian J. Stat. **43**(3), 218–228 (2014)
124. Mishura, Y.S., Shevchenko, G.M.: The rate of convergence for Euler approximations of solutions to of stochastic differential equations driven by fractional Brownian motion. Stochastics **809**(5), 489–511 (2008)
125. Mishura, Y.S., Shevchenko, G.M.: Existence and uniqueness of the solution of stochastic differential equation involving Wiener process and fractional Brownian motion with Hurst index $H > 1/2$. Commun. Stat. Theory Methods **40**(19–20), 3492–3508 (2011)
126. Mishura, Y., Shevchenko, G.: Mixed stochastic differential equations with long-range dependence: existence, uniqueness and convergence of solutions. Comput. Math. Appl. **64**(10), 3217–3227 (2012)
127. Mishura, Y., Shevchenko, G.: Theoretical and Statistical Aspects of Stochastic Processes. Elsevier, Amsterdam (2017)
128. Mishura, Y., Voronov, I.: Construction of maximum likelihood estimator in the mixed fractional–fractional Brownian motion model with double long-range dependence. Mod. Stoch. Theory Appl. **2**(2), 147–164 (2015)
129. Mishura, Y., Ralchenko, K., Seleznev, O., Shevchenko, G.: Asymptotic properties of drift parameter estimator based on discrete observations of stochastic differential equation driven by fractional Brownian motion. In: Modern Stochastics and Applications. Springer Optimization and Applications, vol. 90, pp. 303–318. Springer, Cham (2014)
130. Moers, M.: Hypothesis testing in a fractional Ornstein-Uhlenbeck model. Int. J. Stoch. Anal. **2012**, 23 (2012). Art. ID 268568
131. Norros, I., Valkeila, E., Virtamo, J.: An elementary approach to a Girsanov formula and other analytical results on fractional Brownian motions. Bernoulli **5(4)**, 571–587 (1999)
132. Norvaiša, R.: A complement to Gladyshev's theorem. Lith. Math. J. **51**(1), 26–35 (2011)
133. Norvaiša, R.: Gladyshev's theorem for integrals with respect to a Gaussian process. Preprint (2011). arXiv:1105.1503v1
134. Norvaiša, R., Salopek, D.: Estimating the Orey index of a Gaussian stochastic process with stationary increments: An application to financial data set. In: Stochastic Models. Proceedings of the International Conference, In Honour of Prof. Donald A. Dawson, Ottawa, June 10–13, 1998, pp. 353–374. American Mathematical Society for the Canadian Mathematical Society, Providence, RI (2000)
135. Nourdin, I.: Asymptotic behavior of weighted quadratic and cubic variations of fractional Brownian motion. Ann. Probab. **36**(6), 2159–2175 (2008)
136. Nourdin, I.: Selected Aspects of Fractional Brownian Motion. Bocconi & Springer Series, vol. 4. Springer, Bocconi University Press, Milan (2012)

137. Nourdin, I., Viens, F.G.: Density formula and concentration inequalities with Malliavin calculus. Electron. J. Probab. **14**, 2287–2309 (2009)
138. Nourdin, I., Nualart, D., Tudor, C.A.: Central and non-central limit theorems for weighted power variations of fractional Brownian motion. Ann. Inst. Henri Poincaré, Probab. Stat. **46**(4), 1055–1079 (2010)
139. Nualart, D.: The Malliavin Calculus and Related Topics, 2nd edn. Probability and Its Applications. Springer, Berlin (2006)
140. Nualart, D., Răşcanu, A.: Differential equations driven by fractional Brownian motion. Collect. Math. **53**(1), 55–81 (2002)
141. Orey, S.: Gaussian sample functions and the Hausdorff dimension of level crossings. Z. Wahrscheinlichkeitstheor. Verw. Geb. **15**, 249–256 (1970)
142. Peltier, R.F., Lévy Véhel, J.: Multifractional Brownian motion: definition and preliminary results. INRIA Research Report, vol. 2645 (1995)
143. Perestyuk, M., Mishura, Y., Shevchenko, G.: On the distribution of integral functionals of a homogeneous diffusion process. Mod. Stoch. Theory Appl. **1**(2), 109–116 (2014)
144. Pitman, J., Yor, M.: Hitting, occupation and inverse local times of one-dimensional diffusions: martingale and excursion approaches. Bernoulli **9**(1), 1–24 (2003)
145. Prakasa Rao, B.L.S.: Asymptotic Theory of Statistical Inference. Wiley, New York (1987)
146. Prakasa Rao, B.L.S.: Statistical Inference for Fractional Diffusion Processes. Wiley, New York (2010)
147. R Core Team: A language and environment for statistical computing. R Foundation for Statistical Computing. Vienna (2014). http://www.R-project.org/
148. Ralchenko, K.V.: Approximation of multifractional Brownian motion by absolutely continuous processes. Theory Probab. Math. Stat. **82**, 115–127 (2011)
149. Ralchenko, K.V., Shevchenko, G.M.: Paths properties of multifractal Brownian motion. Theory Probab. Math. Stat. **80**, 119–130 (2010)
150. Resnick, S.I.: Heavy-Tail Phenomena. Probabilistic and Statistical Modeling. Springer Series in Operations Research and Financial Engineering. Springer, Berlin (2007)
151. Rogers, L.C.G., Williams, D.: Diffusions, Markov Processes, and Martingales, vol. 2: Itô Calculus. Cambridge University Press, Cambridge (2000)
152. Russo, F., Tudor, C.A.: On bifractional Brownian motion. Stoch. Process. Appl. **116**(5), 830–856 (2006)
153. Salminen, P., Yor, M.: Properties of perpetual integral functionals of Brownian motion with drift. Ann. Inst. H. Poincaré Probab. Stat. **41**(3), 335–347 (2005)
154. Salopek, D.: Tolerance to Arbitrage: Inclusion of fractional Brownian motion to model stock price fluctuations. Ph.D. thesis, Ottawa-Carleton Institute (1997)
155. Samko, S., Kilbas, A., Marichev, O.: Fractional Integrals and Derivatives: Theory and Applications. Transl. from the Russian. Gordon and Breach, New York, NY (1993)
156. Shevchenko, G.: Mixed stochastic delay differential equations. Theory Probab. Math. Stat. (89), 181–195 (2014)
157. Shiryaev, A.: Essentials of Stochastic Finance: Facts, Models, Theory, 1st edn. Advanced Series on Statistical Science & Applied Probability, vol. 3. World Scientific, Singapore (1999)
158. Skorokhod, A.V.: Studies in the Theory of Random Processes. Addison-Wesley, Reading (1965)
159. Stoev, S.A., Taqqu, M.S.: How rich is the class of multifractional Brownian motions? Stoch. Process. Appl. **116**(2), 200–221 (2006)
160. Stroock, D.W., Varadhan, S.R.S.: Diffusion processes with continuous coefficients. I, II. Commun. Pure Appl. Math. **22**, 345–400, 479–530 (1969)
161. Tanaka, K.: Distributions of the maximum likelihood and minimum contrast estimators associated with the fractional Ornstein-Uhlenbeck process. Stat. Infer. Stoch. Process. **16**, 173–192 (2013)
162. Tanaka, K.: Maximum likelihood estimation for the non-ergodic fractional Ornstein–Uhlenbeck process. Stat. Infer. Stoch. Process. **18**(3), 315–332 (2015)

163. Taqqu, M.S.: Weak convergence to fractional Brownian motion and to the Rosenblatt process. Z. Wahrscheinlichkeitstheor. Verw. Geb. **31**, 287–302 (1975). https://doi.org/10.1007/BF00532868
164. Tricot, C.: Curves and Fractal Dimension. With a Foreword by Michel Mendès France. Transl. from the French. Springer, New York, NY (1995)
165. Tudor, C.: Some properties of the sub-fractional Brownian motion. Stochastics **79**(5), 431–448 (2007)
166. Tudor, C.A., Viens, F.G.: Statistical aspects of the fractional stochastic calculus. Ann. Stat. **35**(3), 1183–1212 (2007)
167. van Zanten, H.: When is a linear combination of independent fBm's equivalent to a single fBm? Stoch. Process. Appl. **117**(1), 57–70 (2007)
168. Wood, A.T.A., Chan, G.: Simulation of stationary Gaussian processes in $[0, 1]^d$. J. Comput. Graph. Stat. **3**(4), 409–432 (1994)
169. Xiao, W.L., Zhang, W.G., Zhang, X.L.: Maximum-likelihood estimators in the mixed fractional Brownian motion. Statistics **45**, 73–85 (2011)
170. Xiao, W., Zhang, W., Xu, W.: Parameter estimation for fractional Ornstein-Uhlenbeck processes at discrete observation. Appl. Math. Model. **35**, 4196–4207 (2011)
171. Yamada, T., Watanabe, S.: On the uniqueness of solutions of stochastic differential equations. J. Math. Kyoto Univ. **11**(1), 155–167 (1971)
172. Zähle, M.: Integration with respect to fractal functions and stochastic calculus. I. Probab. Theory Relat. Fields **111**(3), 333–374 (1998)
173. Zähle, M.: On the link between fractional and stochastic calculus. In: Stochastic Dynamics. Conference on Random Dynamical Systems, Bremen, April 28–May 2, 1997. Dedicated to Ludwig Arnold on the Occasion of his 60th birthday, pp. 305–325. Springer, New York, NY (1999)
174. Zähle, M.: Integration with respect to fractal functions and stochastic calculus. II. Math. Nachr. **225**, 145–183 (2001)
175. Zvonkin, A.K.: A transformation of the phase space of a diffusion process that will remove the drift. Math. USSR Sb. **93**, 129–149 (1974)

Index

asymptotic ratios, 117

bifractional Brownian motion, 40
boxplot, 379

CLT
  for quadratic variations of fBm, 58
  for stationary Gaussian sequences, 330
concentration inequality, 47, 305

delta method, 328

ergodic theorem, 327
Euler–Peano approximation, 82

fractional
  Brownian bridge, 279
  Brownian motion, 3
    mixed, 126
  derivative, 8
  Gaussian noise, 128
  Gompertz diffusion process, 35
  integral, 7
  Ornstein–Uhlenbeck process, 24
    generalized, 230
    stationary, 28

Gauss hypergeometric function, 321
generalized Lebesgue–Stieltjes integral, 9

Hölder continuous function, 3
Helly–Bray theorem, 328
Hermite polynomial, 329

incremental variance function, 271
invariant function, 117
isonormal Gaussian process, 333
Isserlis theorem, 129

Kantorovich–Rubinstein duality theorem, 325
Kolmogorov continuity theorem, 3

Langevin equation, 24
Lipschitz seminorm, 324
local time, 14
long-range dependence, 4
Love–Young inequality, 10

metric massiveness, 339
mixed power variations, 128
Molchan martingale, 7
multifractional
  Brownian motion, 41
  Ornstein–Uhlenbeck process, 262
multiple Wiener–Itô integral, 334

non-degeneracy condition, 79

Orey index, 270
  extended, 272
  of bifBm, 285

© Springer International Publishing AG 2017

K. Kubilius et al., *Parameter Estimation in Fractional Diffusion Models*,
B&SS – Bocconi & Springer Series 8,
https://doi.org/10.1007/978-3-319-71030-3

of fO-U, 282
of fractional Brownian bridge, 280
of sfBm, 283

p-variation, 323
potential, 21

quadratic variation, 46
normalized, 46
second-order, 116

recurrent diffusion process, 14
regularly varying function, 322
regulated function, 117
Rosenblatt distribution, 337

SDE
mixed, 38
with polynomial drift, 31
sequence of partitions, 117
slowly varying function, 322
Slutsky theorem, 328
step function, 117
strong law of large numbers for martingales, 327
sub-fractional Brownian motion, 39

transferred MLE, 179
transient diffusion process, 14

Wasserstein distance, 324
Wiener chaos, 333
Wiener integration, 5

MIX
Papier aus verantwortungsvollen Quellen
Paper from responsible sources
FSC® C105338

If you have any concerns about our products,
you can contact us on
ProductSafety@springernature.com

In case Publisher is established outside the EU,
the EU authorized representative is:
Springer Nature Customer Service Center GmbH
Europaplatz 3, 69115 Heidelberg, Germany

Printed by Libri Plureos GmbH
in Hamburg, Germany